Plasma Physics for Astrophysics

PRINCETON SERIES IN ASTROPHYSICS

EDITED BY DAVID N. SPERGEL

Theory of Rotating Stars, *by Jean-Louis Tassoul*

Theory of Stellar Pulsation, *by John P. Cox*

Galactic Dynamics, *by James Binney and Scott Tremaine*

Dynamical Evolution of Globular Clusters, *by Lyman Spitzer Jr.*

Supernovae and Nucleosynthesis: An Investigation of the History of Matter, from the Big Bang to the Present, *by David Arnett*

Unsolved Problems in Astrophysics, *edited by John N. Bahcall and Jeremiah P. Ostriker*

Galactic Astronomy, *by James Binney and Michael Merrifield*

Active Galactic Nuclei: From the Central Black Hole to the Galactic Environment, *by Julian H. Krolik*

Plasma Physics for Astrophysics, *by Russell M. Kulsrud*

Plasma Physics for Astrophysics

RUSSELL M. KULSRUD

PRINCETON UNIVERSITY PRESS

PRINCETON AND OXFORD

Published by Princeton University Press, 41 William Street, Princeton,
New Jersey 08540
In the United Kingdom: Princeton University Press, 3 Market Place,
Woodstock, Oxfordshire OX20 1SY

Library of Congress Cataloging-in-Publication Data
Kulsrud, R.M.
Plasma physics for astrophysics / Russell M. Kulsrud.
p. cm. — (Princeton series in astrophysics)
Includes bibliographical references and index.
ISBN 0-691-10267-8 (acid-free paper) — ISBN 0-691-12073-0 (pbk. : acid-free paper)
1. Plasma astrophysics. I. Title. II. Series.
QB462.7.K85 2005
523.01–dc22 2004044512

British Library Cataloging-in-Publication Data is available

This book has been composed in Sabon

Printed on acid-free paper. ∞

www.pup.princeton.edu

Printed in the United States of America

1 3 5 7 9 10 8 6 4 2

DEDICATED TO HELENE

CONTENTS

LIST OF FIGURES

FOREWORD

NEARLY ALL THE UNIVERSE is a plasma. If you want to understand the Universe, you need to understand plasmas. Study this book, and you will be well equipped to investigate many of the most fascinating problems in modern astronomy and astrophysics. You will also have the tools and the introduction to the relevant literature that will enable you to begin to work in such diverse areas as, for example, geophysics, fusion physics, and oceanography.

Russell Kulsrud is one of the leading theoretical plasma physicists of his era. In addition, he is a professional astrophysicist, who has made important contributions to both astrophysics and fusion research. If you had to design an ideal author to write a book on plasmas and astrophysics, you would probably have invented someone like Kulsrud.

If you are a student in a physics department, or in an astronomy or engineering-physics department, you should go to the department chair and insist that a course based on this book should be included in the curriculum. It is a scandal, in my view, that many talented physicists and astronomers are being trained today without a proper introduction to plasma physics. Much of what is interesting in the world, and much of what is important for applications to various fields, involves plasma physics. You absolutely need to understand the basics of such ubiquitous concepts as the Fokker–Planck equation, magnetohydodynamic theory, shock waves, and turbulence. For a student in the engineering or physical sciences, you can learn most of what you will need to know about plasma physics from this book.

Fortunately, if you are a professional chemist, or physicist, or astronomer who wants to understand more about the large-scale universe, or if you are just interested in mastering the fundamentals of plasma physics, you can learn what you want to know by studying this book by yourself. The presentation is pedagogical. The discussions are clear and well organized. Moreover, Kulsrud makes frequent use of his exceptional physical insight to motivate qualitatively the results. You are guided step-by-step through the derivations, but all the time your hand is being held by Kulsrud who whispers reasons why you should expect the results that eventually emerge. And there are informative problems at the end of each chapter that you can use to test and deepen your understanding.

You don't even have to read the chapters in order. If you have a good background in electromagnetism, at the level of an advanced undergraduate physics course, then you can jump directly to studying the material in the chapter that has the application of greatest interest to you.

This book is a treasure. You should not put it on your bookshelf. You should keep it on your desk.

John N. Bahcall
Professor of Natural Science
Institute for Advanced Study, Princeton

PREFACE

THIS BOOK IS AN OUTGROWTH of a course in plasma astrophysics which I taught for many years. During that time I was unable to find an appropriate textbook for this course, so I thought I would try to write one myself. I must confess that this was to a large extent at John Bahcall's urging.

I set my goal in this book to present plasma physics as a comprehensible field of physics that could be grasped largely on the basis of physical intuition and qualitative reasoning, in a way similar to other fields of physics. Plasma physics appears to consist of a large number of topics arising independently from astrophysics, fusion physics, and other practical applications. But, in fact, most of these topics can be approached from the same point of view. Namely, they are all comprehensible by simple physical intuition.

I have made no attempt to be comprehensive in my choice of topics. Instead, I have limited myself to those topics with which I am familiar, and that are well established and no longer controversial. In attempting to learn a field such as plasma physics, it is an advantage to have definite answers in which one has confidence, and by and large the topics I chose are unambiguously accepted by the plasma community. However, I do venture into some speculative topics in the last two chapters on dynamos and magnetic reconnection.

I treat each topic in more or less the same way. I first approach it through a physical and qualitative discussion. Then I select a simple example and treat this example with considerable rigor and analytic detail. I recommend that the readers of this book follow through these derivations with care. They will lead to a better understanding of the subject. If the readers cannot verify these calculations on their own, this failure should lead them to uncover possible misunderstandings of the physics. Lastly, I describe the more general results and give the references for them with which I am most familiar, and that I know present the results with clarity and completeness.

I present five or six problems at the end of each chapter. I regard the solving of these problems as very important in the learning process and urge the readers to attempt their solution. From my teaching experience, I found that students gained at least as much benefit from the problems as from my lectures. Further, I include the bulk of the astrophysical applications in these problems. They are useful for gaining a good idea as to the numerical magnitude of plasma effects in astrophysics.

The book is divided into two unequal parts. In the first eleven chapters the primary emphasis is on plasma physical ideas, with astrophysical applications serving as illustrations. The last three chapters are specifically devoted to astrophysical topics that draw on the knowledge presented in the first part of the book.

My thanks go out to many people who have helped me immensely with the preparation of this book. First, I want to thank John Bahcall for needling me into writing the book. I thank Peter Goldreich for his efforts in enabling me to understand MHD turbulence and for a careful and thoughtful reading of the section of the book devoted to it. I found the book to be very time-consuming, and I have grievously neglected my interactions with my colleagues, who have been very patient with me during the past two years. In particular, I thank Masaaki Yamada, Hantao Ji, Eugene Parker, and Steve Cowley. I most particularly thank my wife, Helene, my daughter, Pamela, and my son, Peter, who have been especially patient with my almost total immersion in this book.

Plasma Physics for Astrophysics

Chapter 1

INTRODUCTION

ASTROPHYSICS MAKES USE of nearly every branch of physics: mechanics, electromagnetism, quantum mechanics, fluid mechanics, nuclear physics and general relativity. In addition, the subject of plasma physics has been playing an increasing role in astrophysics. For a large part of astrophysics, we can get along without plasma physics except in a passive role. For example, magnetic fields in a fluid exert forces that modify its behavior in a way that is easy to grasp. But many of the more extreme events in astrophysics involve plasma physics in an essential way, and to comprehend them a deeper understanding of the more subtle and surprising properties of plasma physics is necessary.

Plasmas in the laboratory have many striking properties, but the laboratory involves sizes and other parameters with rather mundane values. When we pass to astrophysics these sizes and parameters take on extreme values, which bring out these properties more forcefully.

A particular interesting example is the so-called Langmuir paradox. In the late 1920s when Tonks and Langmuir were first creating and examining plasmas, they shot a beam of energetic electrons into a plasma and found that the beam came to a screeching halt, in a very short distance, much shorter than the collisional length over which the beam is expected to decelerate (Tonks and Langmuir 1929). A similar phenomenon occurs in astrophysics for cosmic rays. They are injected into the interstellar medium by acceleration around a supernova and rather quickly lose their directed motion becoming isotropic in their velocity space. Both phenomena are caused by strong instabilities on small scales. The intense electric fields in these instabilities interact with the beams and decelerate them or scatter them. The instabilities arise from the free energy in the beams themselves. Such plasma phenomena occur on a small scale that is not visible to the observer, and thus appear hidden, so that the results of their action often appear mysterious.

The solar flare is another important consequence of deeper plasma phenomena. The sudden appearance far above the sun's surface of energy in the form of radiation, particle acceleration, and heating at first seemed very remarkable and incomprehensible, until it was appreciated that the source of energy was the sun's magnetic field. How this conversion occurs is still unclear, but it is believed to be associated with the reconnection of magnetic field lines.

The ideas behind this are rooted in the concept that magnetic field lines have a reality much deeper than the mere vector representation of the magnetic field. Once these lines are present in a plasma they can be considered as real lines bodily transported by macroscopic plasma motions. This bodily transport can be destroyed only by resistivity. However, because of the large scale of astrophysical plasmas this occurs only very slowly. As a consequence, magnetic field lines can be wound up, much as a spring can, and the tightly wound lines represent a great increase in the magnetic energy. In the solar flare, for example, this continues until the lines suddenly break, allowing the rapid conversion of the excess magnetic energy into thermal and radiated energy. How this happens is still not fully understood and is currently the subject of intense research.

The remarkable freezing of magnetic field lines into a plasma can be traced to the large induction $\mathcal{L}$ of a typical large-scale astrophysical plasma and a correspondingly very small resistance $\mathcal{R}$. In electrical circuits the timescale for decay of currents is $\mathcal{L}/\mathcal{R}$ and, correspondingly, the timescale during which the flux freezing holds is $\mathcal{L}/\mathcal{R}$ and in astrophysics is generally very large.

This freezing is one of the outstanding characteristics of plasmas on large scales. As an example, it easily explains why a galaxy can have a long-lasting magnetic field, a field whose decay time is longer than the age of the universe by many orders of magnitude. On the other hand, it makes it difficult to see how a magnetic field on this scale can arise in a time shorter than the life of the galaxy. In fact, such a start-up of a field can happen only while conserving the actual number of magnetic field lines. It is thought to be accomplished in the galaxy by a mechanism in which helical motions fold the lines in a complex topological manner so that a large amount of positive and negative flux results. Then the observed large-scale field of constant sign results by the removal of the negative flux from the galaxy. Such processes are called dynamos. Similar processes seem to be present in the sun, which could explain the mysterious reversal of its magnetic field every eleven years. Accretion disks and compact objects have dynamo mechanisms that generate and amplify magnetic fields. Here the field is amplified to a point where it can provide the viscous force needed for the matter to fall onto the compact body at the center of the disk.

Plasma processes play many other roles, which are not immediately obvious and are detected only through their consequences, such as in particle acceleration, generation of intense radiation, confinement of energetic particles such as cosmic rays, and stellar formation.

Why does a plasma have these remarkable properties? When we inspect a plasma closely, we appreciate that it can have a very large number of degrees of freedom, which manifest themselves as waves with a large range of wavelengths, extending from microscopic to macroscopic length scales, The amplitudes of these waves are generally lifted above thermal levels by specific plasma instabilities. The collective motions of particles in these waves and the resulting strong fields are the agents by which these properties are produced.

A plasma consists of enumerable particles of either sign and the particles carry large amounts of charge and electrical currents. In a normal situation the charges cancel to an extremely high degree of approximation. However, even the slightest imbalance in the numbers of particles of opposite sign, ions and electrons, can produce intense electric fields. Also, a slight anisotropy in their motions can produce extremely strong magnetic fields.

As an example, there are roughly $N = 10^{57}$ electrons and $N = 10^{57}$ ions in the sun. If these numbers were seriously out of balance, say only half the number of electrons as ions, there would be an electric field at the surface of the sun of $Ne/R^2 = 10^{57} \times 4.8 \times 10^{-10}/(6 \times 10^{10})^2 \approx 1.5 \times 10^{26}$ statvolts/cm $= 4.5 \times 10^{28}$ volts/centimeter (V/cm). Such a field would easily pull the sun apart. In fact, the surface electric field would actually be outward and of such a strength so as to balance one-half of the inward gravitational force on an ion ($\approx 10^{-8}$V/cm). (This is the field necessary to keep the electrons from evaporating from the solar surface.) Actually, there is such a surface field and it satisfies $eE = m_H g/2 \approx \nabla p_e/n$. What imbalance between N_i and N_e is necessary to produce this field? We have $(N_i - N_e)e = 4\pi R^2 E$, and if we use $4\pi R^2 g = GN_i m_H$ we easily get $(N_i - N_e)/N_i = Gm_H^2/e^2 \approx 10^{-36}$ so the cancellation is extremely high. The same degree of cancellation holds at every point in the sun.

The cancellation is general and represents the extremely large amount of charge that a plasma can potentially supply if necessary. In fact, suppose we have a plasma slab of thickness L. If there is an imbalance of charge density Δn, then there is an electric field of magnitude $4\pi \Delta neL$ and an electric potential $\Delta\phi = 4\pi\Delta neL^2/2$. The potential energy of an ion $e\Delta\phi$ is comparable to its one-dimensional thermal energy $k_B T/2$ when $\Delta n \approx T/4\pi ne^2 L^2$ or when $\Delta n/n \approx \lambda_D^2/L^2$, where

$$\lambda_D = \sqrt{\frac{T}{4\pi ne^2}} \approx 7\sqrt{\frac{T}{n}} \text{ cm} \tag{1}$$

This expression has the dimensions of length and is called the Debye length. It is a fundamental length in plasma physics. For the central region of the sun of radius $R/10$, $T \approx 10^7$ kelvins (K) and $n \approx 10^{26}$, so that $\lambda_D \approx 2 \times 10^{-9}$ cm and $(\lambda_D/10R)^2 \approx 10^{-37}$, which is consistent with our first estimate. ($\Delta n/n \approx 1$, can hold only when $L \approx \lambda_D$. Since this is the case for electron plasma oscillations, this is also the scale of electron plasma oscillations.)

Similarly, we can find how much potential current the sun could have by taking all the N electrons to move in the same direction. It is $I = Nev/c$, where v is the thermal velocity $v \approx 10^9$ cm/sec at $T \approx 100$ eV. Thus, $I \approx 10^{57} \times 4.8 \times 10^{-10} \times 10^9/(3 \times 10^{10}) = 10^{46}$ abamps ($\approx 10^{47}$ amps). Such a current would produce a magnetic field strength of order $B \approx 2I/R \approx 3 \times 10^{35}$ gauss (G). This enormous field derived from the assumption of totally directed motion of the electrons indicates that even the smallest anisotropy in velocities will produce almost any conceivable field. For example, an anisotropy of $\delta \approx 10^{-29}$ would produce a field strength of about a megagauss.

The large quantity of potential charge and current available in an astrophysical body is quite general. For this reason, we generally calculate the electric and magnetic field first, and afterward check that the required charge and current is quite small compared to what is available. It almost always is.

Two other extreme quantities typical in an astrophysical body, which we have already mentioned, are very large electrical inductance and very small resistance. Inductance $\mathcal{L}$ is the number of lines produced by a unit amount of current, so since $B = 2I/R$ and since the number of lines is BR^2, the inductance is of order R. The resistivity is the reciprocal of the conductivity σ and the total resistance is $\mathcal{R} = \eta c/R$. Thus, the $\mathcal{L}/\mathcal{R}$ decay time of the currents and magnetic field is proportional to $R^2/\eta c$. Since η is about the same in the laboratory as in astrophysics, it is the very large size of R that makes the decay time so long. η is inversely proportional to the electron temperature to the three halves power and independent of density, so we find on establishing the correct constant that the decay time is

$$T_{\text{decay}} = \frac{\mathcal{L}}{\mathcal{R}} = \frac{R^2}{10^7} T_{\text{eV}}^{3/2} \quad \text{sec} \tag{2}$$

where the temperature is measured in electron volts. (One electron volt is the same as 10^4 K.) For the sun $R \approx 10^{11}$ cm, and $T_{\text{eV}} \approx 100$, so that $T_{\text{decay}} \approx 10^{17}$ sec. For a piece of the interstellar medium of size 1 parsec, $T_{\text{eV}} \approx 1$, we get $T_{\text{decay}} \approx 10^{30}$ sec. Thus, we see that resistivity cannot destroy a magnetic field in a static astrophysical plasma of a size greater than the size of a star in a Hubble time.

We can imagine what might happen. The electron motion need only be a very small mean drift v_D to produce the field ($v_D/v_{\text{th}} \approx 10^{-30}$ in the sun and $v_D/v_{\text{th}} \approx 10^{-15}$ in the interstellar medium). Our first impression is that a single collision can easily wipe out this incredibly small anisotropy. But if it did, the magnetic field would collapse to zero in the same short time. The resulting rapid change in the flux would create an enormous electric field, which would very quickly reaccelerate the electrons, recreating the necessary anisotropy to produce the magnetic field. Actually, there is always a very small electric field present that sustains the necessary anisotropy against collisions, and this electric field is produced by induction from the magnetic field decaying on the extremely long $\mathcal{L}/\mathcal{R}$ time mentioned above. In fact, this is an alternative way to arrive at the decay time of the magnetic field.

However, the above picture applies to the grossly simplified case of a static plasma and a magnetic field of scale R. If the plasma is moving, the same considerations apply locally in the moving frame. This brings us back to the flux-freezing idea and actually shows that flux-freezing holds for the $\mathcal{L}/\mathcal{R}$ timescale. It is useful to note that flux freezing holds for any astrophysical plasma in which the magnetic field varies on a scale larger than $10^{12}/T_{\text{eV}}^{3/2}$ cm, since at this scale the $\mathcal{L}/\mathcal{R}$ time is the Hubble time.

The above highly qualitative discussion indicates some of the powerful effects that plasma physics can exert on astrophysical phenomena.

In general, we can most simply understand plasma physics as a double reaction between particles and electromagnetic fields. The plasma moves and generates currents, which produce and change the electromagnetic field. The electromagnetic field reacts back on the ionized plasma controlling its motion.

For example, in the corona the magnetic fields force the particles to move along their field lines. Electric fields accelerate the ions and electrons, causing currents to flow, which twist the field lines and change the paths along which the particles flow. Waves arising from below the photosphere propagate into the corona. These waves consist of perturbed magnetic fields accelerating the plasma particles. The accelerated plasma particles then oscillate the perturbed fields, resulting in the wave motion.

We will see that nearly all plasma physics can best be grasped in this manner: the electromagnetic field influences the plasma, and the plasma influences the electromagnetic field. This parallels what we learned in electromagnetism, where we were taught that charged particles generate fields, and fields exert forces on charged particles controlling their motions.

As is well known, a plasma consists of electrons and of ions of various species and charge states. However, it is not so universally appreciated that a plasma also consists of waves or collective motions. Of course, the waves are nothing more than electrons and ions coherently moving. But the wave motion is so coherent that waves can be thought of as independent entities. When we give the number density of ions and electrons in various states of motion, we should also specify the intensities of the waves, i.e., their energies at different wave numbers. Plasma physics also concerns itself with electric and magnetic fields some of whose structure is bound up in waves.

The subject of plasma physics is, on the face of it, a very complex subject. Yet a lot of progress has been made in systematizing it and laying down a number of fundamental principles and equations. We can divide the subject of plasma physics into one part that is well understood and in which reliable results can be obtained theoretically and demonstrated experimentally, and a second part that is still puzzling, in which results are still very much in doubt and over which there is much debate. The second part consists of tough problems important to astrophysics, such as dynamo theory, magnetic reconnection, and particle acceleration. These problems are slowly beginning to yield to intensive attack by theorists, computationalists, experimentalists, and observers.

Magnetic reconnection bears directly on the mechanism of solar flares, the magnetospheric interaction with the solar wind, and magnetic storms and aurora. Dynamo theory is important for the origin and evolution of cosmic magnetic fields, the understanding of the solar cycle, of accretion disks, and the sustainment and behavior of the earth's magnetic field.

These problems are hard because they are primarily three dimensional and they contain no small dimensionless parameters that provide useful limiting cases that are simple to handle, as is the case with the better understood part of plasma physics. The small parameter for both magnetic reconnection and dynamo problems is the resistivity. In proper dimensionless units, it is a very small parameter. But as it becomes small the difficulties of solution become harder and more intractable, rather than easier to handle.

Plasma physics abounds in quantities, such as the Debye length, the gyroradius, and the collision time, which when expressed in the correct units of the problem are very small. These small parameters allow us to approach, through limiting cases, the simplification of a large number of problems. For example, the mean free path is often much smaller than the scale of variation of physical quantities. This allows the plasma to be treated as a fluid for most macroscopic problems. Where this is not a valid approximation, use can then be made of the smallness of the gyroradius, which leads to almost fluid-like equations.

1.1 How Do We Describe a Plasma and Its Electromagnetic Fields?

To completely describe a plasma at any time requires knowing the position and velocity of each particle of each type, for example, the position and velocity of each ion and electron. Each particle position and velocity changes under the influence of the electromagnetic fields, and we must know these fields at every point. Further, these fields change according to Maxwell's equations in which the current is obtained from the particle positions and velocities.

As a practical matter, we must give up such infinite precision and deal with the statistical distribution of the particle positions and velocities. Then from these distributions we can obtain a reasonable value for the charge and current densities that produce the **E** and **B** fields that move the particles. Similarly, the statistical distribution of the particle densities and currents is influenced by the electric and magnetic fields. To carry out this approach requires two steps in level of description.

First, we consider the distribution function f_j, where

$$f_j(r, v, t)d^3rd^3v \tag{3}$$

gives the mean number of particles of type j in a small volume in phase space $d^3\mathbf{r}d^3\mathbf{v}$:

$$d^6N_j = f_j(\mathbf{r_j}, \mathbf{v}, \mathbf{t})\mathbf{d}^3\mathbf{r}\mathbf{d}^3\mathbf{v} \tag{4}$$

Then from the f_j we obtain the mean current density

$$\mathrm{j}(r, t) = \sum_k \frac{e_k}{c} \int f_k(\mathbf{r}, \mathbf{v}, t)\mathbf{v}d^3\mathbf{r} \tag{5}$$

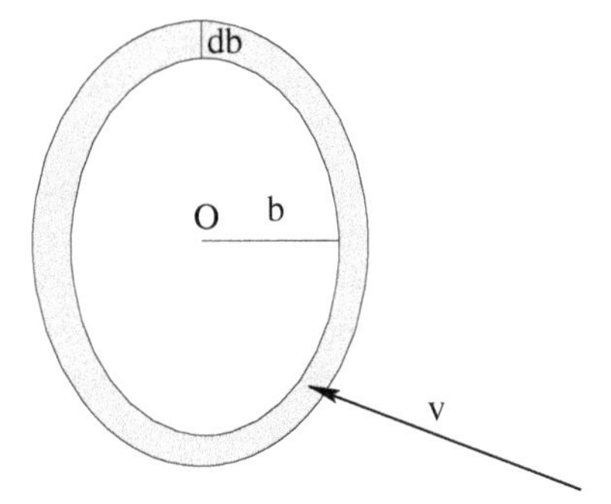

Figure 1.1. Collisions and the stosszahl ansatz

and the mean charge density

$$\rho(r, t) = \sum_k e_k \int f_K(\mathbf{v}, \mathbf{r}, t) d^3\mathbf{r} \tag{6}$$

These can be combined with Maxwell's equations to find **E** and **B**.

However, there is an a important correction to this picture associated with microscopic collisions. If a particle is at **r**, the probability that there is another particle nearby is not uniform, i.e., not exactly equal to $\int f d^3\mathbf{r}$ but on small scales varies erratically. Usually for microscopically stable plasmas this two-particle distribution function is most conveniently treated by Boltzmann's stosszahl ansatz principle. This principle can be roughly expressed as follows.

First, imagine a test particle at rest at O in Figure 1.1. Then imagine an annular zone about O of radius b and thickness db, perpendicular to some velocity **v**. The number of particles with velocity **v** that strike this zone in time dt is taken to be just the average number of particles in a cylinder with this zonal cross section and with height vdt. This number is equal to f times the volume $2\pi b db\, vdt$.

In general, the effect of the collisions between the "test" particle and the incoming particles in this zone is usually small enough so that the cumulative effect of these collisions can be estimated statistically by adding them together independently. The gradual evolution of the velocity of the test particle can thus be calculated. Since this collisional effect is random, the probability distribution for the particle's velocity will diffuse in velocity space. For the case of a test particle initially at rest, the particle will develop a velocity in a somewhat random direction. (If the test particle is moving with velocity **v**, we can apply the above picture by transferring to a frame moving with velocity **v**.) It turns out that there can also be a systematic change in the test particle velocity due to the effect of the many collisional encounters (dynamical friction).

For electric fields on this microscopic scale we must realize that the rapidly varying electric field at a particle is due to the Coulomb fields of these incoming particles passing close by it.

Thus, the electric field (and magnetic field) should be considered as a sum of at least two parts: the large-scale smooth field produced by the large-scale distribution of particles, and the small-scale fields produced by the erratic encounters discussed above. In general, only the first field is discussed explicitly, the small-scale fields and their effect on the particles are treated statistically by collision theory. The large-scale field is predictive, while the small-scale fields are treated statistically.

There are also intermediate-scale fields and particle densities in phase space that are also treated statistically. These are the fields associated with various plasma waves. In a quiescent plasma these waves have very small amplitudes and they have little effect on the particles, However, in certain cases these waves can be excited to larger amplitudes, and their effect on the particles can dominate over normal collisional effects and lead to striking and important results.

To illustrate the behavior of these waves in an important example, consider the propagation of cosmic rays through interstellar medium. If a burst of cosmic rays is suddenly produced by a supernova, the cosmic rays will move away from the supernova at high speed along the interstellar magnetic field. This bulk motion of the cosmic rays will excite Alfven waves, which will grow rapidly. The current in these waves is intense and produces small-scale magnetic fields that interact with the cosmic rays and change the direction of their velocities, forcing their bulk velocity to slow down enough that they no longer excite the Alfven waves. As a consequence, a balance is reached between the waves and the bulk velocity of the cosmic rays, which in a steady state forces the bulk velocity to be approximately equal to the Alfven speed.

The waves themselves consist of coherent electromagnetic fields and the charge and current concentration that produce these fields. Their study is still at the forefront of plasma research and they are generally responsible for nonintuitive phenomena such as this reduction of the motion of cosmic rays to the Alfven speed. They are also responsible for most particle acceleration methods, for the entropy production in collisionless shocks, for the increase of plasma resistivity, and for the rapid conversion of the magnetic energy to heating and bulk particle motion in solar flares.

Although in each wave or wave packet the fields and perturbed particle densities and currents are quite coherent, the actual position and amplitudes of the waves themselves are usually quite random so that they can also be treated statistically.

Thus, the most convenient description of a plasma breaks down first into separate descriptions of the various particle distribution functions in phase space, $f_j(\mathbf{r}, \mathbf{v}, t)$ (six dimensional), and the electromagnetic field, $E(\mathbf{r}, t)$, $B(\mathbf{r}, t)$ (three dimensional). As usual, the fields affect the particles through the equation of motion and the particles affect the fields through Maxwell's equations.

This general description is simplified by breaking each of these descriptions into three parts. The distribution of particles is divided into three parts, a large-scale smooth part, $f(\mathbf{r}, \mathbf{v}, t)$, a smaller scale part on an intermediate scale associated with waves, and a very small-scale part associated with collisions. Similarly, the electric and magnetic fields are divided into the same three parts, a large-scale smooth part, a smaller intermediate-scale part associated with waves, and a very small-scale part associated with collisions.

The smallest scale part, the part that is important for collisional evolution, can be expressed in terms of $f(\mathbf{r}, \mathbf{v}, t)$ by means of Boltzmann's stosszahl ansatz and leads to Boltzmann's collision integral, or for plasmas, essentially the Fokker–Planck equation.

The intermediate part associated with waves can also be treated randomly since the positions of the waves are of no real consequence. However, their amplitudes are of importance. For each possible plasma wave of type j, the intensity $I_j(\mathbf{r}, \mathbf{k})$ as a function of its position $\mathbf{r}$, and its wave number $\mathbf{k}$ must be given.

Thus, a full description of a plasma requires the distribution function for each species j, $f_j(\mathbf{r}, \mathbf{v}, t)$, specified on the large scale, the large-scale electric and magnetic field $\mathbf{E}(\mathbf{r}, t)$ and $\mathbf{B}(\mathbf{r}, t)$, and the mode intensity $I_j(\mathbf{r}, \mathbf{k}, t)$ for the waves. The evolution of each of these quantities is given by its own equations, which involves the other quantities as well. It will be found that this mode of describing the plasma is general enough to encompass most astrophysical plasma phenomena of interest.

It should be noted that the waves themselves can occur on large scales, but these waves are included in the description of the large-scale distributions and fields.

The intermediate-scale waves are described by $I_j(\mathbf{r}, \mathbf{k}, t)$, which depends on the wave number $\mathbf{k}$ as well as position. The wave numbers are large enough that the waves propagate in regions of local homogeneity of the large-scale quantities. The evolution of I will be described by the wave kinetic equation.

Although these matters will be discussed in detail in later chapters we here list the accepted evolution equations for the quantities f_j, E, B, and I. First, $f(\mathbf{r}, \mathbf{v}, t)$ is given by the Fokker–Planck equation

$$\frac{\partial f_j}{\partial t} + \mathbf{v} \cdot \nabla f_j + \frac{e_j}{m_j}\left(\mathbf{E}(\mathbf{r}, t) + \frac{\mathbf{v} \times \mathbf{B}}{c}\right) \cdot \frac{\partial f_j}{\partial \mathbf{v}} + \mathbf{g} \cdot \frac{\partial f_j}{\partial \mathbf{v}} = \left(\frac{\partial f_j}{\partial t}\right)_c + \left(\frac{\partial f_j}{\partial t}\right)_w \tag{7}$$

where e_j and m_j are the charge and mass of the jth species. The first term on the right represents the time rate of change of f_j due to collisions. For plasma particles it can be expressed in terms of the Fokker–Planck operator which in the Landau form reads

$$\left(\frac{\partial f_j}{\partial t}\right)_c =$$

$$\sum_\ell \left[\frac{2\pi(e_j e_\ell)^2 \ln\Lambda}{m_j}\right] \frac{\partial}{\partial \mathbf{v}_j} \cdot \int \left(\frac{f_\ell}{m_j}\frac{\partial f_j}{\partial \mathbf{v}_j} - \frac{f_j}{m_\ell}\frac{\partial f_\ell}{\partial \mathbf{v}_\ell}\right) \cdot \left(\frac{\mathbf{I}}{g} - \frac{\mathbf{g}_{\ell,j}\mathbf{g}_{\ell,j}}{g^3}\right) d^3\mathbf{v}_\ell \tag{8}$$

where $\mathbf{I}$ is the unit dyadic and the collisional evolution of f_j is the sum of encounters with the ℓth species. $f_\ell \equiv f_\ell(\mathbf{r}, \mathbf{v}_\ell, t)$, $f_j = f_j(\mathbf{r}, \mathbf{v}_j, t)$, and $g_{\ell,j} = v_\ell - v_j$. This form of the collision operator was first given by Landau in 1937. It has the pleasing properties that it conserves number, momentum, and energy between the ℓth and jth species.

The last term, $(\partial f/\partial t)_{\text{waves}}$, represents the effect of the waves on the evolution of the distribution function. This effect occurs mainly through the interaction of particles resonant with the waves. The resonance can be either a Cerenkov resonance, in which case the particles are traveling close to phase velocity of the wave $\omega - \mathbf{k}\cdot\mathbf{v} \approx 0$, or, in the presence of a magnetic field, a cyclotron resonance, where the Doppler-shifted frequency of the wave seen by the particle is close to a harmonic of the cyclotron frequency $\omega - \mathbf{k}\cdot\mathbf{v} \approx n\Omega$, where n is an integer (possibly negative). Higher order resonances in an expansion in the amplitude of the waves can also occur between the particle and two waves $\omega_1 - \omega_2 = (\mathbf{k}_1 - \mathbf{k}_2)\cdot\mathbf{v}$. In addition, there is spontaneous emission of the waves. Because of the multiplicity of the particle interactions we will not attempt to write a general formula for the wave–particle interactions, but will treat these interactions in detail in specific examples.

For illustrative purposes, we give the quasilinear form of the wave–particle collision term when the waves are electrostatic and the magnetic field is absent:

$$\left(\frac{\partial f}{\partial t}\right)_{\text{waves}} = \frac{\partial}{\partial \mathbf{v}} \cdot \left[\pi\frac{e^2}{m}\int d^3\mathbf{k}\hat{\mathbf{k}}\hat{\mathbf{k}}I(\mathbf{k},\mathbf{r},t)\delta(\omega - \mathbf{k}\cdot\mathbf{v}) \cdot \frac{\partial f}{\partial \mathbf{v}}\right] \tag{9}$$

where I is the spectrum of the electric field normalized so that

$$\left\langle\frac{E^2}{8\pi}\right\rangle = \int I d^3\mathbf{k} \tag{10}$$

and $\hat{\mathbf{k}} = \mathbf{k}/k$ is the unit vector in the $\mathbf{k}$ direction.

Let us turn to the smooth large-scale parts of the electromagnetic fields $\mathbf{E}(\mathbf{r}, t)$ and $\mathbf{B}(\mathbf{r}, \mathrm{t})$. Their evolution is governed by Maxwell's equations:

$$\nabla\cdot\mathbf{B} = 0 \tag{11}$$

$$-\frac{1}{c}\frac{\partial \mathbf{B}}{\partial t} = \nabla\times\mathbf{E} \tag{12}$$

$$\nabla \cdot \mathbf{E} = 4\pi q \tag{13}$$

$$\nabla \times \mathbf{B} = 4\pi \mathbf{j} + \frac{1}{c}\frac{\partial \mathbf{E}}{\partial t} \tag{14}$$

where q, the net charge density, and $\mathbf{j}$ are given by equations 5 and 6. These Maxwell equations, (11)–(14), represent the total effect of the smooth plasma on the smooth fields. The remainder of the fields, on smaller scales, is bound up with the Landau collisional term in equation 8, and with the waves described by the wave kinetic equation below.

Next, let us consider the intermediate scale and the waves. There are different types of waves in different wave number regions, but in any given wave number region $\mathbf{k}$ there are generally only one or two types. The behavior of these waves, their dispersion relation $\omega(\mathbf{r}, \mathbf{k}, t)$, and their mode structure or polarization depend on the large-scale fields and distributions of $\mathbf{E}$, $\mathbf{B}$, and f. Further, the change in their amplitude, growth, or damping rate γ also depends on the distribution of resonant particles given by f. If the waves are unstable, their amplitude will be limited by various nonlinear interactions among themselves.

The evolution of the waves is governed by the wave kinetic equation,

$$\frac{\partial I}{\partial t} - \frac{\partial \omega}{\partial \mathbf{r}}\frac{\partial I}{\partial \mathbf{k}} + \frac{\partial \omega}{\partial \mathbf{k}}\frac{\partial I}{\partial \mathbf{r}} = 2\gamma I + \left(\frac{\partial I}{\partial t}\right)_{\text{coll}} + S \tag{15}$$

The second term on the left-hand side refers to the refraction of the wave and the third to its group velocity. The first term on the right refers to the damping or growth of the waves. The second term on the right represents the nonlinear wave–wave interaction, while the last term S refers to the various ways in which the waves can get started, such as bremsstrahlung or Cerenkov emission.

These processes, represented by γ and S, are complementary to the wave–particle interaction terms in the kinetic equation (7) for the particle distribution function. Whenever waves gain energy due to an instability ($\gamma > 0$), the $(\partial f/\partial t)_{\text{wave}}$ term in the Fokker–Planck equation reduces the energy of the smoothed parts of the particle distribution function. There is, in addition, a more subtle effect related to the refraction of the waves. If the medium is being compressed, the waves gain coherent energy, which exerts a backpressure force on the smoothed plasma to reduce its momentum and energy. This is included as a nonresonant part $(\partial f/\partial t)_{\text{waves}}$ in (7). In a fluid description it is generally expressed in terms of a wave pressure.

In summary, a plasma and its electric and magnetic field interactions break down into six components. The smallest scale chaotic components are taken care of by collision theory in terms of the other four components. The more coherent intermediate-scale parts of $\mathbf{E}$, $\mathbf{B}$, and f are described by the distribution of energies over the wave spectrum of the various wave modes, I. Thus, we need only keep track of three components: $f(\mathbf{r}, \mathbf{v}, t)$, the one-particle distribution functions on large scales; $\mathbf{E}$ and $\mathbf{B}$, the electromagnetic fields on large scales; and $I(k)$, the intemediate-scale wave component. These quantities are governed by the kinetic or Fokker–Planck equations, Maxwell's equations, and the wave kinetic equation, respectively.

The above method of describing a plasma is quite general. It covers most of the phenomena encountered in astrophysics. However, because of its generality it is difficult to handle except in limiting cases. In specific examples these limiting cases are provided by some parameter being very small, which allows a limiting case to be taken. In this book we will frequently make use of these limiting cases when dealing with plasma phenomena.

For example, because the Debye length is very small, the positive and negative charge densities of the plasma very nearly cancel. Because each separate charge density is enormous, this cancellation is essential, otherwise the electric fields generated would be enormous and lead to rapid motions of the electrons, which would immediately short out these fields and restore this quasi-neutral balance. However, because charge densities of the individual species are so large, even after substantial cancellation there is more than enough net charge to produce the electric fields to enforce this near cancellation. This cancellation is termed "charge neutrality" and means no more than that the positive and negative charge densities are nearly equal.

Another equally important plasma limit is the fluid limit that is valid when the collision rate of each species with itself is large compared to the macroscopic rates of change. At the same time, the mean free paths are short compared to macroscopic dimensions. As a consequence, before plasma conditions change significantly each species collisionally relaxes to a local Maxwellian with its own mean density ρ, temperature T, and mean velocity V. These parameters, then, are sufficient to characterize the plasma state and they are governed by the fluid equations. These equations are the magnetohydrodynamic or MHD equations. They consist of

$$\frac{\partial \rho}{\partial t} + \nabla \cdot (\rho \mathbf{V}) = 0$$

the continuity equation for the total density ρ,

$$\rho \left(\frac{\partial \mathbf{V}}{\partial t} + \mathbf{V} \cdot \nabla \mathbf{V} \right) = -\nabla p + \mathbf{j} \times \mathbf{B} + \rho \mathbf{g}$$

the equation of motion for the velocity $\mathbf{V}$, and some sort of entropy equation for the pressure p. If collisions are very strong, then the entropy per unit mass is conserved following the fluid:

$$\frac{\partial}{\partial t} \left(\frac{p}{\rho^{\gamma}} \right) + \mathbf{V} \cdot \nabla \left(\frac{p}{\rho^{\gamma}} \right) = 0$$

In addition to these equations, the equation of the motion of the electron fluid

$$\rho_e \frac{d\mathbf{V}_e}{dt} = -n_e e \left(\mathbf{E} + \frac{\mathbf{V}_e \mathbf{x} \mathbf{B}}{c} \right) - \nabla p_e + \mathbf{F}_{ei} + \rho_e \mathbf{g}$$

is also important. $\mathbf{F}_{ei}$ is the force exerted on the electrons by all the other species ions and neutrals. Except at very high frequencies the inertial and gravitational terms are negligible. The resulting equation is called the generalized Ohm's law. It couples the fluid properties to the electromagnetic fields. This fluid description is not as universally valid as the "charge neutrality" condition, but when applicable it leads to enormous simplification of plasma problems.

It is almost always the case in astrophysics that most of the plasma species are properly described by fluid equations. However, a single component, such as the cosmic rays, may be present, which is collisionless and must be described by the kinetic equation (7). For this species, binary collisions are negligible and the term $(\partial f/\partial t)_{\mathrm{waves}}$ is very important. This species is coupled through waves to the rest of the plasma by its pressure, which must be included in the main fluid equation of motion.

References

Tonks, L., and I. Langmuir 1929. *Physical Review* **33**, 195.

Landau, L.D. 1937. *ZhETF (Journal Experimental and Theoretical Physics USSR)* 7, 203.

Chapter 2

PARTICLE MOTIONS

AS MENTIONED IN THE INTRODUCTION, the subject of plasma physics can be thought of as the interaction between particles and electromagnetic fields. This chapter is concerned with how these fields affect particles.

For the treatment of a large numbers of particles, the statistical approach involving the smoothed distribution functions, i.e., the kinetic description in phase space, or the more reduced fluid description of the mean density of particles is best. However, plasmas do actually consist of individual particles, electrons and ions, and a better appreciation of plasmas is gained by understanding the individual motions of these particles in electromagnetic fields.

In almost all cases, the motion of individual particles is primarily controlled by a magnetic field, **B**. Any single particle will, to lowest order, execute a helical motion along some field line. The helical motion consists of the combination of a circular motion about the field line together with a uniform motion along the field line. This circular motion involves a circle, of radius ρ, perpendicular to the field line, which is generally very small compared to the scale of variation of the magnetic field.

2.1 Motion in a Uniform Magnetic Field

We consider the motion of a particle with charge e_j and mass m_j about a line of force of a uniform magnetic field, B_0 (and take the electric field to be zero). The equation of motion is

$$m_j \frac{d\mathbf{v}}{dt} = e_j \frac{\mathbf{v} \times \mathbf{B}_0}{c} \tag{1}$$

Take Cartesian coordinates with $\mathbf{B}_0$ along the z axis. The solution of this equation is then

$$\begin{aligned} v_x &= v_\perp \sin(\Omega_j t + \phi) \\ v_y &= v_\perp \cos(\Omega_j t + \phi) \\ v_z &= v_\parallel \end{aligned} \tag{2}$$

where $\Omega_j \equiv e_j B_0 / m_j c$, and $v_\perp$, $v_\parallel$ and ϕ are constants. The position $\mathbf{r}$ is obtained from $d\mathbf{r}/dt = \mathbf{v}$ and is

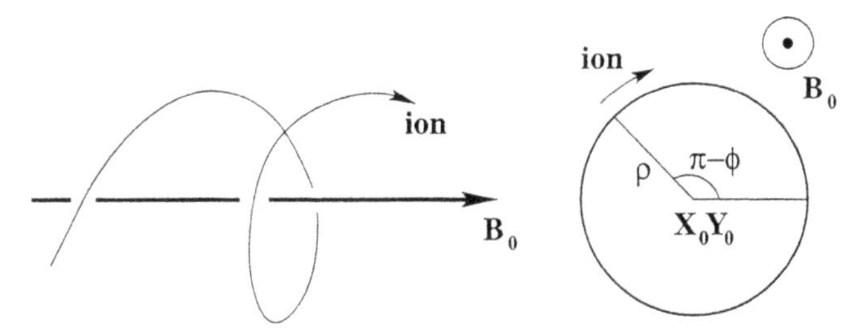

Figure 2.1. Left-handed circular motion of an ion

$$
\begin{aligned}
x &= -\rho_j \cos(\Omega_j t + \phi) + X_o \\
y &= \rho_j \sin(\Omega_j t + \phi) + Y_0 \\
z &= v_\parallel t + Z_0
\end{aligned}
\tag{3}
$$

where $\rho_j \equiv v_\perp/|\Omega_j|$, and X_0, Y_0, and Z_0 are additional constants. Thus, the solution of the particle motion is specified by six constants: $v_\perp$, $v_\parallel$, ϕ, and the vector $\mathbf{R}_0 = (X_0, Y_0, Z_0)$. $\mathbf{R}_0$ is the center of gyration at $t = 0$.

Let us first consider the helical motion of the proton, $e_j = +e$, $m_j = M$, where e the electronic charge, and M the proton mass. Throughout this book the proton mass will be denoted by a large M and the electron mass by a small m and the subscript j will be generally disregarded. Further, when I refer to an ion it should be understood to be a proton unless otherwise specified.

The perpendicular motion, i.e., the motion projected onto the x–y plane is indicated in Figure 2.1. The motion is circular with radius ρ and is in the left-handed sense relative to B. (Putting the thumb of your *left* hand along B, your fingers give the sense of motion.) ϕ gives the phase of the motion at time $t = 0$. At $t = 0$ the center of the circle is at X_0, Y_0, Z_0, and so if we allow this center to move along the line $x = X_0$, $y = Y_0$ at constant velocity $v_\parallel$, then the ion rotates left-handedly around this moving circle of radius ρ with angular velocity Ω.

In summary, the motion of any ion is given by specifying the motion of its gyrocircle. The latter is specified by its radius $\rho_i = v_\perp/\Omega_i$. The line along which its center moves is $x = X_0$, $y = Y_0$. The third equation in (3) specifies where along the line its center is.

In plasmas, these quantities have standardized names,

- $v_\perp$ is the perpendicular velocity
- $v_\parallel$ is the parallel velocity
- ρ is the gyration or cyclotron radius
- Ω is the cyclotron frequency
- The center of the circle is the gyrocenter of the particle

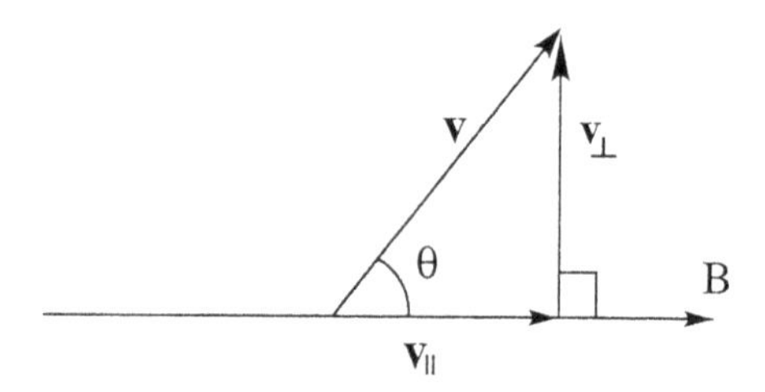

Figure 2.2. The definition of the pitch angle θ

In addition, $v_\parallel$ and $v_\perp$ are given by

$$\begin{aligned} v_\perp &= v \sin\theta \\ v_\parallel &= v \cos\theta \end{aligned} \tag{4}$$

where θ is called the pitch angle. $\theta = 90°$ corresponds to pure circular motion, $\theta = 0$ corresponds to pure straight-line motion, and the size of θ gives the flatness of the helical orbit.

Handy numerical facts that should be memorized are

$$\begin{aligned} \Omega_i &\approx 0.96 \times 10^4 B \text{ sec}^{-1}, & \Omega_e &\approx 1.8 \times 10^7 B \text{ sec}^{-1} \\ v_i &\approx 1.4 \times 10^6 \sqrt{\epsilon_i} \text{ cm/sec} & v_e &\approx 5.9 \times 10^7 \sqrt{\epsilon_e} \text{ cm/sec} \end{aligned} \tag{5}$$

where ϵ_i and ϵ_e are the ion and electron energies in electron volts.

The particle describes a screw-type motion with the pitch of the screw, divided by 2π, equal to $v_\parallel/\Omega = v\cos\theta/\Omega$. For fixed v the pitch of the screw decreases as the "pitch angle" increases, which is a counterintuitive. However, the use of the name "pitch angle" for the angle **v** makes with **B** (figure 2.2) is so well established that it is advisable not to think of the true pitch of the screw.

For electrons, the motion is closely analogous, but with three significant differences:

1. The motion about **B** is right-handed.
2. The cyclotron frequency of the electron, Ω_e, is very large, a factor M/m larger than the ion's cyclotron frequency. (We take Ω_e as positive.)
3. For equal energies $Mv_i^2 = mv_e^2$ its gyroradius is much smaller than the ion gyroradius. $\rho = v_\perp/\Omega$ is smaller by $\sqrt{M/m} \times m/M = \sqrt{m/M}$. (The larger electron velocity partially compensates for the much larger cyclotron frequency.)

The gyromotion of a heavier ion of charge Ze and mass AM is obtained by replacing e/M for a proton by Ze/AM. Thus, for example, a singly charged argon atom has a cyclotron frequency 40 times smaller than that of a proton, and, for the same energy, a gyroradius $\sqrt{40}$ times larger than that of a proton.

It was remarked that the gyroradius of a particle is small compared to the typical scale size for the field. The following three examples should make this plausible:

1. A $\frac{1}{4}$-eV proton in the kilogauss field of a sunspot has cyclotron frequency $10^4 B \approx 10^7$ sec^{-1}, a velocity of $\approx 0.7 \times 10^6$ cm/sec ($\epsilon = \frac{1}{4}$eV), and a gyroradius $\rho = v/\Omega$ of 0.07 cm. In the corona, with a field of ≈ 1 G, a hotter 100-eV proton has a gyroradius of 14 meters.
2. A 1-eV proton in the interstellar medium, where the field strength is 3 μG, has a cyclotron frequency of 3×10^{-2} sec^{-1} and a gyroradius of $v/\Omega = 1.4 \times 10^6/3 \times 10^{-2} \approx 5 \times 10^7$ cm.
3. A keV proton in the earth's dipole field at 5 earth radii, where $B \approx 0.5/(5)^3 \approx 4 \times 10^{-3}$ G, has a cyclotron frequency of 40 sec^{-1}, a velocity of 4.4×10^7 cm/sec, and a gyroradius of $\approx 1.1 \times 10^6$ cm or ≈ 10 km.

All of these are microscopically small lengths compared to the scales of the relevant fields. These examples make the assumptions that ions have a "small gyroradius" very plausible. Of course, the electron gyroradius is even smaller.

We have estimated the gyroradius for particles with a 90° pitch angle. If $\theta = 0$, then the corresponding lengths refer to the "pitch" of the screw motion. Since these are maximum values, we can say the motion of the particle along the field line during one cyclotron period is also very small compared to the scale of the field.

Thus, the correct way to view the particle is by using a "microscope" to look at it on lengths small compared to the scale of variation of the field. On this scale, the field looks very uniform and the particle motion is very closely a cyclotron motion. Any nonuniformity of the field on the gyroradius scale produces only small deviations from pure helical motion.

To reiterate, any particle's orbit can be considered as a circle locally perpendicular to the field and moving along it. Its center is the gyrocenter. Keeping track of the position of this gyrocenter, the gyroradius of the circle, and the parallel motion is essentially all we need to properly describe the motion in a nonuniform field.

Before considering the corrections to the motion of a proton in more general nonuniform fields, let us pause to discuss the relativistic motion of a charged ion in a uniform magnetic field.

$$\frac{d}{dt}(M\gamma \mathbf{v}) = e\frac{\mathbf{v} \times \mathbf{B}}{c} \tag{6}$$

where $\gamma = (1 - v^2/c^2)^{-1/2}$. Dotting this equation with $\mathbf{v}$ shows that

$$\begin{aligned} \mathbf{v} \cdot \frac{d}{dt} M(\gamma \mathbf{v}) &= \mathbf{M}\dot{\mathbf{v}} \cdot \gamma \mathbf{v} + \mathbf{M}\dot{\gamma}\mathbf{v}^2 \\ &= M\left(\frac{1}{2}\gamma + v^2 \frac{d\gamma}{dv^2}\right)\frac{dv^2}{dt} = 0 \end{aligned} \tag{7}$$

or v^2 is a constant. (Dots here mean time derivatives.) This means that γMc^2, the particle energy, is also constant. The magnetic field cannot change the kinetic energy of the particle. (If E is zero, this is true in a general nonuniform magnetic field as well.) Note that γ involves $v_\parallel$ and $v_\perp$. Thus, if we set $\gamma M = M_{\text{rel}}$, the relativistic equation 6 reduces to equation 1 with M replaced by the relativistic mass M_{rel}. Hence, in a uniform field, all the results about nonrelativistic motion of a proton can easily be generalized to relativistic motion, and clearly the same applies to electrons.

Let us now collect together the basic facts about relativistic as well as nonrelativistic particle motion in a uniform magnetic field. Ions and electrons move in a helix. The perpendicular velocity rotates with gyrofrequency

$$\begin{aligned} \Omega_i &= \frac{eB}{\gamma mc} \approx \frac{10^4 ZB}{A\gamma} \quad (\text{sec})^{-1}(\text{ions}) \\ \Omega_e &\approx 1.8 \times 10^7 \frac{B}{\gamma} \quad (\text{sec})^{-1}(\text{electrons}) \end{aligned} \tag{8}$$

These gyrofrequencies are independent of energy for nonrelativistic particles and inversely proportional to the total energy for ultrarelativistic particles. For nonrelativistic particles, the electron gyrofrequency is much larger than the ion gyrofrequency. However, if the electrons have a relativistic energy greater than Mc^2, then their gyrofrequency is essentially the same as for protons of the same relativistic energy. In the range 1 MeV to 1 GeV the ratio drops from 2000 to 1.

The perpendicular gyroradius is

$$\rho_\perp = \frac{v_\perp}{\Omega} = \frac{\gamma m v_\perp c}{eB} = \frac{p_\perp c}{eB} = \frac{pc}{eB} \sin\theta \tag{9}$$

where $p_\perp$ is the perpendicular momentum. pc is often called the rigidity of the particle and reduces to its energy for ultrarelativistic motion. Relativistic particles have the rather peculiar property that for fixed $v_\perp$ their gyroradius increases as $v_\parallel$ increases because of the γ factor.

Again for nonrelativistic particles with the same v the electrons have a much smaller gyroradii, smaller by the mass ratio $m/M \approx 1/2000$ than the proton gyroradius. For the same energy, $v_\perp$ is larger for electrons by $\sqrt{M/m}$ so the electron gyration radius is smaller than the proton gyroradius only by the more modest factor of the square root of the mass ratio $\sqrt{m/M} \approx 1/40$. For GeV electrons the ratio is $\sqrt{m_{\text{rel}}/M_{\text{rel}}}$ and the radii are the same for equal relativistic energies.

All the above facts are very simple but important and it is useful to keep them in mind.

2.2 Motion of a Particle in a Nonuniform Magnetic Field

Let us now consider the motion of ions and electrons in more complicated fields. First, take a magnetic field in the z direction whose strength varies in the x direction. Let the magnitude B_z vary only slightly over the gyroradius. Then to lowest order the motion is still a uniform circular motion. According to equation 9 the curvature $1/\rho$ is slightly greater at the top of the orbit and slightly smaller at the bottom, so the orbit doesn't close but moves to the left as in figure 2.3 with a "drift" velocity of order

$$V_D \approx \frac{\rho}{L} v_{\text{th}} \tag{10}$$

The first factor, ρ/L, represents the nonuniformity of the field and the lack of cancellation of the y motion. The scale length of the field $L \approx B/|\nabla B|$. The actual result has a factor of $\frac{1}{2}$.

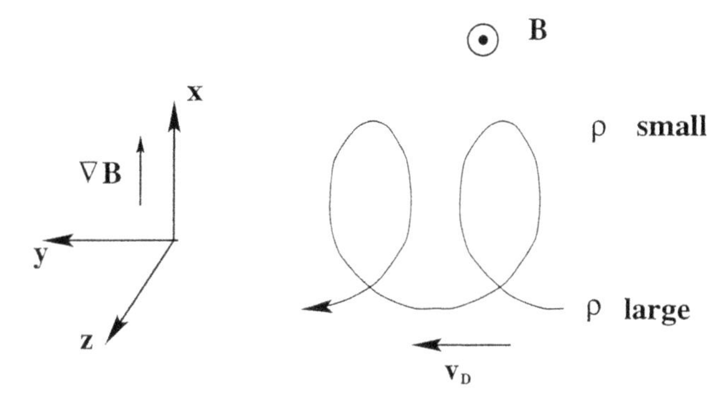

Figure 2.3. Ion motion in a nonuniform magnetic field

Let us understand this result from a different, more physical, perspective. Let us consider why there must be any y drift at all. Assume that there is none and that the circular orbit is closed. The x force on the ion averaged over one orbit is

$$\begin{aligned} \frac{e}{c}\left(\overline{v_y \delta x}\right)\frac{\partial B}{\partial x} &= -\frac{e}{c}\rho v_\perp \overline{\cos^2 \Phi}\frac{\partial B}{\partial x} \\ &= -\frac{1}{2}\frac{ev_\perp^2}{c\Omega}\frac{\partial B}{\partial x} = -\frac{1}{2}\frac{Mv_\perp^2}{B}\nabla B \end{aligned} \tag{11}$$

where $\Phi = \Omega_i t + \phi$ is the phase factor in equations 2 and 3 and the average is taken over one period. (The force is in a direction away from the stronger field, B.)

The factor in front of ∇B,

$$\mu = M\frac{v_\perp^2}{2B} \tag{12}$$

is called the magnetic moment of the spiraling particle. Thus, without the y drift of the particle, there is an unbalanced force in the negative x direction. But if there is a V_D drift velocity in the y direction, as indicated above, then there is an additional x force eV_DB/c, which will balance the force due to the action of the inhomogeneous field on the magnetic moment. For an electron we get the same result with M replaced by m. The force depends on the particle energy and not its charge or mass, so an electron with the same energy as an ion will see exactly the same force.

Equating these two forces, we find the y drift velocity of an ion is

$$V_y = \frac{\mu \nabla B}{eB/c} = \frac{v_\perp^2}{2\Omega_i}\frac{\nabla B}{B} \approx \frac{\rho}{2L} v_\perp \tag{13}$$

as derived above (equation 10). A simple examination of the forces in the diagram shows that for an ion the motion must be in the leftward direction, so the $\mathbf{V}_D \times \mathbf{B}$ force is upward. If the particle is an electron, the $\mu \nabla B$ force is still downward, but the Lorentz force is reversed, so the electron must drift in the rightward direction and with a velocity $\approx \rho v_\perp / L$, as a simple diagram with the electron spiraling in the opposite direction shows. A way to remember the direction of the $\mu \nabla B$ force is that it is always away from the region of stronger field strength.

For a relativistic particle the magnetic moment is $\mu = v_\perp p_\perp / 2B$ and the force is μ times the negative of the gradient in B. For an ultrarelativistic particle moving perpendicular to $\mathbf{B}$, μ is just one-half of its relativistic energy divided by B. The results for the drift of a relativistic particle can still be written as $\rho v_\perp / L$ as we can see both from inspection of the diagram or by replacing m by m_{rel} in the general formula. The sign of the direction of the drift in the y direction can be obtained either by this force balance or by picturing the nonuniformity of curvature of the particle, as in figure 2.3.

The spiraling ion, when we average its motion over a gyration period, is very similar to a coil of wire with radius ρ and current $\Omega e / 2\pi c$ flowing in the left-handed sense about $\mathbf{B}$. The above force argument thus can come directly from an elementary electromagnetic calculation. Further, because such a wire acts like a small magnetic dipole with magnetic moment

$$\mu = IA = \left(\frac{\Omega e}{2\pi c}\right) \pi \rho^2 = M \frac{v_\perp^2}{2B} \tag{14}$$

but in the direction opposite to $\mathbf{B}$, our gyrating particle also acts as such a dipole and produces a dipole magnetic field about itself. A plasma consisting of a large number of electrons and ions possesses a magnetization that is the sum of these dipoles, and that perturbs any magnetic field in which it is embedded, so as to decrease its strength. That is to say, the plasma is diamagnetic. Notice further that the ions gyrate left-handedly, producing a left-circulating current about B, and that the electrons, which gyrate right-handedly, have negative charge, so they also produce a left-circulating current about B. Note also from equation 14 that, if the electron and ion energies are equal, the magnetic moments are equal. The area of gyration is smaller for the electron by m/M but its current is larger by M/m.

Let us next consider a charged ion in a uniform field $\mathbf{B}$ but with a uniform transverse force $\mathbf{F}$, for example, an electric force $e\mathbf{E}$, perpendicular to B. The orbit will be distorted by the force. Consider $\mathbf{B}$ in the z direction and $\mathbf{F}$ in the x direction, as in figure 2.4. Then the ion has an increased velocity

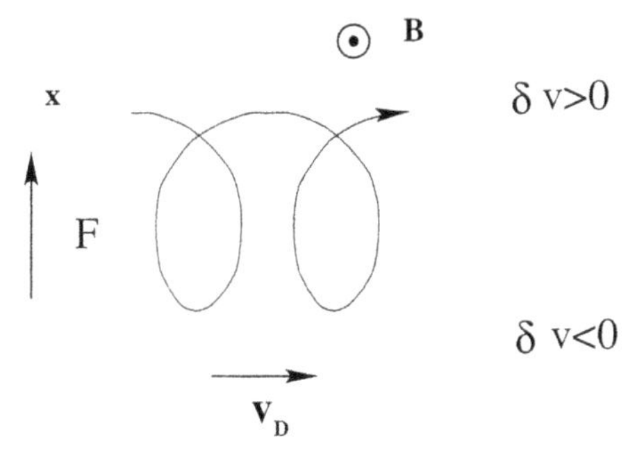

Figure 2.4. Motion of an ion in the presence of a transverse force, F

$\delta v \approx F\rho/Mv$ and radius of curvature at $\delta x > 0$, so it will drift to the right with drift velocity

$$V_D \sim \delta v \approx \frac{F\rho}{Mv_\perp} = \frac{F}{M\Omega} \tag{15}$$

The correct factor again comes from balancing forces

$$e\mathbf{V}_D \times \frac{\mathbf{B}}{c} + \mathbf{F} = 0 \tag{16}$$

or

$$\mathbf{V}_D = c\frac{\mathbf{F} \times \mathbf{B}}{eB^2} \tag{17}$$

V_D has magnitude $cF/eB = F/M\Omega$ and here there is no missing factor of one-half in equation 15. Formula 17 is really the general formula for transverse drifts if we include forces such as $\mu\nabla B$.

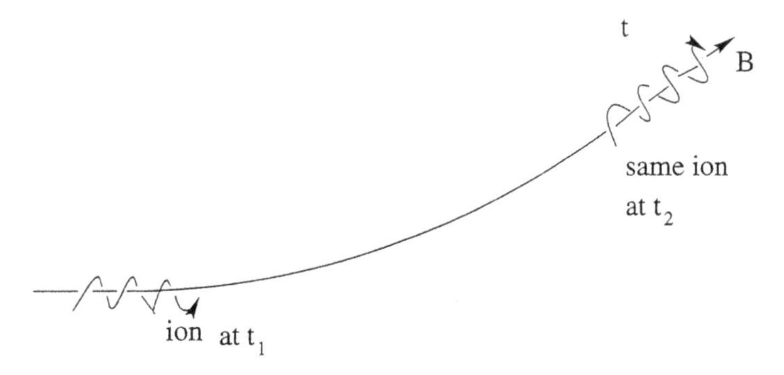

Figure 2.5. Motion of an ion on a curved magnetic field line

There is one further specific transverse drift: the centrifugal, or curvature, drift. Suppose an ion is moving on a curved magnetic field line with radius of curvature R and curvature κ, as in figure 2.5. It does not change its energy or parallel velocity. At the two times, t_1 and t_2, it is spiraling about the magnetic field at two different locations. Its momentum averaged over a gyration period is a parallel momentum along **B**. But **B** changes direction between t_1 and t_2 by an angle $v_\parallel(t_2 - t_1)/R$. A force is required to produce this change and this force is provided by the transverse drift:

$$e\mathbf{V}_D \times \frac{\mathbf{B}}{c} = \boldsymbol{\kappa} M v_\parallel^2 = -\mathbf{F}_c \tag{18}$$

Thus, the curvature drift comes from the general equation 17 with a centrifugal force, $\mathbf{F}_c = -Mv_\parallel^2 \boldsymbol{\kappa}$.

It turns out that these are the only essential deviations of a magnetic field from uniformity that produce transverse drifts, so the total drift velocity of a particle can be written vectorially as

$$\mathbf{V}_D = \left(-\mu\nabla B - \boldsymbol{\kappa} m_j v_\parallel^2 + e_j\mathbf{E} + m_j\mathbf{g}\right) \times \frac{c\mathbf{B}}{e_j B^2} \tag{19}$$

where the subscript j distinguishes whether the particle is an ion or an electron, and $\mathbf{g}$ represents the effective gravitational acceleration resulting from any forces other than the electric force that may act on the particle, such as frictional forces with other particles due to collisions, radiation pressure, inertia, true gravity, etc.

In practice, we find that all drifts except those due to the electric field are small, of order $(\rho/L)v$, where L is the scale size of the field variation, or the radius of curvature, or the gravitational scale height, etc. If we are only concerned with the rough position of the particle, these small drifts can be ignored. For this purpose we need only consider the drift due to the electric field:

$$\mathbf{V}_E = c\frac{\mathbf{E} \times \mathbf{B}}{B^2} \tag{20}$$

Because of its form it is generally referred to as "the E cross B drift."

On the other hand, if we want the electric current due to the particle motions from these additional drifts, we need the more precise formula of equation 19 because of the very large charges in a plasma.

However, as we shall see, there are better ways to find the current. The above discussion is given to make you aware of the origin of these currents. Thus, the above considerations are mostly of use for gaining insight into the microbehavior of the plasma in terms of individual particles. We will not make much use them in the calculations in this book.

2.3 Magnetic Mirrors

We now consider the orbit of an ion in a magnetic field that varies along itself. Consider a magnetic field in the z direction that increases in the z direction. Because the magnetic field is divergence free, it will have components other than the z component on neighboring lines. We therefore consider a field that is purely in the z direction only on the z axis, $x = y = 0$, but has x and y components near the z axis.

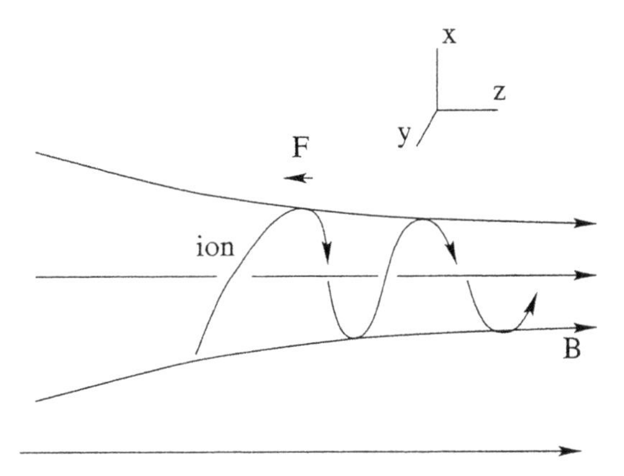

Figure 2.6. Motion in a magnetic field increasing along itself (the mirror field)

Take an ion spiraling about the z axis, as in figure 2.6. Off the axis at $x = \rho$, $y = 0$, a field $B_x < 0$ exerts a force on $v_y < 0$, which in the diagram is in the negative direction. Adding a similar force from B_y, acting on v_x, at $x = 0$, $y = \rho$, and taking it equal to the force due to B_x, we find that the magnitude of the sum of the forces averaged over a gyroorbit is $e\rho v_\perp B_x/c$. But $B_x \approx -\rho\partial B_z/\partial z$, so that we have

$$M\frac{dv_Z}{dt} \approx -\frac{e}{c}\rho v_\perp \frac{\partial B_z}{\partial z} \approx -\mu \nabla B_z \tag{21}$$

In fact, a more careful calculation shows that the order of unity factor in this calculation is actually unity and we get the familiar result that the force on a particle is just $-\mu\nabla|B|$ as before, but now with $|B|$ varying along **B** instead of across **B**. (Again the force is away from the region of stronger B.) Thus, as a particle attempts to move into a stronger field, it is repelled by the increasing field. Further, a closer examination of figure 2.6 shows that the $v_z B_x$ and $-v_z B_y$ forces increase $v_\perp$ as $v_\|$ decreases. This is, of course, necessary so that the ion energy, $M(v_\perp^2 + v_\|^2)/2$, remains constant.

Does the increase in $v_\perp$ change μ? The answer is no. In fact, using the conservation of energy, valid in a constant magnetic field $v_\perp^2 + v_\|^2 =$ constant, we have

$$\begin{aligned} 0 = v_\perp \dot{v}_\perp + v_\| \dot{v}_\| &= \left(\frac{v_\perp^2}{2}\right)^{\cdot} - \frac{v_\perp^2}{2B} v_z \frac{dB}{dz} \\ &= \left(\frac{v_\perp^2}{2}\right)^{\cdot} - \frac{v_\perp^2}{2}\frac{\dot{B}}{B} \end{aligned} \tag{22}$$

where dots indicate time derivatives. Thus,

$$\frac{d\mu}{dt} = \frac{M}{2}\frac{d}{dt}\left(\frac{v_\perp^2}{B}\right) = 0 \tag{23}$$

Thus, in this specific case, the magnetic moment is a constant. We will soon show that the constancy of the magnetic moment is a very general result, (see problem 4).

Let us again consider the motion of a particle into a strong field. We have just seen that the parallel motion continually decreases, and may go to zero and change sign because of the backward force produced by the transverse components of the converging field. The particle is then reflected by an increasing field just as though it were bouncing off of a mirror. This is the general way of expressing this result: If a field increases along itself, it is said to have a magnetic mirror, and the force that produces the deceleration and reflection of the parallel motion is called the magnetic mirror force.

If a particle moves into an increasing magnetic field that goes from B_0 to $B_{\max}$, and then decreases, will it necessarily be reflected? The answer is: not necessarily. If it is moving strictly parallel to B with zero pitch angle, then $v_\perp$ is zero, $\mu = 0$, and there is no mirror force on it at all. If it is moving very slowly along **B**, so that $v_\parallel$ is small and $v_\perp$ large it will certainly be reflected. Thus, at any point on the line there is a range of pitch angles $0 < \theta < \theta_R$, such that a particle with a pitch angle inside this range is able to pass over $B_{\max}$, and a particle outside of it will be reflected.

If the pitch angle is θ, when $B = B_0$ its magnetic moment is $\mu = mv^2 \sin^2\theta/2B_0$ and is constant. $mv^2/2$ is also constant, so

$$\mu' = \frac{\sin^2\theta}{B_0} = \text{constant} \tag{24}$$

If it can reach $B_{\max}$, with pitch angle $\theta_{\max} < \pi/2$, then

$$\mu' = \frac{\sin^2\theta_{\max}}{B_{\max}} < \frac{1}{B_{\max}} \tag{25}$$

Thus, the necessary and sufficient condition for the particle to reach $B_{\max}$ from a point where the field strength is B_0 is that its pitch angle satisfies

$$\sin^2\theta < \sin^2\theta_R = \frac{B_0}{B_{\max}} \tag{26}$$

If $\theta > \theta_R$, and the particle could reach $B_{\max}$, we would have $\sin\theta_{\max} > 1$ which is impossible. (It will be reflected from a somewhat weaker field before it reaches $B_{\max}$.)

The possibility that a particle can be reflected by an increasing mirror has been made use of in an attempt to confine a plasma in an open-ended chamber for fusion. This is the mirror machine (see figure 2.7). In the example shown in figure 2.7, the field strength is uniform in the chamber, and increases its strength at the ends to $B = B_{\max}$. Particles in the central region with pitch angles smaller than θ_R, where

$$\sin^2\theta_R = \frac{B}{B_{\max}} \tag{27}$$

will escape, while those with $\theta > \theta_R$ will be trapped and confined. Thus, particles in the $\theta < \theta_R$ region of velocity space are lost and are said to be

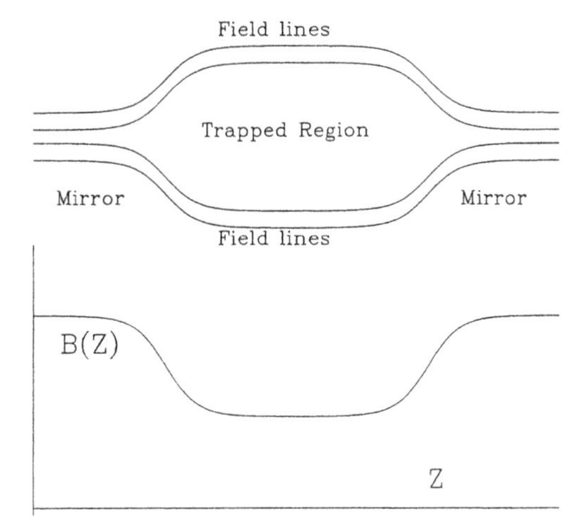

Figure 2.7. Confinement in a magnetic mirror

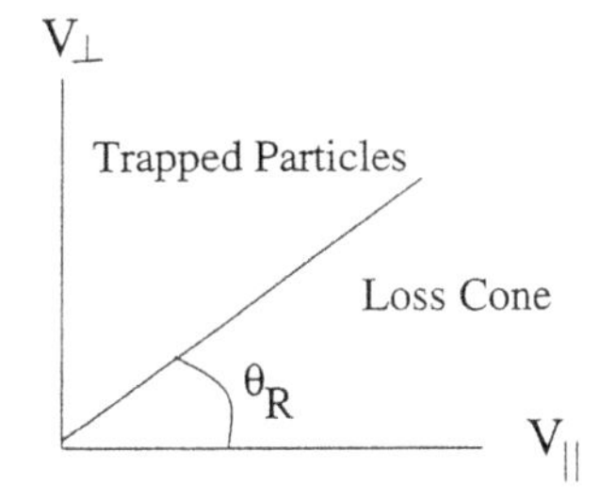

Figure 2.8. Trapped particles and particles in the loss cone

in the "loss cone" (see figure 2.8). After the time for particles to transit the confinement device, all these particles are lost and this part of velocity space becomes empty.

Even before the idea for a mirror machine was proposed, Fermi (1949) suggested that reflection of cosmic rays by the mirror force from moving magnetic clouds in interstellar space could accelerate or decelerate them. This led to his early theory for the origin of cosmic rays. Unfortunately, it was later found that cosmic rays occupy the galactic disk for only a million years or so. From the parameters for interstellar clouds, their velocity and numbers, we can show that Fermi's mechanism takes much longer than this, $\approx 10^8$ years (see problem 5 in Chapter 12). However, the real acceleration of galactic cosmic rays seems to be in shocks and here Fermi's mechanism is vital.

2.4 Polarization Drift

We have now essentially found all the important corrections to particle motions in static inhomogeneous magnetic fields. There are $\mathbf{F} \times \mathbf{B}$ drifts, grad

B drifts, curvature drifts, and changes in $v_{\|}$ due to mirror reflections. All effects except the $\mathbf{E} \times \mathbf{B}$ drifts and mirror reflection are small and essentially negligible as far as keeping track of the particles. The $\mathbf{E} \times \mathbf{B}$ drifts can be large enough that the inertial forces associated with them can produce drifts comparable with the first-order drifts. Because of the complexity of these inertial drifts we consider only the simplest one of them.

The important drift we keep is the "polarization drift." To derive this drift take a static magnetic field $\mathbf{B}$ uniform in the z direction and a time-dependent electric field $\mathbf{E}$ uniform in the x direction. Then the $\mathbf{E} \times \mathbf{B}$ drift in the y direction is

$$V_y(t) = -c\frac{E_x(t)}{B_z} \tag{28}$$

There must be a time-dependent force to produce the acceleration dV_y/dt, $F_y = -MdV_y/dt$. This force arises from a drift in the x direction

$$\begin{aligned} V_{Dx} &= -c\frac{F_y}{eB_z} = +\frac{cM}{eB_z}\frac{d}{dt}\frac{cE_x}{B_z} \\ &= \frac{Mc^2}{eB_z^2}\frac{dE_x}{dt} \end{aligned} \tag{29}$$

This drift is termed the polarization drift, for the following reason: If the plasma were a solid with an electric susceptibility, χ, then a change in E would produce a current $j = \chi dE/dt$. Thus, because the current associated with V_{Dx} is neV_{Dx}, where n is the density of ions, the susceptibility of the plasma by comparison is

$$\chi = \frac{neV_{Dx}}{dE_x/dt} = \frac{nMc^2}{B_z^2} \tag{30}$$

The contribution to χ from the electrons is negligible because of their very small mass.

The dielectric constant of this plasma is thus

$$\varepsilon = 1 + 4\pi\chi = 1 + \frac{4\pi nMc^2}{B^2} \tag{31}$$

The second term is generally quite large. It is one-half of the ratio of the relativistic rest energy of the plasma to the magnetic energy density.

Thus, if a uniform plasma is placed between two plates of a condenser, then as the charge on the condenser plates is increased an electric field is produced in the plasma. But this electric field is much smaller then if there were a vacuum between the plates. The electric field in the plasma produces, by the polarization current, a surface charge that nearly cancels the charge on the condenser plates and shields out most of the charge, as in figure 2.9.

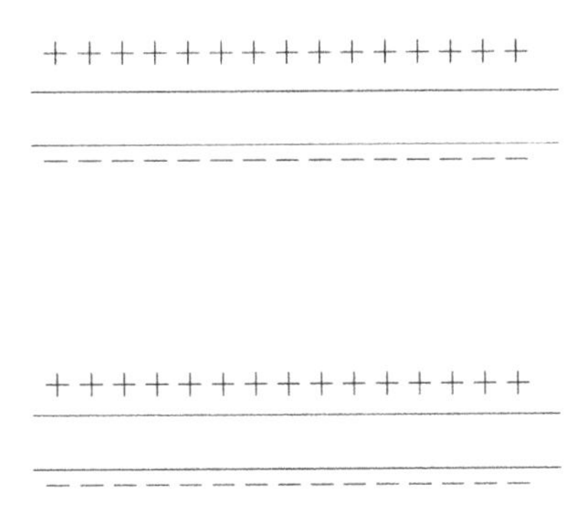

Figure 2.9. A uniform plasma between two condenser plates

The electric field inside of the plasma is just

$$E = \frac{E_0}{\varepsilon} = \frac{E_0}{1 + 4\pi M c^2/B^2} \tag{32}$$

It seen from this example that it is difficult to get an electric field to penetrate the plasma across a magnetic field by applying external charges on the plates. The plasma is very good at shielding out electric fields.

The reason for the term polarization drift is now clear. However, the mechanism for the polarization drift of a plasma is different from that of a solid.

We know that light propagating in a dielectric medium with dielectric constant ε, has a phase velocity $c/\sqrt{\varepsilon}$. This must be true in a plasma as well. What is the velocity of a light wave propagating parallel to B in a plasma? From equation 31 the phase velocity is

$$V_\phi = \frac{c}{\sqrt{1 + 4\pi n M c^2/B^2}} \approx \frac{B}{\sqrt{4\pi\rho}} \tag{33}$$

where in the last equality we neglect the 1 under the square root, assuming that $4\pi n M c^2 \gg B^2$. This speed is termed the Alfven speed because this wave was first proposed by Alfven. This result is correct only for frequencies well below the ion cyclotron frequency. This is because the polarization of the plasma takes an ion cyclotron period to be set up. In fact, all of the considerations of this chapter have involved an average over at least one cyclotron period.

In chapter 5, we will treat the Alfven wave in more detail from an MHD point of view.

2.5 Adiabatic Invariants

The motion of ions and electrons in a plasma is made much simpler by the existence of quasi-constants of the motion, the adiabatic invariants, among

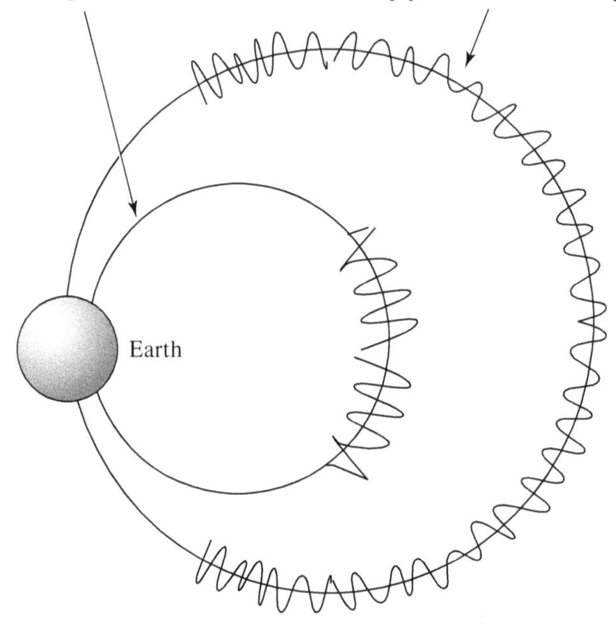

Figure 2.10. Small and large pitch-angle ions trapped in the earth's magnetic field

which the magnetic moment is one of the most important. The applications and importance of adiabatic invariants of particle motions is most clearly understood when we discuss the motion of trapped particles in the earth's magnetosphere.

A charged particle in the magnetosphere executes a variety of motions on different timescales. Each of these motions is associated with an adiabatic invariant. First, there is the cyclotron motion about the local magnetic field.

Second, there is the parallel motion along the magnetic field line on which the particle is located. This motion carries it down toward the earth's poles, where the field strength greatly increases, and then back along the field line. This very strong increase in field strength mirror reflects the particle forcing it to oscillate or bounce between the points of mirror reflection. The point at which the particle mirrors depends on the pitch angle of the particle as it passes the magnetic equator. If the pitch angle θ is finite and not too close to zero, it requires only a moderate increase in the field strength $B_0/B = (\sin^2\theta)^{-1}$ to mirror it, and because of the rapid increase in B, the particle travels only a short distance along the field line before it mirrors. If the pitch angle is very small, a large increase is required and the particle travels much farther along the field line before it reflects (figure 2.10).

As has been shown earlier, the gyroradii of the trapped particles are small and this is exactly the condition for the magnetic moment to be well preserved.

Consider an ion of energy ε in the magnetosphere. The motion of this ion's guiding center along the line can be described by the equation

$$\frac{d\ell}{dt} = v_{\parallel} = \sqrt{\frac{2\left[\varepsilon - Mv_{\perp}^2\right]}{M}} = \sqrt{\frac{2\left[\varepsilon - \mu|B|(\ell)\right]}{M}} \tag{34}$$

where ℓ is the distance measured along the line. For the parallel velocity of the ion, the derivative of ℓ, with respect to time, is equal to the square root and varies with $B(\ell)$.

In addition to the gyration about the line and the bounce motion along the line, the ion has a third motion, the drift motion across the line v_D, which is made up of the $E \times B$, gradient B, and centrifugal drifts. The $E \times B$ drift velocity is essentially the rotation of the earth. When the ion has drifted off the field line on which it starts it will attach itself to a new line horizontally displaced with respect to the old line, and it will bounce between new mirrors on this line. The point is that the particle moves along the line so rapidly compared to its drift motion that it will appear to translate its bounce motion rigidly from line to line. We will not notice the slight unevenness in its drift motion, but see only the mean motion.

Let us now consider other adiabatic invariants associated with the particle motion. The large discrepancy in timescales for the gyration bounce and drift motions leads to the constancy of two more adiabatic invariants in addition to the magnetic moment. Before introducing them let us us first ask, when the particle moves to a new line does the magnetic moment μ change. The answer is no? It will have the same value as on the old line, and for a good reason.

The constancy of the adiabatic invariant follows from two facts. First, the motion of an ion in an electromagnetic field is given exactly by the Hamiltonian

$$H = \frac{1}{2M}\left(\mathrm{p} - \frac{e\mathbf{A}}{c}\right)^2 + e\phi \tag{35}$$

where $\mathbf{A}$ is the vector potential giving $\mathbf{B}$ by $\mathbf{B} = \nabla \times \mathbf{A}$, and $\mathbf{E} = -\nabla\phi - (\partial\mathbf{A}/\partial t)/c$. $\boldsymbol{p}$ is the canonical momentum of the particle. The second important fact is that the cyclotron motion is very large, so that after one gyration of the particle it has hardly moved compared to the scale of the fields, and its energy and pitch angle are nearly the same while the fields at the particle are essentially the same. (This is true even if they are time dependent.)

The constancy of the magnetic moment (and of the other two invariants also) comes from an exact invariant for Hamiltonian motion, the Poincaré invariant. However, it must be appreciated that Poincaré's constant involves many particles rather than single one. The point is that because a single ion hardly moves in phase space during one cycle it can be simulated by a ring of ions arranged around a gyration orbit and the Poincaré constant associated with them leads to the magnetic moment constant of a single ion. The goodness of the constancy depends on how well the ring of ions simulates the single ion.

The Poincaré invariant is defined as follows (Goldstein 1980). Consider a closed circle of gyration about the guiding center of the ion and place a collection of ions (labeled by a parameter λ) about this circle. Each of these ions has the same $v_\perp$ and $v_\parallel$ as the ion they are simulating but they differ from it in gyration angle. As these ions move, the integral over λ,

$$P = \oint \mathrm{p} \cdot \frac{dq}{d\lambda} d\lambda \tag{36}$$

is invariant. In fact, if we differentiate this integral with respect to time making use of the Hamiltonian equations

$$\frac{dp}{dt} = -\frac{\partial H}{\partial q}$$

$$\frac{dq}{dt} = \frac{\partial H}{\partial p} \tag{37}$$

for each particle of fixed λ, we find that the time derivative of the quantity P, which is based on this class of ion orbits, is zero. (This is true for any loop of particles in phase space, not just those initially disposed on a circle of gyration.)

Let us evaluate P as though the ions were in a uniform field. Then the canonical momentum is $M(v - eA/Mc)$ and

$$\begin{aligned} \int p \frac{dq}{d\lambda} d\lambda &= \int M v_\perp dq_\perp - \frac{e}{c} \int A \, dq_\perp \\ &= 2\pi M v_\perp \rho - \frac{e}{c} B \left(\pi \rho^2\right) \end{aligned} \tag{38}$$

where we have used Stoke's theorem to reduce the second integral to the magnetic flux. The first term is $(Mv_\perp^2/2B)(4\pi \; Mc/e)$, while the second integral is one-half of this. Thus, the Poincaré integral is just $(2\pi \; Mc/e)\mu$ and so μ is constant.

By assumption the nonuniform magnetic field is always nearly constant over a gyroorbit. The ions representing P can be chosen to be in the same place as the ion would be if it were slightly delayed in its orbit and the resultant position projected onto the circle perpendicular to $\mathbf{B}$. Then at a later time the ring of ions could be similarly defined. The value of P at the later time is also equal to $2\pi Mc/e$ times $Mv_\perp^2/2B$, where $v_\perp$ is the new value of the perpendicular velocity of the ion. Since P is constant, so is $Mv_\perp^2/2B$.

Because of the way the ring is chosen there is no guarantee that the ring of ions will form an exact circle as the slight inhomogeneity of the field will produce a slight distortion of the circle so there will be a first-order difference between P and $2\pi Mc/e$ times $Mv_{\perp}^2 2B$. The point is that the distortion can never get large, so μ always differs by less than a first-order quantity from a constant no matter how long a time elapses.

2.6 The Motion of Trapped Particles in the Magnetosphere

Before discussing the other two invariants, let us examine the general motion of an ion in the magnetosphere in more detail. Let us consider a numerical example: a 10-keV proton in the earth's magnetosphere placed about 5 earth radii away from the earth. Let us further take the earth's field as a dipole field. The field at the earth's surface at the equator is about 0.3 G and at $5R_E$ it is smaller by a factor of 125, and equal to 0.0024 G and the gyrofrequency is $\Omega = 24$ radians/sec. The velocity is 1.6×10^8 cm/sec, so its gyroradius is 67 km. If the pitch angle is 45°, it will travel about $R_E = 6 \times 10^8$ cm from the equator along the line, so its bounce time is $24R_E/v \approx 100$ sec. Its drift velocity is about $(5\rho/5R_E)v \approx 6$ km/sec so it takes $2\pi \times 5R_E/(5 \text{ km/sec}) \approx 3.6 \times 10^4$ sec to circle the earth. (If we include the $E \times B$ motion associated with the earth's rotation, this is decreased by 50 percent, since the drift motion and the earth's motion are in the same direction.)

Thus, for this particle there are three disparate timescales: the cyclotron time $\approx 2\pi/24$ sec ≈ 0.2 sec, the bounce time ≈ 100 sec, and the drift time $\approx 3.6 \times 10^4$ sec. For example, after one bounce, the line on which the particle moves is displaced only about 20 arc minutes to the east of its initial position. Thus, the particle's line of force appears to be almost stationary after one bounce.

What happens to the particle after it has circled the earth? Will it be on the same line of force or will it move outward or inward? If the earth's field were a perfect dipole field aligned parallel to the earth's rotation axis, then, of course, by symmetry it will come back to the same line. However, this is not the case. The earth's field is distorted by the solar wind so that it is compressed on the sunward side and expands into the magnetic tail on the antisun side. In addition, the dipole, even undistorted by the solar wind, is not aligned with the rotation axis. Thus, the earth's field differs from an axisymmetric field, and it is a nontrivial question whether a particle will return to the same magnetic line after drifting around the earth.

If their orbits are not closed, we would expect the 10-keV protons to be lost in a period of days and we would not expect any density at all of energetic protons in the magnetosphere. However, fortunately for the magnetospheric plasma, it is the case that the particle drift orbits are closed. That is to say, at any time the particle is bouncing back and forth on some line of force. After the particle drifts around the earth to the same longitude relative to the earth–sun direction, it will find itself bouncing on the identical line of force on which it started bouncing. This happens because of a second adiabatic invariant, the longitudinal invariant (Northrop and Teller 1960; Morosov and Solovev 1966).

This second invariant is defined as follows. Consider the bounce motion of a proton along a line of force at some time. Because its energy $\epsilon = M(v_\perp^2 + v_\parallel^2)/2$ and its magnetic moment, $\mu = Mv_1^2/2eB$, are constant its parallel velocity

$$v_\parallel = \sqrt{\frac{2[\epsilon - \mu B]}{M}} \tag{39}$$

is a known function of position. Because B varies along the line, so does the parallel velocity, but for a given particle, when B reaches a certain value, $v_\parallel$ vanishes and the particle reflects at that point. Now integrate $v_\parallel$ between the two mirror reflection points

$$J = \oint v_\parallel d\ell = \oint \sqrt{\frac{2[\epsilon - \mu B(\ell)]}{M}}\, d\ell \tag{40}$$

where the integral is to be taken back and forth over the entire bounce orbit. For any given particle it turns out that J remains constant as long as the drift motion during the bounce time, $\oint d\ell/v_\parallel$, does not take the particle to a drastically different line. In fact, we arrive at this constancy by identifying J with the Poincaré invariant of a ring of particles strung out along the line of force in the same manner as was done for the first invariant. Namely, consider the motion of a single ion bouncing along the magnetic field line. Then place the ions in the ring by delaying the orbit of the ion and projecting it back onto the given line of force on which the ion is question is bouncing. (It may be necessary to close the ring in phase space by adding a few more ions, but it is clear that this will not change the value of P from J of equation 40 by more than a minute amount.)

It turns out that once the magnetic moment is taken constant and gyrophase is ignored, the equation of motion for the remaining variables, $v_\parallel$ and R, the guiding center position, are also Hamiltonian. The proof is of this result is given in Kruskal (1962).

The Poincaré invariant for the ring of such particles is

$$\oint P_\parallel d\ell = \oint M\left(v_\parallel - \frac{eA_\parallel}{c}\right) d\ell = \oint Mv_\parallel d\ell \tag{41}$$

where the $A_\parallel$ integral vanishes by symmetry. This integral is MJ. Indeed, when the particle is at some position ℓ, $A_\parallel$ is the same whether the particle is moving forward $d\ell > 0$ or backward $d\ell < 0$. $v_\parallel$ obviously changes sign, so its integral does not vanish.

Now let us apply our adiabatic invariant to the containment problem. Let L denote two coordinates specifying a given line in the earth's magnetosphere (say the radial and the longitudinal coordinates where it crosses the earth's equator). Remember that μ and ε are fixed. Then on any line L we can calculate J from equation 40. That is, J is a function of μ, ε, L

$$J = J(\varepsilon, \mu, L) \tag{42}$$

and it is plausible that, for a fixed longitude, J is monotonic in the radial coordinate of L. Now select any particle, ion or electron, in the magnetosphere. It will have a definite value for J, say J_0. Then at any longitude we have

$$J(\varepsilon, \mu, L) = J_0 \tag{43}$$

and because J is assumed monotonic there will be only one line crossing the equatorial plane with this longitude, which has the value J_0. (ε and μ are fixed.) Thus, if the particle *starts* with this longitude and this same line, after it drifts around the earth and comes back to this longitude it must come back to this line. Its orbit is closed and it is confined. This is true, independent of any distortion of the earth's field from azimuthal symmetry. In fact, the particle will always lie on a closed topological cylinder.

For simplicity in this discussion we have neglected the rotation of the earth. This can easily be included by introducing a static electric potential ϕ and writing $\varepsilon = Mv^2/2 + e\phi$. With this change the argument goes through as before.

Now, it is true that due to any time dependence of the earth's field, ε will change, but if the magnetic field changes slowly compared to the rate of drift around the earth, then ε will change by a small amount after one drift around the earth. J will still remain an adiabatic invariant of the motions. As a consequence, the particle will always move near to a cylindrical surface specified by J equal to a constant.

In this case the energy is not a constant but is constrained by the existence of a third invariant rather than the energy constant. The second invariant reduces the particle's motion to just two degrees of freedom to a parameter J_0 characterizing the radius of the cylinder $J(\varepsilon, \mu, L) = J_0$ and the energy ε. The relevant Poincaré invariant for the third invariant is obtained by placing particles around the intersection of the cylinder with the equator and is

$$-\frac{e}{c}\Phi = -\frac{e}{c}\oint \mathbf{B} \cdot d\mathbf{A} \tag{44}$$

where $d\mathbf{A}$ is the element of area and the integral is taken over this curve. Thus, if the energy changes slowly, due to a slowly changing $\mathbf{B}(\mathbf{r}, t)$, the corresponding cylinder on which the particle moves also changes slowly in a way determined by the evolution of the energy. As a practical matter, $\mathbf{B}$ usually does not change sufficiently slowly that the third invariant is constant. This third invariant is actually the magnetic flux enclosed by the cylinder, so the constancy of Φ represents in some sense a general form of Lenz's law.

To summarize:

1. μ is constant of the motion and the gyrophase is ignorable. This reduces the six-dimensional phase space for the particle motion to four dimensions with the gyrophase ignored.

2. The particle is trapped by magnetic mirrors, and its bounce motion can be reduced to a two-dimensional problem with the position along the line ignored. The reduction is accomplished by the second or longitudinal invariant. Thus, at this stage, the particle is characterized by μ and J, and all we care about is its energy and the longitude on which it is moving.
3. If B changes slowly compared to the time it takes the particle to drift around the cylinder there is a third invariant, Φ, which reduces the degrees of freedom of the particle to zero. Its motion is completely integrable once the magnetic field $B(x,t)$ is given. Here the particular line on the cylinder on which the particle is bouncing is ignored.

For example, give μ, J, and Φ. For a given ε, J chooses the cylinder on which the particle drifts. Then, ε has to be chosen so that the cylinder picked by J encloses the correct flux, Φ. Φ chooses the energy since, for $\mathbf{B}(\mathbf{r},t)$ at some time, and for each ε, and J, Φ selects a cylinder and this cylinder has to enclose the given flux Φ, and this can happen for only one ε.

The particle drifts around the cylinder at some rate so that any time during this drift the line L on which the particle is bouncing can be found. Further, the velocity along the line is $d\ell/dt = v_\parallel = \sqrt{2(\varepsilon - \mu B)/M}$ so the position along the line can be followed in time. Finally, the gyrophase can be followed locally from $\Omega = eB(\ell, L)/Mc$.

2.7 Particle Motion and Macroscopic Force Balance

As particles move they produce an electric current. For example, due to the ∇B drift of an ion in an inhomogeneous field,

$$V_D = \frac{(B \times \nabla B)Mcv_\perp^2}{2eB^2} \approx \frac{\rho}{H} v_\perp \tag{45}$$

there is an electric current eV_D/c. As has been previously pointed out, this current crossed with $\mathbf{B}$ produces the force that just balances the $\mu\nabla B$ force. If we add up all these elementary currents due to the drifts of a collection of ions whose guiding centers are in a unit box to get a total current density, $\mathbf{j}$, we would then expect that the $\mathbf{j} \times \mathbf{B}$ force would represent the total macroscopic magnetic force density on the plasma. In equilibrium we might expect this to balance the other forces: pressure forces ∇p, inertial forces $\rho\partial\mathbf{v}/\partial t$, etc. Is this actually the case?

To check this let us consider a purely static situation. Let $\mathbf{B}$ be in the z direction and vary in the x direction only, as in figure 2.11. Let us first concentrate on the ions. Then the ions move in the y direction. ($j_y B_z$ is a force in the x direction, which balances the $-\mu\nabla B$ force on the individual

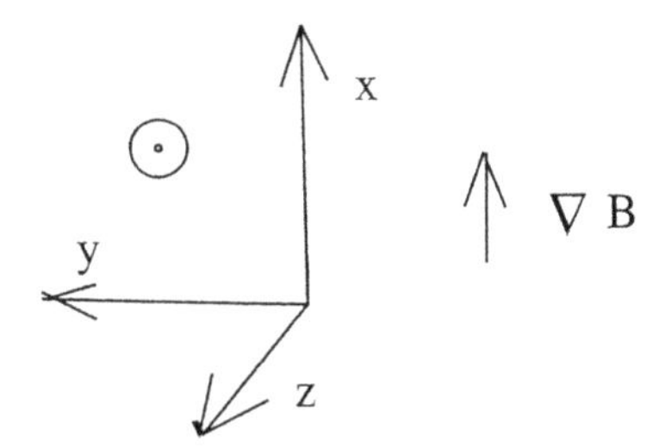

Figure 2.11. The coordinates for macroscopic balance

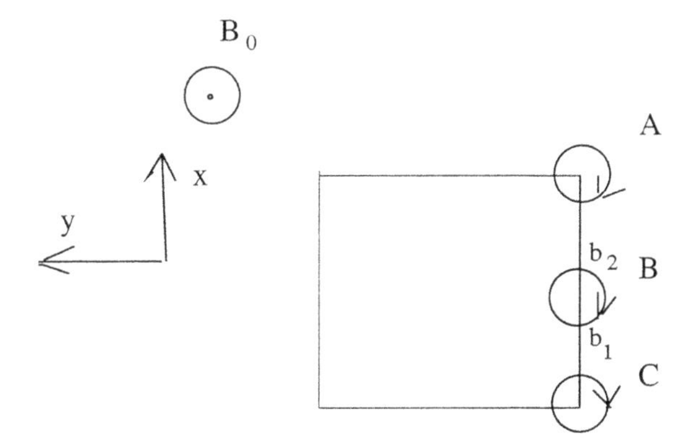

Figure 2.12. The ion magnetization current

ions.) That is, if there are n_i ions in a unit volume, then

$$j_y B_z = n_i e \frac{V_D}{c} B_z = \frac{n_i e}{2c}\left(\rho \frac{\nabla B_z}{B_z}\right) v_\perp B_z = n_i \frac{M v_\perp^2}{2B} \nabla B$$
$$= \frac{p_i}{B} \nabla B \tag{46}$$

and this force is in the x direction. However, the pressure force is ∇p and these two forces do not balance in general.

The problem is resolved when we appreciate that there is another current in addition to that associated with the drifts alone (Longmire 1963). Let us examine a surface $y = 0$, of unit area, on the right side of the unit box in figure 2.12 and for the moment ignore the guiding center drift of the ions.

Note that the some of the gyrating ions intersect this surface. Ions of type A in the figure contribute a current $e\Omega/2\pi c$ to the total current crossing the surface in the y direction (to the left). Those of type B contribute no current, since their leftward motion across the surface at b_1 is canceled by their rightward motion at b_2. Those of type C contribute a negative current. Now there are $n_A \pi \rho_A^2 = \pi n_A v_\perp^2 / \Omega_A^2$ particles of type A, which give a current

$$\frac{n_A v_\perp^2 \pi}{\Omega_A^2} \frac{e}{c} \frac{\Omega_A}{2\pi} = \frac{n_A M v_\perp^2}{2B} \tag{47}$$

Summing over $v_\perp^2$ the total electric current for ions of type A is

$$j_A = \frac{p_{iA}}{B_A} \tag{48}$$

Similarly, the ion current for C particles is

$$j_C = -\frac{p_{iC}}{B} \tag{49}$$

Thus, the additional current due to these surface ions is

$$\frac{\partial}{\partial x}\left(\frac{p_i}{B}\right) \tag{50}$$

This is the ion magnetization current. Adding this to the ion drift current, (see equation 46) we get

$$j = \frac{\partial}{\partial x}\left(\frac{p_i}{B}\right) + \frac{p_i}{B^2}\frac{\partial B}{\partial x} = \frac{1}{B}\frac{\partial p_i}{\partial x} \tag{51}$$

so that the drift current plus this magnetization current when crossed with **B** just balances the ion pressure gradient. The same holds for the electron current, and then for the total current. Thus, this must be the correct magnetic force.

Note that the sum of all the drift currents crossed with the magnetic field actually balances the sum of all the forces on the guiding centers, but this is not the correct total force, since some of the gyrating orbits stick out of the box and the forces on these parts of the orbits should not be included when evaluating the force on the box.

Two special cases emphasize this. Suppose that the plasma pressure is constant. Then, since the pressure gradient force is absent, there should be no current. Clearly, the drift velocity and its current are the same as above. But this current is canceled by the magnetization current $\nabla(p/B)$. Also, if B is constant and p is not, then there is no drift velocity or a drift velocity current, but there is magnetization current $\nabla(p/B) = \nabla p/B$ and this current times B balances the ion pressure gradient.

It is important to understand the origin of the magnetization electric currents from the particle point of view. Each particle carries a large current in its circular motion but because of the circular motion this current generally cancels, as in orbits of type B above. Just a few orbits of type A or C times the large current per ion give a finite current. It is not always simple to evaluate the current from the particle point of view when the geometry is complicated.

In this book, we will turn the problem around and find the electrical currents j from the macroscopic point of view. The current j must flow so that $j \times B$ balances the perpendicular macroscopic forces. Because of the abundance of current in the plasma, the plasma can easily supply this net current. Further, the particles do not move smoothly according to the orbit equations since collisions with other particles interfere. In this case, it is essential to take the macroscopic point of view to find the current. This remark is important since collisions are almost always dominant in astrophysics, as we shall see.

2.8 Problems

1. (Escape of cosmic rays) Consider the possibility of the escape of cosmic rays from the galactic disk by a drift due to inhomogeneities in the 3-μG magnetic field B. Let the horizontal scale of variation of the magnetic field be 100 parsecs. Let the vertical half-thickness of the disk be 200 parsecs. Above what energy do you expect a cosmic-ray proton to drift this half-thickness in a time of three million years. Assume a constant vertical drift.

2. (Ultrarelativistic cosmic rays from a gamma-ray burst) Assume that there is a uniform intergalactic field with field strength $B = 10^{-13}$ G. Assume that a 10^{20}-eV cosmic-ray proton is produced by a gamma-ray burst a distance of 30 megaparsecs away in a direction perpendicular to $\mathbf{B}$. How long after the gamma rays arrive will the cosmic ray arrive? How much longer will it take for a 10^{19}-eV cosmic-ray proton to arrive? (Note that conditions for this problem are such that the cosmic ray does not execute a complete gyroorbit.)

3. (Confinement of pressure by a weak magnetic field) Suppose we have a slab of protons and helium ions. Let the protons have a pressure of $p_{\mathrm{H}}(z)$ and the helium ions have a pressure of $p_{\mathrm{He}}(z)$, where z is the coordinate perpendicular to the slab. Assume that there is a magnetic field in the x direction. What are the proton current and helium ion current in the y direction? Express them in terms of the pressure gradients. If the sum of the pressures is constant, how does the magnetic field B vary with z? Does it have a gradient or is it a constant? If it is a constant how does it manage to balance the proton pressure gradient? Explain in this case where the proton current comes from—magnetization current, grad B current, or what? Consider the same problem replacing the helium ions by cosmic rays. Is it possible for a very weak magnetic field to confine a very high pressure of cosmic rays? How strong must the magnetic field be to accomplish this? Note that if we picture the cosmic rays as transferring momentum to the field we can reasonably imagine that the field is immediately transferring this momentum to the protons in a cyclotron time.

4. (Constancy of the magnetic moment) Consider a uniform magnetic field $B(t)\hat{\mathbf{z}}$ whose strength is time dependent. Show that Faraday's law, $\partial \mathbf{B}/\partial t = -c\nabla \times \mathbf{E}$, can be satisfied by an associated electric field

$$\mathbf{E} = -\frac{\dot{B}}{2c}(x\hat{\mathbf{y}} - y\hat{\mathbf{x}})$$

where the dot indicates a time derivative. Let any additional electrostatic field be zero.

Show that the equation of motion of an ion can be written in the form

$$\ddot{z} = -i\Omega\dot{z} - i\frac{\dot{\Omega}}{2}z$$

where $\Omega(t) = eB(t)/Mc$ and $z = x + iy$. Set

$$z(t) = w(t)\exp -i\int[\Omega(t)/2]dt$$

and show that w satisfies the harmonic oscillator equation

$$\ddot{w} + \frac{\Omega^2(t)}{4}w = 0$$

Let B be linear in time and show that one solution of this equation can be expressed in terms of ordinary Bessel function J. Take the limits $t \to \pm\infty$, and show that the magnetic moment of the ion is asymptoticly constant and the same in both limits, even though B has passed through zero and violated the usual condition for the conservation of the magnetic moment. If B is a constant, show that the motion is a circle in the x–y plane.

5. (Increase in plasma energy by the increase in an ultraweak magnetic field) Consider the same time-dependent magnetic field as in the last problem. Let the variation be so slow that the magnetic moment is constant. Let B be made to decrease by a factor of two—say, due to plasma motions. If the field is in a uniform Maxwellian plasma and they are both in a cylinder of radius r, then the energy of the plasma must decrease by a factor of two. This is true even if the energy of the field is infinitesimal compared to the energy of the plasma, provided that $B(t)$ decreases slowly compared to the cyclotron period. Where does the extra energy go? Such a situation may occur in the *early universe* when the field is weak but time dependent.

6. (Particle acceleration in equal-strength, perpendicular electric and magnetic fields) Consider an ion starting from rest at the origin in a uniform magnetic field of strength B in the z direction and a uniform perpendicular electric field E in the y direction and take $E = B$. The ion quickly becomes relativistic. Show that the equations of motion for it can be written as

$$\frac{du_x}{d\tau} = \Omega u_y$$

$$\frac{du_y}{d\tau} = \Omega(\gamma c - u_x)$$

$$\frac{du_z}{d\tau} = 0$$

$$\frac{d\gamma c}{d\tau} = \Omega u_y$$

where $\Omega = eB/Mc$, $\mathbf{u} = \gamma \mathbf{v}$, and γMc^2 is the relativistic energy where γ is the usual relativistic factor. τ is the proper time given by $dt/d\tau = \gamma$.

Show that during the motion $\gamma c - u_x = c$ and and integrate the remaining equations for $\mathbf{u}$ and $\mathbf{x}$ as a functions of τ (see Landau and Lifshitz (1994).

Consider a particle starting from rest at the light cylinder from a pulsar and take x in the radial direction. Assume that E and B are the fields that the pulsar electromagnetic wave would have but that they are constant in time and for a radial distance c/ω, where ω is the angular velocity of the pulsar. The E and B of this wave are to be taken in the y and z direction as above. Solve for τ when the particle increase its x by c/ω and show that its energy ε is

$$\varepsilon = \gamma mc^2 = \left[1 + \frac{1}{2}\left(\frac{6\Omega}{\omega}\right)^{2/3}\right] mc^2$$

Take the value of B to be that of of a dipole at the light cylinder for the Crab pulsar, say $\approx 10^6$ G, and take $\omega = 200$ Hz. Find the accelerated energy of the particle if it is a proton, and if it is an electron (see Landau and Lifshitz 1994; Ostriker and Gunn 1969).

References

Fermi, E. 1949. *Physical Review* **75**, 1169.

Goldstein, H. 1980. *Classical Mechanics*, 2nd ed., Addison & Wesley, Reading, MA

Kruskal, M. D. 1962. *Journal of Mathematical Physics* **3**, 806.

Landau, L. D., and E. M. Lifshitz. 1994. *Classical Theory of Fields*, 4th ed., p. 58, trans. M. Hamermesh, Pergamen Press, Oxford, UK.

Longmire, C. 1963. *Elementary Plasma Physics*, Interscience, New York.

Morosov, A. I., and I. S. Solovev. 1966. *Reviews in Plasma Physics*, vol. 2, p. 221, ed. E. M. Leontovitch, vol 2, trans. H. Lashinsky, Consultant Bureau Enterprise, New York.

Northrop, T., and E. Teller. 1960. *Physical Review* **117**, 215.

Ostriker, J. P., and J. Gunn. 1969. *Astrophysical Journal* **157**, 1395.

Chapter 3

MAGNETOHYDRODYNAMICS

3.1 The Basic Equations

If a plasma is sufficiently collisional, it can be described macroscopically by the reduced description of magnetohydrodynamics. The basic state of a collisional plasma is given locally by specifying its mass density ρ, its momentum density $\rho\mathbf{V}$, its pressure p, the electric field $\mathbf{E}$, and the magnetic field $\mathbf{B}$. These are all functions of position $\mathbf{r}$ and time t. If collisions are strong, the ion and electron pressures are isotropic and the heat flow is small. In addition, if the dimensionless quantity, the magnetic Reynold's number,

$$R_M = \frac{VL}{\eta c/4\pi} \tag{1}$$

is large, resistivity can be neglected (Cowling 1976).

In this case, the plasma is described by a simple set of equations, the "ideal MHD equations." Although, frequently, the conditions for isotropy and negligible heat flow are not so well satisfied, it will be found that the ideal equations possess so many attractive properties that it is useful to employ them to describe the plasma. Then we can consider the deviations from them as corrections. These corrections are generally inferred from the transport or dissipative terms, the viscous terms, thermal conduction terms, and resistive terms. A more complete description of a plasma is given by the two-fluid description in which the ion and election fluids are considered separately. The standard and best equations for describing the two fluids are the Braginski equations, which are presented in chapter 8.

In ideal MHD, the quantities that describe the plasma are its mean density ρ, its mean velocity $\mathbf{V}$, its mean pressure p, and the magnetic field $\mathbf{B}$. Each of these quantities requires a time-dependent equation to advance it in time. The ideal equations for them are as follows.

For density, the continuity equation:

$$\frac{\partial \rho}{\partial t} + \nabla \cdot (\rho \mathbf{V}) = 0 \tag{2}$$

For velocity, the Euler equation, commonly referred to as the equation of motion:

$$\rho\left(\frac{\partial \mathbf{V}}{\partial t} + \mathbf{V} \cdot \nabla \mathbf{V}\right) = -\nabla p + \mathbf{j} \times \mathbf{B} + \rho \mathbf{g} \tag{3}$$

For pressure, the constant entropy equation:

$$\frac{\partial}{\partial t}\left(\frac{p}{\rho^{\gamma}}\right)+\mathbf{V}\cdot\nabla\left(\frac{p}{\rho^{\gamma}}\right)=0 \tag{4}$$

We will shortly come to the differential equation for **B**.

The plasma consists primarily of charged ions and electrons, but it can also include neutrals, in which case it is called partially ionized. If there are no neutrals, it is called fully ionized. The ions are usually protons, but can be of a mixed composition. ρ is, of course, the total mass density, i.e., the sum of masses of all species, neutrals and ions, in a unit volume. The velocity **V** is chosen so that the ρV is the total momentum in a unit volume. It is possible that different species have substantially different mean velocities, but in this case the total pressure is not isotropic (that is, not a scalar), so that in this case the ideal equations do not represent the fluid motion very well.

Now, the functions $\rho(\mathbf{r},\mathbf{t})$ and $\mathbf{V}(\mathbf{r},\mathbf{t})$ give the density and velocity in the Eulerian frame. They specify the density and velocity at a fixed position. Because plasma passes through any given point, with velocity **V**, $\mathbf{V}(\mathbf{r},t)$ represents the velocity of *different* elements of fluid at different times. For this reason the left-hand side of the equation of motion, which is the instantaneous acceleration of an element of fluid times its mass, is not simply $\rho\partial\mathbf{V}/\partial t$ but must include a term to compensate for the changing fluid element at the fixed position **r**. In fact, the increment in the velocity of a fixed fluid element is its velocity at $t+\delta dt$ and $\mathbf{r}+\mathbf{v}dt$, i.e., $\mathbf{V}(\mathbf{r}+\mathbf{V}\delta t, t+\delta t)$, minus its velocity at **r** and t. To find the velocity at $t+\delta t$ we must look at its displaced position at $t+\delta t$, i.e., $\mathbf{r}+\mathbf{V}\delta t$. Taylor expanding $\mathbf{V}(\mathbf{r}+\mathbf{V}\delta t, t+\delta t)$ in δt, subtracting $\mathbf{V}(\mathbf{r},t)$, and dividing by δt gives the left-hand side of equation 3.

The right-hand side of equation 3 is the force on the unit volume. It consists, first, of the magnetic force, $\mathbf{j}\times\mathbf{B}$, where $\mathbf{j}\times\mathbf{B}$ is the sum over all the species of the Lorentz forces, $\Sigma q(\mathbf{V}\times\mathbf{B})/c$, where $\mathbf{j}=\Sigma q\mathbf{V}/c$ is the electric current density. Second, there is a pressure force that is made of the different pressure stresses on the faces of unit volume. Taylor expanding the pressure leads to $-\nabla p$. (If the pressure were a tensor, P_{ij}, the ith component of the force would be $\partial P_{ij}/\partial x_j$.) We have not included the electric force $q\mathbf{E}$. As we shall show shortly, the electric force $q\mathbf{E}$ is negligible.

It should be remembered that the force due to the pressure stress on any face is the rate at which particles carry momentum across that face. Perhaps, since no physical force is involved, it could be included in the left-hand, inertial, side of the equation of motion, but it is conventional to treat it as a force and keep it on the right side. This picture makes it clear that the pressure must be treated as the sum of the individual pressures of the individual species and the pressure forces of each of these species act independently of each other.

There are cases, such as very partially ionized gases, where it is not safe to treat the total gas, ionized species plus neutrals, as a single fluid with a single fluid velocity, since the ions may move a large distance through the predominant neutral species. In this case, it is appropriate to introduce two momentum equations, one for the neutrals and one for the plasma (electrons plus ions). This occurs in stellar formation and is generally referred to as ambipolar diffusion. (The name is presumably lifted from an analogous laboratory case where ions and electrons diffuse independently but must do so in a manner to preserve charge neutrality.)

A general rule of thumb that we will follow is to treat the fluid as a single fluid, with a single velocity, and later check whether ion-neutral collisions are adequate to balance the difference between the ion and neutral fluid forces. For example, there is a magnetic force on the plasma and no magnetic force on the neutrals. However, collisions can overcome this difference and force the ions and neutrals to move together. If this is the case, then the single-fluid picture is justified.

We mentioned in the introduction that waves are excited to substantial amplitude. These waves then should be treated as a separate component. They interact with the main plasma through refraction and particle resonance, and either extract energy from it to grow, or give up energy to it to damp. The consequence of this is that the waves exert a wave pressure on the fluid, often called a ponderomotive force. In fact, if the waves grow they must extract energy from the rest of the plasma, and the mechanism for this requires an additional force term on the plasma, $-\nabla \cdot \mathbf{P}_w$, that must be added to the pressure gradient term (see chapter 5).

The above remarks illustrate the ways in which the ideal fluid equations can be extended to cover various complex cases of importance to astrophysics, as well as to show the limitations of applicability of the ideal equations.

We now turn to the entropy equation 4. The entropy of a perfect gas per unit mass

$$S = C_V \ln \left(\frac{p}{\rho^\gamma} \right) \tag{5}$$

where $C_V = \frac{3}{2} k_B / \mu M$ is the specific heat per unit mass. (μ is the mean molecular weight per particle. It is $\frac{1}{2}$ for a fully ionized hydrogenic plasma. k_B is Boltzmann's constant.) If heat flow is small and there is little frictional heating or radiative heating or cooling, then the entropy of any given fluid element is a constant. This is equation 4. In astrophysics, it is usually the case that frictional heating is negligible except in shocks and current layers. Also, radiative cooling is usually unimportant for rapid phenomena. On the other hand, heat flow of electrons along lines of force is almost always large. Because, as we know from orbit theory, electrons are strongly constrained to move along lines of force, there is scarcely any heat flow across the lines. Therefore, the temperature is usually nearly constant along lines, or, at any rate, varies slowly.

The consequences of this suppression of perpendicular heat flow can easily be seen in the sun. The streamers seen in the corona are actually lines of force along which the temperature is high. (The high temperature allows the plasma to rise high above the sun's surface.) Thus, high temperature leads to high plasma density and, consequently, electron scattering of the solar radiation, making these field lines visible. On the other hand, the neighboring lines can have a much cooler temperature and lower electron density so that they are not seen. In this way the corona, seen in reflected light, appears to consist of long streamers that trace out the field lines.

In summary, it is only strictly appropriate to apply the entropy equation to MHD structures when the temperature varies slowly along the magnetic field lines. In turn, this condition is enforced by the large heat conduction.

Equations 2–4 describe the ideal equations for the plasma. However, we must include Maxwell's equations to get the time evolution of the fields **E** and **B**, which enter the equation of motion. These equations are

Faraday's induction equation
$$\frac{\partial \mathbf{B}}{\partial t} = -c(\nabla \times \mathbf{E}) \tag{6}$$

Ampere's law
$$\nabla \times \mathbf{B} = 4\pi \mathbf{j} + \frac{1}{c}\frac{\partial \mathbf{E}}{\partial t} \tag{7}$$

The divergence **B** equation
$$\nabla \cdot \mathbf{B} = 0 \tag{8}$$

Poisson's equation
$$\nabla \cdot \mathbf{E} = 4\pi q \tag{9}$$

where q is the net electric charge density. In these equations we use Gaussian units (gauss for **B**, statvolts/cm for **E**, and abamperes/cm^2 for **j**; 1 statvolt = 300 volts, 1 abampere = 10 amperes). We use the electromagnetic unit of abamperes for current because this conveniently removes c from the ideal equations.

Maxwell's equations must be coupled to the plasma equations and this is accomplished through Ohm's law. Ohm's law, in its simplest form, states that $\mathbf{E}'$, the electric field seen in the moving frame of the plasma, i.e., $\mathbf{E}' = \mathbf{E} + \mathbf{V} \times \mathbf{B}/c$, is proportional to the resistivity times the current density. In our units, Ohm's law is

$$\mathbf{E} + \frac{\mathbf{V} \times \mathbf{B}}{c} = \eta \mathbf{j} \tag{10}$$

where η is the resistivity. Ohm's law is only an approximation. The real, or generalized, Ohm's law includes several corrections: the Hall term, the electron pressure gradient (responsible for the Biermann battery, chapter 13), and the fact that η is not always a scalar. If the electron–ion collision rate is less than the electron cyclotron frequency, then the resistivity perpendicular to the magnetic field is 1.9 times that parallel to the magnetic field. In most cases, the extra terms are small and the resistivity itself can be neglected, so in our ideal approximation we take Ohm's law with scalar η in the form equation 10.

We get the equation for $\partial \mathbf{B}/\partial t$ by substituting equation 10 in equation 6:

$$\frac{\partial \mathbf{B}}{\partial t} = \nabla \times (\mathbf{V} \times \mathbf{B}) - c\nabla \times (\eta \mathbf{j}) \tag{11}$$

We next eliminate $\mathbf{E}$ entirely from the MHD equations by first showing that the displacement current in Ampere's law, equation 7, is small if $V \ll c$, i.e., the fluid motions are nonrelativistic. We accomplish this by solving equation 7 for $\mathbf{j}$ and substituting the result in the equation of motion (3). The result is

$$\rho\left(\frac{\partial \mathbf{V}}{\partial t} + \mathbf{V}\cdot\nabla\mathbf{V}\right) = -\nabla p + \frac{(\nabla\times\mathbf{B})\times\mathbf{B}}{4\pi} - \frac{1}{4\pi c}\frac{\partial \mathbf{E}}{\partial t}\times B$$
$$+\rho\mathbf{g}$$

The displacement current produces an extra $\partial\mathbf{E}/\partial t$ term. We then estimate $\mathbf{E}$ from Ohm's law by dropping the resistivity term to get $\mathbf{E} \approx -\mathbf{V}\times\mathbf{B}/c$. Hence,

$$-\frac{1}{4\pi c}\frac{\partial \mathbf{E}}{\partial t}\times\mathbf{B} = -\frac{\partial}{\partial t}\left[\frac{(\mathbf{V}\times\mathbf{B})\times\mathbf{B}}{4\pi c^2}\right]$$
$$\approx \frac{B^2}{4\pi c^2}\frac{\partial \mathbf{V}}{\partial t}$$

This is of order $B^2/4\pi\rho c^2$ times the $\rho\partial\mathbf{V}/\partial t$ term on the left so it is smaller than this inertial term by the ratio of the magnetic energy density $B^2/8\pi$ to one-half the rest mass energy density of the plasma, $\rho c^2/2$. Keeping this term can be thought of as adding the "mass" of the magnetic field to the plasma mass.

Next, we show that the electric force term in the equation of motion $q\mathbf{E}$ is small compared to the other, $\rho\mathbf{V}\cdot\nabla\mathbf{V}$, inertial term by the same factor, and by a similar argument. Indeed, with $\mathbf{E} \approx -\mathbf{V}\times\mathbf{B}/c$ we have, using Poisson's equation 9,

$$q\mathbf{E} = \frac{\mathbf{E}\nabla\cdot\mathbf{E}}{4\pi} \approx \frac{(\mathbf{V}\times\mathbf{B})\nabla\cdot(\mathbf{V}\times\mathbf{B})}{4\pi c^2} \approx \frac{B^2\mathbf{V}\cdot\nabla\mathbf{V}}{4\pi c^2}$$

This is smaller than $\rho(\mathbf{V}\cdot\nabla\mathbf{V})$ by the same ratio $B^2/\rho c^2$. In making the approximation in this equation, we assume that the length scale of $\mathbf{V}\times\mathbf{B}$ in the third term to be comparable to the length scale on the right-hand side. However, it is almost always the case that these length scales are at least of the same order of the magnitude. Hence, dropping the electric force $q\mathbf{E}$ is justified for $v_A^2 \ll c^2$.

Let us check our estimate, $\mathbf{E} \approx \mathbf{V}\times\mathbf{B}/c$. We do this after the fact. If this estimate is correct, then we can estimate $\mathbf{j}$ from Ampere's law without displacement current:

$$\eta\mathbf{j} = \eta\frac{\nabla\times\mathbf{B}}{4\pi} \approx \eta\frac{\mathbf{B}}{4\pi L}$$

and this is small compared to $\mathbf{V}\times\mathbf{B}/c$ if

$$\frac{\eta c}{4\pi L} \ll V.$$

This is commonly expressed in terms of the magnetic Reynold's number. Dividing this inequality by the left-hand side gives

$$1 \ll \frac{4\pi V L}{\eta c} \equiv R_M \tag{12}$$

Thus, if R_M is large, then the resistivity term is small,

$$\mathbf{E} + \frac{\mathbf{V} \times \mathbf{B}}{c} = 0 \tag{13}$$

and our above estimate for $\mathbf{E}$, used in dropping the displacement current and electric force, is justified. Now, because L in astrophysics is large, the magnetic Reynold's number in astrophysics is always large. This fact is quite important.

In this argument it might be objected that there are situations in astrophysics that are static, and therefore $R_M = 0$. However, in practice if this is the case, then all three terms in Ohm's law are small and $\mathbf{B}$ will evolve so slowly that, before it changes at all, due to the resistivity, a velocity will develop that will make the $\mathbf{V} \times \mathbf{B}$ term dominant again. Therefore, $\mathbf{V}$ should be taken as a typical velocity that occurs in the system, say the Alfven velocity, when justifying the neglect of resistivity. Then $R_M = 4\pi L v_A/\eta c$. When V is replaced by v_A, the number is called the Lundquist number and is denoted by S.

We now gather our equations together in the most convenient and compact form to display the generally accepted ideal MHD equations.

Continuity equation
$$\frac{\partial \rho}{\partial t} + \nabla \cdot (\rho \mathbf{V}) = 0 \tag{14}$$

Equation of motion (with $q\mathbf{E}$ force dropped)
$$\rho \frac{d\mathbf{V}}{dt} = \mathbf{j} \times \mathbf{B} - \nabla p + \rho \mathbf{g} \tag{15}$$

Entropy or pressure equation (again)
$$\frac{d}{dt}\left(\frac{p}{\rho^\gamma}\right) = 0 \tag{16}$$

Magnetic differential equation without resistivity
$$\frac{\partial \mathbf{B}}{\partial t} = \nabla \times (\mathbf{V} \times \mathbf{B}) \tag{17}$$

Magnetic differential equation with resistivity (assuming η to be constant)
$$\frac{\partial \mathbf{B}}{\partial t} = \nabla \times (\mathbf{V} \times \mathbf{B}) + \frac{\eta c}{4\pi} \nabla^2 \mathbf{B} \tag{18}$$

Equation 18 follows from

$$\nabla \times \mathbf{B} = 4\pi \mathbf{j} \tag{19}$$

and $\nabla \times (\nabla \times \mathbf{B}) = -\nabla^2 \mathbf{B}$. Equation 17 is equation 18 without resistivity. Equation 19 is Ampere's law without the displacement current.

The basic ideal equations are equations 14–17. If we replace (17) by (18), we refer to these equations as resistive ideal equations and the fluid satisfying them as a resistive ideal fluid. (Without the adjective resistive we always mean the true or perfect ideal equations and the ideal fluid satisfying them.)

Recognizing that j can be considered as an abbreviation for $(\nabla \times \mathbf{B})/4\pi$ and $\mathbf{j} \times \mathbf{B} \equiv (\nabla \times \mathbf{B}) \times \mathbf{B}/4\pi$, we see that these four equations involve only the fundamental quantities ρ, $\mathbf{V}$, p, and $\mathbf{B}$. At the same time these equations give the time evolution of these four quantities so that they are essentially complete.

There is one slight addendum to this completeness statement. Maxwell's divergence B equation 8, $\nabla \cdot \mathbf{B} = 0$, must be satisfied. Equation 17 or 18 guarantees that this will continue to be satisfied if it is satisfied initially.

We now imagine a perfect ideal fluid with density ρ, velocity $\mathbf{V}$, pressure p, and magnetic field $\mathbf{B}$ satisfying the ideal set of equations, 14–17. The properties of this plasma are said to be those of an ideal MHD fluid. Since its properties approximate those of a real plasma very well, when we consider macroscopic phenomena, the ideal plasma is well worth studying. These equations have so many nice properties that this approach to macroscopic phenomena is extremely useful. Throughout a large part of the book we will adapt this approach. Once a macroscopic phenomena has been well described in this way, we will be in a position to check the extent to which its approximation to a real plasma is accurate.

However, we will find that one of the properties of the plasma, flux freezing, can sometimes lead to such tangled or stressed fields that intense small-scale structures emerge and the neglect of resistivity is no longer valid. In this case, we must replace equation 17 with the resistive magnetic differential equation 18, at least for the small-scale structures.

3.2 Flux Freezing

The most important property of an ideal plasma is flux freezing. Consider equation 17 governing the evolution of the magnetic field. This equation leads immediately to the very important concept of flux freezing. Flux freezing says that when the magnetic field is described by its field lines, these lines must possess a reality beyond that usually attributed to them in a vacuum. The number of such lines through any given moving fluid element does not change. From this it follows that magnetic field lines can be considered to be bodily transported by the plasma in which they are embedded. If this were not the case at some point P, then an electric field would be induced in the local moving frame at P, and this would produce a current proportional to $1/\eta$, which would therefore be nearly infinite. This follows essentially from a modification of Lenz's law, which states that in a static system a current system resists any attempt to modify it. We will shortly give a formal proof of these facts.

Granted this flux freezing result, we see that our usual intuitive picture of a magnetic field as a set of field lines is given great force. Start initially with any field **B** and its lines of force. Then the plasma as it moves drags the lines along with it. If we can know where the plasma goes, we know where the lines go, and thus know the **B** field. Further, since the plasma moves continuously, the lines move continuously and cannot break. Thus, the topology of the lines is preserved and this brings topological properties into the study of plasma physics. This provides a powerful tool for understanding much of macroscopic plasma physics.

Let us consider a simple example to make this clear, that of the interaction of the solar corona with the solar photosphere. Observationally we know that coronal field lines leave or return to the photosphere. Many lines that leave the photosphere rise into the corona and then return to it, although a few enter into the solar wind and may not return. Ignore these solar wind lines. Now, as the plasma in the photosphere moves along its surface, it carries the foot points of these lines along in such a way that the two plasma elements situated initially at the foot points of any line of force stay connected by this line of force. Since the photospheric motions are somewhat random, the lines in the corona get more and more twisted without breaking, the magnetic field in the corona becomes more and more stressed, and its magnetic energy increases without bound. Clearly, this cannot happen indefinitely. Either the energy becomes so large that the photospheric motions no longer have the power to continue to amplify it and these motions are reduced, or the ideal equations fail to describe the solar corona and flux freezing breaks down. The latter seems to be the case. Further, the breakdown seem to occur in narrow regions (probably sheets), and can occur suddenly. This phenomenon is believed by many people to be responsible for coronal heating, and also underlies the remarkable solar flare event.

This example illustrates how, with the idea of flux freezing the imagination, without actually solving any equations, can conceive of situations where important physical phenomena may arise.

Let us now establish the idea of flux freezing on a firm mathematical basis, showing how it follows from the magnetic differential equation 17. Consider an arbitrary moving surface S and the flux through it:

$$\Phi(t) = \int_{S(t)} \mathbf{B}(\mathbf{r}, t) \cdot d\mathbf{A} \tag{20}$$

We expect $\Phi(t)$ to change in time by both the dependence of $\mathbf{B}(\mathbf{r}, \mathbf{t})$ on time and by the changing surface $S(t)$ carried by the plasma motions. The

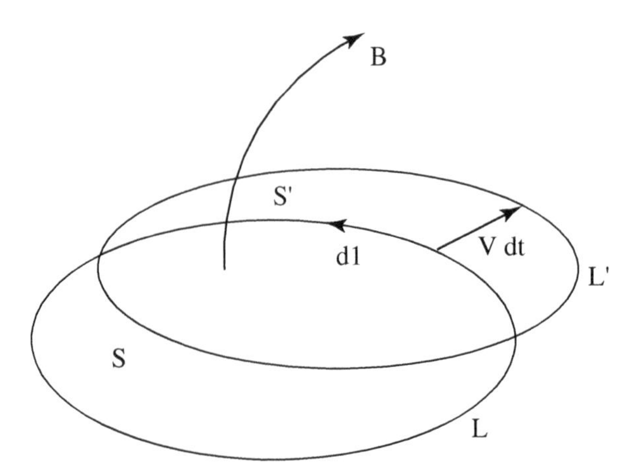

Figure 3.1. The ribbon between the two surfaces at t and $t + dt$

contribution from the first effect is

$$\int_S \frac{\partial \mathbf{B}}{\partial t} \cdot dA = \int_S \nabla \times (\mathbf{V} \times \mathbf{B}) \cdot d\mathbf{A} = \int_L \mathbf{V} \times \mathbf{B} \cdot d\boldsymbol{\ell} \tag{21}$$

where L is the curve bounding the surface S.

The contribution from the second effect is obtained by inspection of figure 3.1 which displays the surface at two different times t and $t+dt$. The second change in the flux is the flux through the ribbon between the two bounding curves L and L' of the surface $S(t)$ and $S(t + dt)$ (surface S' in figure 3.1). The elementary area of the ribbon is

$$\mathbf{V}dt \times d\boldsymbol{\ell} \tag{22}$$

and the flux through the ribbon divided by dt is

$$-\int_L (d\boldsymbol{\ell} \times \mathbf{V}) \cdot \mathbf{B} = -\int d\boldsymbol{\ell} \cdot (\mathbf{V} \times \mathbf{B}) \tag{23}$$

This change in Φ (from the moving surface) just cancels the previous change (from the changing field) through the fixed surface S. Thus, the flux through any moving surface of any size or orientation or shape in three-dimensional space is constant. That is, the number of lines of force through the surface (which is the flux Φ) is conserved.

From this theorem we can now show that the lines are dragged by the matter. Consider a given line of force L at time t. Enclose it in a thin flux tube whose surface is made up of neighboring lines of force, as in figure 3.2. Consequently, at this instant the flux through any part of this surface is zero. Assume that we follow the plasma in the physical tube to a later time t' and that the field $\mathbf{B}$ at some point along the new tube is no longer parallel to it. Then about this point we can find a patch of the tube through which there is a nonzero flux, contrary to flux conservation for this patch. Therefore, this

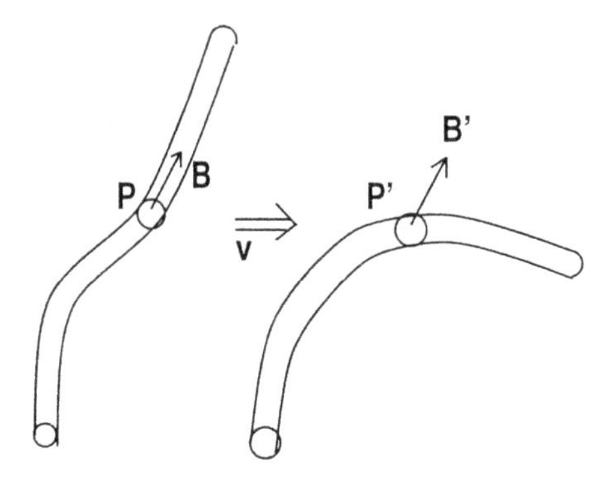

Figure 3.2. Demonstration that lines must be bodily carried along by the plasma

cannot happen and the tube of plasma at the new position $\mathbf{r}'$ must remain parallel to $\mathbf{B}(\mathbf{r}', t')$ at this new position.

Note that this proof holds for more general flows. In fact, for an imaginary fluid to preserve flux, it is sufficient for its velocity $\mathbf{U}$ to satisfy $\partial \mathbf{B}/\partial t = \nabla \times (\mathbf{U} \times \mathbf{B})$ or from the induction equation

$$\nabla \times \left(\mathbf{E} + \frac{\mathbf{U} \times \mathbf{B}}{c} \right) = 0 \tag{24}$$

This, in turn, is a necessary and sufficient condition for flux conservation by $\mathbf{U}$.

The weaker theorem, that field lines of plasma follow the plasma, can reasonably be called the " line preservation theorem." In fact, if we inspect our proof of the line preservation theorem, we see that we only need flux conservation for surfaces (the patch on our tube) whose normal is perpendicular to B. The general necessary and sufficient condition for line preservation is (Newcomb 1958)

$$\mathbf{B} \times \left[\nabla \times \left(\mathbf{E} \times \frac{\mathbf{U} \times \mathbf{B}}{c} \right) \right] = 0 \tag{25}$$

This is clearly a weaker condition than the flux conservation condition (24).

Even in the presence of a real plasma with velocity $\mathbf{V}$ we can conceive of an imaginary fluid with velocity $\mathbf{U}$, not necessarily equal to $\mathbf{V}$, but such that it preserves the flux and lines. This is often convenient for understanding plasma instabilities such as the interchange instability (see chapter 7).

We see from this that there is no absolute way to say how magnetic lines of force move. We can only say that there is at least one way to assign a motion to them, which is simply to give them the velocity of the plasma.

If resistivity is important, magnetic field lines can slip relative to the plasma. For axisymmetric situations we can distinguish the poloidal and toroidal fields and fluxes. It is possible to introduce two poloidal velocities for two imaginary fluids that slip relative to the real fluid, but in which the

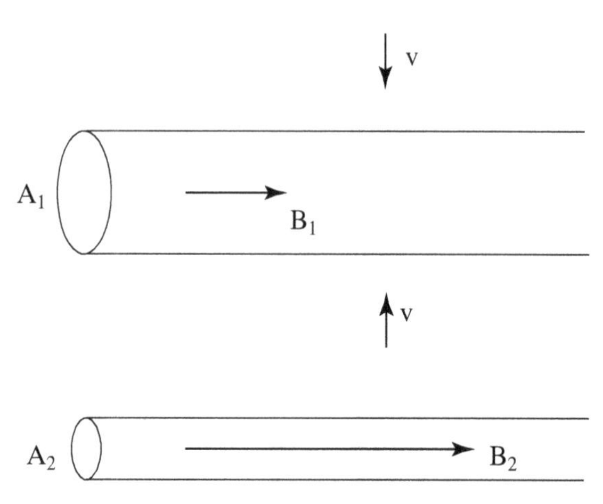

Figure 3.3. Compression perpendicular to B

two different components of the field, the poloidal and toroidal fluxes are respectively frozen.

Our description and proofs of flux freezing are semiquantitative. Later we shall introduce a more analytical description for the change in **B** (problem 2 and Lundquist's identity, equation 75 in chapter 4. For the moment let us discuss some applications that bring out the importance and usefulness of the flux freezing.

3.3 Applications of Flux Freezing

3.3.1 The Symmetric Cases

We now take up some simple examples of flux freezing. First consider a uniform plasma and a uniform magnetic field, taken in the z direction, and let there be motion of the plasma perpendicular to **B** that is independent of z as in figure 3.3. Follow a cylinder of cross-sectional area A. Both the mass and the flux in this cylinder are conserved. Therefore, ρA and BA are conserved and we have

$$\frac{B}{\rho} = \text{constant} \tag{26}$$

Next consider motion parallel to **B** but now dependent on z. In this case the cross-sectional area of a cylinder is preserved so $\mathbf{B} = \text{constant}$. This is valid, even though ρ for some plasma of length ℓ changes as $1/\ell$. $\rho\ell$ is constant so $\rho \sim 1/\ell$ and changes. We can say in this case that for such motions,

$$\frac{B}{\rho\ell} = \text{constant} \tag{27}$$

Next consider a uniform isotropic contraction of the plasma toward the origin. The plasma in a sphere of radius r not necessarily at the origin conserves mass and flux so that both ρr^3 and Br^2 are constant and so

$$\frac{B}{\rho^{2/3}} = \text{constant} \tag{28}$$

Collecting these three cases together we have

$$\begin{aligned} &\frac{B}{\rho^n} = \text{constant} \\ &n = 1 \quad \text{Perpendicular contraction or expansion} \\ &n = 0 \quad \text{Parallel contraction or expansion} \\ &n = \tfrac{2}{3} \quad \text{Isotropic, contraction or expansion} \end{aligned} \tag{29}$$

Thus, in these symmetric cases there exists a simple relation between field strength B and density ρ, but the relation depends on the particular symmetry. Such relations, which are occasionally invoked in astrophysics, must be handled with care.

3.3.2 Stellar Collapse

Now, let us consider some more specific cases. First a collapsing star. Let a star of radius $\approx 10^{11}$ cm collapse to a white dwarf of radius 10^9 cm. If this collapse is isotropic, $B \sim \rho^{2/3}$ and ρ increases by $(10^{11}/10^9)^3 = 10^6$ so the magnetic field increases by $(10^6)^{2/3} = 10^4$. For example, an initial field of 10^2 G increases to 10^6 G. Such fields are frequently observed in white dwarfs and flux freezing makes clear why they should be present.

For a more extreme case consider the isotropic collapse of the same star to a neutron star of radius 10^6 cm and density 10^{15} g/cm^3. The radial collapse of r from 10^{11} to 10^6 means a decrease of r by 10^{-5} and area $\sim r^2$ by 10^{-10}, so the field would be increased by 10^{10}, say from 100 G to 10^{12} G (a typical field for a neutron star).

Let us ask whether flux conservation is a reliable assumption. To check this, let us go to the magnetic field equation with resistivity, equation 18:

$$\frac{\partial \mathbf{B}}{\partial t} = \nabla \times (\mathbf{V} \times \mathbf{B}) + \frac{\eta c}{4\pi}\nabla^2 \mathbf{B}$$

The factor $\eta c/4\pi$, the resistivity of the plasma, is very weakly (logarithmically) dependent on density and is proportional to the electron temperature to the minus three-halves power, $T_e^{-3/2}$. Its value at $T = 10^4$ K is $\approx 10^7$ cm^2/sec, so

$$\frac{\eta c}{4\pi} \approx \frac{0.42 \times 10^7}{T_{\text{eV}}^{3/2}} \ \text{cm}^2/\text{sec} \tag{30}$$

where T_{eV} is the temperature in electron volts. Its dimensions are clear from equation 18. If we estimate $\nabla^2 \sim 1/L^2$, where L is the scale length over which B varies, then the last term is of order B/T_{decay}, where (without the 0.42 factor)

$$T_{\text{decay}} \approx 10^{-7} T_{\text{eV}}^{3/2} L^2 \text{sec} \tag{31}$$

For the white dwarf with $T_{\text{eV}} = 10^2$ (10^6 K) and $L \sim R \sim 10^9$ cm we find $T_{\text{decay}} \approx 10^{-7+3}(10^9)^2 = 10^{14}\text{sec}^{-1}$. The time-derivative term in equation 18 is B/t, where t is the timescale during which a star collapses to reach the white dwarf stage. Thus, the resistive term is negligible compared to the time derivative term if

$$t \ll T_{\text{decay}} = \frac{L^2}{\eta c/4\pi} \tag{32}$$

(Cowling 1976).

Flux freezing results from balancing the $\partial B/\partial t$ and the $\nabla \times (V \times B)$ term. If t is shorter than 10^{14} sec $= 3 \times 10^6$ years, then the resistivity term is negligible, these other two terms must balance, and we can safely assume flux freezing.

Our discussion has been very rough. η varies with T_{eV} and it and L^2 clearly change by large factors as the star collapses, so the effective decay time T_{decay} is probably larger than the above estimate. Also, the later stages of collapse occur faster. Since the total collapse term is probably shorter than 10^{14} sec, our criteria for flux freezing is even more strongly satisfied.

A similar calculation for increase in the magnetic field during the collapse of a neutron star performed at its smallest radius with $T_{\text{eV}} \gg 10^4_{\text{eV}}$ $L = 10^6$ cm leads to a decay time of $T_{\text{decay}} > 10^{6-7+12} \approx 10^{11}$ sec. Neutron star collapse occurs on a timescale of a few seconds, so from this calculation flux freezing is also well satisfied in this case.

Let us turn to an example where flux freezing, at least in it simplest form, definitely cannot hold. This example concerns the initial formation of a star. Again, for simplicity, assume that the collapse is isotropic. Let the protostar start with a density of 1 atom/cm^3 and collapse to a uniform sphere with a density of 1 g/cm^3 corresponding to $\approx 10^{24}$ atoms/cm^3. B increases by $(10^{24})^{2/3} = 10^{16}$. Starting with a field of 3×10^{-6} G, characteristic of the interstellar field, its final value would be 3×10^{10} G. Such a magnetic field has a magnetic pressure $B^2/8\pi = 4 \times 10^{19}\text{ergs/cm}^3$. Compare this with the internal pressure of a normal star, $nT_{\text{eV}} \approx 10^{24-10} \approx 10^{14}$ ergs/cm. (Here and in the rest of this book we include the Boltzmann constant intrinsically in the temperature, although when we refer to the temperature we continue to mean its value in units of electron volts. This is convenient and never seems to cause a problem.)

We see that such a field could not be present in stars. In fact, such a field would be strong enough to resist gravitational collapse altogether. Also, for a typical star, $T = 10^2$ eV, $L = 10^{11}$ cm, and $T_{\text{decay}} \approx 10^{18}$ sec. Because this is longer than the Hubble time, flux freezing should be valid.

To understand what must happen, we have to appreciate that before the star can form, the interstellar matter must have a very low degree of ionization, and the neutrals and the ionized component need not move together. In fact, the magnetic $\mathbf{j} \times \mathbf{B}$ forces act only on the ionized component. The ionized component cannot collapse at all because of this strong force. The neutral component can collapse, but its collapse is limited by the collision of the neutrals with the stationary ions. The balance of forces give

$$n_0 n_i m_{\mathrm{H}} V_r \langle \sigma v \rangle = n_o m_{\mathrm{H}} g = n_0 m_{\mathrm{H}} \left(\frac{GM}{R^2} \right) \tag{33}$$

where m_{H} is the mass of atomic hydrogen, n_i is the ion density, V_r is the radial velocity of the collapse, $n_i \langle \sigma v \rangle$ is the collision rate, g is the (inward) gravitational force, and n_0 is the neutral density. $\langle \sigma v \rangle$ is of appreciable size, but when n_i becomes very small, V_r can become large enough to lead to collapse of a star in a reasonable time.

Note that the same forces on the neutrals that occur in this equation are also exerted on the ionized component, since the total collisional forces are equal (and opposite), by momentum conservation of each collision. Since the collisional force on the ions is balanced mostly by the Lorentz force, the Lorentz force $\mathbf{j} \times \mathbf{B}$ must balance the right-hand side. Since any appreciable distortion of the field leads to a Lorentz force much larger then the gravitational force on the neutrals, the distortion must be negligible.

To summarize: During collapse, the magnetic field remains frozen in the ions that are held essentially motionless by the large Lorentz force. The neutrals collapse moving through the ions at a velocity such that the collisional force with the ions just balances the gravitational collapse force. This process of stellar formation is called ambipolar diffusion (Elmegreen 1985).

3.3.3 The Solar Wind and the Magnetosphere

Next we consider the solar wind and its interaction with the earth's magnetic field. The solar wind cannot enter the region of the earth's field because to do so would violate flux freezing. The solar wind is "frozen off of" the earth's magnetic field lines. Alternatively, if the plasma did enter the field we could consider it frozen onto the earth's field lines. Turning time around the plasma could not leave and return to the sun. Note that equation 17, from which flux freezing derives, is time reversible.

Thus, insofar as flux freezing holds, the solar wind plasma cannot enter the earth's magnetic field. When it encounters the earth it must flow around on either side of the earth's field. This deflection of the solar wind requires a considerable force. This leads to a compression of the earth's magnetic field into a closed cavity, called the magnetosphere. By balancing the force resulting from the compression of the earth's field with the pressure exerted by the solar wind on it we find that the size of the earth's magnetic cavity is between 8 and 10 earth radii. Indeed, from measurements of this cavity

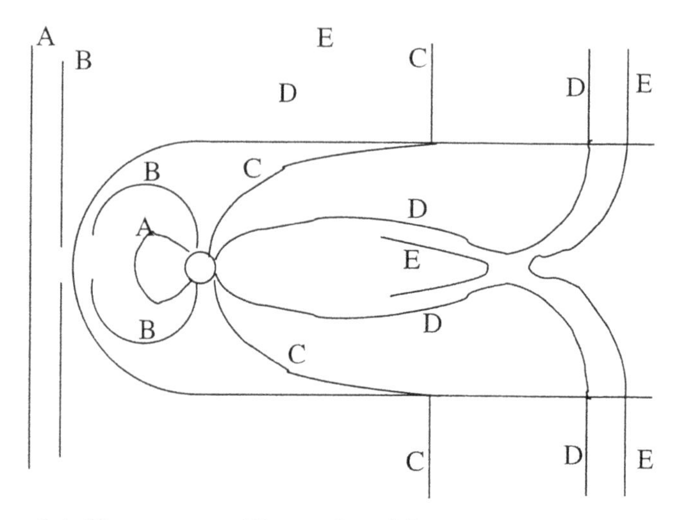

Figure 3.4. The reconnected lines at four different positions, *A*, *B*, *C*, *D*, *E*

by satellites, we find that this is its approximate size in the direction toward the sun, and also that as the speed and density of the solar wind change, the radius of the cavity varies between these two scales. It is smaller when the solar wind is more intense.

On the downstream side of the earth, away from the sun, the picture is very different from what would be expected from perfect flux freezing. With flux freezing, we would expect that the shape would be a teardrop, with the cavity ending in a point where the solar wind from the different transverse sides of the cavity come together. Instead, the earth's field fills a long, cylindrical cavity extending many hundreds of earth radii downstream. The inside of this cylinder is called the earth's magnetotail, perhaps in analogy with a comet tail.

The explanation for this is due to Dungey (1961). He postulated that when the magnetic lines carried by the solar wind come in contact with the earth's magnetic lines of force, not all of the solar wind's lines pass smoothly around the earth. He supposed that 5 to 10 percent of the solar wind's magnetic lines break. Simultaneously, an equal number of the earth's magnetic lines break. The broken lines of different types join, or reconnect, the earth's broken lines connecting to the solar wind's broken lines (Hughes 1995). The result of this breaking and reconnecting can be visualized from figure 3.4, where the positions of five different pairs of lines, *A*, *B*, *C*, *D*, and *E*, at a given time are shown. If the situation is in a steady state, we can visualize these five positions as indicating the change in time of a single pair of lines.

Start with pair *A*. The left line of the pair is in the solar wind and is approaching the earth with the solar wind velocity. The right line of the pair is connected to the earth. It will move to touch the solar wind. The pair will then occupy the position of the pair *B*. The two lines break and the northern parts of the lines in the pair will reconnect, as will the southern parts. The two parts of the pair are now each connected to the earth and embedded in the solar wind. The solar wind parts of the lines continue to move with the solar wind velocity.

At a later time the lines are at position C, where the solar wind part of the line is being carried rapidly downstream, pulling the earth's part of the line far downstream. The stretched part of our line and other such lines form the magnetotail of the earth. This stretching continues until a second breaking of the line occurs at the position of the pair D. The solar wind line again reattaches to its original part at position E.

Subsequently, the restored solar wind is swept downstream without further disturbance. However, after reconnection the earth's part of the line is greatly stretched and is expected to snap back to the earth, delivering the considerable energy it has obtained from the energy of the solar wind velocity to the inner part of the earth's magnetosphere. This can generally lead to a large disturbance, which is often credited with being responsible for magnetic substorms and aurora.

This crude picture seems to correspond remarkably well to what is observed. That breaking of the flux-freezing condition occurs here is not so surprising when it is appreciated that the contact surface between the solar wind and the magnetosphere is very thin and, because the magnetic field is quite different across it, the current density is very large. Thus, the ηj term in Ohm's law, equation (10), or, alternatively, the diffusion term $(\eta c/4\pi)\nabla^2 B$ in the magnetic differential equation 18 need no longer be negligible. However, as shown in chapter 14, a closer look at the *reconnection* event that occurs in the narrow layer shows that only a tiny number of lines should reconnect on the basis of the resistive ideal equations, that is to say, only a much smaller fraction than 5–10 percent of the lines reconnect. The resolution of this discrepancy is a subject of great current interest. The subject of the reconnection of lines over this layer is called magnetic reconnection and is very important. We devote chapter 14 entirely to magnetic reconnection.

The conditions on the surface of the magnetosphere are so extreme that even the resistive ideal equations cannot adequately describe them. Much microscopic physics comes in. Shocks and intense microscopic waves appear that may enhance the resistivity. It is probable that these phenomena are responsible for the substantial amount of reconnection that is actually observed. It is interesting that after a line reconnects it stays reconnected for a good portion of a day before it separates back to its previous state. During this time it can be drawn downstream about 1000 earth radii.

In the Dungey picture the concept of flux freezing is valid everywhere except in a tiny region on the sunward side of the surface of the magnetosphere, and at one point far down the magnetotail. But these tiny regions change the topology of the lines and this has a profound effect in changing the whole picture of the solar wind interaction with the magnetosphere, producing the magnetotail, the magnetic storms, and the aurora near the earth's polar caps.

Another remarkable phenomenon that can be easily understood in terms of flux freezing is the spiral structure of the lines of force in the solar wind (Parker 1963). Beyond about ten solar radii the solar wind velocity is found to be nearly radial. Let us imagine a blob of solar wind plasma with azimuthal angle $\theta = 0$ near the sun that moves out radially, starting at time $t = 0$. If the sun were not rotating, the blob would draw the line of force out of the sun in the strictly radial direction. But because of rotation, the position of the next blob, which was at $\theta = 0$ at $t = 0$, is at $\theta = \Omega t$ at time t. If these two blobs were on the same line of force, the line must connect the first blob at $\theta = 0$, $r_1 = V_0 t_1$ with the second blob at $r = 0$, $\theta_2 = \Omega t_1 = (r_1/V_0)\Omega$. It is easily seen that at later times the points that are connected lie along the curve

$$\theta = r\frac{\Omega}{V} \quad \text{or} \quad r = \frac{V\theta}{\Omega} \tag{34}$$

the equation of an Archimedian spiral. The angle α this curve makes with the azimuthal direction satisfies

$$\tan\alpha = \frac{dr}{rd\theta} = \frac{V}{\Omega r} \tag{35}$$

with $\Omega = 3 \times 10^{-6}$. At the earth $r = 1$ a.u. $= 1.5 \times 10^{13}$ cm and taking $V = 3.5 \times 10^7$ cm/sec, we find that $\alpha \approx 45°$ at the earth, which is close to the observed angle. Note that for large r, $\alpha \sim 1/r$ and the spiral becomes flatter at large distances, also in agreement with observations.

3.3.4 Stellar Formation and the Angular Momentum Problem

Let us return to star formation for another important example where flux freezing plays a crucial role, We earlier showed how magnetic fields inhibit star formation since the interstellar magnetic field is too strong to be drawn into the newly forming protostar. In exactly the same way, angular momentum inhibits star formation. Conservation of angular momentum Ωr^2 means that as the star collapses its angular velocity increases. In fact, we have $\Omega \sim r^{-2} = \rho^{2/3}$. Thus, if ρ increases from $\rho = 10^{-24}$ g/cm^3 to $\rho = 1$ g/cm^3, Ω increases by $(10^{24})^{2/3} \approx 10^{16}$. The initial blob of plasma has at least the angular velocity of the galaxy $\Omega \approx 2\pi/(200 \times 10^8 \text{ yr}) = 10^{-15}$radians/sec, so the collapsed star would have $\Omega = 10\,\text{sec}^{-1}$, much too fast too allow the star to collapse.

This angular momentum barrier can be removed by the interstellar field (Elmegreen 1985). The field lines in the collapsing star are also connected to the neighboring part of the interstellar medium not directly involved in the collapse. See figure 3.5, where B is parallel to Ω. As the star collapses the lines becomes twisted about the rotational axis Ω.

The twisted field lines passing through both the star and the interstellar medium have a current j that leads to magnetic forces that exert a backward torque on the star and a forward torque in the cylindrical column not

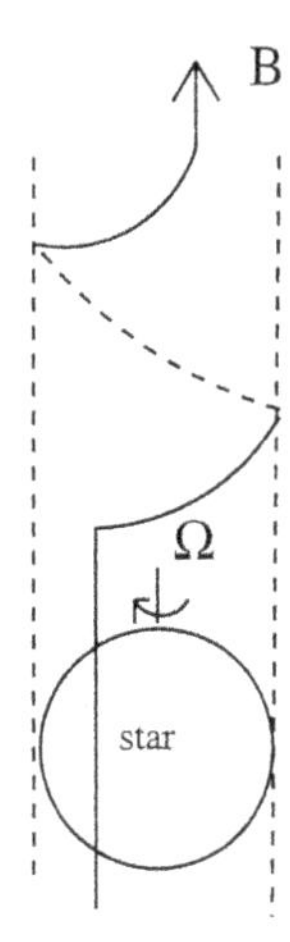

Figure 3.5. The transfer of angular momentum from a rotating protostar to the interstellar medium by the interstellar field

included in the sphere. The result is that the angular momentum of the collapsing star is shared with a piece of the column. The length of the column that is forced to corotate is $v_A t$, since the twist in the line is actually an Alfven wave propagating up the interstellar field (see problem 4). Assuming initially that the density of the column and the sphere are the same, we see that the angular velocity Ω of the protostar decreases as

$$\Omega = \frac{\Omega_0}{1 + v_A t / R_0} \tag{36}$$

where R_0 and Ω_0 are the initial values of the protostellar radius and angular velocity, and the Alfven speed is $v_A = B/\sqrt{4\pi\rho}$. This decrease continues until the lines starts to slip, by ambipolar diffusion, through the neutral component of the sphere. Generally, a large enough angular momentum is removed to allow collapse. (It is now appreciated that in the later stages of collapse further angular momentum is removed by formation of a disk and the existence of anomalous viscosity in the disk.) The magnetic field has a twofold role in the stellar formation. First it removes a sizable fraction of angular momentum, helping to break through the angular momentum barrier. Then through its radial stress it resists further collapse until ambipolar slippage occurs.

3.3.5 Magnetic Fields in Turbulence

Now let us consider a turbulent flow of a weakly magnetized infinitely conducting plasma. Let us follow a short piece of a magnetic tube of force of cross-sectional area A and of length ℓ.

By flux conservation,

$$\Phi = BA = \text{constant} \tag{37}$$

and by mass conservation,

$$M = \rho A \ell = \text{constant} \tag{38}$$

Dividing the first of these two equations by the second we have

$$\frac{B}{\rho \ell} = \text{constant} \tag{39}$$

If the density is constant, we have, more simply,

$$\frac{B}{\ell} = \text{constant} \tag{40}$$

or the magnetic field strength is proportional to its length ℓ, where ℓ refers to the length of a particular short piece of plasma along the line.

Now if the field is weak compared to the motions in the plasma (the condition for this is $B^2/8\pi \ll \rho V^2/2$), then the motions proceed unaffected by the magnetic forces. In a general turbulent medium, two fluid elements separated by a length ℓ, short compared to the decorrelation distance of the turbulence, d_c, separate at a rate proportional to ℓ itself. That is, their rms distance increases exponentially with an exponentiation time $\tau \sim d_c/V$. This remark, together with the line stretching property of the field, implies that the square of the magnetic field strength, B^2, grows as

$$B^2 = B_0^2 \exp 2\gamma t \tag{41}$$

where B_0 is the initial field strength and γ is of order the "eddy turnover rate" V/d_c. (Of course, any given initial length would quickly grow to be larger than d_c, but we can work backward for any time t and pick the initial length shorter than $d_c e^{-t/\tau}$.)

This exponential growth of B cannot proceed indefinitely. One of the two assumptions, the flux-freezing assumption or the weak-field assumption, eventually breaks down. When $B^2/8\pi \sim \rho V^2/2$ the second assumption will break down. But the magnetic field scale, the length over which B changes appreciably, will also decrease exponentially. In fact, over some scale the mean field can be considered constant (because of flux conservation). Thus, as the mean square field increases exponentially, the decorrelation scale for the magnetic field must decrease exponentially. These very important results are discussed much more carefully in the chapter on dynamo theory (Chapter 13).

The above remarks tacitly assume a constant plasma density. If the density varies, B/ρ is the quantity that increases exponentially. But if the density varies within finite bounds, it is clear that B still must increase exponentially.

3.4 Io and Jupiter

An extreme example of flux freezing is provided by Io, the inner moon of Jupiter. Io is conducting and threaded by the magnetic field of Jupiter. It

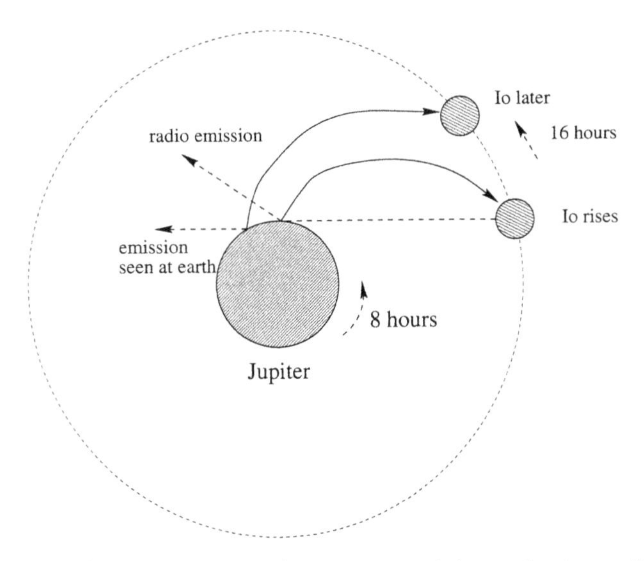

Figure 3.6. The relative orientation of Io, Jupiter, and the earth when radio emission is detected

revolves around Jupiter in roughly 43 hours. Jupiter itself rotates faster, and in the same direction, with an equatorial period of 10 hours.

If the flux lines were completely frozen, both in Io and in Jupiter's ionosphere, and this ionosphere is rigidly attached to Jupiter itself, then the lines would be wrapped around Jupiter once every 13 hours and, thus, the field strength would quickly be amplified to an enormous value, roughly 12 G times the number of rotations. It is clear that flux freezing cannot hold, and that the lines must slip, either through Io or through Jupiter's ionosphere. In any event, the slippage gives rise to a large voltage. Thus, it is not surprising that the emission of synchrotron radiation, characteristic of electrons with energies of many kilovolts, is detected from the surface of Jupiter. This radiation tracks the relative motion of Io across the face of Jupiter (Duncan 1965). Being synchrotron radiation from nearly perpendicular-pitch-angle electrons, it is emitted primarily perpendicular to the magnetic field lines connecting Io and Jupiter. As a result, the emission is observed at the earth shortly after Io emerges from eclipse by Jupiter.

The slight delay can be understood from the geometry of the lines of force dragged by Io. Emission should be seen when the plane perpendicular to the line of force at its foot intersects the earth. The geometry of these dragged field lines when Io emerges from behind Jupiter is sketched in figure 3.6.

To understand what goes on during line slippage let us consider the grossly simplified model shown in figure 3.7. In this model assume that there are two horizontal infinitely conducting walls at $y = \pm(d+D')$, and that initially there is a vertical uniform field $B_y = B_0$. Let there be a horizontal rigid slab of thickness $2d$ moving with constant velocity $V_x = V_0$ in the x direction and centered about $y = 0$ (between $y = \pm d$). Let the open spaces between

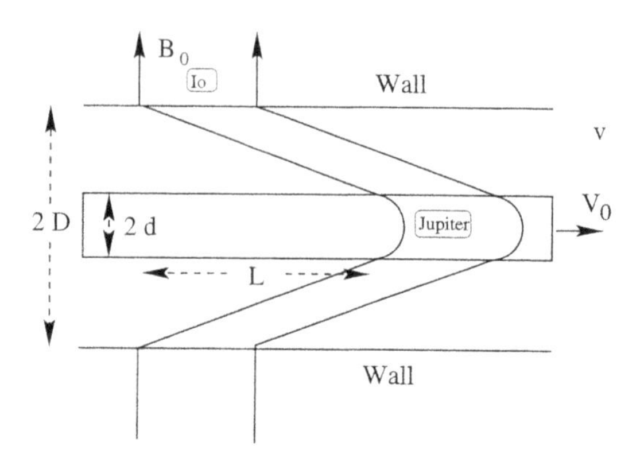

Figure 3.7. A conducting slab passing between two conducting plates as a model for the Io–Jupiter interaction

the slab and the walls be filled with a very low pressure infinitely conducting plasma. Let the walls, the slab, and the field extend infinitely in the x and z direction. Let the slab have finite conductivity. What will happen?

If η is very small, the slab will first draw the lines to the right, satisfying $E_z = -V_x B_y/c$ inside itself. By symmetry B_y does not change, The dragged field lines will have a B_x component negative above and positive below $y = 0$. Consequently, at time t there will be a current density in the slab $4\pi j_z = -\partial B_x/\partial y \approx 2B_x/2d = -(L/dD')B_0$, where $L = V_0 t$ is the displacement of the slab. (Thus, $B_x \approx B_0 D'/L$.) As L increases, j_z increases until ηj_z becomes comparable to $V_0 B/c$ and the electric field approaches zero. That is to say, a steady state is reached.

To derive the steady state let us take as given that the current in the rarefied plasmas, in $d < y < d + D'$ and $-(d + D') < y < -d$, is zero, so $B_x =$ constant there. Then, above the slab

$$\frac{B_x}{B_0} = -\frac{L}{D'} \tag{42}$$

and below $B_x = LB_0/D'$. In this slab the z component of Ohm's law gives

$$\left(\frac{V \times B}{c}\right)_z = \frac{V_0 B_0}{c} = \eta j_z = -\frac{\eta}{4\pi}\frac{\partial B_x}{\partial y} \tag{43}$$

This is a constant because $V_x B_0/c$ is. Integrating this equation and matching to $B_x = \mp(L/D')B_0$ at $y = \pm d$, we find for $0 < y < d$

$$B_x = -\frac{V_x B_0}{\eta c/4\pi} y$$

and when $y = d$

$$B_x = \frac{V_x B_0}{\eta c/4\pi} d = -\frac{L}{D'} \tag{44}$$

and thus,

$$L = \frac{V_0 d D'}{\eta c/4\pi} \tag{45}$$

This can be related to T_{decay}, given by the numerical resistive time-decay formula, equation 32, if we let the length in that formula be $\sqrt{dD'}$. In fact, we get $L = V_0 T_{\text{decay}}$ this way. (Using this result and with the parameters of Io, we find that Jupiter's magnetic lines of force are stretched through only a small angle.)

From $L = T_{\text{decay}} V_0$, we can see that the lines will be stretched at the velocity V_0 for the time it takes the plasma to resistively diffuse through $\sqrt{dD'}$, after which the stretching stops and a steady state is reached. At this point the electric field required to drive the current in the frame of the slab motion (to produce the stretched lines) against resistivity is just $V_0 B_0/c$. This electric field is produced by the dynamo term.

Now let us apply our model to the Jupiter–Io problem. It turns out that the ionosphere of Jupiter is considerably more resistive than that of Io. Thus, in our model Jupiter is represented by a piece of the slab and Io is represented by the piece of the wall connected to the slab, which is connected to it by the magnetic field lines. V_0 represents Jupiter's rotational velocity, relative to Io's orbital motion. As Jupiter rotates, the lines slip through its ionosphere. In the frame of Jupiter there is a north–south electric field $V_0 B_y/c$ perpendicular to the direction of its motion. This electric field gives rise to a potential voltage of about one million volts, which is presumably responsible for the particle acceleration and the radio emission (Goldreich and Lynden-Bell 1969).

In our infinite slab model, we would like to follow how the lines of force move even in the presence of resistivity. In the outer regions where the conductivity is infinite the lines of force should not move. Similarly, inside the slab where the plasma moves we see that the lines of force should not move even though the fluid velocity is finite. We generally interpret this by saying that the lines slip backward relative to the slab due to the finite resistivity. To quantify this slippage in the case of our slab symmetry, let us write Ohm's law as

$$\mathbf{E} = \frac{-\mathbf{V} \times \mathbf{B}}{c} + \eta \mathbf{j} = -\frac{\mathbf{V} \times \mathbf{B}}{c} - \frac{\mathbf{V}_\eta \times \mathbf{B}}{c} \tag{46}$$

with $\mathbf{V}_\eta \times \mathbf{B}/c = -\eta \mathbf{j}$, or

$$\mathbf{V}_\eta = \frac{\mathbf{B} \times \eta c \mathbf{j}}{B^2} \tag{47}$$

where $\mathbf{V}_\eta$ is the slippage velocity. By the argument given for flux freezing we see that we should consider the lines to move with velocity

$$\mathbf{V} = \mathbf{V}_{\text{plasma}} + \mathbf{V}_\eta \tag{48}$$

For our simple model, $\mathbf{V}_\eta = -\mathbf{V}_{\text{plasma}}$ and the lines do not move in the slab, once a steady state has been reached. In Jupiter's frame V_η is the actual

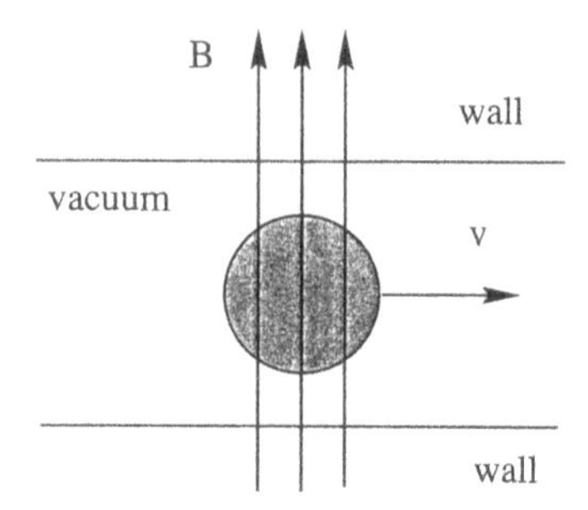

Figure 3.8. A sphere moving through a vacuum across a magnetic field

velocity of the lines as they are dragged through Jupiter's ionosphere by Io's backward motion.

It should be remarked that a great deal more research has been done on this famous problem, both observationally by satellite flybys and theoretically. We will not describe this research, which is still not conclusive, but merely content ourselves with this employment of the Jupiter–Io connection to show how flux freezing can be limited.

3.5 Motions of Lines of Force in a Vacuum

Let us next consider the rarefied regions between the slab and the walls to be a vacuum, and let the slab be replaced by a sphere as in figure 3.8. Since there is no current in the vacuum or the sphere, there must be an electric field in the sphere such that $\mathbf{E} + \mathbf{V} \times \mathbf{B}/c = 0$. This field is uniform in the sphere and extends into the vacuum region. This electric field is an electrostatic field produced by charges on the surface of the sphere. The electrostatic field in the vacuum has a component along the magnetic field, but, unlike the case of the rarefied plasma, this field does not produce any currents. Because of the absence of any current, the field lines are straight and undisturbed. This electric field cannot exist when there is a rarefied plasma because it would drain off the charge on the sphere, and the current that would be set up to restore this charge would continue to flow, leading us back to the situation somewhat like that of the plasma slab in the Io–Jupiter model.

This example is a good illustration of how flux freezing might break down in the presence of vacuum or finite resistivity plasmas. Finally, suppose we shoot a spherical ball into an inhomogeneous increasing vacuum field. In this case, the electric field in the sphere increases, again producing a transient current in the sphere. This transient current gives a reverse impulse to the sphere, slowing it down by an amount that we would get by equating the drop in the kinetic energy of the sphere to the increase of the electrostatic energy in the sphere $\approx (V/c)^2$ times the magnetic energy in the sphere.

3.6 The Validity of the MHD Equations

In the preceding sections we have considered general properties of MHD plasmas. Remember that this MHD description of a plasma is valid only if there are sufficient collisions to give it fluid properties. Let us estimate the required collision rate. For example, we regard a given cube of plasma with its mass energy and momentum as a unit. If too many particles leave the cube, it will not preserve its unity and the equations for it will not be valid.

Take such a cube and let it have a scale L comparable to or a somewhat smaller than the scale of variation of the plasma quantities ρ, p, and $\mathbf{V}$. If particles leak out of this cube in a time short compared to T, the time for the plasma quantities to change, then we cannot regard the cube as a unit of plasma, and the MHD equations fail. Let the typical mean free path for the plasma particles be λ, their thermal velocity be v_T, their collision rate be ν, and their collision time be $\tau = 1/\nu$. Then in a time T a particular plasma particle will random walk a distance ℓ with

$$\ell^2 = \frac{T}{\tau}\lambda^2 \approx T v_T \lambda \tag{49}$$

since it will make T/τ collisions, and between each collision it will move a distance λ in a random direction with the direction of the steps uncorrelated. (More formally, $\ell^2 = \langle(\Sigma x_i)^2\rangle$ and $\langle x_i^2\rangle = \lambda^2$, $\langle x_i x_j\rangle = 0$, where angle brackets denote averages. We use the fact that $\lambda = v_T \tau$.)

Thus, our treatment of the plasma as a fluid is justified if $\ell^2 \ll L^2$ or

$$v_T \lambda \ll \frac{L^2}{T} \tag{50}$$

This condition is satisfied if

$$\nu \ll \frac{1}{T} \quad \text{and} \quad \lambda \ll L \tag{51}$$

since $v_T \lambda = \lambda^2 \nu$.

There are further conditions that must be satisfied, such as a sufficiently small thermal conductivity that entropy is conserved and a small enough viscosity that the pressure remains isotropic. These conditions are discussed in full detail in chapter 8. For the moment let us restrict ourselves to the criteria of equation 51 for judging the validity of the MHD equations.

To apply these rough criteria, we give the prescription for a qualitative determination of ν and λ. For these quantities we need an estimate of the appropriate Coulomb collision cross section, which is the effective cross section for scattering of a particle through 90°. (This will be pursued in more detail in chapter 8.) Electrons scatter off of both ions and electrons and the cross section is of order

$$\sigma \approx \frac{10^{-12}}{T_{\text{eV}}^2}\text{cm}^2 \tag{52}$$

where T_{eV} is the electron temperature in electron volts. A temperature of one electron volt is equivalent to a temperature of 10^4 K. Ions primarily scatter off only ions and not off electrons. Their cross section is given by equation (52), with the ion temperature instead of the electron temperature. In addition, we might multiply it by a factor of order one-half because there are only half as many collisions. (But see the last section of chapter 8.)

For equal temperatures, the mean free path is roughly the same for electrons and ions and is

$$\lambda = \frac{1}{n\sigma} \tag{53}$$

where n is either the ion or electron density. (They are the about same for a hydrogenic plasma.) The collision rate is

$$\nu = \frac{v_T}{\lambda} \tag{54}$$

where v_T is the thermal velocity of the appropriate species. Because for the same temperature the thermal speed is larger for electrons than for ions by a factor of 40 (the square root of the mass ratio), ν is larger for electrons than for ions by the same factor. At 1 eV $\approx 10^4$ K, $v_T \approx 7 \times 10^7$ cm/sec for electrons, and $v_T \approx 1.5 \times 10^6$ cm/sec for ions.

If the plasma is partially ionized, then there will also be collisions with neutrals. However, the atomic scattering cross section is the gas atomic cross section of about 10^{-16} cm^2. Because this is much smaller than the Coulomb cross section at 1 eV, collision rates with neutrals are much smaller than collision rates with other ions and can generally be neglected. At higher temperatures, when the Coulomb cross section is smaller, the plasma will be fully ionized, while at lower temperatures the Coulomb cross section is even larger. Thus, as a general rule, electrons and ions are tightly coupled compared to the coupling of electrons with neutrals.

3.7 Pulsar Magnetospheres

The plasmas we have discussed so far have all consisted of ions and electrons with nearly equal densities. As remarked in the introduction, the relative balance between these densities is of order $(\lambda_D/L)^2$. However, there is one noteworthy example, the magnetosphere around a rotating neutron star or pulsar, where charges of only one sign occur. A neutron star is a compact star of solar mass with a radius of 10^6 cm, so its surface gravitational acceleration is 10^{14} cm^2/sec and the ion scale height for a temperature of 100 eV is about 1 cm. Thus, it was at first supposed that the density in its atmosphere must be near that of a vacuum, $\sim 10^{-10^6}$ cm^{-3}.

However, it was pointed out by Goldreich and Julian in 1969 that a magnetized neutron star rotating with an angular velocity ω must have a very strong electric field. For a one-second-period pulsar with a magnetic field of 10^{13} G, $\mathbf{E} = -\mathbf{v} \times \mathbf{B}/c$ and $E = \omega r B_z/c \approx 10^{12}$ V/cm, where z is along the rotational axis so that the electrical force on an ion is 1 dyne, a factor of 10^{10} times the gravitational force.

Only the component of $\mathbf{E}$ along $\mathbf{B}$ actually lifts the ion away from the pulsar against the gravitational force, because the force across $\mathbf{B}$ is balanced by the Lorentz force. However, we can show (see problem 7) that if the atmosphere were actually uncharged, the parallel electric field would be comparable to the total field. This force may be up or down, depending on the relative orientation of $\mathbf{B}$ and $\boldsymbol{\omega}$. Where they are parallel an electron would be pulled out of the crust, and where they are antiparallel an ion would be.

The ions (or electrons) are pulled out into the magnetosphere until the parallel electric field vanishes at the surface and indeed vanishes everywhere except near the poles. (Actually, $E_{\parallel}$ does not completely vanish, but is greatly reduced until at any given point the parallel electric force just balances the gravitational force on either the electrons or the ions, whichever is present at that point.) Then we have

$$\mathbf{E} + \frac{\mathbf{V} \times \mathbf{B}}{c} = 0 \tag{55}$$

neglecting the tiny value of $E_{\parallel}$. We do not know the velocity $\mathbf{V}$, ab initio, but whatever it is the magnetic lines of force are frozen into it.

Now assume that $\mathbf{B}$ is static and axisymmetric. Then, by flux freezing, the lines must rotate rigidly into each other, and since the the lines are tied to the rotating neutron star they must rotate with the same angular velocity ω, carrying the plasma with them. That is, we may replace $\mathbf{V}$ in equation 55 by $\boldsymbol{\omega} \times \mathbf{r}$. Therefore, the electric field is given by

$$\mathbf{E} = -\frac{\mathbf{V} \times \mathbf{B}}{c} = -\frac{(\boldsymbol{\omega} \times \mathbf{r}) \times \mathbf{B}}{c} \tag{56}$$

As remarked above, $\mathbf{E}$ is produced electrostatically by charges in the neutron star and charges in the magnetosphere. The charge density q of the latter is given by

$$4\pi q = \nabla \cdot \mathbf{E} = -\frac{\mathbf{B} \cdot (\nabla \times \mathbf{V})}{c} + \frac{\mathbf{V} \cdot \nabla \times \mathbf{B}}{c} \tag{57}$$

But $\nabla \times \mathbf{B} = 4\pi \mathbf{j} = 4\pi q \mathbf{V}/c$ and $\nabla \times \mathbf{V} = \boldsymbol{\omega}$, so we get

$$q = -\frac{\omega \cdot \mathbf{B}}{4\pi c} \frac{1}{1 - v^2/c^2} \tag{58}$$

Now, $|q| = ne$ because only particles of one sign are present at any point, since the $E_{\parallel}$ cannot support both types of particles with opposite sign against gravity. We thus arrive at the particle density

$$n = \frac{\boldsymbol{\omega} \cdot \mathbf{B}}{4\pi ec(1 - v^2/c^2)} = \frac{B_{13}}{P} \frac{1}{1 - v^2/c^2} 3 \times 10^{11}/\mathrm{cm}^3 \tag{59}$$

where B_{13} is the field strength in units of 10^{13} G and P is the rotational period in seconds. This density is known as the Goldreich–Julian density (generally

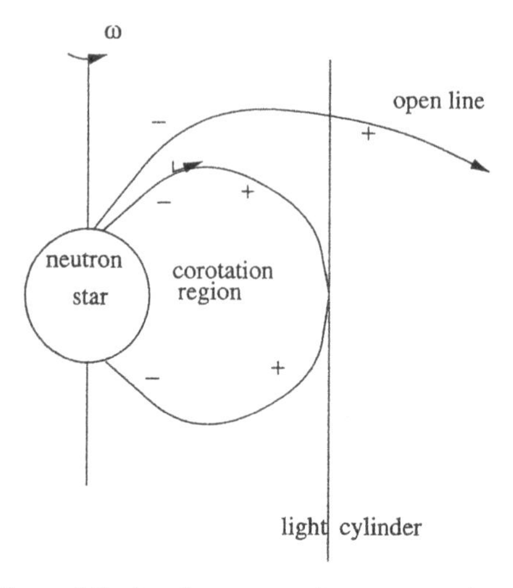

Figure 3.9. A pulsar magnetic magnetosphere

without the factor in the denominator). As a convenient mnemonic device it can be written in terms of the local cyclotron frequency Ω and the plasma frequency as

$$\omega_{pe}^2 = \frac{\boldsymbol{\omega} \cdot \boldsymbol{\Omega}}{1 - v^2/c^2} \tag{60}$$

Thus, we see that the density around a rotating neutron star is much larger than predicted by unmagnetized hydrostatic theory. The sign of the charge of the particles at any point is given by the sign of $-\boldsymbol{\omega} \cdot \mathbf{B}$, so regions where $\mathbf{B}$ is parallel to $\boldsymbol{\omega}$ are occupied by electrons, and regions where they are antiparallel are occupied by ions.

Note that as the cylindrical coordinate r approaches the radius $r_L = c/\omega$, the rotational velocity of the lines approaches c and the simple picture of a purely poloidal field with plasma on it corotating at ω must break down. The cylinder with radius r_L is called the light cylinder.

The situation can be appreciated if we temporarily assume that the field is a purely poloidal dipole field. As seen from figure 3.9, there are two distinct regions. The region occupied by lines that close inside the light cylinder is called the corotation region. It is well described by the above equations except near $r = r_L$. The bounding line between the two regions reaches the neutron star at a colatitude of $\sqrt{R/r_L}$. Lines outside of this region cross the light cylinder. In this region particles stream across the light cylinder. Because they cannot move faster than the speed of light, they bend the poloidal field lines back into a toroidal field. The pattern of lines still rotates at the fixed angular velocity ω, but beyond the light cylinder the corotation speed is compensated by the opposite toroidal velocity of the particles streaming along the toroidal component of the field lines.

The actual velocity of the particles is

$$\mathbf{V} = \kappa \mathbf{B} + \boldsymbol{\omega} \times \mathbf{r} \tag{61}$$

where κ is a scalar that depends on position. The cancellation results because the toroidal part of $\mathbf{B}$ is in the opposite direction to the toroidal coordinate ϕ.

The particles on the open lines of force (those that pass beyond the light cylinder) stream out through the light cylinder and continue until they reach the nebula that surrounds the neutron star. The toroidal angular momentum in their flow plus the tension in the backward-stretched toroidal lines removes angular momentum from the rotating neutron star slowing it down. The particles far from the light cylinder are accelerated to high energy and their outflowing energy balances the decreasing of the rotational energy of the star. This energy goes to energize the nebula. The energy continues to supply the nebula, keeping it radiating a long time after the initial supernova that produced the neutron star has decayed. This is essentially the Goldreich-Julian picture of the rotational dynamics of the pulsar rotating neutron star. It is based on an axisymmetric model in which the magnetic axis of the pulsar is aligned with the rotational axis and therefore cannot be the exact picture of a pulsar, since, by symmetry, there is nothing to pulse. However, most of the physics in the Goldreich–Julian model persists in the nonaligned rotating neutron stars (pulsars).

3.8 Problems

1. (Flux freezing) Consider a small-amplitude disturbance in a uniform plasma in a uniform magnetic field $\mathbf{B}$ in the z direction. Set $\rho =$ constant, $p =$ constant, and

$$\mathbf{V} = \epsilon V_A \sin(kz - \omega t)\hat{\mathbf{x}}$$

$$\mathbf{B} = B_0 - \epsilon B_0 \sin(kz - \omega t)\hat{\mathbf{x}}$$

Show that, if $\omega = kv_A$, these equations satisfy the ideal MHD equations to first order in ϵ. Calculate the displacement current and show that it is small compared to the $4\pi j$, if $v_A \ll c$. Calculate the resistive correction to $\mathbf{B}$ and show that it is small if $\omega T_{\text{decay}} \gg 1$, where $T_{\text{decay}} = (k^2 \eta c/4\pi)^{-1}$.

Find the equation for a magnetic line of force to order ϵ by integrating

$$\frac{dx}{dz} = \frac{B_x}{B_0}$$

and show that we get the same equation from the displacement of the fluid $\delta x = \int V_x dt$, assuming flux freezing of the lines in the plasma.

2. (Clebsch coordinates) It is well known that, because $\nabla \cdot \mathbf{B} = 0$, we can always find functions $\alpha(\mathbf{r})$, $\beta(\mathbf{r})$ such that $\mathbf{B}$ can be written in the form

$$\mathbf{B} = \nabla\alpha \times \nabla\beta$$

Show that such a form is always divergence free. Show that α and β are constant along any line, so that any line can be written as $\alpha =$ constant, $\beta =$ constant. Show that the flux Φ through any area in the x–y plane is

$$\Phi = \int \nabla\alpha \times \nabla\beta \cdot \hat{\mathbf{z}}\, dS = \int d\alpha d\beta$$

where we change coordinates from x, y to $\alpha(x, y, 0)$, $\beta(x, y, 0)$. Note that the first integral involves the Jacobian of the transformation.

Assume that we know α and β at time $t = t_0$ and that we advance them in time according to

$$\frac{d\alpha}{dt} = \frac{\partial\alpha}{\partial t} + \mathbf{V} \cdot \nabla\alpha = 0$$

$$\frac{d\beta}{dt} = \frac{\partial\beta}{\partial t} + \mathbf{V} \cdot \nabla\beta = 0$$

Show that

$$\mathbf{B}(\mathbf{r}, t) = \nabla\alpha(\mathbf{r}, t) \times \beta(\mathbf{r}, t)$$

satisfies equation 17. Use the fact that $d\alpha d\beta$ is related to the flux and the fact that $\alpha =$ constant, $\beta =$ constant is related to the lines to show that this mathematically expresses flux freezing and line preservation.

3. (Helicity) Consider a closed, simply connected volume and let **B** be tangent to its surface. Introduce a vector potential **A** for **B**:

$$\mathbf{B} = \nabla \times \mathbf{A}$$

Show that the quantity known as helicity

$$H = \int \mathbf{B} \cdot \mathbf{A}\, d^3x$$

is gauge invariant. That is, change **A** by $\nabla\chi$ for an arbitrary function χ and show that H does not change.

To find how H changes in time, take some vector potential at $t = 0$, and let it change in time according to

$$\frac{\partial \mathbf{A}}{\partial t} = -c\mathbf{E}$$

Show that such an **A** continues to represent **B**.

Next show that

$$\frac{dH}{dt} = -2c \int \mathbf{E} \cdot \mathbf{B}\, d^3x - c \int \mathbf{A} \cdot (\mathbf{n} \times \mathbf{E})\, dS$$

Now, if $\mathbf{E}+\mathbf{V} \times \mathbf{B}/c = 0$, then the first term vanishes. Show that the second term vanishes if $\mathbf{V} \cdot \mathbf{n} = 0$ on the surface. Thus, under these conditions helicity is a conserved quantity in a ideal plasma.

4. (Rotational damping of a protostar) Show that in cylindrical coordinates the toroidal Alfven wave

$$\mathbf{V} = \epsilon V_A g(r) f(z - V_A t)\hat{\boldsymbol{\theta}}$$

$$\mathbf{B} = \mathbf{B}_0 - \epsilon \mathbf{B}_0 g(r) f(z - V_A t)\hat{\boldsymbol{\theta}}$$

satisfies the ideal equations to order ϵ, where f and g are arbitrary functions.

Start the plasma at rest and introduce an infinitely conducting, thin disk of radius a at the origin in the x–y plane. Suddenly start the disk rotating with the small constant angular velocity ω. Use the above solution and the condition that the tangential velocity at the disk is continuous to show that at any later time t all the plasma with $r < a$, $|z| < V_A t$ is rotating with this velocity ω and all the rest of the plasma is still at rest. If the disk has a surface density of σ, how long will it be before it "starts" to slow down? (*Hint*: Compare the rotational angular momentum of the plasma with that of the disk.)

5. (Flux freezing in a general velocity) Consider a constant magnetic field **B** in the z direction. Introduce a neutral fluid that does not interact with the plasma. While the plasma and field remain undisturbed let the neutrals move with an arbitrary velocity $\mathbf{U} = \hat{\mathbf{r}} U_r$ in the direction perpendicular to **B**, where U_r depends only on the radius r. Follow the flux contained in the neutral medium. Show that for this radial flow equation 25, with this velocity **U**, is satisfied so that the lines are preserved in the neutral fluid. Show that if the radial flow is compressible then equation 24 is not satisfied. Does this agree with a simple intuitive picture of flux freezing in the neutrals? This shows that flux freezing is definitely a stronger constraint than line preservation. If **U** is not strictly radial, or U_r depends on z, need the lines be preserved in the neutral fluid?

6. (The solar cycle) Consider the poloidal flux inside the sun. Assume that the solar plasma above the surface is infinitely conducting and that there is no plasma flow through the surface. By integrating the electric field around the equator determine the rate of change of the flux threading the equator inside the sun. Make no particular assumption as to the magnetic field behavior inside the sun, i.e., calculate the rate independent of the details of the flow, or the resistivity, or, in general, whatever processes determine the magnetic field evolution. Show that the poloidal flux cannot change if B_r is zero on the equator, but that if it is nonzero the flux may change. Can you interpret this result geometrically? The above results imply that that the solar activity must be deeply involved in the reversal of the magnetic field in the solar cycle.

7. (A vacuum about a neutron star) Assume that the region outside a rotating neutron star of radius R is a vacuum. Inside the star take **B** to be constant with magnitude B_0 and aligned along the rotational axis. Assume that $\mathbf{E} + (\mathbf{V} \times \mathbf{B})/c = 0$ inside the star. Show that at the surface $E_\theta = -\omega r B_0 \sin\theta \cos\theta / c$. Using the continuity of E_θ and taking $\mathbf{E} = -\nabla\phi$ in the

vacuum show that

$$\phi = \frac{-B_0 \omega R^5}{3cr^3} \left(\cos^2\theta - \frac{1}{3} \right)$$

Show that just outside the star the resulting **E** field has a component along **B** at the surface. Take **B** to be a dipole field with $B_r = B_0 \cos\theta$, $B_\theta = \frac{1}{2} B_0 \sin\theta$ on the surface of the star.

References

Cowling, T. G. 1976. *Magnetohydrodynamics*, Monographs in Astronomical Subjects 2, Adam Hilger, Bristol, UK.
Duncan, R. A. 1965. *Planetary and Space Sciences*, **13**, 997.
Dungey, J. W. 1961. *Physical Review Letters*, **6**, 47.
Elmegreen, B. G. 1985. In *Protostars and Planets II*, p. 33, edited by D. C. Black, University of Arizona Press, Tucson.
Goldreich, P. G., and W. H. Julian. 1969. *Astrophysical Journal*, **157**, 869.
Goldreich, P. G., and D. Lynden-Bell. 1969. *Astrophysical Journal*, **156**, 50.
Hughes, W. 1995. In *Introduction to Space Science*, p. 227, edited by M. G. Kivelson and C. T. Russell, Cambridge University Press, Cambridge, UK.
Newcomb, W. 1958. *Annals of Physics*, **3**, 347.
Parker, E. N. 1963. *Interplanetary and Space Sciences*, p. 137, Interscience, Wiley, New York.

Chapter 4

CONSERVATION RELATIONS

4.1 Introduction

Many MHD processes in plasmas are most easily understood as exchanges of the various forms of energy. These forms are kinetic, thermal, magnetic, and gravitational. For example, waves involve the oscillatory exchange of kinetic energy and the more potential forms of energy such as thermal and magnetic energies. Equilibria involve a balance between gravitational, thermal, and magnetic energy. For a precise understanding of the crucial role energy plays in MHD we must write down its conservation relation.

In addition, there are other quantities that are conserved, such as momentum and angular momentum. Further, there is the mysterious quasi-conservation relation, the virial theorem. While the virial theorem is not in the direct mainstream of physics, it is very useful in interpreting general results in MHD. Further, it puts constraints on equilibria. Last, but not least, the very usefulness of the ideal equations springs from the fact that they arise from a Lagrangian action principle. While these results are all somewhat formal, they will turn out to very helpful in the solution of problems, as well as being useful in arriving at a rough idea of the magnitude of phenomena.

4.2 The Lorentz Force

Before directly presenting these conservation relations we write the equation of motion (equation 15 in chapter 3) in a different form. First, let us examine the magnetic force $\mathbf{j} \times \mathbf{B}$. By employing Ampere's law, equation 19 in chapter 3, regarded as the definition of $\mathbf{j}$ in terms of $\mathbf{B}$, we can write

$$\mathbf{j} \times \mathbf{B} = \frac{(\nabla \times \mathbf{B}) \times \mathbf{B}}{4\pi} = \frac{\mathbf{B} \cdot \nabla \mathbf{B}}{4\pi} - \nabla \frac{B^2}{8\pi} \tag{1}$$

The first term is considered to be the magnetic tension force, and the second the magnetic pressure force. This is not a strictly reasonable interpretation since the tension force, analogous to the tension in a string, is usually perpendicular to the line of force, while $\mathbf{B} \cdot \nabla \mathbf{B}$ is not. Further, the magnetic pressure term $-\nabla B^2/8\pi$ can have a component along $\mathbf{B}$, while from the original form $\mathbf{j} \times \mathbf{B}$, the Lorentz force is always perpendicular to $\mathbf{B}$. This last fact, although obvious, is very important, and we should always be aware of this simple result. The resolution of the above remarks is that the part of

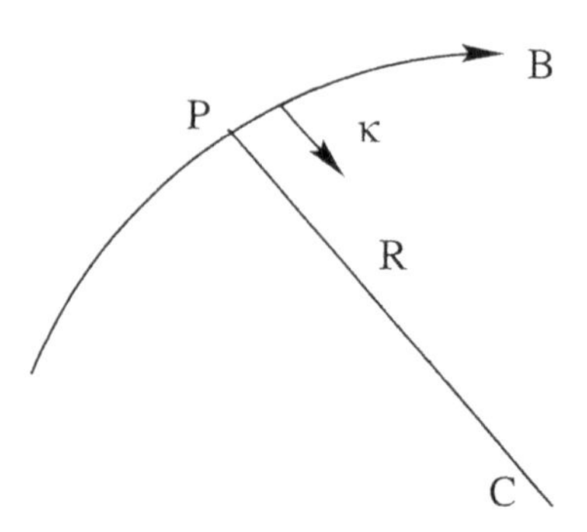

Figure 4.1. The line curvature and magnetic tension

the "tension" term $\mathbf{B} \cdot \nabla \mathbf{B}/4\pi$ parallel to $\mathbf{B}$ just cancels the corresponding part of the "magnetic pressure" term $-\nabla B^2/8\pi$.

Indeed, write the magnetic vector $\mathbf{B}$ as

$$\mathbf{B} = |\mathbf{B}|\mathbf{b} = B\mathbf{b} \tag{2}$$

where $\mathbf{b}$ is a unit vector parallel to $\mathbf{B}$. Then by the Leibnitz rule applied to vectors

$$\mathbf{B} \cdot \nabla \mathbf{B} = |B|\mathbf{b} \cdot \nabla(|B|\mathbf{b}) = |B|^2\mathbf{b} \cdot \nabla \mathbf{b} + \mathbf{b}\mathbf{b} \cdot \nabla |B|^2/2 \tag{3}$$

If we wished to, we could discard the last term because $\mathbf{j} \times \mathbf{B}$ clearly has no $\mathbf{b}$ component, but for the moment we will keep it. In the first term, $\mathbf{b} \cdot \nabla \mathbf{b}$ is the curvature vector $\boldsymbol{\kappa}$ for a line of force, since $\mathbf{b} \cdot \nabla$ is a derivative following the line and $\mathbf{b} \cdot \nabla \mathbf{b}$ is the rate at which the tangent vector, $\mathbf{b}$, turns.

If $\mathbf{R}$ is a vector from the center of curvature C of the line at a point P on the line to P (as in figure 4.1), then

$$\mathbf{b} \cdot \nabla \mathbf{b} = \boldsymbol{\kappa} = -\frac{\mathbf{R}}{R^2} \tag{4}$$

Next, combining equations 1, 3, and 4, we get

$$\begin{aligned} \mathbf{j} \times \mathbf{B} &= \boldsymbol{\kappa}\frac{B^2}{4\pi} + \mathbf{b}\mathbf{b} \cdot \nabla \frac{B^2}{8\pi} - \nabla \left(\frac{B^2}{8\pi}\right) \\ &= \boldsymbol{\kappa}\frac{B^2}{4\pi} - \nabla_\perp \frac{B^2}{8\pi} \end{aligned} \tag{5}$$

where $\nabla_\perp$ means keep only the derivatives of $B^2/8\pi$ perpendicular to $\mathbf{B}$ (formally, $\nabla_\perp = \nabla - \mathbf{b}\,\mathbf{b} \cdot \nabla$). The form for the Lorentz force in equation 5 correctly breaks it into two forces perpendicular to $\mathbf{B}$, a tension force and the perpendicular gradient of the magnetic pressure.

We might ask what happened to the mirror force in all this. We know that particles are reflected from a region where B^2 increases. We know that in a magnetic mirror region all the particles are subject to a force in the same direction away from the mirror, that opposite to the gradient of $\mathbf{B}^2$ and parallel to $\mathbf{B}$.

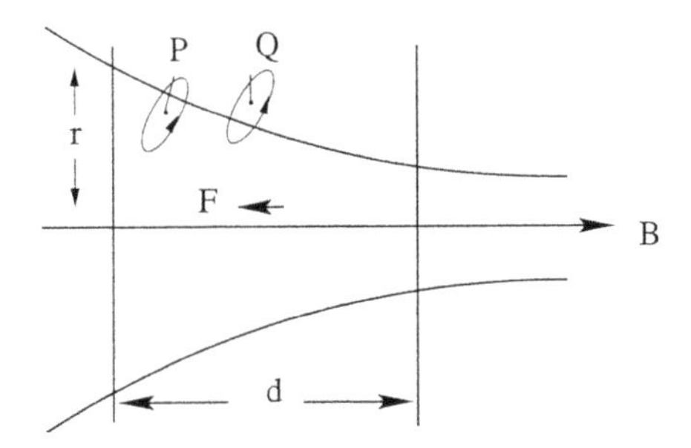

Figure 4.2. The Lorentz force in a mirror field

But MHD refers to the Lorentz force $\mathbf{j} \times \mathbf{B}$ on all the particles in a given small volume, and some particles have their gyroorbit only partially inside this volume. Consider such a small volume aligned along the mean field and bounded by magnetic field lines on the side and by two planes perpendicular to the mean B (figure 4.2).

The total magnetic moment force along $\mathbf{B}$, summed over all the particles with guiding centers in the volume is

$$\mathbf{F} = -\Sigma \left(\frac{M v_\perp^2}{2B} \frac{\delta B}{d} \right) \pi r^2 d \tag{6}$$

This must be corrected for the fringe ions, for which only that part of the Lorentz force which acts on the piece of the gyroorbit inside the volume should be included.

In figure 4.2, P and Q are fringe ions. P has its guiding center in the volume, but part of the Lorentz force evB/c is outside and directed toward the volume. The projection of this force along the mean field must be subtracted. On the other hand, Q has its guiding center outside so it has not been counted in the sum of equation 6. However, it spends part of its orbit in the volume and the Lorentz force on this part is directed outward away from the volume. This force must be added so that the sum of the corrections is a force outward. Taking into account the angle α for the tilt of the orbit we see that the corrections are both opposite to $\mathbf{F}$. Adding up the corrections over all the fringe particles yields an additional force that is equal and opposite to $\mathbf{F}$.

In fact, the angle α of the orbit to the main field is

$$\alpha = \frac{\delta r}{d} = \frac{r \delta r}{dr} = \frac{\delta A r}{2Ad} = \frac{\delta B r}{-2Bd} \tag{7}$$

the total number of fringe particle is $n_{\text{fringe}} = 2 \times 2\pi \rho r d$, and the mean change in the force per fringe particle is $evB/2c$. We easily see that correction $n_{\text{fringe}} \times evB/2c \times \sin\alpha$ from the fringe particles is equal and opposite to F. This above example should assure the reader that the magnetic force $\mathbf{j} \times \mathbf{B}$ never has a component along $\mathbf{B}$.

4.3 Conservation of Linear Momentum

Now we return to the conservation relations. Write the equation of motion as

$$\rho\left(\frac{\partial \mathbf{V}}{\partial t}+\mathbf{V}\cdot\nabla\mathbf{V}\right)=-\nabla\cdot\left(\mathbf{T}_p+\mathbf{T}_B\right)+\rho\mathbf{g} \tag{8}$$

where, introducing the unit dyadic tensor **I** with components δ_{jk}, we have for the two tensors

$$\mathbf{T}_p=\mathbf{I}p=\delta_{ij}p \tag{9}$$

and

$$\mathbf{T}_B=\mathbf{I}\frac{B^2}{8\pi}-\frac{\mathbf{BB}}{4\pi} \tag{10}$$

Note that $\nabla\cdot\mathbf{T}_p=\nabla p$, since $\partial T_{jk}/\partial x_j=\partial(\delta_{jk}p)/\partial x_j=\partial p/\partial x_k=\nabla p$. Similarly, $\nabla\cdot\mathbf{T}_B=(\nabla B^2/8\pi)\mathbf{I}-\nabla\cdot(\mathbf{BB})/4\pi=\nabla(B^2/8\pi)-\mathbf{B}\cdot\nabla\mathbf{B}/4\pi$, since $\nabla\cdot\mathbf{B}=0$. Reference to equation 15 in chapter 3 establishes that equation 8 is equivalent to the equation of motion.

We can make a further reduction of the equation of motions. Multiply the equation of continuity (equation 14 in chapter 3) by **V** and add the result to equation 8. Since

$$\rho\frac{\partial\mathbf{V}}{\partial t}+\mathbf{V}\frac{\partial\rho}{\partial t}=\frac{\partial}{\partial t}(\rho\mathbf{V}) \tag{11}$$

and

$$\mathbf{V}\nabla\cdot(\rho\mathbf{V})+\rho\mathbf{V}\cdot\nabla\mathbf{V}=\nabla\cdot(\rho\mathbf{VV}) \tag{12}$$

we find that the equation of motion can be written

$$\frac{\partial}{\partial t}(\rho\mathbf{V})=-\nabla\cdot\mathbf{T}+\rho\mathbf{g}=-\nabla\cdot(\mathbf{T}_V+\mathbf{T}_p+\mathbf{T}_B)+\rho\mathbf{g} \tag{13}$$

where

$$\mathbf{T}_V=\rho\mathbf{VV} \tag{14}$$

Now consider in figure 4.3, a general volume V with surface S, integrate equation 13 over it, and make use of Gauss's theorem on the first component of **T**. We see that the momentum **P** inside the volume should be

$$\mathbf{P}=\int_V\rho\mathbf{V}d^3x \tag{15}$$

and

$$\begin{aligned}\frac{d\mathbf{P}}{dt}&=\int_V\frac{\partial}{\partial t}(\rho\mathbf{V})d^3x=-\int_V(\nabla\cdot\mathbf{T})d^3x+\int_V\rho\mathbf{g}d^3x\\&=-\int_S\mathbf{T}\cdot d\mathbf{S}+\int_V\rho\mathbf{g}d^3x\end{aligned} \tag{16}$$

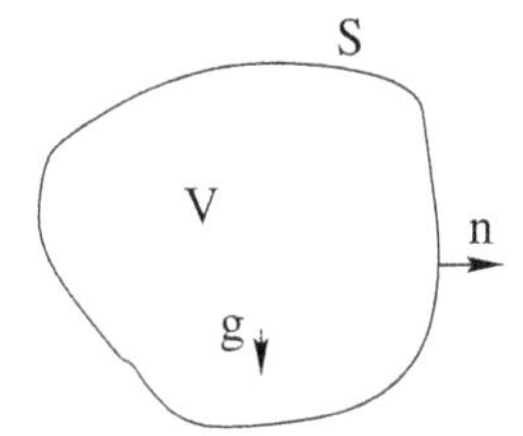

Figure 4.3. Rate of change of momentum in a general volume

P clearly represents the fluid momentum. In electromagnetism there is also a momentum associated with the electromagnetic field $\mathbf{E} \times \mathbf{B}/4\pi c$, but, since $E \approx VB/c$, this momentum is of order $\rho\mathbf{V}(B^2/4\pi\rho c^2)$ and is small compared to the fluid momentum $\rho\mathbf{V}$, if again $B^2/4\pi \ll \rho c^2$. This condition, which is usually well satisfied, is assumed in the derivation of the MHD equations.

The conservation of **P** shows that for an ideal MHD plasma there is no other momentum. **P** changes in time due to forces on the surface, and a gravitational volume force $\rho\mathbf{g}$. These forces on the surface are the sum of Reynold's stresses, pressure stresses, and magnetic stresses. Explicitly, the Reynold's stress

$$-\mathbf{T}_V \cdot dS = -\rho\mathbf{V}\mathbf{V} \cdot d\mathbf{S} \tag{17}$$

represents the loss of momentum due to a fluid crossing an elementary surface element $d\mathbf{S}$ at a volume rate $\mathbf{V} \cdot d\mathbf{S}$ per unit time, carrying $\rho\mathbf{V}$ per unit volume, and, thereby, if $\mathbf{V} \cdot \mathbf{dS}$ is positive, reducing the momentum inside.

The pressure stress on $d\mathbf{S}$

$$-\mathbf{T}_p \cdot dS = -p d\mathbf{S} \tag{18}$$

represents the pressure force on a surface and the force is in the $-d\mathbf{S}$ direction. (Microscopically, it arises by single particles crossing the surface, those moving inward increasing the momentum in the $-d\mathbf{S}$ direction, while those moving outward decreasing the momentum in the $+d\mathbf{S}$ direction, so that the contribution of the two streams add rather than cancel. In MHD, the pressure is considered a macroscopic force.)

The magnetic stress on $d\mathbf{S}$ is

$$-\mathbf{T}_B \cdot d\mathbf{S} = -\frac{B^2}{8\pi} d\mathbf{S} + \frac{\mathbf{B}\mathbf{B}}{4\pi} \cdot d\mathbf{S} \tag{19}$$

It is made up of a magnetic pressure force in the $-d\mathbf{S}$ direction and a magnetic tension force in the $+\mathbf{B}$ direction. The latter is proportional to the number of lines crossing $d\mathbf{S}$ in the outward direction, $\mathbf{B} \cdot d\mathbf{S}$, times a tension force $\mathbf{B}/4\pi$ per each line. Note that this breakup of the magnetic force is analogous to our original break up of the Lorentz force given in equation 5 into the perpendicular pressure and the tension.

In the breakup of the Lorentz force into stress per unit area, our restriction that there be no force along **B** no longer holds. The tension force across the

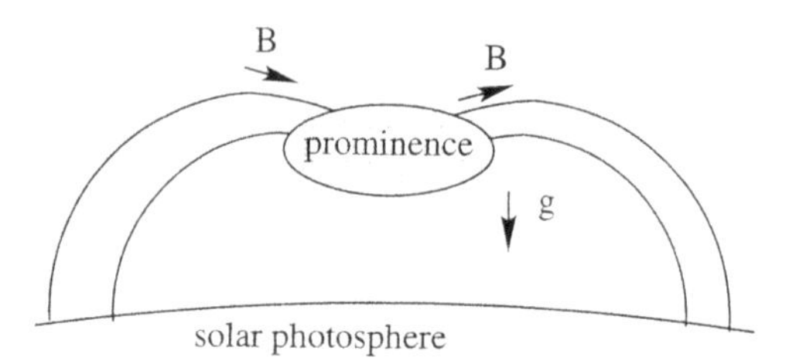

Figure 4.4. A prominence supported by magnetic tension

surface is actually along **B**, just as if the field lines acted as strings under tension. Similarly, the pressure force acts as a force perpendicular to the surface, and even the sum of the two forces on a given $d\mathbf{S}$ can have a component along **B**. This is because, for the sum of the Lorentz force over a total volume, there is no particular direction of **B**, which it need be perpendicular to. The theorem that the volume force be perpendicular to **B** at each point has no bearing on the force on the total volume. Gathering the results of equations 16–19 together, we have the explicit version for the evolution of the momentum **P** in a volume without the **g** term:

$$\frac{d}{dt}\int_V \rho \mathbf{V} d^3x = -\int_V \rho \mathbf{V}\mathbf{V} \cdot d\mathbf{S} + \int_V \left(p + \frac{B^2}{8\pi}\right)(-d\mathbf{S}) + \int_S \mathbf{B} \cdot d\mathbf{S}\frac{\mathbf{B}}{4\pi} \quad (20)$$

As an example of the tension force, consider the prominence in figure 4.4 above the solar surface. The downward gravitational force is balanced by the magnetic stresses on the boundary of the prominence. On the right, the number of lines per unit area is positive and the force per line is toward the upper right side. For the surface on the left, the number of lines per unit area is negative $\mathbf{B} \cdot d\mathbf{S} < 0$ and the force per line is toward the lower right, so the mean force per unit area is toward the upper left. The sum of the tension forces on the right and left surfaces each has an upward component, while the horizontal components cancel by symmetry. Thus, the prominence can be supported by the tension forces.

In reality, there is also some support by magnetic pressure. We expect the field lines to be crushed together below the prominence, and the magnetic pressure increased there, and similarly to be decreased above the prominence. Thus, on the bottom this pressure has an upward component $-(B^2/8\pi)d\mathbf{S}$, and a downward component on the top. Since the latter is weaker, there is also some net support from the magnetic pressure.

The tension part of the magnetic stress is independent of the sign of B, so we could equally well describe it by taking the number of lines crossing the surface per unit area as always positive and taking the tension force of each line as always outward. This is much closer to the analogy of the string, since the number of strings crossing an area is always positive and the tension in the string is always a pull outward. (The pull is outward since

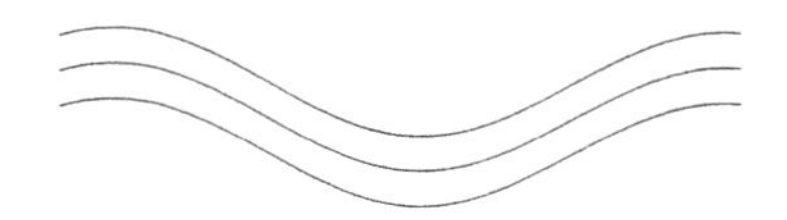

Figure 4.5. Magnetic field lines in an alfven wave

we are calculating the force on the inside volume. If we were calculating the force on the outside volume, the pull would then be inward.)

Another important illustration of the magnetic stress concept is the Alfven wave. A parallel propagating Alfven wave is a transverse sinusoidal perturbation in an initially uniform magnetic field. Replacing the field lines by strings with tension $T = B/4\pi$, as in figure 4.5, and taking the number of strings per unit area as B, we can assign the mass of the plasma to the strings, with each string having a mass $\sigma = \rho/B$ per unit length. The strings will vibrate together, carrying a wave with phase velocity $\sqrt{T/\sigma} = \sqrt{(B/4\pi)/(\rho/B)} = \sqrt{B^2/4\pi\rho}$, which is the velocity of an Alfven wave in MHD. There is no compression of the field lines because the wave is transverse.

We have left the gravitational force as a body force. This is appropriate if $\mathbf{g}$ is the gravitational field of a neighboring massive body. For example, if we are studying the plasma phenomena in the outer layers of a star or in the coronal region of the sun, $\mathbf{g}$ is given by internal massive part of the sun and is constant in time.

On the other hand, if we are studying the motion of a self-gravitating system, such as the internal motion of a star or the large-scale behavior of a galaxy, $\mathbf{g}$ will change with time according to the gravitational Poisson equation. If $\mathbf{g} = -\nabla\phi$, then

$$\nabla^2\phi = 4\pi G\rho \tag{21}$$

In this case, we can also write the gravitational force as the divergence of the stress tensor

$$\mathbf{T}_g = \frac{(\nabla\phi)^2\mathbf{I}}{8\pi G} - \frac{\nabla\phi\nabla\phi}{4\pi G} \tag{22}$$

A simple calculation shows that

$$\rho\mathbf{g} = -\nabla\cdot\mathbf{T}_g \tag{23}$$

and the equation of motion can be written with all forces as divergences of stresses

$$\frac{\partial(\rho\mathbf{V})}{\partial t} - \nabla\cdot\mathbf{T}_{\text{total}} = -\nabla\cdot(\mathbf{T}_K + \mathbf{T}_p + \mathbf{T}_B + \mathbf{T}_g) \tag{24}$$

On the other hand, to determine the rate of change of momentum in a finite volume in terms of surface terms is difficult because it hard to find $\mathbf{T}_g$ and it is usually better to treat $\rho\mathbf{g}$ as a body force. Only when we consider an isolated system and includes all space, in which the gravitational stress goes to zero at infinity, is it convenient to introduce the gravitational stress tensor. We will treat $\mathbf{g}$ as the divergence of a tensor when we discuss the virial theorem.

4.4 Conservation of Angular Momentum

The angular momentum $\mathbf{H}$ for an MHD system and its conservation law is a simple extension of the discussion of linear momentum $\mathbf{P}$. For a general volume V the angular momentum is again all in the plasma motion $\rho\mathbf{V}$, so

$$\mathbf{H} = \int_V \mathbf{r} \times \rho \mathbf{V} d^3x \tag{25}$$

where $\mathbf{r}$ is the radius vector from some point O about which the angular momentum is defined. Differentiating $\mathbf{H}$ with respect to time, we differentiate those quantities in the integrand that depend on t, i.e., ρ and $\mathbf{V}$. Thus, making use of the equation of motion in the form of equation 13 we have

$$\frac{d\mathbf{H}}{dt} = \int_V \mathbf{r} \times \frac{\partial(\rho\mathbf{V})}{\partial t} d^3x = \int_V \left(-\mathbf{r} \times \nabla \cdot \mathbf{T} + \mathbf{r} \times \rho \mathbf{g}\right) d^3x \tag{26}$$

The $\mathbf{T}$ term in the integrand can be transformed. A simple vector calculation shows that, because $\mathbf{T}$ is a symmetric tensor,

$$-\mathbf{r} \times (\nabla \cdot \mathbf{T}) = -\nabla \cdot (\mathbf{r} \times \mathbf{T}) \tag{27}$$

Thus,

$$\begin{aligned}
\frac{d\mathbf{H}}{dt} &= -\int_S \mathbf{r} \times \mathbf{T} \cdot d\mathbf{S} + \int \mathbf{r} \times \rho \mathbf{g} d^3x \\
&= -\int_S \mathbf{V} \cdot d\mathbf{S}(\mathbf{r} \times \rho\mathbf{V}) + \int_S \mathbf{r} \times (-d\mathbf{S}) \left(p + \frac{B^2}{8\pi}\right) \\
&\quad + \int_S \mathbf{r} \times \mathbf{B} \frac{\mathbf{B}}{4\pi} \cdot d\mathbf{S} + \int \mathbf{r} \times \rho \mathbf{g} d^3x.
\end{aligned} \tag{28}$$

The first term is the loss of angular momentum carried by the plasma flowing through $d\mathbf{S}$. The second term is the torque exerted on the surface $-d\mathbf{S}$ by the outside plasma pressure plus magnetic pressure. The third term is the torque exerted on $d\mathbf{S}$ by the $\mathbf{B} \cdot d\mathbf{S}$ lines each exerting the torque $\mathbf{r} \times \mathbf{B}/4\pi$. As has already been noted, the magnetic force along $\mathbf{B}$ can be considered to be outward, if the absolute value of $\mathbf{B} \cdot d\mathbf{S}$ is taken, so the number of lines is always positive. The last term is the volume torque due to the gravitational force.

It is interesting to compare this equation with equation 20. The surface terms in equation 20 are the individual forces on a given surface element $d\mathbf{S}$, while those in equation 28 are the corresponding torques exerted by these forces on the surface element about the center O. Indeed, equation 20 actually follows formally from equation 28 by changing the location of O by an arbitrary dispacement $\mathbf{a}$. The extra terms proportional to $\mathbf{a}$ are just the terms in equation 20. It does not seem possible to derive equation 28 from equation 20 in this way.

The conservation of angular momentum is usually applied to problems where there is a center of symmetry, O. Examples of this are the solar wind, where the center is the sun, or an accretion disk around a central body. In these cases the gravitational torque is zero since $\mathbf{r} \times \mathbf{g} = 0$, and if the surface is taken as a sphere then the pressure torques vanish also, since $\mathbf{r} \times d\mathbf{S} = 0$. If the surface is a cylinder, then the the axial components of the torques vanish. Therefore, for such cases we have

$$\frac{d\mathbf{H}}{dt} = -\int_S \mathbf{V} \cdot d\mathbf{S}(\mathbf{r} \times \rho \mathbf{V}) + \int_S \mathbf{r} \times \mathbf{B} \frac{\mathbf{B}}{4\pi} \cdot d\mathbf{S} \tag{29}$$

The first term still represents the flow of angular momentum carried by the plasma crossing the surface.

For an axisymmetric accretion disk, the flow of angular momentum due to the first term is $\dot{M}J$, where $\dot{M}$ is the inward mass flow into the volume and J is the angular momentum per unit mass. $\dot{M}$ is positive for an accreting object and negative for the solar wind. $J = rV_\theta$.

To interpret the behavior of angular momentum in an accreting disk let us introduce a second smaller cylinder at r_1 and take the outer cylinder at r_2. Consider the angular momentum in the volume in between the spheres. For Keplerian motions about the central object $J_2 > J_1$, while for a steady state $\dot{M}$ is the same at the two spheres. Therefore, the contribution of plasma flow to the angular momentum increase is positive. Since H is constant, a magnetic torque on the two surfaces must balance this. We usually assume that we can neglect the magnetic torque on the inner sphere.

We can say that steady-state accretion cannot occur unless there is magnetic torque that removes angular momentum. For the example of the solar wind, it is the case that near the sun the magnetic field is strong enough to force the plasma to corotate with the sun. This holds out to the so-called Alfven radius r_A, where the Alfven speed is comparable to the corotation speed at this radius, $\Omega_S r_A$. Beyond this radius the field is significantly bent, eventually turning into the Archimedian spiral. The angular momentum per unit mass is $r_A^2 \Omega_S$ at the Alfven radius r_A, approximately $10 r_S$, so the mean angular momentum per unit mass of the solar wind material at r_A is more than 100 times that of the sun $\approx \frac{1}{5} r_S^2 \Omega_S$. Thus, if the sun loses 1 percent of its mass, its angular velocity will decrease by an amount comparable to its initial angular velocity. This is approximately the numerical situation. In fact, the measured magnetic torque at the earth integrated over the lifetime of the sun is of order of the angular momentum of the sun (see problem 1 and Davis and Weber 1967).

It is conceivable that the sun rotated much faster when it formed, and that the solar wind was much more intense. Even if this is the case, it is plausible that the solar rotation could be brought down to its present value. On the other hand, if the torque has always been of the present order of magnitude, then the sun could not have been rotating much faster.

4.5 Conservation of Energy

We now consider the third and most important conservation law, that of energy. While the momentum and angular momentum is in entirely carried by the kinetic motion of the plasma $\rho\mathbf{V}$, this is not the case for energy. Energy occurs in four forms: kinetic, thermal, magnetic, and gravitational.

We will show that for conservation of energy to hold, the energy in any volume V must be

$$\mathcal{E} = \mathcal{E}_V + \mathcal{E}_p + \mathcal{E}_B + \mathcal{E}_g = \int_V \left(\frac{\rho V^2}{2} + \frac{p}{\gamma - 1} + \frac{B^2}{8\pi} + \rho\phi \right) d^3x \qquad (30)$$

where ϕ is the gravitational potential, $\mathbf{g} = -\nabla\phi$. This is the correct form of the energy in the case when $\mathbf{g}$ is a fixed time-independent gravitational field, unaffected by the plasma motion, as is the case for plasma motions in the solar atmosphere, where the sun's gravitational field $\mathbf{g}$ is essentially unaffected by these motions. For larger systems, where gravity is determined self-consistently through the gravitational Poisson equation and *is* affected by the motions, $\rho\phi$ should be replaced by $\rho\phi/2$. We first consider the former case, where $\mathbf{g}$ is fixed.

The four forms of energy, kinetic energy $\mathcal{E}_V$, thermal energy $\mathcal{E}_p$, magnetic energy $\mathcal{E}_B$, and gravitational energy $\mathcal{E}_g$, are familiar.

The form of the thermal energy $\mathcal{E}_p$ can be verified from the $-pdV$ work done on an isolated adiabatic gram of plasma. Let us compress it at constant entropy from zero density to a density ρ_0 and a pressure P_0. The work done on the plasma is

$$dW = -\int pdV = -\int_0^{\rho_0} pd\left(\frac{1}{\rho}\right) = \int_0^{\rho_0} \left(\frac{\rho}{\rho_0}\right)^\gamma p_0 \frac{d\rho}{\rho^2} = \frac{1}{\gamma - 1}\frac{p_0}{\rho_0} \qquad (31)$$

since $p/\rho^\gamma = p_0/\rho_0^\gamma$ is constant. The energy per unit volume comes from multiplying this by ρ_0. Note that for a plasma with no molecules, $\gamma = \frac{5}{3}$, $p = nT$ and the energy per unit volume is $3/2nT$, where, as has been mentioned earlier, Boltzmann's constant is absorbed into the definition of T.

Let us separately investigate the time derivatives of each type of energy in a given volume V. For the kinetic energy

$$\frac{d\mathcal{E}_V}{dt} = \int_V \frac{\partial}{\partial t}\left(\frac{\rho V^2}{2}\right) d^3x = \int_V V^2 \frac{\partial \rho}{\partial t} d^3x + \int_V \rho \frac{\partial \mathbf{V}}{\partial t} \cdot \mathbf{V} d^3x \qquad (32)$$

From equations 14 and 15 in chapter 3,

$$\frac{d\mathcal{E}_V}{dt} = \int_V \left[-\nabla\cdot(\rho\mathbf{V})V^2 - \mathbf{V}\cdot\rho(\mathbf{V}\cdot\nabla\mathbf{V}) - \mathbf{V}\cdot\nabla p + \mathbf{V}\cdot\mathbf{j}\times\mathbf{B} - \rho\mathbf{V}\cdot\nabla\phi\right] d^3x \tag{33}$$

Now consider the change in the thermal energy. For this we rewrite the ideal equation for p, equation 16 in Chapter 3. Divide it by p/ρ^γ to get logarithmic derivatives:

$$\frac{1}{p}\frac{dp}{dt} + \frac{\gamma}{\rho}\frac{d\rho}{dt} = 0 \tag{34}$$

Treating the continuity equation similarly, we have

$$\frac{1}{\rho}\frac{d\rho}{dt} + \nabla\cdot\mathbf{V} = 0 \tag{35}$$

(since $d/dt = \partial/\partial t + \mathbf{V}\cdot\nabla$). Combining these two equations to eliminate $d\rho/dt$, we obtain

$$\frac{\partial p}{\partial t} = -\mathbf{V}\cdot\nabla p - \gamma p\nabla\cdot\mathbf{V} \tag{36}$$

Now, differentiate $\mathcal{E}_p$, taken over the volume V, with respect to time and use this equation:

$$\frac{d\mathcal{E}_p}{dt} = \frac{1}{\gamma-1}\int_V \frac{\partial p}{\partial t} d^3x = \frac{1}{\gamma-1}\int_V (-\mathbf{V}\cdot\nabla p - \gamma p\nabla\cdot\mathbf{V})\, d^3x \tag{37}$$

This can be written

$$\frac{d\mathcal{E}_p}{dt} = -\int_V \left[\frac{\gamma}{\gamma-1}\nabla\cdot(p\mathbf{V}) - \mathbf{V}\cdot\nabla p\right] d^3x \tag{38}$$

The $\mathbf{V}\cdot\nabla p$ term cancels the corresponding term in equation 33 and represents the energy that must be taken from the thermal energy to increase the kinetic energy by the pressure gradient term, $-\nabla p$. The other, divergence, term converts to a surface term, by Gauss's theorem. We come back to it when we collect all the terms.

The $\mathbf{j}\times\mathbf{B}$ term in equation 33 should be compensated by a change in magnetic energy, since as it accelerates the plasma, increasing its kinetic energy, the magnetic energy should be lowered. Taking the time derivative of $\mathcal{E}_B$ and using equation 17 in chapter 3 we have

$$\frac{d\mathcal{E}_B}{dt} = \frac{1}{4\pi}\int_V \mathbf{B}\cdot\frac{\partial\mathbf{B}}{\partial t} d^3x = \int_V \frac{\mathbf{B}\cdot\nabla\times(\mathbf{V}\times\mathbf{B})}{4\pi} d^3x \tag{39}$$

By a vector identity this can be written

$$\frac{d\mathcal{E}_B}{dt} = -\int_V \left\{\nabla\cdot\left[\frac{\mathbf{B}\times(\mathbf{V}\times\mathbf{B})}{4\pi}\right] - \frac{\nabla\times\mathbf{B}}{4\pi}\cdot(\mathbf{V}\times\mathbf{B})\right\} d^3x \tag{40}$$

The second term is just $\mathbf{j} \cdot (\mathbf{V} \times \mathbf{B}) = -\mathbf{V} \cdot (\mathbf{j} \times \mathbf{B})$ by Ampere's law, equation 19 in chapter 3. This term is the opposite of the magnetic acceleration term in equation 33, and does indeed represent the reduction in magnetic energy associated with an increase in kinetic energy by magnetic acceleration. The divergence term will be treated later.

Finally, differentiating the gravitational energy with respect to time, we get

$$\frac{d\mathcal{E}_g}{dt} = \int_V \frac{\partial \rho}{\partial t} \phi d^3x = -\int_V \phi \nabla \cdot (\rho \mathbf{V}) = -\int_V [\nabla \cdot (\rho \mathbf{V} \phi) - \rho \mathbf{V} \cdot \nabla \phi] d^3x \quad (41)$$

and we obtain the term $\rho \mathbf{V} \cdot \nabla \phi$ corresponding to the gravitational acceleration term in equation 33.

Adding equations 33, 38, 40, and 41, we find that all but the divergence terms cancel, and with Gauss's theorem we get

$$\frac{d\mathcal{E}}{dt} = -\int_S \left[\rho \mathbf{V} \frac{V^2}{2} + \frac{\gamma}{\gamma - 1} p \mathbf{V} - \frac{(\mathbf{V} \times \mathbf{B}) \times \mathbf{B}}{4\pi} + \rho \mathbf{V} \phi \right] \cdot d\mathbf{S} \quad (42)$$

where

$$\frac{d\mathcal{E}}{dt} = \frac{d\mathcal{E}_V}{dt} + \frac{d\mathcal{E}_p}{dt} + \frac{d\mathcal{E}_B}{dt} + \frac{d\mathcal{E}_g}{dt} \quad (43)$$

Similarly to the momentum and angular momentum loss through the surface, $-(\rho V^2/2)\mathbf{V} \cdot d\mathbf{S}$ represents the kinetic energy lost by plasma crossing the surface element $d\mathbf{S}$.

The pressure term $-\gamma p \mathbf{V} \cdot d\mathbf{S}/(\gamma - 1)$ is actually the outward flow of enthalpy. If we write it as

$$-\frac{\gamma p \mathbf{V} \cdot d\mathbf{S}}{\gamma - 1} = -\frac{p}{\gamma - 1} \mathbf{V} \cdot d\mathbf{S} - p \mathbf{V} \cdot d\mathbf{S} \quad (44)$$

then we see that the first term is the outward flow of thermal energy through the surface, while the second term is the "pdV" work that the internal pressure does on the external surroundings.

The magnetic term $-(\mathbf{V} \times \mathbf{B}) \times \mathbf{B} \cdot d\mathbf{S}/4\pi = c(\mathbf{E} \times \mathbf{B}) \cdot d\mathbf{S}/4\pi$ is the energy carried away by the Poynting vector across the surface element and is the usual expression for the flow of electromagnetic energy. By expanding the triple product, it can also be written as

$$-\frac{B^2}{8\pi} \mathbf{V} \cdot d\mathbf{S} - \mathbf{V} \cdot \left(\frac{B^2}{8\pi} \mathbf{I} - \frac{\mathbf{BB}}{4\pi} \right) \cdot d\mathbf{S} = -\frac{B^2}{8\pi} \mathbf{V} \cdot d\mathbf{S} - \mathbf{V} \cdot \mathbf{T}_B \cdot d\mathbf{S} \quad (45)$$

where $\mathbf{T}_B$ is the Maxwell stress tensor. In this form it can be interpreted as the sum of the flow of magnetic energy outward plus the work the magnetic field does on its surroundings. If $\mathbf{B} \cdot d\mathbf{S} = 0$, the sum is $-(B^2/8\pi)\mathbf{V} \cdot d\mathbf{S} - (B^2/8\pi)\mathbf{V} \cdot d\mathbf{S}$, and is the exact analogy of enthalpy for the magnetic field. In fact, the magnetic field often is said to act as a $\gamma_B = 2$ gas, and $\gamma_B/(\gamma_B - 1) = 2$, so the surface term is equal to $\gamma_B/(\gamma_B - 1)(B^2/8\pi)$.

Finally, it is easily shown directly from Maxwell's equations (without displacement current) that

$$\frac{1}{8\pi}\int_V \frac{\partial B^2}{\partial t} d^3x = -c\int_V \mathbf{j}\cdot\mathbf{E}d^3x - c\int_V \frac{\mathbf{E}\times\mathbf{B}}{4\pi}\cdot d\mathbf{S} \tag{46}$$

so $\mathbf{j}\cdot\mathbf{E}$ represents the destruction of that part the magnetic field energy that heats and accelerates the plasma in the volume, while the Poynting vector represents the transfer of the remainder out of the volume. In fact, making use of Ohm's law, $-c\mathbf{j}\cdot\mathbf{E} = \mathbf{j}\cdot\mathbf{V}\times\mathbf{B} = -\mathbf{V}\cdot(\mathbf{j}\times\mathbf{B})$, we see that the destruction of the magnetic energy is directly related to the acceleration.

The $-\rho\phi\mathbf{V}$ term in equation 42 at a point P on the surface can be interpreted as the loss of gravitational energy associated with the gravitational potential energy per unit mass ϕ at P being carried outward with the plasma flow through the surface at P.

For completeness, let us consider gravitational energy in the self-gravitational case, with the factor half, $\mathcal{E}_g' = \int_V \frac{1}{2}\rho\phi$, and with ϕ given by

$$\nabla^2\phi = 4\pi G\rho \tag{47}$$

Differentiating $\mathcal{E}_g'$ with respect to time gives

$$\begin{aligned}
\frac{d\mathcal{E}_g'}{dt} &= \frac{1}{2}\int_V \left(\phi\frac{\partial\rho}{\partial t} + \rho\frac{\partial\phi}{\partial t}\right) d^3x \\
&= \frac{1}{2}\int_V \phi\frac{\partial\rho}{\partial t} d^3x + \int_V \frac{\nabla^2\phi}{8\pi G}\frac{\partial\phi}{\partial t} d^3x \\
&= \frac{1}{2}\int_V \phi\frac{\partial\rho}{\partial t} d^3x + \int_V \frac{\phi}{8\pi G}\nabla^2\frac{\partial\phi}{\partial t} d^3x \\
&= \frac{1}{2}\int_V \phi\frac{\partial\rho}{\partial t} d^3x + \frac{1}{2}\int_V \phi\frac{\partial\rho}{\partial t} d^3x = \int_V \phi\frac{\partial\rho}{\partial t} d^3x
\end{aligned} \tag{48}$$

where here we integrate the Laplacian term twice by parts and ignore the surface terms. Then, using the continuity equation, we get $\int_V \rho\mathbf{V}\cdot\nabla\phi d^3x$, which cancels the gravitational acceleration term in equation 33. For this case it is necessary to take the surface to infinity to get rid of the surface terms.

The factor of two difference in the two expressions for the gravitational energy may seem strange. The explanation for this may be made clear by an example.

Consider a prominence of mass M supported against solar gravity $\mathbf{g}$ a height h above the surface of the sun. Let the prominence be raised δh. During this shift $\mathbf{g}$ does not change because the mass of the prominence is too small to move the sun. If we take the volume V' to be outside of the sun, then $\delta\mathcal{E}_g = Mg\delta h$. On the other hand, if the volume includes the entire sun, then we must include $\delta\phi$, produced by the prominence shift, at each element of the sun, $\delta\phi GM\delta\mathbf{h}\cdot\mathbf{r}/r^2$, where $\mathbf{r}$ is the radius vector from the solar element to the prominence. Summing this over the entire sun we get a contribution to $\delta\mathcal{E}_g$ of $\Sigma GM\rho\delta\mathbf{h}\cdot\mathbf{r}/r^2$. But this expression by symmetry, is just $Mg\delta h$. Thus, taking the sum of these two terms, $2g\delta hM$ divided by 2, we again get $g\delta hM$, the same answer as above, when we took the volume V' and the first expression for $\delta\mathcal{E}_g$. In other words, $\rho\phi/2$ is the correct integrand for the gravitational energy taken over the bigger volume, but half of the energy change is outside the volume V', so $\int_{V'}\rho\phi$ does give the correct answer for the gravitational energy.

4.6 The Virial Theorem

A theorem closely related to the conservation of energy is the virial theorem, which is often of use in checking the validity of theoretical results (Chandrasekhar 1961). First take $\mathbf{g}=0$ and a finite volume. Then choose an origin O inside the volume and define a radius vector $\mathbf{r}$ relative to it. Consider the scalar moment of inertia,

$$I=\frac{1}{2}\int_V \rho r^2 d^3x \tag{49}$$

and differentiate it with respect to time:

$$\frac{dI}{dt}=\frac{1}{2}\int_V \frac{\partial\rho}{\partial t}r^2 d^3x=-\frac{1}{2}\int_V \nabla\cdot(\rho\mathbf{V})r^2 d^3x$$

$$=-\frac{1}{2}\int_V \nabla\cdot(\rho\mathbf{V}r^2)d^3x+\frac{1}{2}\int_V \rho\mathbf{V}\cdot\nabla r^2 d^3x$$

$$=-\frac{1}{2}\int_S r^2\rho\mathbf{V}\cdot d\mathbf{S}+\int_V \rho\mathbf{V}\cdot\mathbf{r}d^3x \tag{50}$$

Let us assume that either the volume is large enough that no plasma escapes through the surface, or that the surface is rigid, so that $\mathbf{V}\cdot\mathbf{n}=0$ on it, so that the surface term is zero. The second derivative of I gives

$$\frac{d^2I}{dt^2}=\int_V \mathbf{r}\cdot\frac{\partial}{\partial t}(\rho\mathbf{V})d^3x=-\int_V(\nabla\cdot\mathbf{T})\cdot\mathbf{r}d^3x$$

$$=-\int_V \nabla\cdot(\mathbf{T}\cdot\mathbf{r})d^3x+\int_V \mathbf{T}:\nabla\mathbf{r}d^3x$$

$$=-\int d\mathbf{S}\cdot\mathbf{T}\cdot\mathbf{r}+\int \mathrm{trace}(\mathbf{T})d^3x \tag{51}$$

In subscript notation $-(\nabla\cdot\mathbf{T})\cdot\mathbf{r}=-(\partial/\partial r_\alpha)(T_{\alpha\beta})r_\beta=-\partial/\partial r_\alpha(T_{\alpha\beta}r_\beta)+T_{\alpha\beta}(\partial r_\alpha/\partial r_\beta)$. The second term is $T_{\alpha\beta}\delta_{\alpha\beta}=\mathrm{trace}\ \mathbf{T}$. Hence, using the definition of T, equations 9, 10, and 14, and dropping **g**, we get

$$\frac{d^2 I}{dt^2} = \int_V \left(\rho v^2 + 3p + \frac{3B^2}{8\pi} - \frac{B^2}{4\pi}\right) d^3x + \int d\mathbf{S} \cdot \mathbf{T} \cdot \mathbf{r}$$

$$= 2\mathcal{E}_V + 3(\gamma - 1)\mathcal{E}_p + \mathcal{E}_B \quad + \int d\mathbf{S} \cdot \mathbf{T} \cdot \mathbf{r} \tag{52}$$

If the system is stationary, or I does not grow on the average, we can take the time average of equation 51 and see that a plasma cannot be confined inside a volume without a corresponding stress on the walls, $\mathbf{r} \cdot \mathbf{T} \cdot d\mathbf{S}$.

Suppose that the plasma is static, $\mathbf{V} = 0$ and $\mathbf{B} \cdot d\mathbf{S} = 0$. Then $\mathbf{T} \cdot d\mathbf{S} = (p + B^2/8\pi)d\mathbf{S}$. Let the maximum value of $p + B^2/8\pi$ be $(p + B^2/8\pi)_{\max}$. Then the surface integral in equation 51 is less than $(p+B^2/8\pi)_{\max} \int_S \mathbf{r} \cdot d\mathbf{S} = 3 \times \text{volume} \times (p + B^2/8\pi)_{\max}$ and the sum of the energies in equation 52 must be less than this. Further, if the mean values of p and $B^2/8\pi$ are $\bar{p}$ and $\bar{B}^2/8\pi$ and we take $\gamma = \frac{5}{3}$, $3(\gamma - 1) = 2$, then we have as a necessary condition for confinement

$$2\bar{p} + \frac{\bar{B}^2}{8\pi} < 3\left(p + \frac{B^2}{8\pi}\right)_{\max} \tag{53}$$

Put another way, we see that a finite system without walls must explode, d^2I/dt^2 must always be positive. The relation of the energy of the body to its surface forces, e.g., the pressure, was the first application of the virial theorem, which was introduced by Clausius for this purpose when he calculated the first correction to the ideal equation of state of a real gas (Milne 1925; Fowler and Guggenheim 1939).

The virial theorem becomes much more interesting when we include gravity. To include gravity we must add a term

$$\mathbf{r} \cdot \rho\mathbf{g} = -\mathbf{r} \cdot (\nabla \cdot \mathbf{T}_g) \tag{54}$$

to the integrand of equation 51. That is to say, we must include the gravitational stress tensor $\mathbf{T}_g$ of equation 22 in $\mathbf{T}$. Taking the system so large that we can neglect the surface terms we get an additional contribution to the trace of the total stress tensor:

$$\text{trace } \mathbf{T}_g = -\frac{1}{8\pi G}(\nabla\phi)^2 \tag{55}$$

and this is the self-gravitational energy of the system, as can be seen by a simple integration by parts. Thus, we can rewrite equation 52:

$$\frac{d^2 I}{dt^2} = 2\mathcal{E}_V + 3(\gamma - 1)\mathcal{E}_p + \mathcal{E}_B + \mathcal{E}_g \tag{56}$$

If the system remains confined, so that the time average

$$\frac{1}{T}\int_0^T \frac{d^2 I}{dt^2} = \frac{1}{T}\frac{dI}{dt}\bigg|_0^T \to 0 \tag{57}$$

as $T \to \infty$, we find that the sum of the average energies on the right-hand side of equation 56 must vanish. If $\gamma = \frac{5}{3}$, then we get

$$2\bar{\mathcal{E}}_V + 2\bar{\mathcal{E}}_p + \bar{\mathcal{E}}_B + \bar{\mathcal{E}}_g = 0 \tag{58}$$

Note that the integrals that involve motion, $\mathcal{E}_V$ and $\mathcal{E}_p$, enter with a factor 2, while the others, $\mathcal{E}_B$ and $\mathcal{E}_g$, have a factor unity. The total energy $\mathcal{E}_{\text{total}} = \mathcal{E}_V + \mathcal{E}_p + \mathcal{E}_B + \mathcal{E}_g$, which is constant, then satisfies

$$\mathcal{E}_{\text{total}} = -\bar{\mathcal{E}}_V - \bar{\mathcal{E}}_p = \tfrac{1}{2}(\bar{\mathcal{E}}_B + \bar{\mathcal{E}}_g) : \qquad \gamma = \tfrac{5}{3} \tag{59}$$

Now, the energy $\mathcal{E}$ is equal to minus the sum of the two positive kinetic and thermal energies and, therefore, is negative. This is not surprising when we remember that the gravitational energy $\mathcal{E}_g$ is negative (see equation 55). From this equation we get the familiar result that as a star gives off energy by radiation, $\mathcal{E}_{\text{total}}$ decreases, and, by equation 59, $\mathcal{E}_p + \mathcal{E}_B$ increases. That is, if $\mathcal{E}_B = 0$, then the star heats up as it cools! (This ignores the nuclear energy, of course.) When $\mathcal{E}_B \neq 0$, then the thermal plus the magnetic energy increases as the star cools.

For a static star, $\mathcal{E}_V = 0$, and for an arbitrary but constant γ the binding energy of the star is $(4-3\gamma)\mathcal{E}_p$. If the thermal motions are relativistic, $\gamma \approx \frac{4}{3}$, and the binding energy of a static star can be very small. The pressure energy is very nearly balanced by the sum of the gravitational and magnetic energies. Such a star can change its structure drastically by a very little loss of energy. This is the state of a star just prior to a supernova explosion.

4.7 The Action Principle for MHD

An ideal MHD fluid is really a very tight system. Suppose that we know the properties of the system at time t_0, its magnetic field $\mathbf{B}(\mathbf{r}_0, t)$, and its pressure $p(\mathbf{r}_0, t)$. If at any later time we know the displacement $\boldsymbol{\xi}(\mathbf{r}_0, t)$ of any fluid element from its initial position, $\mathbf{r}_0$ at time t_0, as in figure 4.6, then we know the mass density at the displaced point $\mathbf{r} = \mathbf{r}_0 + \boldsymbol{\xi}(\mathbf{r}_0, t)$ (simply from the Jacobian $|\partial\mathbf{r}/\partial\mathbf{r}_0|$). Then from the entropy constant of the fluid element, or alternatively from the constancy of p/ρ^γ, we know the pressure at the displaced point. Finally, from flux freezing we know the direction and density of the field lines at the displaced point, and, thereby, we know the magnetic field as well. In other words, we know the entire state of the MHD fluid at time t if we know the displacement $\boldsymbol{\xi}(\mathbf{r}_0, t)$ everywhere as a function of $\mathbf{r}_0$ at time t. Putting all this together we can find the force at any point $\mathbf{r}$. The velocity and acceleration of the plasma element at time t are, of course, given simply by taking the derivative of $\boldsymbol{\xi}(\mathbf{r}_0, t)$ with respect to t at fixed $\mathbf{r}_0$.

All this is independent of the history of how the displacement is carried out. It depends only on the initial and final position of the displacement.

The plasma actually consists of an innumerable number of particles, which altogether, with the general electromagnetic field, satisfy an action principle

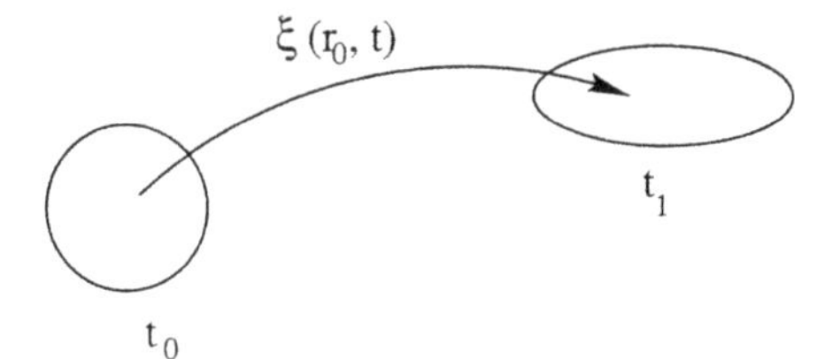

Figure 4.6. The displacement ξ in the action principle

that takes into account everything—collisions, mass motions, Ampere's law, etc. It is not surprising that the MHD fluid itself, whose equations are a limiting form of the particle equations, also satisfies an action principle (Eckart 1960; Newcomb 1962).

Before turning to this action principle, let us examine how we might determine the displacement function $\boldsymbol{\xi}(\mathbf{r}_0, t)$ from the Eulerian form of the velocity V(r,t). During the interval δt the fluid element moves from the position $\mathbf{r} = \mathbf{r}_0 + \boldsymbol{\xi}(\mathbf{r}_0, t)$ to $\mathbf{r}_0 + \boldsymbol{\xi}(\mathbf{r}_0, t) + (\partial\boldsymbol{\xi}/\partial t)\delta t$. This change must also be equal to $\mathbf{V}(\mathbf{r}, t)\delta t$, so we have

$$\frac{\partial \boldsymbol{\xi}}{\partial t}(\mathbf{r}_0, t) = \mathbf{V}[\mathbf{r}_0 + \boldsymbol{\xi}(\mathbf{r}_0, t), t] \tag{60}$$

For a given velocity field $\mathbf{V}(\mathbf{r}, t)$, the three components of $\boldsymbol{\xi}$ satisfy these three nonlinear ordinary differential equations with initial conditions $\boldsymbol{\xi}(\mathbf{r}_0, t_0) = 0$. In general, it is difficult to solve these equations except in systems with one-dimensional symmetry. However, because of the existence and uniqueness properties of ordinary differential equations, we are assured that $\boldsymbol{\xi}(\mathbf{r}_0, t)$ is defined and unique. For the general discussion we make below this is all we need.

The Lagrangian for the MHD action principle should be

$$\mathcal{L} = \int_V \rho \frac{V^2}{2} d^3x - W \tag{61}$$

where W is the "potential energy" of the plasma,

$$W = \int_V \left(\frac{p}{\gamma - 1} + \frac{B^2}{8\pi} + \rho\phi \right) d^3x \tag{62}$$

The action principle states that of all the fluid motions $\mathbf{V}(\mathbf{r}, t)$ between time t_1 and time t_2, each of which has a definite $\rho(\mathbf{r}, t)$, $p(\mathbf{r}, t)$, $\mathbf{B}(\mathbf{r}, t)$ associated with it through equations 14, 16, and 17 in chapter 3, respectively, the fluid motion that minimizes the action

$$S = \int_{t_1}^{t_2} \mathcal{L} dt \tag{63}$$

(or at least makes it stationary) is the one that satisfies the equation of motion, equation 15 in chapter 3, and is thus the real velocity. The choice of motions over which we minimize is restricted to those motions that have the same displacements at t_1 and t_2. Note that the other ideal equations (14, 16, and 17 in chapter 3) are assumed satisfied for each of the possible choice of motion of the plasma and that this makes sense. For each motion we see that the plasma is forced by these equations to have a unique ρ, p, and $\mathbf{B}$.

We denote the full set of physical quantities $\mathbf{V}$, ρ, p, $\mathbf{B}$, considered as functions of $\mathbf{r}$ and t, as an ideal motion. Then only ideal motions will be considered, but these need not satisfy the equation of motion.

Conversely, the ideal motion that *does* satisfy the equation of motion makes the action stationary to all other possible ideal motions, at least those that satisfy the displacement constraint at t_1 and t_2. On the other hand, if different displacements are specified at t_1 and t_2, then we get a different real ideal motion that satisfies the equation of motion. The specification of displacements at t_1 and t_2 in this way determines all the possible ideal motions.

Note that the action principle works both ways. That ideal motion that makes S stationary is the true one, and the true one is that which makes S stationary. This always with the stricture that the displacements are properly restricted at t_1 and t_2.

We now prove the first part of the principle. Let us assume that some ideal motion characterized by its displacement $\boldsymbol{\xi}(\mathbf{r}_0, t)$, makes the action S stationary. We denote this motion the "trial motion." Select some interval of time between t_1 and t_2 and let $\delta\boldsymbol{\xi}(\mathbf{r}_0, t)$ be a small displacement of the fluid element from the trial displacement $\boldsymbol{\xi}(\mathbf{r}_0, t)$, and let it be zero outside of this interval. Bear in mind that the function over which we are varying the action S is the displacement $\boldsymbol{\xi}(\mathbf{r}_0, t)$, regarded as a function of $\mathbf{r}_0$ and t.

First, what is the change in the density at t and $\mathbf{r} = \mathbf{r}_0 + \boldsymbol{\xi}(\mathbf{r}_0, t)$ induced by the change $\delta\boldsymbol{\xi}(\mathbf{r}_0, t)$? We may carry out the changed displacement in any order we wish. Let us do it in two stages. First, carry out the unperturbed displacement $\boldsymbol{\xi}(\mathbf{r}_0, t)$, and then carry out the perturbed displacement from $\mathbf{r}$ to $\mathbf{r} + \delta\boldsymbol{\xi}(\mathbf{r}_0, t)$. (This need not be carried out in time at the same rate as the trial displacement.) The way the two displacements are carried out has no effect on the resulting density. The displacement must be carried out not just at the point $\mathbf{r}$, but also in the neighborhood of $\mathbf{r}$.

The first displacement leads to the density $\rho(\mathbf{r})$ of the trial ideal motion. The second displacement is obtained from integrating the continuity (equation 14 in chapter 3) for the perturbed density $\delta\rho$ with the artificial time t', from $t' = 0$ to $t' = 1$:

$$\frac{\partial \delta\rho}{\partial t'} = -\nabla \cdot [(\rho + \delta\rho)\mathbf{V}'] \tag{64}$$

where $\mathbf{V}'$ is some artificial velocity that carries the fluid element from $\mathbf{r}$ to $\mathbf{r} + \delta\boldsymbol{\xi}(\mathbf{r}_0, t)$. The boundary condition is $\delta\rho(\mathbf{r}, t') = 0$ at $t' = 0$. The result is

$$\delta\rho(\mathbf{r}, t' = 1) = -\nabla \cdot [\rho\, \delta\boldsymbol{\xi}(\mathbf{r}, t)] \tag{65}$$

Two remarks are in order. First, we have integrated the change in density at a fixed point $\mathbf{r}$ rather than following the fluid. Second, we have found the perturbed density as a function of $\delta\boldsymbol{\xi}(\mathbf{r}, t)$ rather than $\delta\boldsymbol{\xi}(\mathbf{r}_0, t)$, the fundamental function over which we are supposed to vary the action. But, for a fixed trial displacement and a fixed time t, we can regard $\mathbf{r}$ as a definite function of $\mathbf{r}_0$, that is $\mathbf{r} = \mathbf{r}(\mathbf{r}_0)$, so after substitution of this in the argument, $\mathbf{r}_0$ of $\delta\boldsymbol{\xi}$ we can regard $\delta\boldsymbol{\xi}$ as also a function of $\mathbf{r}_0$ and t.

In summary, we start with a displacement function $\boldsymbol{\xi}(\mathbf{r}_0, t)$ and perturb it by $\delta\boldsymbol{\xi}(\mathbf{r}_0, t)$ which by the above transformation can be expressed at t in terms of a $\mathbf{r}$ as $\delta\boldsymbol{\xi}(\mathbf{r}, t)$. Then the perturbed density at t is

$$\delta\rho(\mathbf{r}, t) = -\nabla \cdot [\rho(\mathbf{r}, t)\delta\boldsymbol{\xi}(\mathbf{r}, t)] \tag{66}$$

(We carefully write out all the independent variables to emphasize which quantities are to be thought of as functions of $\mathbf{r}_0$ and which as functions of $\mathbf{r}$. Note that the functional relation between them depends on t as a parameter.)

Exactly the same argument leads to the variation of pressure at $\mathbf{r}$ and at t. The differential equation for the second step in the displacement in terms of the artificial time t' is from equation 36:

$$\frac{\partial \delta p(\mathbf{r}, t')}{\partial t'} = -\gamma p(\mathbf{r}, t)\left[\nabla \cdot \frac{\partial \delta\boldsymbol{\xi}(\mathbf{r}, t')}{\partial t'}\right] - \frac{\partial \delta\boldsymbol{\xi}(\mathbf{r}, t')}{\partial t'} \cdot \nabla p(\mathbf{r}, t) \tag{67}$$

and we get, similarly to the result for the perturbed density,

$$\delta p(\mathbf{r}, t) = -\gamma p(\mathbf{r}, t)\nabla \cdot [\delta\boldsymbol{\xi}(\mathbf{r}, t)] - \delta\boldsymbol{\xi}(\mathbf{r}, t) \cdot \nabla p(\mathbf{r}, t) \tag{68}$$

The treatment of the magnetic field is identical and using equation 17 in chapter 3 we get

$$\delta\mathbf{B}(\mathbf{r}, t) = \nabla \times [\delta\boldsymbol{\xi}(\mathbf{r}, t) \times \mathbf{B}(\mathbf{r}, t)] \tag{69}$$

The perturbed velocity of the fluid at $\mathbf{r}, t$ is

$$\delta\mathbf{V}(\mathbf{r}, t) = \mathbf{V}(\mathbf{r}, t) \cdot \nabla\delta\boldsymbol{\xi}(\mathbf{r}, t) - \delta\boldsymbol{\xi}(\mathbf{r}, t) \cdot \nabla\mathbf{V}(\mathbf{r}, t) + \frac{\partial \delta\boldsymbol{\xi}(\mathbf{r}, t)}{\partial t} \tag{70}$$

where the partial derivative in the last term is to be carried out at fixed $\mathbf{r}$, that is, with $\delta\boldsymbol{\xi}(\mathbf{r}, t)$ treated as a function of $\mathbf{r}$. We can see this by first considering the fluid element that would have been at $\mathbf{r}$ from its unperturbed motion, and following it from its perturbed position at $\mathbf{r}+\delta\boldsymbol{\xi}(\mathbf{r}, t)$ to its perturbed position at $\mathbf{r}+\mathbf{V}(\mathbf{r}, t)\delta t+\delta\boldsymbol{\xi}(\mathbf{r}+\mathbf{V}\delta t, t+\delta t)$. The resulting displacement is thus $(\mathbf{V}+\mathbf{V}\cdot\nabla\delta\boldsymbol{\xi}+\partial\delta\boldsymbol{\xi}/\partial t)\delta t$. This displacement also is equal to $[\mathbf{V}(\mathbf{r}, t)+\delta\boldsymbol{\xi}\cdot\nabla\mathbf{V}+\delta\mathbf{V}]\delta t$. Comparing these, we get the expession for $\delta\mathbf{V}$, equation 70. (Alternatively, we can also note that its perturbed velocity at $\mathbf{r}+\delta\boldsymbol{\xi}$, is $\mathbf{V}\cdot\nabla\delta\boldsymbol{\xi} + \partial\delta\boldsymbol{\xi}/\partial t$ and the perturbed velocity of the fluid element at $\mathbf{r}$ is this minus $\delta\boldsymbol{\xi} \cdot \nabla\mathbf{V}$.)

Now, the perturbed action is

$$\begin{aligned}\delta S &= \int_{t_1}^{t_2} dt \int \delta\mathcal{L} dt d^3x \\ &= \int_{t_1}^{t_2} dt \int d^3x \left(\delta\rho\frac{V^2}{2} + \rho\mathbf{V}\cdot\delta\mathbf{V} - \frac{\delta p}{\gamma - 1} - \frac{\mathbf{B}\cdot\delta\mathbf{B}}{4\pi} - \delta\rho\phi\right) \qquad (71)\end{aligned}$$

Substituting the values of the perturbed quantities from equations 66, 68, 69, and 70 into the perturbed action we get

$$\delta S = \int_{t_1}^{t_2} dt \int d^3x \left\{ \nabla \cdot (-\rho\delta\boldsymbol{\xi})\frac{V^2}{2} + \rho\mathbf{V} \cdot \left(\mathbf{V} \cdot \nabla\delta\boldsymbol{\xi} - \delta\boldsymbol{\xi} \cdot \nabla\mathbf{V} + \frac{\partial \delta\boldsymbol{\xi}}{\partial t} \right) \right\}$$

$$+ \int_{t_1}^{t_2} dt \int d^3x \frac{\gamma p \nabla \cdot \delta\boldsymbol{\xi} + \delta\boldsymbol{\xi} \cdot \nabla p}{\gamma - 1} - \int_{t_1}^{t_2} dt \int d^3x \frac{\mathbf{B} \cdot \nabla \times (\delta\boldsymbol{\xi} \times \mathbf{B})}{4\pi}$$

$$+ \int_{t_1}^{t_2} dt \int d^3x \nabla \cdot (\rho\delta\boldsymbol{\xi})\phi \qquad (72)$$

Integrating by parts those terms containing $\delta\boldsymbol{\xi}$ under the derivatives, and for the moment ignoring the integrated terms, we find (term by term)

$$\delta S = \int_{t_1}^{t_2} dt \int d^3x \left\{ \delta\boldsymbol{\xi} \cdot \left[\rho\nabla\frac{V^2}{2} - \nabla \cdot (\rho\mathbf{V}\mathbf{V}) - \rho\nabla\frac{V^2}{2} - \frac{\partial}{\partial t}(\rho\mathbf{V}) \right] \right.$$

$$\left. + \frac{-\delta\boldsymbol{\xi} \cdot \gamma\nabla p + \delta\boldsymbol{\xi} \cdot \nabla p}{\gamma - 1} + \frac{\delta\boldsymbol{\xi} \cdot (\nabla \times \mathbf{B}) \times \mathbf{B}}{4\pi} - \delta\boldsymbol{\xi} \cdot \rho\nabla\phi \right\} \qquad (73)$$

The integrated terms vanish if we choose $\delta\boldsymbol{\xi}$ to vanish on the boundaries and on the time surfaces t_1 and t_2. This is all we need. It is clear that since $\boldsymbol{\xi}$ as a function of $\mathbf{r}_0$ is arbitrary, so is $\boldsymbol{\xi}$ as a function of $\mathbf{r}$. Collecting terms and making use of Ampere's law, equation 19 in chapter 3, we have

$$\delta S = -\int_{t_1}^{t_2} dt \int d^3x \delta\boldsymbol{\xi} \cdot \left[\frac{\partial(\rho\mathbf{V})}{\partial t} + \nabla \cdot (\rho\mathbf{V}\mathbf{V}) + \nabla p - \mathbf{j} \times \mathbf{B} + \rho\nabla\phi \right] \qquad (74)$$

Now if $\delta\boldsymbol{\xi}$ vanishes everywhere except in a space–time neighborhood of a point P at t, then the bracket must vanish at P, and this vanishing is equivalent to the equation of motion (equation 15 in chapter 3). Thus, $\delta S = 0$ for all the possible ideal motions implies the ideal MHD equations.

Conversely, if the trial ideal motion is a solution of the actual equation of motion, then $\delta S = 0$ for all perturbations $\delta\boldsymbol{\xi}$ that vanish at t_1 and t_2 and that vanish at the surface of the volume.

Let our ideal MHD system be enclosed by a boundary at which $\mathbf{B} \cdot d\mathbf{S} = 0$ $\mathbf{V} \cdot d\mathbf{S} = 0$; that is, $\mathbf{B}$ is tangent to the boundary, and the boundary is a rigid wall. In this case, the requirement that the surface terms vanish is simply $\delta\boldsymbol{\xi} \cdot d\mathbf{S} = 0$. This is the same condition that must be applied if energy is to be conserved. If, on the other hand, $\mathbf{B} \cdot d\mathbf{S}$, or $\mathbf{V} \cdot dS \neq 0$, then we need the full $\delta\boldsymbol{\xi}$ to vanish at the boundary, not just the normal component. These remarks apply to only the second part of the action principle, which states that for the ideal motion that satisfies the equation of motion δS all perturbed displacements vanish. These perturbations must be restricted as above according to the boundaries of the system.

We might wonder what we gain by formulating the ideal MHD equation as an action principle. The answer is twofold. Most of the general results come out more simply from the action principle. But the real power gained from it is the fact that the action principle holds for almost any general description of the plasma. For example, it will be shown in chapter 5 that the number of ideal waves is conserved, where this number is defined in terms of wave action, i.e., energy divided by frequency. The simplest proof of this comes from the action principle (Whitham 1965). Further, the reaction of the waves on the rest of the medium can be reduced to a tensor wave pressure most simply by use of the action principle (Whitham 1965; Dewar 1970) (see equation 72 in chapter 5). It turns out that most of the subtle and surprising properties of ideal MHD will emerge in this way.

4.8 Lundquist's Identity

Before leaving the subject of the action principle it is of interest to give an analytic expression for the evolution of the field. We have stated rather pictorially that given the displacement $\boldsymbol{\xi}(\mathbf{r}_0, t)$ we know where the plasma fluid element at initial position $\mathbf{r}_0$ ends up at time t. Then from flux freezing we know the position of the lines of force, and thus the magnetic field, both in its magnitude and its direction.

This can all be encapsulated in one formula due to Lundquist, and called the Lundquist identity (Lundquist 1951; Roberto 1967).

$$\frac{\mathbf{B}(\mathbf{r}, t)}{\rho(\mathbf{r}, t)} = \frac{\mathbf{B}_0 + \mathbf{B}_0 \cdot \nabla_0 \boldsymbol{\xi}(\mathbf{r}_0, t)}{\rho_0} \tag{75}$$

where $\mathbf{B}_0$ is the magnetic field at time t_0 and position $\mathbf{r}_0$, and ρ_0 is the density there. The subscript zero on ∇ indicates that the gradient is to be taken with respect to $\mathbf{r}_0$. Examination of this formula shows that it should roughly give the correct answer consistent with flux freezing, since it has the correct features. For example, the gradient of $\boldsymbol{\xi}(\mathbf{r}_0, t)$ with respect to $\mathbf{r}_0$ gives the shearing or turning of the field lines as in figure 4.7.

Further, the stretching along the lines, also included in the equation, tends to increase the volume due to the increase in the length ℓ, along the line. Thus, division by the density, which represents the increase in volume equal to its area $\times$ length, yields the decrease in area, and thus the increase in field strength that the field should have. This is the analytic expression of the amplification of the magnetic field strength by line stretching.

A more formal derivation is as follows. By vector algebra we can rewrite equation 17 in chapter 3 as

$$\frac{\partial \mathbf{B}}{\partial t} = \nabla \times (\mathbf{V} \times \mathbf{B}) = \mathbf{B} \cdot \nabla \mathbf{V} - \mathbf{V} \cdot \nabla \mathbf{B} - \mathbf{B} \nabla \cdot \mathbf{V} \tag{76}$$

where the $\nabla \cdot \mathbf{B}$ term is dropped. Making use of the continuity equation

$$\frac{\partial \rho}{\partial t} = -\mathbf{V} \cdot \nabla \rho - \rho \nabla \cdot \mathbf{V} \tag{77}$$

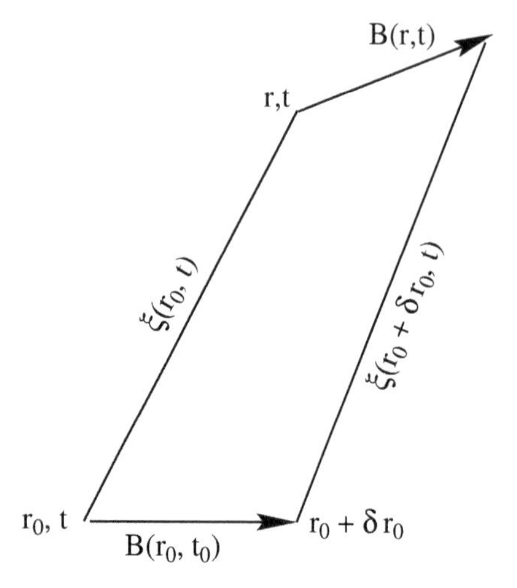

Figure 4.7. Diagram illustrating Lundquist's identity

equation 76 can rewritten as follows. Bringing the $\mathbf{V}\cdot\nabla$ term to the left in both equations, replacing $\partial/\partial t+\mathbf{V}\cdot\nabla$ by the convective derivative d/dt, and combining the two equations, we get

$$\frac{d}{dt}\left(\frac{\mathbf{B}}{\rho}\right)=\frac{1}{\rho}\frac{d\mathbf{B}}{dt}-\frac{\mathbf{B}}{\rho^2}\frac{d\rho}{dt}=\frac{\mathbf{B}}{\rho}\cdot\nabla\mathbf{V} \tag{78}$$

the other terms canceling.

Now $\mathbf{r}=\mathbf{r}_0+\boldsymbol{\xi}(\mathbf{r}_0,t)$, which allows us to write Lundquist's identity, equation 75, as

$$\frac{\mathbf{B}}{\rho}=\frac{\mathbf{B}_0}{\rho_0}\cdot\frac{\partial}{\partial\mathbf{r}_0}[\mathbf{r}(\mathbf{r}_0,t)] \tag{79}$$

Substituting this into the left-hand side of equation 78 gives

$$\frac{\mathbf{B}_0}{\rho_0}\cdot\frac{\partial}{\partial\mathbf{r}_0}\left(\frac{d}{dt}\mathbf{r}\right)=\frac{\mathbf{B}_0}{\rho_0}\cdot\frac{\partial}{\partial\mathbf{r}_0}\mathbf{V} \tag{80}$$

Substituting this on the right-hand side of equation 78 gives

$$\frac{\mathbf{B}_0}{\rho_0}\cdot\frac{\partial\mathbf{r}}{\partial\mathbf{r}_0}\cdot\nabla\mathbf{V}=\frac{\mathbf{B}_0}{\rho_0}\cdot\frac{\partial}{\partial\mathbf{r}_0}\mathbf{V} \tag{81}$$

by the chain rule of differentiation.

The identity of the left and right sides shows that Lundquist's form for $\mathbf{B}/\rho$ satisfies the differential equation. Since it also satisfies it initially at $t=t_0$, it must give the correct result for $\mathbf{B}/\rho$, and, thus, give $\mathbf{B}$ as it is supposed to. Incidentally, Lundquist's identity is identical in form with that for the evolution of vorticity ω in an ideal fluid (Lamb 1945, p. 205). This is because vorticity satisfies the same equation as the magnetic field (see equations 52 and 56 in chapter 13).

4.9 Axisymmetry

The ideas of MHD are well illustrated by axisymmetric situations. We define this symmetry by taking cylindrical coordinates r, θ, z and insisting that the components of all vectors (e.g., B_r, B_θ, B_z) and scalars (e.g., p, ρ) are independent of θ. We can reduce the number of independent components of **B** by using the divergence condition

$$\nabla \cdot \mathbf{B} = 0 \tag{82}$$

to introduce a flux function ψ. In cylindrical coordinates equation 82 can be written

$$\frac{1}{r}\frac{\partial}{\partial r}(rB_r) + \frac{\partial}{\partial z}B_z = 0 \tag{83}$$

or

$$-\frac{\partial}{\partial r}(-rB_r) + \frac{\partial}{\partial z}(rB_z) = 0 \tag{84}$$

Treating r and z as Cartesian coordinates, this is the two-dimensional curl of the vector

$$\mathbf{A} = rB_z\hat{\mathbf{r}} - rB_r\hat{\mathbf{z}} \tag{85}$$

so in r, z coordinates **A** is the gradient of a scalar,

$$\mathbf{A} = \nabla\psi \tag{86}$$

or, explicitly,

$$B_r = -\frac{1}{r}\frac{\partial\psi}{\partial z}$$

$$B_z = \frac{1}{r}\frac{\partial\psi}{\partial r} \tag{87}$$

Thus, provided $\nabla \cdot \mathbf{B} = 0$, the two components of **B**, B_r and B_z, can be replaced by the single scalar function $\psi(r, z)$. Conversely, any $\psi(r, z)$ generates a magnetic field **B** that is divergence free. B_θ still remains as the other component of **B**.

In axisymmetry, the magnetic field is conventionally broken up into a poloidal field,

$$\mathbf{B}_p = B_r\hat{\mathbf{r}} + B_z\hat{\mathbf{z}} = -\frac{\hat{\boldsymbol{\theta}} \times \nabla\psi}{r} \tag{88}$$

and a toroidal field $\mathbf{B}_T = B_\theta\hat{\boldsymbol{\theta}}$, so that

$$\mathbf{B} = \mathbf{B}_p + \mathbf{B}_T \tag{89}$$

The reason for this terminology is clear. If **B** had only a toroidal component, $\mathbf{B}_T$, then the lines of force would be closed circles, perpendicular to and

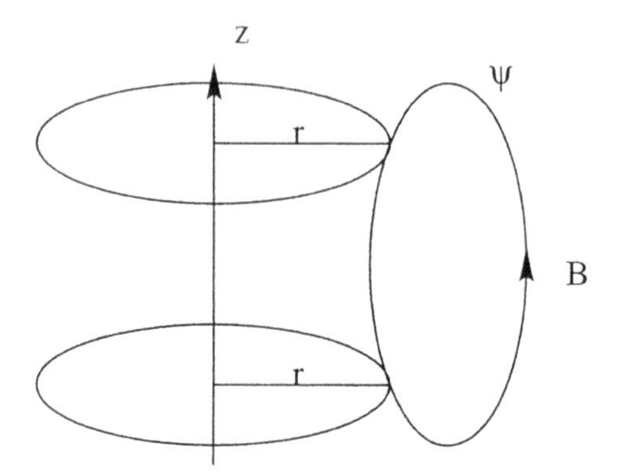

Figure 4.8. The poloidal flux function

centered about the z axis. Similarly, if the magnetic field had only a poloidal component, then the lines of force would lie in planes θ = constant and would appear similar to the magnetic field of the earth, which is dipolar.

The scalar ψ is determined only up to an additive constant, which could be a function of time. From equation 87 multiplied by r, $rB_r = -\partial\psi/\partial z$, we see that ψ is constant on the z axis. Thus, we can choose the constant so that ψ vanishes on the axis. With this choice we can see that $2\pi\psi(r, z)$ is just the flux threading the circle through r, z. Namely, from equation 87

$$2\pi\psi = 2\pi \int_0^r B_z r dr \tag{90}$$

which is just this flux. The flux must be constant on a line of force, that is to say, poloidal flux encompassed by each of two circles passing through two points on the same line of force are equal (see the figure 4.8). Needless to say, the flux we refer to can be either the total magnetic flux or the poloidal flux.

Thus, in axisymmetry the equation of a poloidal line of force that lies in a θ = constant plane is

$$\psi(r, z) = \text{ const} \tag{91}$$

and a total line of force also lies on the two-dimensional surface, ψ = constant. These last two things are true no matter how the free constant in ψ is chosen.

Finally, it should be noted that $\hat{\theta}\psi/r$ is a vector potential $\mathbf{A} = A_\theta\hat{\theta}$ for the poloidal field. The constant in ψ here corresponds to the gauge freedom in $\mathbf{A}$. (This is a different $\mathbf{A}$ from that in equation 85.)

Now let us consider the time evolution of the magnetic field due to an axisymmetric velocity

$$\mathbf{V} = \mathbf{V}_p + \mathbf{V}_T = (V_r\hat{\mathbf{r}} + V_z\hat{\mathbf{z}}) + \Omega r\hat{\boldsymbol{\theta}} \tag{92}$$

where we break the velocity field into a poloidal velocity and a toroidal velocity in analogy with the breakup of the magnetic field. The toroidal velocity has been expressed as an angular velocity, $V_\theta = \Omega r$. Then we have for the toroidal part of equation 17 in chapter 3

$$\frac{\partial}{\partial t}\left(B_\theta \hat{\boldsymbol{\theta}}\right) = \nabla \times (\mathbf{V} \times \mathbf{B})_p \tag{93}$$

since the curl of a toroidal vector is poloidal. The poloidal part of $\mathbf{V} \times \mathbf{B}$ is

$$(\mathbf{V} \times \mathbf{B})_p = \mathbf{V}_p \times \mathbf{B}_T + \mathbf{V}_T \times \mathbf{B}_p = \mathbf{V}_p \times \mathbf{B}_T - \Omega \nabla \psi \tag{94}$$

The curl of $\mathbf{V}_p \times \mathbf{B}_T$ is

$$\begin{aligned} \nabla \times (\mathbf{V}_p \times \mathbf{B}_T) &= \mathbf{B}_T \cdot \nabla \mathbf{V}_p - \mathbf{V}_p \cdot \mathbf{B}_T - \mathbf{B}_T \nabla \cdot \mathbf{V}_p \\ &= -\mathbf{V}_p \cdot \nabla \mathbf{B}_T - \mathbf{B}_T (\nabla_2 \cdot \mathbf{V}_p) \end{aligned} \tag{95}$$

where the subscript 2 on ∇ indicates a two-dimensional (Cartesian) divergence. We get this from $\mathbf{B}_T \cdot \nabla \mathbf{V}_P = -\mathbf{B}_T V_r / r$, which cancels the middle term of $\mathbf{B}_T \nabla \cdot \mathbf{V}_p$. The curl of the $\Omega \nabla \psi$ term is

$$\nabla \times (\Omega \nabla \psi) = \nabla \Omega \times \nabla \psi = -r \mathbf{B}_p \cdot \nabla \Omega \hat{\boldsymbol{\theta}} \tag{96}$$

Combining these results, we have

$$\frac{\partial B_T}{\partial t} + \mathbf{V}_p \cdot \nabla B_T + (\nabla_2 \cdot \mathbf{V}_p) B_T = r \mathbf{B}_p \cdot \nabla \Omega \tag{97}$$

where in this equation B_T is a scalar, and where, explicitly,

$$\nabla_2 \cdot \mathbf{V}_p = \frac{\partial V_r}{\partial r} + \frac{\partial V_z}{\partial z} \tag{98}$$

Equation 97 is easy to interpret. The left side gives the evolution of B_T as a two-dimensional continuity equation for B_T. A given poloidal area S contains $B_T S$ lines of force and in the absence of $\mathbf{B}_p$ will continue to contain these lines as it moves in a poloidal surface under the poloidal flow velocity $\mathbf{V}_p$. The right-hand side represents the shearing of poloidal field lines if Ω varies along the constant ψ line.

Note that if $\mathbf{B}_p \cdot \nabla \Omega = 0$, there is no production of toroidal field flux. If $\mathbf{V}_p$ were zero, then $\mathbf{B}_p \cdot \nabla \Omega = 0$ is the required condition that B_T be time independent. This is known as Ferraro's isorotation law (Alfven and Falthammar 1963). It is obvious, physically, from flux freezing. To understand the rate at which differential rotation produces toroidal flux, start with $B_T = 0$. Consider a short line ℓ lying on a flux surface. After a time δt, the top of ℓ will advance an angle $\delta\theta = \ell \cdot \nabla \Omega \delta t$ further than the bottom. Therefore, $\delta B_T / r \delta\theta = B_p / \ell$, so that $\delta B_T = r \mathbf{B}_p \cdot \nabla \Omega \delta t$, in accordance the right-hand side of equation 97.

Now consider the evolution of the poloidal field, which is given by the poloidal part of equation 17 in chapter 3. This part can be written

$$\frac{\partial}{\partial t}\left(\nabla \times \hat{\boldsymbol{\theta}} \frac{\psi}{r}\right) = \nabla \times \left[\frac{\mathbf{V}_p \times (\nabla \psi \times \hat{\boldsymbol{\theta}})}{r}\right] \tag{99}$$

We can see this from the fact that in a curl term, only the toroidal vector gives a poloidal vector, and the bracket represents the complete toroidal part of $\mathbf{V} \times \mathbf{B}$. Since the right bracket reduces to $-\mathbf{V}_p \cdot \nabla\psi\hat{\boldsymbol{\theta}}/r$, equation 99 reduces to

$$\nabla \times \left(\frac{1}{r}\frac{d\psi}{dt}\hat{\boldsymbol{\theta}}\right) = 0 \tag{100}$$

where $d\psi/dt = \partial\psi/\partial t + \mathbf{V}_p \cdot \nabla\psi$. Therefore, the r and z derivatives of $d\psi/dt$ are zero and $d\psi/dt$ is a constant in space. This constant is the time derivative of the free constant in the definition of ψ. If we choose the constant to make ψ vanish on the z axis at all times, then we have

$$\frac{d\psi}{dt} = 0 \tag{101}$$

and ψ remains the flux function for all times. This corresponds to flux freezing of the field, since, clearly, poloidal flux is conserved by this equation.

Finally, for this gauge choice, $\psi = 0$ on axis, we can relate it to the toroidal component of the electric field, E_θ. In fact,

$$\left(\mathbf{E} + \frac{\mathbf{V} \times \mathbf{B}}{c}\right)_\theta = E_\theta - \frac{1}{c}\frac{\mathbf{V} \cdot \nabla\psi}{r} = 0 \tag{102}$$

So, because $d\psi/dt = 0$, we have

$$E_\theta = -\frac{1}{cr}\frac{\partial\psi}{\partial t} \tag{103}$$

a result consistent with the relationship, already mentioned, of ψ to the vector potential, and valid for the proper choice of gauge.

If the region does not include the axis but is bounded by a conducting wall on which the normal component of the magnetic field is zero, and the tangential component of the electric field is zero, then we can choose the constant in ψ to make it zero on this wall. In this case, equation 103 is also valid. Even if the wall is moving but if the tangential component of E is zero in the moving frame, equation 103 will still hold.

Let us consider the plasma equations for the other axisymmetric plasma quantities. It is easy to calculate the current from equation 80. The result is

$$\begin{aligned}
-4\pi j_r &= \frac{1}{r}\frac{\partial}{\partial z}(rB_\theta) \\
-4\pi j_z &= -\frac{1}{r}\frac{\partial}{\partial r}(rB_\theta) \\
-4\pi j_\theta &= \frac{1}{r}\left(r\frac{\partial}{\partial r}\frac{1}{r}\frac{\partial\psi}{\partial r} + \frac{\partial^2\psi}{\partial z^2}\right) \\
&= \frac{\Delta^*\psi}{r}
\end{aligned} \tag{104}$$

The symbol Δ^* is introduced because the operator is close to the Laplacian operator in cylindrical coordinates.

We write down the equation of motion only in the case of static equilibrium, with no gravity. The equation is

$$\mathbf{j} \times \mathbf{B} = \nabla p \tag{105}$$

Its θ component is

$$j_z B_r - j_r B_z = 0 \tag{106}$$

since $\partial p / \partial \theta = 0$ by the symmetry. With equations 87 and 104 this becomes

$$-\frac{1}{4\pi r^2}\left[\frac{\partial}{\partial r}(r B_\theta)\frac{\partial \psi}{\partial z} - \frac{\partial}{\partial z}(r B_\theta)\frac{\partial \psi}{\partial r}\right] = 0 \tag{107}$$

But this equation is equivalent to the Jacobian of $r B_\theta$ and ψ vanishing, so that $r B_\theta$ must be a function of ψ alone:

$$r B_\theta = F(\psi) \tag{108}$$

Further, we can derive directly from equation 105 that

$$\mathbf{B} \cdot \nabla p = 0 \tag{109}$$

Making use of equation 87 again, we find that this equation is equivalent to the Jacobian of p and ψ being zero, so that p must be a function of ψ alone:

$$p = p(\psi) \tag{110}$$

This, of course, can be seen directly, since equation 109 says that p is constant along $\mathbf{B}$.

Now, the r component of equation 105 is

$$j_\theta B_z - j_z B_\theta = \frac{\partial p}{\partial r} \tag{111}$$

so from equations 104, 87, 108, and 110 we obtain

$$-\frac{\Delta^*\psi}{4\pi r}\frac{1}{r}\frac{\partial \psi}{\partial r} - \frac{1}{4\pi r}\frac{dF}{d\psi}\frac{\partial \psi}{\partial r}\frac{F}{r} = \frac{dp}{d\psi}\frac{\partial \psi}{\partial r} \tag{112}$$

or

$$\Delta^*\psi + FF' = -4\pi r^2 p' \tag{113}$$

where primes denote derivatives with respect to ψ. This is the renowned Grad–Shafranov equation for static MHD equilibria in cylindrical symmetry (Shafranov 1966).

There are several important points to made about this equation. All the terms in this equation have a factor $\partial \psi / \partial r$, which we cancel. If we write out the z component of equation 105, we find almost the same equation with the factor $\partial \psi / \partial z$ instead of $\partial \psi / \partial r$, and canceling this, we arrive at the Grad–Shafranov equation again. This is not surprising since we made use of equation 110 in deriving equation 113, so these equations are not independent. In any event, any solution of the equilibrium equation 105 is a solution of the Grad–Shafranov equation and conversely any solution of the Grad–Shafranov equation gives a solution of the equilibrium equation 105.

The Grad–Shafranov equation is a nonlinear partial differential equation and the quantities $F(\psi)$ and $p(\psi)$ are arbitrary functions. For each choice of them we should get an equilibrium, provided the proper boundary conditions are added. These boundary conditions are usually specified by making ψ constant on some boundary and giving the total amount of poloidal flux. Thus, we see that the freedom of axisymmetric equilibrium solutions is given by the two functions: $p(\psi)$, which gives the distribution of pressure with poloidal flux ψ, and $F(\psi)$, which gives the distribution of toroidal flux with ψ.

In a certain sense, this is natural. Suppose we start off with some poloidal field lines, say periodic in z. Then, in between two flux surfaces we place some plasma with pressure $P(\psi)$ and some toroidal flux $d\Phi = G(\psi)d\psi$. We would expect that by adjusting the shape of the flux surfaces, keeping the flux frozen in the plasma, and keeping the "entropy" p/ρ^{γ} of each fluid element constant, we would arrive at an equilibrium. This equilibrium has a pressure p that is constant on ψ surfaces (but will be a different function than that initially imposed, since it has to correspond to the same mass between the two flux tubes and the same entropy.) Also, $rB_\theta = F(\psi)$ will be constant on ψ surfaces and $F(\psi)$ will be related to $\Phi(\psi)$ in some complicated way that need not concern us, except that the toroidal flux between the surfaces must also be the same. The point is that changing the functions $P(\psi)$ and $G(\psi)$ will lead to changed functions $p(\psi)$ and $F(\psi)$. This indicates why we might expect the class of all axisymmetric equilibrium solutions to be at least that of two arbitrary functions of ψ. In fact, this is actually a special case of a much more general result that does not depend on axisymmetry. It is always the case that two functions give the correct degree of freedom for general topologically toroidal static equilibrium (see problem 5).

For a trivial example of a solution of the Grad–Shafranov equation, let us assume that $F(\psi) = 0$, and look for an equilibrium independent of z. Then the Grad–Shafranov equation 113 becomes

$$\Delta^*\psi = \frac{d^2\psi}{dr^2} - \frac{1}{r}\frac{d\psi}{dr} = -4\pi r^2 \frac{dp}{d\psi} \tag{114}$$

where ψ is a function of r alone. Multiplying by $d\psi/dr$ and dividing by $4\pi r^2$ reduces this equation to

$$\frac{1}{4\pi}\frac{1}{r}\frac{d\psi}{dr}\frac{d}{dr}\left(\frac{1}{r}\frac{d\psi}{dr}\right) = -\frac{dp}{dr} \tag{115}$$

or, since $B_z = d\psi/rdr$,

$$\frac{d}{dr}\left(p + \frac{B_z^2}{8\pi}\right) = 0 \tag{116}$$

a well-known solution, easily verified directly from equation 105.

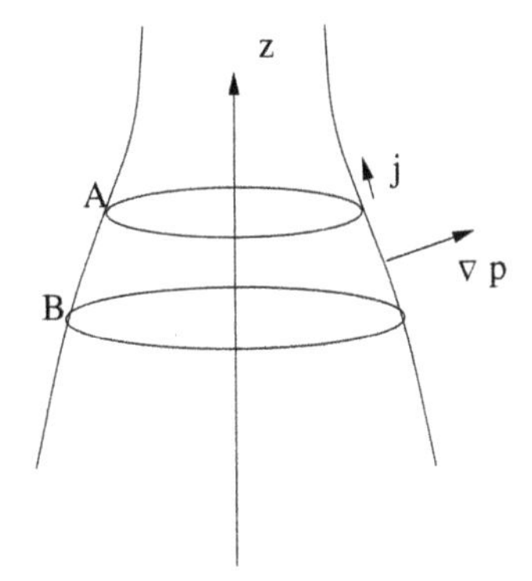

Figure 4.9. Proof that rB_θ is a function of ψ

A slightly more complicated solution is that of the screw pinch. Again take the equilibrium independent of z, i.e., $\partial/\partial z = 0$, but now take $F \neq 0$. Then, again multiplying the Grad–Shafranov equation by $(1/4\pi r^2)d\psi/dr$, we get

$$\frac{B_z}{4\pi}\frac{dB_z}{dr} + \frac{F}{4\pi r^2}\frac{dF}{dr} = -\frac{dp}{dr} \tag{117}$$

or, since $F = rB_\theta$,

$$\frac{d}{dr}\left(\frac{B_\theta^2 + B_z^2}{8\pi}\right) + \frac{B_\theta^2}{4\pi r} = -\frac{dp}{dr} \tag{118}$$

This equation is easy to interpret in terms of the concepts of magnetic tension and pressure. The first term on the left is the magnetic pressure force, the second is the inward magnetic tension force, and the right-hand side is the pressure force.

What is the physical reason that rB_θ is a function of ψ alone? By dotting the general force equation 105 with j we find that

$$\mathrm{j} \cdot \nabla p = 0 \tag{119}$$

which shows that j lies in the flux surfaces (see figure 4.9).

Now consider two circles centered on the z axis and passing through the points A and B on a flux surface. $2\pi rB_\theta$ at A is the current passing through circle A and $2\pi rB_\theta$ at B is the current passing through circle B (by a simple integration of Ampere's law). But, because j cannot cross the flux surface, these currents must be the same, so that the two values of rB_θ must also be equal and, thus, rB_θ must be a function of ψ alone as given by equation 108.

Let us now consider the resistive evolution of ψ in the axisymmetric case. We assume that $V = 0$. This would be appropriate inside a star, even if we include gravity but neglect convection. Let us further neglect the toroidal component of the field. Then it can easily be shown from equation 18 in chapter 3 that the equation for $\partial\psi/\partial t$ is

$$\frac{\partial\psi}{\partial t} = \frac{\eta c}{4\pi}\Delta^*\psi \tag{120}$$

for any space-dependent, but axisymmetric, resistivity. Simply add $-\nabla \times (\eta c\mathbf{j})_T$ to the right-hand side of equation 99, noting that only j_θ contributes to the poloidal part. Then argue as before to remove the curl operator, and use equation 104 for j_θ.

Wrubel (1952) has given the eigenvalue solution of equation 120 for a star, with η taken as the Spitzer resistivity inside the star. We can see qualitatively what should happen. In general, the magnetic field does not totally decay during the lifetime of the star. However, since $\eta(r)$ is much larger in the outer regions of a star (because it is proportional to the temperature to the minus three-halves power), we expect that in these outer regions the electrical current will decay faster than in the inner region and j_θ should disappear there. The field in the outer regions should tend to a vacuum dipole solution. On the other hand, in the inner parts the field would decay more slowly, and over the stellar lifetime would hardly change. (From the estimate of T_{decay} given in chapter 3, we see the stellar lifetime of the field is probably longer than the lifetime of the star.) In other words, the solution would be a slowly decaying field surrounded by an essentially dipole solution in the outer layers that would track it as it decayed.

A word of caution about this solution. It was predicated on the assumption that $V = 0$, and this is not true in the outer layers of the sun. Thus, although presumably our solution is valid in the deep layers of the sun, its behavior is quite different in the outer convection zone where $V \neq 0$. Here there seems to be much dynamo activity, which leads to the solar cycle, the 11-year periodic behavior of the solar field.

4.10 Problems

1. (Solar wind) Assume that the magnetic field is stationary in a frame rotating with the angular velocity of the sun. Show that if the solar wind is infinitely conducting, then

$$B_\theta v_r = (v_\theta - \Omega r)B_r$$

where Ω is the angular velocity of the sun. Assume that the angular torque J across a cylinder of height $2r$ is a constant with r. Show that it can written as

$$-J = 2\pi \int_{-r}^{r} \left(r\rho v_r v_\theta - r\frac{B_\theta B_r}{4\pi} \right) r dz$$

Assume that the expression in parentheses is independent of z, and that $4\pi\rho r^2 v_r$, the solar wind flux through the cylinder, is a constant. Show that

$$-\frac{J}{4\pi\rho r^2 v_r} = r v_\theta \left(1 - \frac{B_r^2}{4\pi\rho v_r^2}\right) + \frac{B_r^2}{4\pi\rho v_r^2}\Omega r^2$$

Show that the denominator of the left-hand side is mass outflow rate and therefore a constant. There is a radius, r_A, called the Alfven radius where the radial Alfven speed $B_r/\sqrt{4\pi\rho}$ is equal to v_r. Evaluate the angular momentum loss per unit mass at this radius and show that it is equal to $\Omega^2 r_A^2$

If r_A is 10 solar radii, approximately what fraction of its mass must the sun lose to slow its rotation down? Take the moment of inertia of the sun to be $0.06 M_{\rm sun} R_{\rm sun}^2$. What fraction of its mass has the sun lost in 5 billion years if the solar wind was steady and the density at the earth is 10/cm^3 and the velocity is 300 km/sec (see Davis and Weber 1967)?

2. (Vacuum fields) Show that at any point in a vacuum

$$\nabla_\perp B^2 = -2\kappa B^2$$

where κ is the curvature of the magnetic field line at that point.

3. (Magnetic and gravitational energy) Consider a star of uniform density with a uniform magnetic field. Show that as the star collapses, the ratio of the gravitational energy to the magnetic energy is a constant.

4. (Force free fields) A force-free field is a field that is so strong its force is potentially stronger than any other force if $\mathbf{j}$ and $\mathbf{B}$ make a finite angle with each other. Therefore, the angle must be very small and the equilibrium equation is

$$\mathbf{j} \times \mathbf{B} = 0$$

Show that for a force-free field we have $\mathbf{j} = \alpha \mathbf{B}$, where α is a constant along $\mathbf{B}$. Take α a constant everywhere and show that in a cylindrical equilibrium one has

$$B_r = 0$$

$$B_\theta = B_0 J_1(\alpha r)$$

$$B_a = B_0 J_0(\alpha r)$$

where B_0 is a constant, and J_0 and J_1 are Bessel functions. This equilibrium is known as the Lundquist equilibrium.

5. (Energy principle for equilibrium) Consider a plasma with no velocity or gravitational field, but a pressure p and a magnetic field $\mathbf{B}$. Perturb its energy in a volume V by a displacement $\delta\boldsymbol{\xi}$, and show that the first-order change in its energy is

$$\delta W = -\int d^3x \delta\boldsymbol{\xi} \cdot (\mathbf{j} \times \mathbf{B} - \nabla p)$$

ignoring surface terms. Thus, conclude that the necessary and sufficient condition that the change in the energy vanish to first order in the displacement is that

$$\mathbf{j} \times \mathbf{B} = \nabla p$$

or, in other words, the plasma must be in static ($\mathbf{V} = 0$) equilibrium. Can you specify boundary conditions on the boundary of V so that when they are imposed we have sufficiency? Necessity?

6. (The flux function for a dipole field) The equations for an axisymmetric dipole field in spherical coordinates are

$$B_r = B_0 \frac{\cos\theta}{r^3}$$

$$B_\theta = B_0 \frac{\sin\theta}{2r^3}$$

where B_0 is the field at the north pole.

Show from equation 87 that the flux function for this field is

$$\psi = B_0 \frac{\sin^2\theta}{2r}$$

Does this flux function satisfy the Grad–Shafranov equation 113 with $p = F = 0$? Consider the earth's magnetic field a dipole field. A line that intersects the equatorial plane at $5R_e$ intersects the earth's surface at what latitude?

References

Alfven, H., and C.-G. Falthammar. 1963. *Cosmical Electrodynamics Fundamental Principles*, p. 109, Oxford Clarendon Press, London.

Chandrasekhar, S. 1961. *Hydrodynamic and Hydromagnetic Stability*, p. 577, Oxford University Press, London.

Davis, L., and E. J. Weber. 1967. *Astrophysical Journal* **148**, 217.

Dewar, R. L. 1970. *Physics of Fluids* **13**, 2710.

Eckart, C. 1960. *Physics of Fluids* **3**, 921.

Fowler, K. H., and E. A. Guggenheim. 1939. *Statistical Thermodynamics*, p. 270, Cambridge University Press, London.

Lamb, H. 1945. *Hydrodynamics*, 6th ed., p. 205, Dover, New York.

Lundquist, S. 1951. *Physical Review* **83**, 307.

Milne, E. A. 1925. *Philosophical Magazine* **50**, 409.

Newcomb, W. A. 1962. *Nuclear Fusion Supplement* **2**, 451.

Roberts, P. H. 1967. *Introduction to Magnetohydrodynamic*, p. 46, Longmans, Green, London.

Shafranov, V. D. 1966. In *Reviews of Plasma Physics*, Vol. 2, p. 116, ed. M. A. Leontovitch, trans. H. Lashinsky, Consultants Bureau, 1966.

Whitham, G. B. 1965. *Journal of Fluid Mechanics* **22**, 273.

Wrubel, M. H. 1952. *Astrophysical Journal* **116**, 291.

Chapter 5

MHD WAVES

5.1 The Basic Equations

THE SIMPLEST EXAMPLE OF MHD motions is MHD waves. Here the interplay between motion $\mathbf{V}$ and magnetic field $\mathbf{B}$ is most easily seen. The waves are perturbations of ρ, $\mathbf{V}$, p, and $\mathbf{B}$ from a smooth motion.

In general, we imagine that the unperturbed system is a large-scale MHD motion, and we first study the perturbed wave motion on a smaller scale. If this scale is small enough, ρ, $\mathbf{V}$, p, and $\mathbf{B}$ are uniform, locally homogeneous, and, if we translate to the moving frame, locally static. Then, in zero approximation in the wave scale, we can consider the wave motion to be in an effectively infinite, homogeneous static medium. The unperturbed quantities ρ_0, p_0, and $\mathbf{B}_0$ are constant in space and time and $\mathbf{V}_0 = 0$. It turns out that in this approximation $\mathbf{g}$ can be neglected.

Then we consider small disturbances $\delta\rho$, $\delta\mathbf{V}$, δp, and $\delta\mathbf{B}$ in the medium. If we again introduce the displacement vector $\delta\boldsymbol{\xi}$, as in the discussion of the action principle, then we can express the perturbed quantities in the same way as we did in equations 66 and 68–70 in chapter 4:

$$\begin{aligned}
\delta\rho &= -\rho_0 \nabla \cdot \boldsymbol{\xi} \\
\delta p &= -\gamma p_0 \nabla \cdot \boldsymbol{\xi} \\
\delta\mathbf{B} &= \nabla \times (\boldsymbol{\xi} \times \mathbf{B}_0) \\
\delta\mathbf{V} &= \partial\boldsymbol{\xi}/\partial t
\end{aligned} \tag{1}$$

We have introduced several modifications in the original equations. We have replaced $\delta\boldsymbol{\xi}$ with $\boldsymbol{\xi}$, since the original $\boldsymbol{\xi}$ would have been zero for the static media. Also, we identify $\mathbf{r}_0$ and $\mathbf{r}$ for the same reason, so that our new $\boldsymbol{\xi}$ should be thought of as a function of $\mathbf{r}$. Second, the derivatives of the unperturbed equilibrium quantities are taken to be zero, since the unperturbed equilibrium is homogeneous. Finally, we keep only terms linear in $\boldsymbol{\xi}$, since our wave amplitude is assumed small.

We are mainly interested in wave packets, so that we can localize our waves to a small region. However, at first we pretend that the system is infinite and take our waves to be monochromatic. We take them as sinusoidal in space and time.

Let us assume that all perturbed quantities are the real part of a complex constant times $\exp(i\mathbf{k}\cdot\mathbf{r} - i\omega t)$. For example,

$$\delta p = Re\left[p_1 e^{i(\mathbf{k}\cdot\mathbf{r}-\omega t)}\right] \tag{2}$$

We assume that the complex quantities under the real parts satisfy the equations of motion. The equation of motion (equation 15 in chapter 3) to first order is

$$\rho_0 \frac{\partial^2 \boldsymbol{\xi}}{\partial t^2} = -\nabla \delta p + \delta\mathbf{j} \times \mathbf{B}_0 \tag{3}$$

From Ampere's law, equation 19 in chapter 3, $\delta\mathbf{j}$ is

$$\delta\mathbf{j} = \frac{1}{4\pi} \nabla \times \delta\mathbf{B} \tag{4}$$

Now, clearly, differentiating quantities with respect to space introduces a factor of the wave vector $i\mathbf{k}$ and with respect to time a factor of the angular frequency $-i\omega$. These factors are to be included under the real part operator. Thus, the perturbed quantities are

$$p_1 = -i\gamma p_0(\mathbf{k}\cdot\boldsymbol{\xi}_1)$$
$$\mathbf{B}_1 = i\mathbf{k} \times (\boldsymbol{\xi}_1 \times \mathbf{B}_0)$$
$$\mathbf{j}_1 = \frac{i}{4\pi}\mathbf{k} \times \mathbf{B}_1 \tag{5}$$

Introducing these results into equation 3 we find

$$-\rho_0\omega^2\boldsymbol{\xi} = -\gamma p_0 \mathbf{k}(\mathbf{k}\cdot\boldsymbol{\xi}) - \frac{1}{4\pi}\left\{\mathbf{k} \times [\mathbf{k} \times (\boldsymbol{\xi} \times \mathbf{B}_0)]\right\} \times \mathbf{B}_0 \tag{6}$$

This equation can be written either for $\boldsymbol{\xi}_1$ or for $\boldsymbol{\xi}$, since they differ by the exponential, which is a constant common to all terms.

Equation 6 is essentially three linear equations in the three components of $\boldsymbol{\xi}$, ξ_x, ξ_y, and ξ_z, which are the unknowns. The condition for a nontrivial solution for them is that the determinant vanish. The vanishing of the determinant gives a cubic equation for ω^2.

The parameters of the problem are ρ_0, p_0, $\mathbf{B}_0$, and $\mathbf{k}$. The vector $\mathbf{k}$ gives the direction of wave propagation and the wave number $|\mathbf{k}| = 2\pi/\lambda$. For each choice of this parameter, there are three independent solutions each arising from one solution of the cubic for ω^2. For each value of ω^2, the positive and negative values of ω give propagation parallel or antiparallel to $\mathbf{k}$. Also, for each value of ω^2 we have a nontrivial solution of equation 6 for the three components of $\boldsymbol{\xi}$. This solution gives the polarization of the wave.

It will appear that the three solutions for ω^2 are all real and positive and, in general, not equal. The largest solution corresponds to the fastest wave and is actually termed the "fast MHD wave" in the literature. Similarly, the smallest solution gives the "slow MHD wave," and the intermediate solution

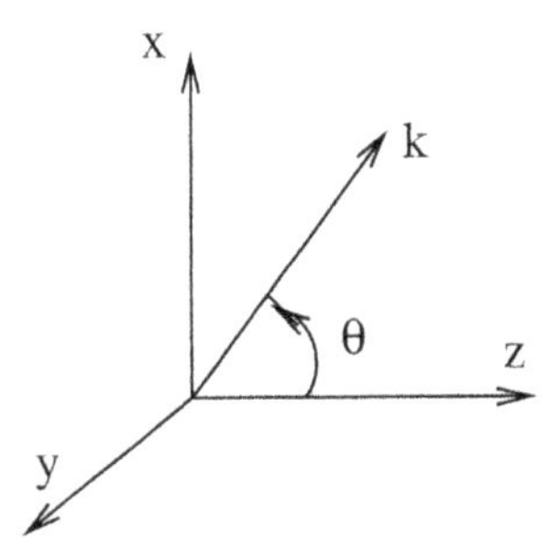

Figure 5.1. Direction of the wave vector **k**

gives the "intermediate wave. The first and last waves are also called the magnetosonic wave and the shear Alfven wave.

Let us now examine these three solutions by writing out equation 6 in component form. Take a coordinate system such that $\mathbf{B}_0$ is in the z direction, and $\mathbf{k}$ is in the x–z plane. Let $\mathbf{k}$ make an angle θ with $\mathbf{B}_0$ as in figure 5.1. It turns out that the three solutions (waves) split into one that has only the y component, $\xi_y \neq 0$, of the displacement, the intermediate wave, and two others that have $\xi_y = 0$ and $\xi_x \neq 0, \xi_z \neq 0$, the fast and slow waves.

5.2 The Intermediate Wave

First, we take up the intermediate wave with $\boldsymbol{\xi}$ in the y direction (Alfven 1942; Alfven and Falthammar 1963; Astron 1950). Then $\boldsymbol{\xi} \times \mathbf{B}_0$ is in the x direction with magnitude $\xi_y B_0 \hat{\mathbf{x}}$. Next, $\mathbf{k} \times (\boldsymbol{\xi} \times \mathbf{B}_0)$ is in the $\hat{\mathbf{y}}$ direction and has a magnitude $k\xi_y B_0 \cos\theta$. The cross product of $\mathbf{k}$ with this vector has a magnitude $k^2 \xi_y B_0 \cos\theta$, is in the x–z plane, and makes an angle θ with the minus $-x$ axis.

The last two quantities give $\mathbf{B}_1$ and the direction of $\mathbf{j}_1$, both of which are perpendicular to $\mathbf{k}$. The last result we need is $\mathbf{j}_1 \times \mathbf{B}$ with magnitude $k^2 \xi_y B_0 \cos\theta$, which is in the y direction. Its magnitude is most easily found by expanding the last term as a triple product, treating $\mathbf{k} \times (\boldsymbol{\xi} \times \mathbf{B}_0)$ as a single vector in the y direction,

$$\frac{1}{4\pi}\left\{\mathbf{k} \times [\mathbf{k} \times (\boldsymbol{\xi} \times \mathbf{B}_0)]\right\} \times \mathbf{B}_0 = \frac{(\mathbf{k} \cdot \mathbf{B}_0)\left[\mathbf{k} \times (\boldsymbol{\xi} \times \mathbf{B}_0)\right]}{4\pi} = \frac{(\mathbf{k} \cdot \mathbf{B}_0)^2}{4\pi}\boldsymbol{\xi} \quad (7)$$

where the second equality comes from expanding $\mathbf{k} \times (\boldsymbol{\xi} \times \mathbf{B}_0)$ as a triple product. (We could have proceeded directly to this result, but we wished to show that $\mathbf{B}_1$ was in the y direction and parallel to $\boldsymbol{\xi}$.)

The pressure term is zero for the intermediate mode because $\mathbf{k} \cdot \boldsymbol{\xi} = 0$. Thus, equation 6 reduces to

$$-\rho\omega^2 \xi_y = -\frac{(\mathbf{k} \cdot \mathbf{B}_0)^2}{4\pi}\xi_y \quad (8)$$

Since $\xi_y \neq 0$, we have

$$\omega^2 = (k\cos\theta)^2 \frac{\mathbf{B}_0^2}{4\pi\rho} = k_z^2 V_A^2 \tag{9}$$

so the phase velocity along the z axis is

$$\frac{\omega}{k_z} = \pm V_A \tag{10}$$

We have found a single equation for the single component ξ_y. We assume that ξ_x and ξ_z are zero, and this produces no more terms in the equation so this assumption legitimate. The wave is polarized in the y direction, perpendicular to both $\mathbf{k}$ and $\mathbf{B}_0$.

Note that the mathematics of the derivation reflects the physical picture of the Alfven wave given in chapter 4. A simple integration of $\mathbf{B} = \mathbf{B}_0 + k_z\xi_y B_0\hat{\mathbf{y}}$ for the line of force shows that it agrees with the result of displacing the line $x = 0$, $y = 0$ by $\boldsymbol{\xi} = \xi_y\hat{\mathbf{y}}$ (see problem 1 in chapter 3). Also, the force $\mathbf{F}_1 = \mathbf{B}_0 k_z \mathbf{B}_1/4\pi$ agrees with the tension argument since the curvature is $-ik_z^2\xi_y\hat{\mathbf{y}}$.

Note also that all the displaced lines lie in the planes x = constant, and the force is also in these planes. Thus, the dynamics of the mode is such that each plane moves independently of the others. We assume that the mode has an x dependence given by the factor $\exp ik_x x$, but this dependence is actually obtained by arranging the displacement of the lines in each plane relative to the others to gain a pure sinusoidal dependence.

This picture can be viewed in a slightly different way by making use of the fact that ω^2 does not depend on k_x. Let us take a general x profile for $\xi_y(x)$ at $t = 0$,

$$\xi_y = f(x) = e^{ik_z z} \int dk_x f_{k_x} e^{ik_x x} \tag{11}$$

Then the solution is obtained by a superposition of different k_x's as

$$\xi_y = \int dk_x f_{k_x} e^{-i\omega t + ik_z z + ik_x x} = f(x) e^{-i\omega t + ik_z z} \tag{12}$$

where we have $\omega = k_z v_A$ (or perhaps $-k_z v_A$), corresponding to a right-propagating wave (or a left one). This is possible since ω does not depend on k_x.

We could have a profile $f_+(x)$ for the right-moving wave and another profile $f_-(x)$ for the left-moving wave. If we now add together the waves for all possible k_z, we obtain the general Dalembertian solution

$$\xi_y = g_+(x, z - v_A t) + g_-(x, z + v_A t) \tag{13}$$

where g_+ and g_- are arbitrary functions. Each wave keeps its shape in x and z as it propagates. It is interesting to show directly that equation 13 satisfies equation 3 (compare with problem 4 in chapter 3).

Finally, we can find the group velocity of the intermediate wave by vectorally differentiating the dispersion relation, say, for the right-moving wave, in the form $\omega = \mathbf{k} \cdot \mathbf{B}_0/\sqrt{4\pi\rho}$, with respect to $\mathbf{k}$,

$$\mathbf{v}_g = \frac{\partial \omega}{\partial \mathbf{k}} = \frac{\mathbf{B}_0}{\sqrt{4\pi\rho}} \tag{14}$$

The group velocity is along $\mathbf{B}_0$ and is a constant. For this reason the x profile does not change. If the wave is localized to a small region in x, the rest of the x space will not be disturbed as time progresses. All this is very simple from the point of view of flux freezing.

5.3 The Fast and Slow Modes

We now turn to the other two solutions of equation 6, the solutions that correspond to the fast and slow modes (Spitzer 1962). Choose $\boldsymbol{\xi}$ to lie in the x–z plane. Consider the last term $\boldsymbol{\xi} \times \mathbf{B}_0$, which lies along the y axis and is perpendicular to $\mathbf{k}$. Then, expanding the triple product (with $\boldsymbol{\xi} \times \mathbf{B}_0$ considered a single vector) $\mathbf{k} \times [\mathbf{k} \times (\boldsymbol{\xi} \times \mathbf{B}_0)] = -k^2(\boldsymbol{\xi} \times \mathbf{B}_0)$, we see that the current is along the y axis and the force $\mathbf{j}_1 \times \mathbf{B}_0$ is in the x direction. Only ξ_x enters, since $\xi_z \hat{\mathbf{z}} \times \mathbf{B}_0 = 0$. Thus,

$$\frac{1}{4\pi}\left\{\mathbf{k} \times [\mathbf{k} \times (\boldsymbol{\xi} \times \mathbf{B}_0)]\right\} \times \mathbf{B}_0 = -k^2 \frac{B_0^2 \xi_x}{4\pi} \hat{\mathbf{x}} \tag{15}$$

Further, we expect no force in the z direction because $\mathbf{F} = \mathbf{j}_1 \times \mathbf{B}_0$. Also, $\mathbf{B}_1$ does not involve ξ_z because $\mathbf{B}_1 = \nabla \times (\boldsymbol{\xi} \times \mathbf{B}_0)$ and the same for $\mathbf{j}_1$. Therefore, the magnetic $\mathbf{F}$ must be in the x direction and independent of ξ_z, which explains the simple form of equation 15.

The pressure term in these coordinates is more complex. It is easily written out using equation 5 and we find

$$-\nabla p_1 = -\gamma p_0 \mathbf{k}\mathbf{k} \cdot \boldsymbol{\xi}$$

$$-\frac{\partial p_1}{\partial x} = -k^2 \gamma p_0 (\sin^2\theta \xi_x + \sin\theta\cos\theta \xi_z)$$

$$-\frac{\partial p_1}{\partial z} = -k^2 \gamma p_0 (\sin\theta\cos\theta \xi_x + \cos^2\theta \xi_z)$$

As a result of these last equations the $\hat{\mathbf{x}}$ and $\hat{\mathbf{z}}$ components of equation 15 are (after dividing by ρ)

$$-\omega^2 \xi_x = -k^2(c_s^2 \sin^2\theta + v_A^2)\xi_x - k^2 c_s^2 \sin\theta\cos\theta \xi_z \tag{16}$$

$$-\omega^2 \xi_z = -k^2 c_s^2 \sin\theta\cos\theta \xi_x - k^2 {c_s}^2 \cos^2\theta \xi_z \tag{17}$$

where $c_s^2 = \gamma p_0/\rho$ and c_s is the speed of sound.

The determinant of these equations, which must be zero, is

$$\begin{vmatrix} k^2 v_A^2 + k^2 c_s^2 \sin^2\theta - \omega^2 & k^2 {c_s}^2 \sin\theta \cos\theta \\ k^2 c_s^2 \sin\theta \cos\theta & k^2 c_s^2 \cos^2\theta - \omega^2 \end{vmatrix} = 0$$

or

$$\omega^4 - k^2 \left(c_s^2 + v_A^2 \right) \omega^2 + k^4 c_s^2 v_A^2 \cos^2\theta = 0 \qquad (18)$$

The solution of this quadratic equation for ω^2/k^2 can be written as

$$\frac{\omega^2}{k^2} = \frac{v_A^2 + c_s^2}{2} \pm \frac{1}{2}\sqrt{(v_A^2 - c_S^2)^2 + 4 c_s^2 v_A^2 \sin^2\theta} \qquad (19)$$

Inspection of this equation reveals that the two roots for ω^2/k^2 are distinct unless $v_A^2 = c_s^2$ and $\sin\theta = 0$ (**k** parallel to $\mathbf{B}_0$, or parallel propagation). The upper root gives the fast mode and the lower root the slow mode.

The intermediate mode, $(\omega/k)^2 = v_A^2 \cos^2\theta$, always lies between the slow and fast modes, which we can see as follows. Consider the left-hand side of equation 18 regarded as a function of ω^2. First, take $cos^2\theta \neq 0$. Then for small ω^2 it is positive. It changes sign at the first root of ω^2 and again at the second where it becomes positive again. Substitution of $\omega^2 = k^2 v_A^2 \cos^2\theta$, the value for the intermediate mode, into the left-hand side makes it negative (or zero if $\cos^2\theta = 1$). Thus, this ω^2 must lie between the two roots, but is equal to one of them if $\cos^2\theta = 1$. If $\cos^2\theta = 0$, then ω^2 vanishes for both the slow mode and the intermediate mode, so the intermediate mode is still in between, justifying its name.

5.3.1 The Nature of the Fast and Slow Modes

To form a picture of the dispersion relation of these three modes let us consider some limiting cases. First, for parallel propagation we have that $(\omega/k)^2 = v_A^2$ for the intermediate mode while the slow and fast modes have $\omega^2/k^2 = v_A^2$ and c_s^2, not necessarily in this order. If c_s^2 is larger than v_A^2, then the sound mode is the fast mode and its phase velocity is c_s^2, while in the reverse case, $v_A^2 > c_s^2$, the fast mode has the phase velocity v_A^2. The modes have separate characteristics: the Alfven mode has only transverse motion and no compression, while the sound mode has no transverse motion and only parallel motion. Because it has no transverse motion, the sound mode does not interact with the field at all. Further, at least in this case, the two motions, i.e., directions of polarization, are perpendicular.

For perpendicular propagation, $\cos\theta = 0$, the slow mode and the intermediate mode have zero phase velocity while the fast mode has phase velocity $\sqrt{c_s^2 + v_A^2}$. Its polarization is perpendicular to B_0 and the force driving the wave is simultaneous compression of both the pressure and the magnetic field leading to the addition (in quadrature) of the sound and Alfven phase velocities. Again, the motion of the slow mode is along B_0 as is seen by setting ω^2 and $\cos\theta = 0$ in equation 16, so again the slow and fast modes have perpendicular polarization vectors $\boldsymbol{\xi}$.

This result, that the two modes are perpendicularly polarized, holds for all θ. This can be seen as follows. Write equations 16 and 17 together as a two-dimensional matrix equation:

$$-\rho\omega^2\boldsymbol{\xi} = M\boldsymbol{\xi} \tag{20}$$

where $\boldsymbol{\xi}$ is a two-component vector with components ξ_x and ξ_y. Note further that the 2×2 matrix M is symmetric in its x and y components. Let $\boldsymbol{\xi}_s$ and $\boldsymbol{\xi}_f$ be the vectors corresponding to the two solutions, and ω_s^2 and ω_f^2 be the corresponding eigenvalues. Then

$$-\rho\omega_s^2\boldsymbol{\xi}_s = M\boldsymbol{\xi}_s \tag{21}$$

$$-\rho\omega_f^2\boldsymbol{\xi}_f = M\boldsymbol{\xi}_f \tag{22}$$

Multiply (21) vectorially on the left by $\boldsymbol{\xi}_f$ and (22) on the left by $\boldsymbol{\xi}_s$ and subtract to get

$$-\rho\left(\omega_s^2 - \omega_f^2\right)\boldsymbol{\xi}_s \cdot \boldsymbol{\xi}_f = \boldsymbol{\xi}_f \cdot M \cdot \boldsymbol{\xi}_s - \boldsymbol{\xi}_s \cdot M\boldsymbol{\xi}_f = 0$$

because of the symmetry of M. This establishes the orthogonality. This orthogonality is a special case of a much more general result that will appear in chapter 7 on the energy principle. Of course, the intermediate mode is polarized in the y direction, which is perpendicular to the x–z plane and therefore to both the fast and the slow polarization directions. All three modes are mutually perpendicular.

Now let $c_s^2 \gg v_A^2$ (and θ arbitrary). In this case the square root in (19) is c_s^2 to lowest order so to lowest order $\omega_f^2 = k^2c_s^2$, $\omega_s^2 = 0$. We can get the correction to ω_f^2 and ω_s^2 by expanding the square root, but it is more easily obtained from the quadratic equation in ω^2, equation 18. For the fast mode divide the equation by ω^2

$$\omega^2 = k^2\left(\mathrm{c}_s^2 + \mathrm{v}_A^2\right) - \frac{k^4c_s^2v_A^2\cos^2\theta}{\omega^2}$$

and substitute the lowest order for ω^2, i.e., $k^2c_s^2$, in the denominator of the last term to get

$$\frac{\omega_f^2}{k^2} = c_s^2 + v_A^2 - v_A^2\cos^2\theta = c_s^2 + v_A^2\sin^2\theta \tag{23}$$

and

$$\frac{\omega_f}{k} \approx c_s + \frac{v_A^2}{2c_s}\sin^2\theta \tag{24}$$

To get the slow mode write the quadratic equation as

$$k^2\left(v_A^2 + c_s^2\right)\omega_s^2 = k^4c_s^2v_A^2\cos^2\theta + \omega_s^4$$

and substitute zero for the ω^4 term, the lowest order value for ω_s^4, to get

$$\frac{\omega_s^2}{k^2} = \frac{c^2 v_A^2 \cos^2\theta}{v_A^2 + c_s^2} \approx v_A^2 \cos^2\theta \tag{25}$$

Note that these results, (24) and (25), agree with the results for the phase velocity for the special cases of perpendicular and parallel propagation $\theta = \pi/2$ and 0.

We can carry out the same operations for the opposite case $c_s \ll v_A$ to obtain

$$\frac{\omega_f^2}{k^2} = v_A^2 + c_s^2 \sin^2\theta$$

$$\frac{\omega_s^2}{k^2} = c_s^2 \cos^2\theta \tag{26}$$

These results are just the reflected images of (24) and (25) with c_s and v_A interchanged. This is because the full dispersion relation is symmetric under this interchange, as are our two limiting conditions ($c_s > v_A$ goes to $v_A > c_s$).

Although the dispersion relations are symmetric under this interchange, for $\theta \neq 0, \pi/2$ the polarization structures are not. For example, if $c_s \gg v_A$, the slow mode must have $\boldsymbol{\xi}$ nearly perpendicular to $\mathbf{k}$; otherwise, the perturbed pressure would lead to a large ω. (This can be checked from equation 17, which gives $\xi_x \sin\theta + \xi_z \cos\theta \approx 0$ since the left-hand side is small.) The fast mode must have $\boldsymbol{\xi}$ nearly parallel to $\mathbf{k}$ since the displacments of the two modes are always perpendicular to each other. We can also see that this must be true since there is no strong force perpendicular to $\mathbf{k}$ to drive perpendicular motions at the large sound frequency. Ignoring the magnetic force, we can easily see that the frequency of the fast mode is nearly kc_s. The frequency of the slow mode is $kv_A \cos\theta$, since the only force along $\boldsymbol{\xi}$ is the projected magnetic force. The projection of the magnetic force along $\mathbf{k}$ is easily balanced by a slight compression of the large pressure due to a very small component of $\boldsymbol{\xi}$ along $\mathbf{k}$.

On the other hand, in the opposite limit, $v_A^2 \gg c_s$, the polarizations of the two modes are different. For the slow mode, ξ_x must be small; otherwise, the strong magnetic field would drive ξ_x at nearly the Alfven frequency. The slow mode is thus polarized parallel to $\mathbf{B}_0$. Since there is no magnetic force in this direction, the mode is driven by the projection of the compressional force along $\mathbf{B}_0$ (a factor of $\cos\theta$) and this compressional force is also reduced by projection of $\boldsymbol{\xi}$ along $\mathbf{k}$ (another factor of $\cos\theta$).

The fast mode must be polarized in the x direction since there is a no large force to drive fast motions in the z direction. The magnetic force consists of two parts: a tension force, reduced from the full magnetic force $k^2 v_A^2 \xi$ by $\cos^2\theta$, and a compressional force, reduced from the full magnetic force $k^2 v_A^2 \xi$ by $\sin^2\theta$, so the total magnetic force is proportional to $k^2 v_A^2 \xi$.

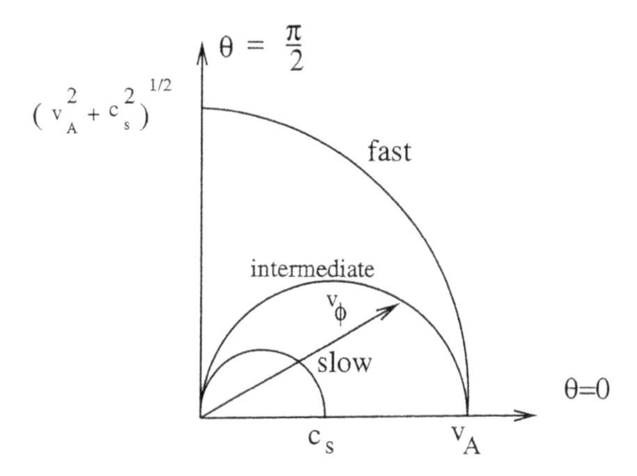

Figure 5.2. Friedricks diagram for MHD waves when c_s is less than v_A

For the intermediate mode only the tension force acts so that ω^2 is decreased from $k^2 v_A^2$ to $k^2 v_A^2 \cos^2\theta$ and the mode is correspondingly slower.

In each of the two limiting cases, the group velocity of the slow mode is along $\mathbf{B}_0$ to the lowest order. That is, when $c_s^2 \gg v_A^2$, $\omega_s = k_z v_A$, and $\mathbf{v}_g = \nabla_k \omega_s = v_A \hat{\mathbf{z}}$. When $v_A^2 \gg c_s^2$, $\omega_s = k_z v_s$ and $\mathbf{v}_g = \nabla_k \omega = c_s \hat{\mathbf{z}}$. In particular, in the second case the magnetic field can be considered so strong that for the slow mode the motions along one line are effectively insulated from those along another and the phase velocity along each line is the sound speed c_s. The cross-field force from any perpendicular variation of the pressure is easily balanced by a small component of ξ_x. Thus, any x profile for the wave packet is allowed just as for the shear mode and at least to lowest order the slow mode allows a Dalembertian solution just as the intermediate mode does.

However, these remarks are valid only to lowest order in $c_s^2/v_A^2 \ll 1$. There is a small component of the group velocity of the slow mode in the x direction of order $-(c_s^3/v_A^2)\cos^3\theta \sin\theta$ that for finite not close to 90 degrees $\theta \approx 1$ spreads any slow mode very localized in a magnetic tube, highly localized in the x direction, in about v_A^2/c_s^2 wave periods. In the opposite limit $v_A \ll c_s$, the x component of the slow mode group velocity is $-(v_A^3/c_s^2)\cos^3\theta \sin\theta$.

I have presented the various details of the mechanism of the MHD modes so that the reader can develop some facility in MHD reasoning. They represent MHD phenomena in the simplest terms.

5.3.2 The Friedricks diagram

The characteristic behavior of the these modes is most simply seen in terms of a polar diagram where $v_\phi = \omega/k$ is plotted against θ. This diagram is called the Friedricks diagram (Friedricks 1957; Kantrovitch and Petschek 1966). For $c_s < v_A$, it is given in figure 5.2.

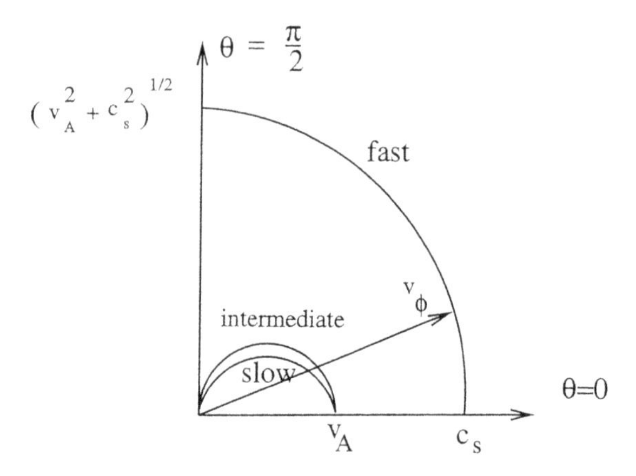

Figure 5.3. Friedricks diagram when c_s is greater than v_A

The various results expressed above are easily seen from this diagram. For parallel propagation ($\theta = 0$) the slow mode has phase velocity c_s and the intermediate and fast modes have phase velocities v_A. For perpendicular propagation the two slower modes have zero phase velocity and the fast-mode phase velocity equals the magnetosonic velocity $\sqrt{v_A^2 + c_s^2}$. The curve for the intermediate velocity $\omega/k = v_A \cos\theta$ is always a circle, and (if $c_s^2 \ll V_A^2$) the slow mode is nearly a circle. The fast mode is a distorted circle slightly larger in the $\theta = \pi/2$ direction.

For $v_A < c_s$, the Friedricks diagram is given in figure 5.3. In this case, the two slower-mode, parallel phase velocities are equal to v_A and the fast mode phase velocity is c_s. In neither diagram do any of the curves ever cross except at $\theta = 0$ or $\pi/2$. The group velocity can be expressed as $\mathbf{v}_\phi + (\partial v_\phi/\partial/\theta)\hat{\boldsymbol{\theta}}$ and it can be seen that it is tilted upward from the radial direction for the fast mode and downward for the slow mode. For the intermediate mode it is exactly along $\mathbf{B}_0$.

One further interesting remark might explain the different structures of the slow and fast mode. Refer back to equation 18. It is seen from the form of this dispersion relation as a quadratic equation that $\omega^2 = k^2 c_s^2 \cos^2\theta$ lies in between ω_s^2 and ω_f^2, so that $\omega_S^2 - k^2 c_s \cos^2\theta$ is negative and $\omega_f^2 - k^2 c_s^2 \cos^2\theta$ is positive. (Simply substitute $\omega^2 - k_s^2 c^2 \cos^2\theta$ in equation 18 and verify that the quadratic expression is negative.) Thus, ξ_x and ξ_z have the same sign for the fast mode and opposite sign for the slow mode. Therefore, ξ_x contributes the same compression to the magnetic field and the pressure for both modes, but ξ_z gives additional compression of pressure for the fast mode but reduces it for the slow mode. This will be important when we discuss shocks.

5.4 The Number of Modes

So far, we have discussed the eigenmodes of the MHD waves. But one of the main uses of eigenmodes is to solve initial value problems. We consider some small deviation away from the uniform equilibrium at some initial time t, say $t = 0$. Let the initial perturbations be $\rho_1(x,t)$, $p_1(x,t)$, $\mathbf{B}_1(x,t)$, and $\mathbf{v}_1(x,t)$. We wish to determine the further evolution of these quantities in time. It is natural to carry out a Fourier analysis of these quantities, e.g.,

$$p_1 = \int \tilde{p}_1(\mathbf{k},t)e^{i\mathbf{k}\cdot\mathbf{r}}d^3\mathbf{k} \tag{27}$$

We find $\tilde{p}_1(\mathbf{k},t)$ etc. at $t = 0$ from our initial perturbations, where $\tilde{p}_1(\mathbf{k},t)$ is complex. Then, for each $\mathbf{k}$ the most general initial condition involves seven complex constants. These are the three components of the velocity $\mathbf{v}_\mathbf{k}$, two components of the magnetic field $\mathbf{B}_1$, and the two scalar perturbations p_1 and ρ_1($\mathbf{k}\cdot\mathbf{B}_1$ must be zero). This corresponds to seven wave amplitudes. (The seven phases are included in the complex amplitudes.) Then in terms of these seven waves we can propagate the perturbations forward in time. However, we have only found six waves so far—two fast waves, two intermediate waves and two slow waves—each going in the $\pm\mathbf{k}$ directions. One wave is missing and this is the so-called entropy wave.

We have assumed that the entropy $S = RT\ln(p/\rho\gamma)$ is the same in the perturbation as in the unperturbed situation, so that $p_1/\mathbf{p}_0 - \gamma\rho_1/\rho_0 = 0$. This couples the initial perturbed pressure to the initial perturbed density. If we wish p_1 and ρ_1 to be independent we must relax this constraint. We then find the entropy wave, which consists of a perturbation in the density alone (all other perturbed quantities being zero) and which has zero frequency,

$$\rho_1 = \hat{\rho}_1 e^{i(\mathbf{k}\cdot\mathbf{r})} \tag{28}$$

It is also a solution of the perturbed MHD equations.

Clearly, this apparently trivial wave involves no forces, since both the pressure and the magnetic field are unperturbed. Thus, this perturbation remains undisturbed. This wave adds one more complex constant (the amplitude and phase) to the six constants of the six standard waves, thus enabling us to solve for the evolution of the most general initial perturbation.

As an example of an initial value problem consider $\mathbf{k}$ parallel to $\mathbf{B}_0$ (parallel propagation) and let only $B_y = B_1\cos kz = \mathrm{Re}\frac{1}{2}(B_1e^{ikz} + B_1e^{-ikz})$ be nonzero initially. Then this initial condition will give rise to two Alfven waves propagating in opposite directions, such that initially their B_y's add but their initial v_y's cancel:

$$v_{y1} = \frac{1}{2}\frac{\omega}{k}\frac{B_1}{B_0}\mathrm{Re}\left(e^{ikz} - e^{-ikz}\right) = 0$$

Closer examination of the counting would suggest a double counting since the waves with $-\mathbf{k}$ and $-\omega$ are really identical to those for $+\mathbf{k}$ and $+\omega$, but because of the reality condition $a_{-\mathbf{k}} = a^*_{\mathbf{k}}$, this is not the case. It is convenient when counting waves to always take $\omega > 0$ in the solutions for the eigenmodes, and the $-\mathbf{k}$ waves independent of the $+\mathbf{k}$ wave, so that all waves then propagate in the $\mathbf{k}$ direction.

5.5 Wave Energy and Momentum

Now that we have enumerated all of the MHD waves let us determine their energy and momentum per unit volume in terms of their amplitudes (Stix 1962). A direct evaluation of the various energies is complicated and a little ambiguous. This is because the energy is clearly second order in the amplitude. For example, we need terms like $(2\mathbf{B}_0 \cdot \mathbf{B}_2/8\pi)$, as well as B_1^2, to evaluate the magnetic energy, and $\mathbf{B}_2$ is a little complicated to define. $\mathbf{B}_2$ has a space-independent part, and an ambiguity arises in second order as to exactly what equilibrium we are perturbing about. $\mathbf{B}_0$ could have an uncontrolled part that might represent a slight shift of the equilibrium we want to perturb about. (This is also true for the pressure.)

For the discussion of wave energy it is best to imagine that the perturbation is set up by applying some external force $\mathbf{F}_{\text{ext}}(\mathbf{t})$ to some specific unperturbed situation and to define the energy of the wave to be simply the work done by this external force. (For example, the external force could be the motion of one of the discrete particles.) To carry this out we shall work in purely real notation. Consider either the fast or slow mode.

Write the eigen equations (16) and (17) together as a vector equation:

$$-\omega_n^2 \rho \boldsymbol{\xi}_n = \mathbf{F}(\boldsymbol{\xi}_n) \tag{29}$$

where ω_n is the eigenfrequency, $\boldsymbol{\xi}_n$ is the eigenmode, and $\mathbf{F}(\boldsymbol{\xi}_n)$ is the force in the eigenmode. Let the spatial part of the external force $\mathbf{F}_{\text{ext}}$ be taken proportional to the displacement eigenvector $\boldsymbol{\xi}_n$. Now start with $\boldsymbol{\xi} = 0$ and introduce this external force $\mathbf{F}_{\text{ext}}(\mathbf{t})$ into equation 29. Then, under the influence of the external force, $\boldsymbol{\xi}$ satisfies

$$\rho \frac{\partial^2 \boldsymbol{\xi}}{\partial t^2} = \mathbf{F}(\boldsymbol{\xi}) + \mathbf{F}_{\text{ext}}(t) \tag{30}$$

Let $\mathbf{F}_{\text{ext}}(t)$ grow exponentially from zero at $t = -\infty$ as

$$\mathbf{F}_{\text{ext}}(t) = \mathbf{F}_0 e^{\gamma t} \cos \mathbf{k} \cdot \mathbf{r}$$

with growth rate γ. Then, since $\mathbf{F}(\boldsymbol{\xi}_n) = -\rho_0 \omega_n^2 \boldsymbol{\xi}_n$, we have

$$\boldsymbol{\xi} = \hat{\boldsymbol{\xi}}_n e^{\gamma t} \cos \mathbf{k} \cdot \mathbf{r},$$

with

$$\rho \gamma^2 \hat{\boldsymbol{\xi}}_n = -\rho \omega_n^2 \hat{\boldsymbol{\xi}}_n + \mathbf{F}_0 \tag{31}$$

and the relation

$$\mathbf{F}_0 = \rho(\gamma^2 + \omega_n^2)\boldsymbol{\xi}_n \tag{32}$$

between $\mathbf{F}_0$ and $\boldsymbol{\xi}_n$. The work per unit volume done by this force from $t' = -\infty$ to $t' = t$ is

$$W = \int_0^t \overline{\frac{\partial \boldsymbol{\xi}}{\partial t} \cdot \mathbf{F}_0 e^{\gamma t'} dt'} = \frac{\rho_0 \tilde{\xi}_n^2}{4} \left(\gamma^2 + \omega_n^2\right)$$

At time t, $\rho_0 \gamma^2 \tilde{\boldsymbol{\xi}}_n^2/4$ is clearly the kinetic energy and $-\rho_0 \omega_n^2 \tilde{\xi}_n^2/4$ is the potential energy per unit volume. The potential energy depends on the amplitude of the wave and is independent of γ. We know this already, since the perturbed pressure and magnetic field also depend only on the displacement, but not on how the displacement was set up.

Now, if the force $\mathbf{F}_{\text{ext}}$ is turned off, two free waves are "launched" propagating in opposite directions, each with an energy equal to one-half of the work W done by the force and with amplitudes $\tilde{\boldsymbol{\xi}}/2$. For simplicity let us take the limit $\gamma \to 0$ so that the energy in each wave is

$$\frac{W}{2} = \frac{\rho_0 \omega_n^2 \tilde{\xi}_n^2}{8} \tag{33}$$

For comparison the kinetic energy of each wave is $K_{\text{wave}} = \rho_0 \omega_n^2 \tilde{\xi}_n^2/16$ so that the energy in either one of the waves satisfies

$$\mathcal{E}_{\text{wave}} = 2 \times K_{\text{wave}} \tag{34}$$

That is to say, for any MHD wave its total energy is just twice its kinetic energy.

Equation 34 is really the simplest expression for the wave energy, since its kinetic energy is always easy to express in terms of its amplitude. (Specifically, its maximum kinetic energy at any point is $\rho_0 \omega_n^2 \boldsymbol{\xi}^2/2$, where $\boldsymbol{\xi}$ is its maximum amplitude. Its mean kinetic energy averaged over space and time is half of this and its total energy is equal to this. Calculating the energy this way will remove any ambiguity between maximum and average kinetic energy.)

We have mentioned the possible ambiguities we might face in a direct calculation of the potential and kinetic energies due to the fact that in setting up the wave we might actually change the parameters of the equilibrium, and since the wave energy might be thought to be defined as the total energy when the wave is present minus the energy when it is absent, we need to know the basic equilibrium from which it is deviating. However, operationally we are really interested in the wave energy defined in terms of the energy exchange between the plasma and the wave source, so the above definition of the work done by the source is the natural one.

We have taken the source as an external force, but it need not be external. In a kinetic picture, the source could be due to individual particles in the plasma itself. The production of wave energy would then be associated with a reduction in the particle energy. For example, if there is a particle beam exciting the wave, its energy could go into wave energy, carried by the rest of the plasma.

In terms of the energy $\mathcal{E}_{\text{wave}}$ and phase velocity v_ϕ of a wave, the momentum of a wave is simply $P_{\text{wave}} = \mathcal{E}_{\text{wave}}/v_\phi$ and it is in the direction of $\mathbf{k}$ (if $\omega > 0$), so that we can write

$$\mathbf{P}_{\text{wave}} = \frac{k}{\omega}\mathcal{E}_{\text{wave}} \tag{35}$$

In fact, we arrive at this unambiguous definition of $\mathbf{P}_{\text{wave}}$ by again applying an external force $\mathbf{F}_{\text{ext}}$ from $t = -\infty$ to $t = 0$ to create a wave in the unperturbed plasma, and calculating the impulse that it imparts to the plasma in setting up this MHD wave. In this case, it is easier to consider a finite but long wave packet made up of waves with nearly the same wave numbers. We let $\mathbf{F}_{\text{ext}}$ depend on space and time as

$$\mathbf{F}_{\text{ext}} = e^{\gamma t}\mathbf{F}_0(x - v_\phi t)\sin(\mathbf{k}_0 \cdot \mathbf{x} - \omega t): \quad t < 0$$

and be directed along $\mathbf{k}$, where $\mathbf{F}_0(\mathbf{x} - v_\phi t)$ is a slowly varying envelope function of $\mathbf{k}_0 \cdot \mathbf{x} - \omega t$. Such a force will set up the appropriate wave packet. (This external force is different from the one we used to calculate the wave energy, which is directed along $\boldsymbol{\xi}_n$.) The momentum delivered to the plasma to create the wave is

$$\mathbf{P} = \int_{-\infty}^{0}\int d^3x\mathbf{F}_{\text{ext}} \tag{36}$$

while the work done and consequently the energy of the wave packet is

$$\mathcal{E} = \int_{\infty}^{0}\int d^3x\mathbf{v}_\phi \cdot \mathbf{F}_{\text{ext}} \tag{37}$$

Thus, the ratio of these equations yields

$$\frac{\mathbf{P}}{\mathcal{E}} = \frac{\mathbf{k}}{v_\phi}$$

If $\mathbf{F}_0$ varies slowly enough, then a wave with $\mathbf{k}$'s concentrated about $\mathbf{k}_0$ and phase velocity ω/k will be excited. The simplest way to view this external disturbance is to consider a collection of wires moving through the plasma. The wires carry currents proportional to $\mathbf{F}_{\text{ext}}$ and interact with the plasma electromagnetically through the current but do not collide with it physically. The currents in the wires start from zero and $t = -\infty$ and terminate at t. If the wires did not travel at the phase velocity v_ϕ, their effect would average out and no wave would be excited.

Again the force need not be external. The principal point is that in defining a momentum for the wave the effective definition is the impulse required to set the wave up, since one important property of the wave is to transfer momentum from one region of the plasma to another and from one class of particles to another.

There is a simple example of the amount of momentum explicitly carried by a wave packet in which it is easy to evaluate this momentum. This is the

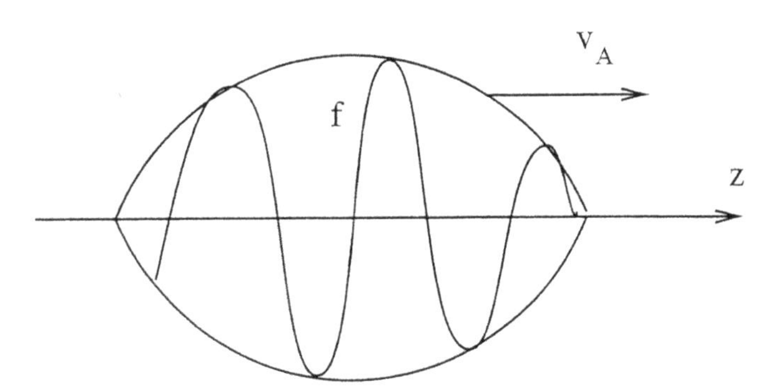

Figure 5.4. A localized wave packet

momentum carried by a parallel propagating Alfven wave packet in a cold, zero-pressure, plasma (see figure 5.4). Let the transverse magnetic field of the Alfven wave be

$$\mathbf{B}_1 = \hat{x} f(z - v_A t) \tag{38}$$

The linearized force in this wave is in the x direction and averages out. The net force in the z direction is quadratic and equal to the negative gradient of $B_x^2/8\pi$ in the wave packet. The z component of the equation of motion is thus

$$\rho \frac{\partial v_z}{\partial t} = -\frac{\partial}{\partial z} \frac{f^2(x - v_A t)}{8\pi} = +\frac{1}{v_A} \frac{\partial}{\partial t} \frac{f^2(z - v_A t)}{8\pi} \tag{39}$$

Let f represent a pulse of finite length propagating to the right. Then for large enough negative t, the velocity of a fluid element that has not been reached by the wave packet is zero, and integration of equation 39 then gives

$$\rho v_z = \frac{f^2}{8\pi v_A} \tag{40}$$

The momentum of the wave per unit area in the wave packet at any time is

$$P = \int \rho v_z dz = \frac{1}{v_A} \int \frac{f^2}{8\pi} dz \tag{41}$$

But

$$\mathcal{E} = \int \left(\frac{B_1^2}{8\pi} + \frac{\rho V_1^2}{2} \right) dz = 2 \int \frac{B_1^2}{8\pi} dz = 2 \int \frac{f^2}{8\pi} dz$$

Unfortunately, this example, which is very easy to calculate, misses the correct answer by a factor of two!

$$P = \frac{1}{2} \frac{\mathcal{E}}{v_A} \tag{42}$$

The explanation for this discrepancy of a factor of $\frac{1}{2}$ in the momentum is that we evaluated the momentum only where the wave packet was at time $t > 0$

after the external force is removed. There is also a second-order motion from the region in which the wave packet was created during $t < 0$.

In creating the wave packet by our force F_{ext} we produced a small amount of motion in the plasma not associated with the wave. This motion has the missing momentum $\mathcal{E}/2v_A$ in it and is left behind the actual wave. In point of fact, according to our prescription for the definition of **P**, we should start with zero amplitude. If we take into account the behavior of B_1 during negative time as $e^{\gamma t} f_1$, where γ is small, the integral of $e^{\gamma t}\partial f/\partial t$ does not cancel, and a wake of moving plasma is left behind the wave. The momentum of this wake is $\mathcal{E}/2v_A$, so the sum of the momentum in this wake and in the wave satisfies equation 35. (The energy of the wake is higher order.)

It is clear that this momentum that is left behind in the cold plasma never goes anywhere. The question is, which is the correct momentum to assign to the wave? Since we are not really interested in the momentum of the wave per se, but only the impulse on the source of the wave, equation 35 without the factor of $\frac{1}{2}$ is the correct operational definition. If at a later time the wave is removed by the action of another external force, the force would receive an impulse given by the explicit expression for the wave but in addition during the absorption another wake of backward moving plasma would be set up, which would have an equal amount of negative momentum, and the forcing agent would receive a total momentum equal to $\mathcal{E}/v_\phi$ without the factor of $\frac{1}{2}$.

5.6 Waves in Nonuniform Media

Up until this point we have treated MHD waves that are perturbations about a uniform background. This leads to the zeroth approximation for waves propagating in a nonuniform, time-dependent plasma of scale length L and timescale T when the wavelength of the waves is very short compared to L and the period of the wave packet is short compared to T. Actually, to get the zeroth approximation we must assume a wave packet of waves whose length is also shorter than L, while, of course, larger than the wavelength. This is analogous to our treatment in chapter 2 of the orbit of a spiraling particle in a nonuniform magnetic field.

At any given time t_a the wave packet is in some region R_a with properties ρ_a, p_a $\mathbf{B}_a$, $\mathbf{V}_a$. At a later time t_b it will be in a different region, R_b, with different properties, ρ_b, p_b, $\mathbf{B}_b$, $\mathbf{V}_b$. Also, it first has wave vectors centered about wave number $\mathbf{k}_a$ and at the later time t_b they are centered about a different wave number $\mathbf{k}_b$. In our zeroth-order treatment we took $\mathbf{V}$ to be zero by Doppler shifting the frame of reference for our calculation. Hence, our previous results involved $\omega' = \omega - \mathbf{k}_a \cdot \mathbf{V}_a$ rather than ω, where ω is the frequency in the lab frame, while ω' is the frequency in the local rest frame of the plasma.

Our first task is twofold: to find the region R_b where the wave is located at t_b, and to find the later value of wave number when the wave reaches this region, $\mathbf{k}_b$. This is accomplished by the wave-normal equations, often referred to as the eikonal equations (Landau and Lifshitz 1975).

In our zero-order treatment we took the phase of the wave as given by $e^{ikx-\omega t}$ in complex notation. Actually, there should be an additional term of ϕ in the exponential to correctly represent the phase. If we set $\mathbf{k}\cdot\mathbf{x}-\omega t+\phi = \Phi/\epsilon$, we would write the wave amplitude as

$$Ae^{i\Phi(\mathbf{x},t)/\epsilon}$$

where

$$\mathrm{k} = \frac{\partial\Phi}{\epsilon\partial\mathbf{x}}, \qquad \omega = \frac{-\partial\Phi}{\epsilon\partial t} \tag{43}$$

The small parameter ϵ represents the fact that the wavelength is short and the wave frequency high. If we are dealing with wave packets, Φ is of interest in a small region of space R_a, at t_a. As the wave propagates to a different region at t_b, Φ will be of interest only in R_b. The factor A in front of the exponential is supposedly nonnegligible only in these regions and, in fact, in a localized strip in space–time.

If we know Φ as a function of $\mathbf{x}$ and t, its $\mathbf{x}$ derivative would give us the desired $\mathbf{k}$, and its t derivative the desired ω. Since Φ arises from the solution of the wave equation, its derivatives must satisfy the dispersion relation

$$\omega = D(\mathbf{k}, \mathbf{x}, t) \tag{44}$$

or

$$-\frac{1}{\epsilon}\frac{\partial\Phi}{\partial t} = D\left(\frac{1}{\epsilon}\frac{\partial\Phi}{\partial\mathbf{x}}, \mathbf{x}, t\right)$$

On the other hand, its mixed second derivatives, $\partial\Phi/\partial\mathbf{x}\partial t$ and $\partial\Phi/\partial t\partial\mathbf{x}$, must be equal. Thus,

$$\frac{\partial^2\Phi/\epsilon}{\partial t\partial\mathbf{x}} = \frac{\partial\mathbf{k}}{\partial t} = \frac{\partial^2\Phi/\epsilon}{\partial\mathbf{x}\partial t} = -\frac{\partial\omega}{\partial\mathbf{x}} \tag{45}$$

But in equation 44 ω depends on $\mathbf{x}$ through both arguments of D, $\mathbf{k}$ and $\mathbf{x}$, so

$$\frac{\partial\omega}{\partial\mathbf{x}} = \frac{\partial D}{\partial\mathbf{k}}\frac{\partial\mathbf{k}}{\partial\mathbf{x}} + \frac{\partial D}{\partial\mathbf{x}} \tag{46}$$

$\partial D/\partial\mathbf{k}$ is $(\partial\omega/\partial\mathbf{k})_x$ at fixed $\mathbf{x}$, and is thus the group velocity of the wave packet. The group velocity is connected with the evolution of the amplitude A of the wave packet, which is nonzero only in a localized region of space. As time evolves this localized region changes and the velocity of this change is the group velocity $\mathbf{v}_g$.

Substituting these results in equation 45, we have for the total derivative of $\mathbf{k}$ following the wave packet:

$$\frac{d\mathbf{k}}{dt} = \frac{\partial\mathbf{k}}{\partial t} + \mathbf{v}_g\cdot\frac{\partial\mathbf{k}}{\partial\mathbf{x}} = -\left(\frac{\partial\omega}{\partial\mathbf{x}}\right)_t + \mathbf{v}_g\cdot\frac{\partial\mathbf{k}}{\partial\mathbf{x}} = -\frac{\partial D}{\partial\mathbf{k}}\cdot\frac{\partial\mathbf{k}}{\partial\mathbf{x}} - \frac{\partial D}{\partial\mathbf{x}} + \mathbf{v}_g\cdot\frac{\partial\mathbf{k}}{\partial\mathbf{x}} \tag{47}$$

or

$$\frac{d\mathbf{k}}{dt} = -\frac{\partial D}{\partial \mathbf{x}} \equiv -\left(\frac{\partial \omega}{\partial \mathbf{x}}\right)_{\mathbf{k}} \tag{48}$$

This equation, known as the wave normal equation, or sometimes as the eikonal equation, gives the rate of change of the wave number of the wave packet (its refraction) following the wave packet. Making use of our result for the group velocity, which gives the rate of change of the position of the wave packet, we have

$$\frac{d\mathbf{x}}{dt} = \left(\frac{\partial \omega}{\partial \mathbf{k}}\right)_{\mathbf{x}} \tag{49}$$

Equations 48 and 49 are the wave normal equations (Landau and Lifshitz 1975).

Finally, we should note that if we regard equation 44 as expressing ω in terms of position $\mathbf{x}$ and wave number $\mathbf{k}$

$$\omega = \omega(\mathbf{k}, \mathbf{x}, t) \tag{50}$$

then $\omega(\mathbf{k}, \mathbf{x}, t)$ becomes the Hamiltonian for the evolution of the $\mathbf{k}$ and $\mathbf{x}$ of a wave packet. The function ω can be determined at any point and time by local analysis, treating the unperturbed system as homogeneous with the values for ρ, p, $\mathbf{B}$, and $\mathbf{V}$ as the parameters the uniform system.

Let us complete this discussion by verifying that the amplitude of the wave packet $\mathbf{A}$ actually moves with the group velocity equal to $(\partial\omega/\partial k)_{x,t}$. We can locally write $\mathbf{A}$ as a Fourier series:

$$\mathbf{A}(\mathbf{x}, t) = \int \mathbf{a}(\Delta\mathbf{k}) \exp\left[-i(\Delta\omega t - \Delta\mathbf{k}\cdot\mathbf{x})\right] d^3\Delta\mathbf{k} \tag{51}$$

where the range of $\Delta\mathbf{k}$ is much smaller than the principal $\mathbf{k}$ of the wave packet $\mathbf{k}_0$. Since the dispersion relation specifies $\omega = \omega(\mathbf{k})$, we can expand it about $\mathbf{k}_0$ to find $\Delta\omega = (\partial\omega/\partial\mathbf{k})_0 \cdot \Delta\mathbf{k}$, where $(\partial\omega/\partial\mathbf{k})_0$ may be considered a constant. Taking $\mathbf{A}(\mathbf{x}, t) = \mathbf{A}(\mathbf{x}) \exp -i(\omega_0 t - \mathbf{k}_0 \cdot \mathbf{x})$, we may substitute in equation 51 to find

$$\mathbf{A}(\mathbf{x}) = \int \mathbf{a}(\Delta\mathbf{k}) \exp\left\{-i[(\partial\omega/\partial\mathbf{k})_0 t - \mathbf{x}]\cdot\Delta\mathbf{k}\right\} d^3\Delta\mathbf{k} \tag{52}$$

from which we see that $\mathbf{A}$ is a function of $\mathbf{x} - (\partial\omega/\partial k)_0 t$ and therefore locally moves with the group velocity $\mathbf{v}_g = (\partial\omega/\partial\mathbf{k})_0$.

$A(x)$, the magnitude of $\mathbf{A}$, is the function that determines the shape of the wave packet, and the faster exponential $e^{-i\omega t - \mathbf{k}_0\cdot\mathbf{x}}$ represents the oscillation in the wave packet. The phase velocity usually differs from the group velocity. If it is bigger, we can picture these oscillations entering the wave packet say from the rear of the wave packet propagating through it, and leaving it through the front, as in figure 5.5.

A closer examination shows that the wave packet actually spreads out with a spread $\Delta x = t\partial^2\omega/\partial k^2(\Delta k)$ in a homogeneous median and a rate

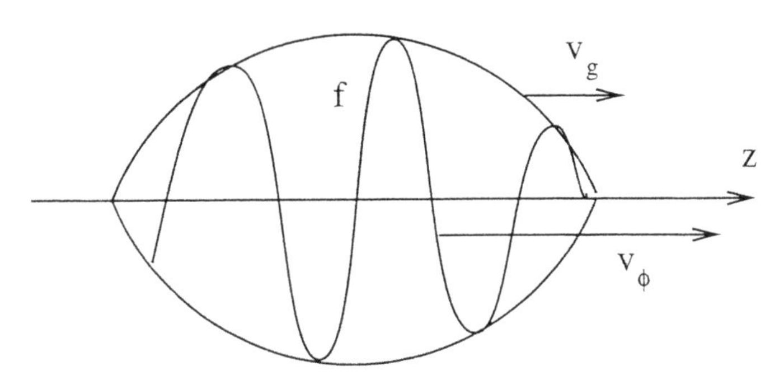

Figure 5.5. Relative motion of phase velocity and group velocity

$\Delta \dot{x} = \ell(\partial v_g/\partial x)t$ in an inhomogeneous median, where ℓ is the length of the wave packet.

Let us consider two simple examples to make the application the wave normal equation 48, and its physical content, clear. For the first example, assume that **B** is uniform in the z direction, ρ varies in the x direction, and **k** is initially in the z direction. Then

$$\omega = \frac{\mathbf{B} \cdot \mathbf{k}}{\sqrt{4\pi\rho}} \tag{53}$$

Differentiating ω at constant **k** we have

$$\frac{d\mathbf{k}}{dt} = -\left(\frac{\partial \omega}{\partial \mathbf{x}}\right)_{\mathbf{k}} = \omega \frac{\partial \rho/\partial x}{2\rho}\hat{\mathbf{x}} \tag{54}$$

If ρ increases with x, this equation says that **k** refracts in the positive $\hat{x}$ direction at the rate $\omega/2L = (v_A/2L)k_z$, where $L = (\ln \rho/\partial x)^{-1}$ is the density scale height. Now, comparing this with a simple physical argument we see that waves at larger x are slower and at smaller x are faster, so we expect the wave to turn upward, because the smaller x part of the wave is passing the larger x part as in figure 5.6. In a time L/v_A the waves at smaller x and a distance L from the larger x waves will have gone one-half wave further in z and the waves will refract finitely.

Second, take the same situation but with uniform ρ and **B** and allow the medium to have a z velocity $V_0(x)$ that varies with x. Then

$$\omega = k_z V_0(x) + k_z v_A$$

and

$$\frac{d\mathbf{k}}{dt} = -\left(\frac{\partial \omega}{\partial \mathbf{x}}\right)_{\mathbf{k}} = -k_z \frac{\partial V_0}{\partial x}\hat{x} \tag{55}$$

Now, if $V_0(x)$ increases with x, this equation says that the waves refract downward, consistent with physical expectations, since the waves at larger x are now moving faster to the right. The rate of change of k is $(k_z V_0/L)\hat{\mathbf{x}}$

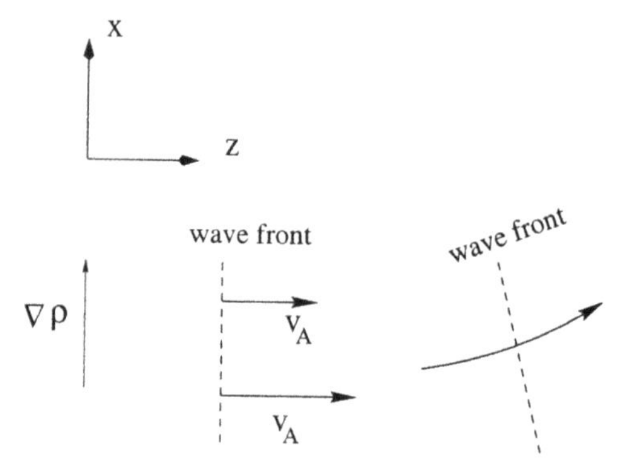

Figure 5.6. A parallel propagating Alfven wave in a density gradient

and the rate of change of the angle θ that the wave vector makes with $\mathbf{B}$ is

$$\dot{\theta} = \frac{\dot{\mathbf{k}}}{k_z} \approx -\frac{V_0}{L} \tag{56}$$

where L is the scale height of V_0. For the first case, of variable density, $\dot{\theta} \approx v_A/2L$.

5.6.1 The Variation in Amplitude

The wave normal equations describe how the centered wave number of the wave packet $\mathbf{k}$ changes and how the region in which the wave packet is nonzero moves. However, to find how the amplitude of the wave packet changes we must make use of the Lagrangian action principle of the last chapter and the concept of wave action. The general ideas behind this approach are due to Whitham (1965) and have been elaborated for MHD waves by Dewar (1970). The basic result is that a quantity called the wave action density, $\partial \mathcal{L}_2/\partial \omega$, satisfies a continuity equation with wave group velocity replacing fluid velocity. ($\mathcal{L}_2$ is the second-order Lagrangian of the wave.) The rate of change of the wave action density in any volume V is equal to the flow of wave action through the surface of this volume with velocity $\mathbf{v}_g$.

To derive the result formally, let us consider a general, nonuniform, time-dependent, unperturbed medium described by the a zero-order Lagrangian, $\mathcal{L}_0$. Then, let there be a small-amplitude wave packet propagating in this medium. The total action for the perturbed system is that of the unperturbed medium plus that of the wave packet,

$$\int d^3x dt \mathcal{L}_{\text{total}} = \int d^3x dt \left(\mathcal{L}_0 + \mathcal{L}_2\right) \tag{57}$$

where $\mathcal{L}_0$ is the Lagrangian density for the unperturbed system and $\mathcal{L}_2$ is the Lagrangian density for the wave. $\mathcal{L}_2$ is proportional to the square of the amplitude of the wave, which we take as described by the displacement vector $\boldsymbol{\xi}$.

The second-order Lagrangian density $\mathcal{L}_2$ is the difference between the kinetic energy of the wave, $\frac{1}{2}\rho(\partial\boldsymbol{\xi}/\partial t)^2$ and the potential energy of the wave, $\frac{1}{2}\rho\omega_n^2\boldsymbol{\xi}^2 - \frac{1}{2}\rho[D(\mathbf{k},\mathbf{x},t)]^2\boldsymbol{\xi}^2$, where $\omega_n = D(k,x,t)$ is the local dispersion relation at the space time position x, t. Take $\boldsymbol{\xi}$ in the form

$$\xi = \mathbf{A}e^{-i\Phi} + \mathbf{A}^* e^{i\Phi} \tag{58}$$

(In this section we set the small parameter ϵ equal to one and simply regard Φ as large.) Then we can write

$$\mathcal{L}_2 = \tfrac{1}{2}\rho\dot{\boldsymbol{\xi}}^2 - \tfrac{1}{2}\rho\left[D(\mathbf{k},\mathbf{x},t)\right]^2\boldsymbol{\xi}^2 \tag{59}$$

or replacing ω by $-\partial\Phi/\partial t$, and $\mathbf{k}$ by $\partial\Phi/\partial\mathbf{x}$, we have on averaging

$$\mathcal{L}_2 = \rho\left(\frac{\partial\Phi}{\partial t}\right)^2|\mathbf{A}|^2 - \rho\left[D\left(\frac{\partial\Phi}{\partial\mathbf{x}},\mathbf{x},t\right)\right]^2|\mathbf{A}|^2 \tag{60}$$

The total action for the MHD system is

$$S = \int(\mathcal{L}_0 + \mathcal{L}_2)d^3x\,dt \tag{61}$$

Varying the action S with respect to A we have

$$\mathbf{A}^*\left(\frac{\partial\Phi}{\partial t}\right)^2 - \mathbf{A}^*D^2 = 0$$

or

$$\left(\frac{\partial\Phi}{\partial t}\right)^2 = D^2\left(\frac{\partial\Phi}{\partial\mathbf{x}},\mathbf{x},t\right) \tag{62}$$

which is the dispersion relation. In fact, with $\partial\Phi/\partial x$ replaced by $\mathbf{x}$ and $\partial\Phi/\partial t$ by ω we have

$$\omega^2 = D^2(\mathbf{k},\mathbf{x},t) \tag{63}$$

as in the previous section.

Now vary the action S with respect to Φ. The Euler–Lagrangian equation becomes

$$\frac{\partial}{\partial t}\left[\frac{\partial\mathcal{L}_2}{\partial(\partial\Phi/\partial t)}\right] + \frac{\partial}{\partial\mathbf{x}}\cdot\left[\frac{\partial\mathcal{L}_s}{\partial(\partial\Phi/\partial\mathbf{x})}\right] = 0 \tag{64}$$

But again replacing $\partial\Phi/\partial\mathbf{x}$ by $\mathbf{k}$ and $\partial\Phi/\partial t$ by $-\omega$ we have $\partial\mathcal{L}_2/\partial\mathbf{k} = -(\partial\mathcal{L}_2/\partial\omega)(\partial\omega/\partial\mathbf{k})$, since $\mathcal{L}_2 = 0$ by equations 60 and 62, and this ties ω to $\mathbf{k}$. But $\partial\omega/\partial\mathbf{k}$ is the group velocity $\mathbf{v}_g$ so equation 64 becomes

$$-\frac{\partial}{\partial t}\left(\frac{\partial\mathcal{L}}{\partial\omega}\right) - \frac{\partial}{\partial\mathbf{x}}\cdot\left(\mathbf{v}_g\frac{\partial\mathcal{L}}{\partial\omega}\right) = 0 \tag{65}$$

Defining the wave action density by

$$N = \frac{\partial\mathcal{L}}{\partial\omega} = \rho\omega|A|^2 = \frac{\mathcal{E}_{\text{wave}}}{\omega} \tag{66}$$

we have

$$\frac{\partial N}{\partial t} + \frac{\partial}{\partial \mathbf{x}} \cdot (\mathbf{v}_g N) = 0 \tag{67}$$

$N = \mathcal{E}_{\text{wave}}/\omega$ is proportional to the density of the number of quantum mechanical "photons" in the wave packet. (We thus see how its amplitude varies as the wave is propagated at the speed of the group velocity.)

(If we examine the derivation, we see that the total situation is expressed in terms of the global function Φ, which is regarded as a coordinate describing the evolution of the wave. The importance of the Lagrangian approach is that it allows the introduction of such a general coordinate that connects the wave at different positions and times.)

Now, if the background medium is time independent, we have, from the wave normal equations (48), (49), and (50),

$$\frac{d\omega}{dt} = \frac{\partial \omega}{\partial \mathbf{x}}\dot{\mathbf{x}} + \frac{\partial \omega}{\partial \mathbf{k}}\dot{\mathbf{k}} = 0 \tag{68}$$

or ω is a constant. Then the energy is conserved, and the wave action equation reduces to

$$\frac{\partial \mathcal{E}}{\partial t} + \nabla \cdot (\mathbf{v}_g \mathcal{E}) = 0 \qquad \text{(time independent case)} \tag{69}$$

As an example of the latter equation consider the vertical propagation of sound waves in the chromosphere (Osterbrock 1961). Take the temperature constant in space. Then $v_g = c_s$ is constant and $\mathcal{E}$ is constant. But

$$\mathcal{E} = \rho \omega^2 \overline{\xi^2} \tag{70}$$

so as ρ decreases as $e^{-z/H}$ (with $H = kT/g$), ξ increases as $e^{+z/2H}$. Eventually, the amplitude is large enough that the waves steepen and shock.

If there is a constant vertical magnetic field and $v_A \gg c_s$, then we have

$$v_A \mathcal{E} = \text{constant} \tag{71}$$

or

$$\frac{B}{\sqrt{\pi \rho}} \rho \omega^2 \xi^2 = \text{constant} \quad \text{or} \quad \xi \sim \rho^{-1/4} \sim e^{z/4H}$$

and the growth in amplitude is slower. Of course, ω is a constant in both of these cases.

5.6.2 Wave Pressure

Consider a time-dependent example of a MHD wave packet in a uniform compressing medium. Let the zero-order field $\mathbf{B}$ be in the z direction and let the compression take place perpendicular to $\mathbf{B}$. Let the wave packet be propagating parallel to $\mathbf{B}$. Then because the dispersion relation is independent of z the wave number $\mathbf{k}$ is constant. Because B is proportional to ρ the Alfven speed will increase as $\rho^{1/2}$ and the frequency will also increase the same way. Thus, because of the conservation of wave action, the energy of the wave packet will increase and extra work must be done to compress the plasma due to the presence of the wave packet.

(If the compression was along **B** we would have to take the velocity of compression into account in the dispersion relation, which would read $\omega - \mathbf{k} \cdot \mathbf{V}(z, t) = B/\sqrt{4\pi\rho}$. Although B would be constant in this case, the energy of the wave packet would still increase (see problem 2). Again, extra work would be needed to compress the plasma.)

From these last examples we see that the presence of waves affects the energy and momentum of the background plasma. The MHD equations need to be supplemented by forces that represent this effect. Dewar (1970) has shown how these can be found from again invoking the action principle. In our derivation of MHD from the action principle in chapter 4, we carried out a virtual displacement and demanded that the corresponding variation of $\int \mathcal{L}_0 d^3x dt$ must vanish. In the presence of waves we must include the variation of the second-order action $\int d^3x dt \mathcal{L}_2$ as will. This introduces an additional term in the equation of motion of the form

$$-\nabla \cdot \mathbf{P}_W \tag{72}$$

that is to say, a wave pressure term must be included in the equation of motion. For MHD waves the wave pressure is a tensor.

Dewar has calculated $\mathbf{P}_W$ for each of these modes—fast, intermediate, and slow—once and for all, in terms of their energy densities. For intermediate waves it is

$$\mathbf{P}_W = \tfrac{1}{2} N(\omega - \mathbf{k} \cdot \mathbf{V})\mathbf{I} \tag{73}$$

For the other waves the pressure tensor is rather complicated to write out and is given in the Dewar (1970) paper.

For the the limiting case $v_A \ll c_s$, the wave pressure for the sound wave is

$$\mathbf{P}_W = N(\omega - \mathbf{k} \cdot \mathbf{V}) \left(\frac{\gamma - 1}{2} \mathbf{I} + \hat{\mathbf{k}}\hat{\mathbf{k}} \right) \tag{74}$$

In the limiting case $c_s \ll v_A$ and for cold plasma, the wave pressure for magnetosonic waves is

$$\mathbf{P}_W = N(\omega - \mathbf{k} \cdot \mathbf{V}) \left(\tfrac{1}{2} \mathbf{I} - \hat{\mathbf{B}}\hat{\mathbf{B}} + \hat{\mathbf{k}}\hat{\mathbf{k}} \right) \tag{75}$$

This reduces to the same expression as the intermediate mode for $\theta = 0$. (If **k** is perpendicular to **B** then the zz component of the wave pressure is negative.) At perpendicular propagation $\theta = \pi/2$ we have

$$\mathbf{P}_W = \tfrac{1}{2} N(\omega - \mathbf{k} \cdot \mathbf{V}) \left(3\hat{\mathbf{k}}\hat{\mathbf{k}} - \hat{\mathbf{B}}\hat{\mathbf{B}} \right) \tag{76}$$

Note that the pressure along B_0 for this last case is negative (a tension force), while the component across B_0 is positive. This can be checked, as in our simple example of compression, by keeping track of the wave energy when N is conserved.

5.7 Problems

1. (Sound wave momentum) Consider a wave packet of one-dimensional sound waves propagating in the z direction in an unmagnetized plasma. Show that an exact linearized solution of the ideal equations is

$$\rho_1 = \epsilon \rho_0 f(z - c_s t)$$
$$v_{1z} = \epsilon c_s f(z - c_s t)$$

for any function f and for $c_s^2 = \gamma p_0/\rho_0$. Assume that the wave packet is localized in space at any give time, i.e., $f(x - c_s t) = 0$ when $|z - c_s t| > L$. Calculate the wave momentum

$$P = \int \rho_1 v_{1z} dz$$

and show that it is equal to the wave energy divided by c_s. (There is a ρv_{2z} term in the wave momentum, which you may disregard. It can be shown that its integral is zero.)

2. (Alfven wave pressure $\mathbf{P}_W$) Consider a collection of polarized Alfven waves propagating parallel to a uniform magnetic field $\mathbf{B}_0$ in the z direction and in a plasma of uniform density ρ_0. Let the waves be confined to a box, $|x| < L/2, |y| < L/2, |z| < L/2$, and let the box contract uniformly in the z direction with velocity $v_z = -(2z/L)v_0$. Show from the wave normal equation and the dispersion relation

$$\omega = k_z \left[\frac{B}{\sqrt{4\pi\rho}} + v_z(z) \right]$$

that $k_z L$ remains constant. From the conservation of wave action show that the total energy of the Alfven waves increases proportional to $1/\sqrt{L}$. Setting the change in energy equal to the $P_W dV$ work done against the wave pressure, show that the zz component of the wave pressure is just $\frac{1}{2}$ times the energy density of the Alfven waves, confirming the expression for the wave pressure given in the text. Repeat for the same parallel propagating waves but allow a contraction in the x direction and show that the xx component of the wave pressure is equal to the zz component.

3. (Support of a protostar by wave pressure) Consider a protostar in a magnetic field $\mathbf{B}_0$ supported transversely by this field against its self-gravity

$$\frac{GM\rho}{R^2} \ll \frac{B_0^2}{8\pi R}$$

Assume that an energy density of Alfven waves is concentrated near the center of the protostar. Estimate what this energy density must be to support the protostar against collapse parallel to the $\mathbf{B}_0$ field.

4. (Ambipolar damping of Alfven waves) Consider a parallel propagating Alfven wave in a partially ionized medium, and let its frequency ω be

large compared to the ion-neutral collision rate ν_{in} and the neutral-ion collision rate ν_{ni}. Show that the equations of motion for the ions and neutrals can be written

$$\rho_i \frac{\partial^2 V_i}{\partial t^2} = -k^2 \frac{B^2}{4\pi} V_i - \rho_i \nu_{in} \frac{\partial (V_i - V_n)}{\partial t}$$

$$\rho_n \frac{\partial V_n}{\partial t} = -\rho_n \nu_{ni} (V_n - V_i)$$

where the subscripts i and n denote plasma and neutral quantities, and v_A^2 is the Alfven speed involving only the plasma density. Derive the dispersion relation for ω in the high-frequency limit $\omega \gg \nu$, and show that the damping rate of the wave is approximately $\nu_{in}/2$. Remembering that only half of the energy of an Alfven wave is kinetic, can you give a physical interpretation for this rate of damping (Kulsrud and Pearce 1969)?

5. (A positron-electron plasma) What is the propagation speed of an Alfven wave in a pure positron–electron plasma and a uniform magnetic field $\mathbf{B}_0$? Suppose that there is a standing Alfven wave and that the positron–electron density is suddenly doubled at a time when the velocity is passing through zero. What happens to the wave? What is its new frequency and amplitude? Show that the energy of the new wave is smaller by a factor of four. Does this satisfy conservation of wave number N? Why or why not?

References

Alfven, H. 1942. *Arkiv Mat. Astron. Fysik* **29B** (2).

Alfven, H., and C.-G. Falthammar. 1963. *Cosmical Electrodynamics Fundamental Principles*, p. 109, Oxford Clarendon Press, London.

Astron, E. O. 1950. *Arkiv Fysik* **2**, 443.

Dewar, R. L. 1970. *Physics of Fluids* **13**, 2710.

Friedricks, K. O. 1957. *Wave Motion in Magnetohydrodynamics*, Los Alamos Report, LAMS-2105.

Kantrovitch, A., and H. E. Petschek. 1966. In *Plasma Physics in Theory and Application*, p. 158, ed. by W. B. Kunkel, McGraw-Hill, New York.

Kulsrud, R. M., and W. P. Pearce. 1969. *Astrophysical Journal* **156**, 445.

Landau, L. D., and E. M. Lifshitz. 1975. *Classical Theory of Fields*, 4th ed., p. 130, trans. M. Hamermesh, Pergamen Press, Oxford, UK.

Osterbrock, D. 1961. *Astrophysical Journal* **134**, 347.

Spitzer, L., Jr. 1962. *Physics of Fully Ionized Gases*, 2nd ed., p. 61, Interscience, Wiley, New York.

Stix, T. H. 1962. *The Theory of Plasma Waves*, p. 45, McGraw-Hill, New York.

Whitham, G. B. 1965. *Journal of Fluid Mechanics* **22**, 273.

Chapter 6

NONLINEAR STEEPENING AND SHOCKS

6.1 Nonlinear Steepening

In the small-amplitude limit the profile of an MHD wave does not change as it propagates, at least for short times. But even a small-amplitude wave will eventually distort due to "steepening." Consider a pure sound wave propagating in a uniform medium with zero magnetic field. In this wave there is a longitudinal perturbed velocity, δV. On the other hand, we have shown that waves in a medium moving with velocity $\mathbf{V}_0$ have phase velocities that include $\mathbf{V}_0$ as an additional term. Should the perturbed longitudinal velocity be included in the phase velocity? Not directly, since the perturbed velocity varies sinusoidally over the wave. But it does have the effect of increasing the local phase velocity where $\delta V > 0$, and decreasing it where $\delta V < 0$ (see figure 6.1). As a result, one part of the wave catches up with the other part, leading to a steepening of the profile.

In a time t such that $\delta V t \approx \lambda$, the wave will be finitely distorted, and the gradient at the point where δV vanishes will approach infinity. This will result in a shock. This effect is easily demonstrated analytically in the one-dimensional nonlinear case in terms of the idea of a "simple wave," which is defined as follows.

Let p/ρ^γ be a constant in the perturbed situation. Then the nonlinear one-dimensional equations are

$$\frac{\partial \rho}{\partial t} + \frac{\partial}{\partial x}(\rho V) = 0$$

$$\rho\left(\frac{\partial V}{\partial t} + V\frac{\partial V}{\partial x}\right) = -\frac{\partial}{\partial x}\left(\frac{p_0}{\rho_0^\gamma}\rho^\gamma\right) \tag{1}$$

where all quantities depend only on x and t, and p_0 and ρ_0 are constants representing the pressure and the density in some reference equilibrium.

A "simple" solution of these equations is defined as one in which a function $\theta(x,t)$ exists such that ρ and V are functions only of θ (Courant and Friedricks 1948).

Let us assume that such a function, $\theta(x,t)$, exists. Then for $\rho[\theta(x,t)]$ and $V[\theta(x,t)]$ to satisfy equation 1 we must have

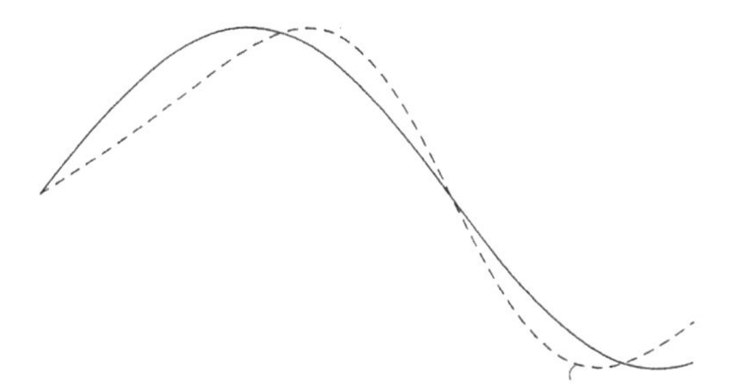

Figure 6.1. Non linear steepening of a wave

$$\left(\frac{d\rho}{d\theta}\right)\frac{\partial\theta}{\partial t}+\left(\rho\frac{dV}{d\theta}+V\frac{d\rho}{d\theta}\right)\frac{\partial\theta}{\partial x}=0$$

$$\left(\rho\frac{dV}{d\theta}\right)\frac{d\theta}{dt}+\left(\rho V\frac{dV}{d\theta}+c_s^2\frac{d\rho}{d\theta}\right)\frac{\partial\theta}{\partial x}=0 \tag{2}$$

where we use $dp/d\theta=(p_0/\rho_0^\gamma)d\rho^\gamma/d\theta=(\gamma p/\rho)d\rho/d\theta=c_s^2 d\rho/d\theta$, where c_s is the space-dependent speed of sound, $c_s^2=\gamma p(x,t)/\rho(x,t)$.

Now this implies that the determinant of the coefficients of $\partial\theta/\partial t$ and $\partial\theta/\partial x$ must vanish. The vanishing of this determinant reduces to

$$\frac{c_s^2}{\rho^2}\left(\frac{d\rho}{d\theta}\right)^2=\left(\frac{dV}{d\theta}\right)^2 \tag{3}$$

Let us assume that the square roots are positive so that this becomes

$$\frac{c_s}{\rho}\frac{d\rho}{d\theta}=\frac{dV}{d\theta} \tag{4}$$

Then, substituting this into equation 2, we get that both equations reduce to

$$\frac{\partial\theta}{\partial t}=-(V+c_s)\frac{\partial\theta}{\partial x} \tag{5}$$

Now, taking into account that V and c_s are functions of θ alone, we see that θ, V and c_s must be constant along the *straight* lines in the x,t plane:

$$\frac{dx}{dt}=c_s+V \tag{6}$$

the lines along which a right-moving, small-amplitude sound wave propagates. More explicitly, if c_s and V are known at $t=0$, say $c_s=c_s^0(x)$, $V=V^0(x)$, then we draw straight lines in the x,t plane through x_0 with slopes $c_s^0(x_0)+V^0(x_0)$, and a necessary condition for our simple wave is the constancy of θ, p, ρ, and V along these lines. If the opposite sign is chosen in equation 4, then we would obtain the left-moving simple wave and the relevant straight lines would have slopes $dx/dt=V-c_s$, instead of equation 6.

Are the functions thus obtained necessarily solutions of the fluid equations 1? The answer is no, since V and ρ must also satisfy equation 4, which says that

$$\frac{2c_s}{\gamma - 1} - V \tag{7}$$

must be everywhere constant, since this quantity is a function of θ alone and also constant in θ. Therefore, equation 7 must be constant at $t = 0$. However, if it is so constant, then the above procedure of taking straight lines through x_0 with slopes $c_s^0(x_0) + V^0(x_0)$ and with ρ and V constant along these lines does yield a solution of equation 1.

In fact, we may choose θ anyway we please at t_0, and extend its definition for $t > 0$ by making it constant along the same lines. Then ρ and V are functions of this θ alone, and, obviously, make equation 7 constant everywhere, and also satisfy equations 5 and 6. This, in turn, guarantees that equations 1 and 2 are satisfied.

We can find the most general right-moving simple wave by taking V to be any function of x, picking ρ to make equation 7 constant, and carrying out the above construction. (Note that θ has dropped out of our construction.) If we choose ρ to make $2c_s/(\gamma - 1) + V$ to be constant, instead, and take the straight lines with slope $-c_s + V$, we get the most general left-moving simple wave.

These simple waves solutions are the natural extension of the linearized Dalembertian solutions of chapter 5. They are exact solutions of the nonlinear fluid equations.

Now it is easy to see that eventually neighboring straight lines with neighboring values of x^0 will cross. As an example, let us take the initial conditions for a simple wave, c_s^0 and V^0, to be those of a linearized right-moving wave. Let

$$c_s^0 + V^0 = \epsilon \sin kx^0 \tag{8}$$

Then our straight lines converge around one part of the wave and diverge around the other as in figure 6.2. We can see the steepening of the profile of the perturbed quantities discussed in the opening paragraph. Now differentiate x (obtained from equation 6) with respect to x^0 at some later time t,

$$\frac{\partial x}{\partial x^0} = 1 + k\epsilon t \cos kx^0 \tag{9}$$

or

$$\frac{\partial x^0}{\partial x} = \frac{1}{1 + k\epsilon t \cos kx^0} \tag{10}$$

and the latter becomes infinite when $kx^0 = \pi, k\epsilon t = 1$. The line emerging from this point is the first line on which this happens.

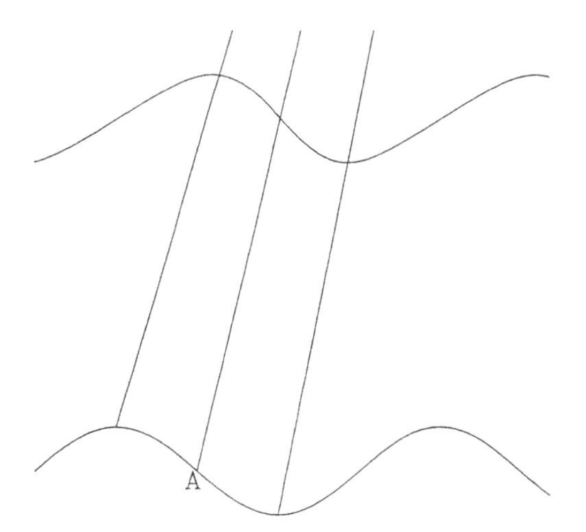

Figure 6.2. Constant θ lines for a simple wave

Let $V = \hat{V} \sin kx^0$. The derivative of V with respect to x at t is then

$$\frac{\partial V}{\partial x} = \frac{\partial V}{\partial x_0}\frac{\partial x^0}{\partial x} = \frac{k\hat{V}^0 \cos kx^0}{1 + k\epsilon t \cos kx^0} \to -\infty \tag{11}$$

for $\cos kx^0 = -1$, as $k\epsilon t \to 1$.

This leads to a breakdown in the fluid picture, for as the derivative of V becomes very large the dissipative terms become important. When the V derivative becomes so large, the steepening is balanced by the velocity diffusion introduced by the dissipative terms. This happens when the scale of variation of the velocity is comparable to the mean free path. This is actually the condition for shock formation. Thus, all undisturbed propagating waves eventually lead to shock formation.

The straight lines that underlie the simple wave solution are called characteristics. Strictly, the definition of characteristics is that they are the curves along which smaller perturbations of the wave propagate. There are generally two sets of characteristics. One class of perturbations propagates along one set and the other along the other set. Perturbations propagating to the right always satisfy $2c_S/(\gamma - 1) - V =$ constant, and these are propagated along the curves $dx/dt = c_S + V$, the local sound speed in the laboratory frame. These are the perturbations of interest for our right-moving simple wave.

We can view the simple wave in a clearer and more physical way, which lends itself to a natural generalization to MHD simple waves (Kantrovitch and Petschek 1966). Suppose that our simple wave is not a smooth function but a function that is defined by a series of steps in V and c_s, across which $2c_s/(\gamma - 1) - V$ is constant, and that separates small regions in space and time in which V and c_s are constant. Now, we know that linearized propagating solutions allow any profile, and, in particular, a step function is allowed.

After each constant region, the step profile will propagate purely to the right at the speed $V + c_s$, of the small uniform region and this step keeps $2c_s/(\gamma - 1) + V$ constant. Thus, each of the steps propagates strictly to the right with the speed of $c_s + V$ created by the previous steps. The jumps of c_s and V across the step remain constant so that c_s and V are constant in the regions between the steps. Thus, after merging these steps together to make a continuous function, we arrive at the properties of our simple wave solution.

Now, MHD waves also propagate without dispersion, so we can also construct simple waves for them. This is more complex than the simple sound waves, because there are six MHD modes and their propagation properties vary with the angle ϕ that $\mathbf{k}$ makes with the background field $\mathbf{B}_0$. We will sketch how to construct a simple wave out of small steps, as in the last paragraph.

Let us start with a uniform magnetic field in the z direction, into which the simple wave will propagate. Let us chose a direction in the x–z plane making an angle ϕ with respect to the z axis. Let us choose one of the three modes to be our simple wave solution. For definiteness let it be the fast mode.

Then for the six independent variables, $\rho, \mathbf{V}(3), \mathbf{B}(2)$, we can find a set of small step function perturbations, $\delta\rho, \delta\mathbf{V}, \delta\mathbf{B}$, that correspond to a small-amplitude, right-moving fast wave, propagating in the ϕ direction. To satisfy this condition we chose an arbitrary value for $\delta\rho$, and then the other five quantities, $\delta\mathbf{V}, \delta\mathbf{B}$, are fixed. Next, in the small uniform region that is moving just behind the step, we can choose another set of step functions that give a second small-amplitude fast mode that propagates parallel to the first but makes a different angle ϕ' with the magnetic field $\mathbf{B}$. This entails slightly different conditions on the steps in the six quantities. (The presence of velocities makes complications with the actual calculations but not with the principle of the algorithm.)

Proceeding in this way we see that again by merging our series of steps we can arrive at a simple one-dimensional MHD wave (see Andersen 1963; Shercliff 1960; Kulikovskii and Lyubimov 1965). Since all quantities, $\rho, \mathbf{V}, \mathbf{B}$, are sums of previous steps, they are all constant in the little uniform regions. Thus, the direction and speed of propagation of the steps are constants and the propagation of each in space–time is again a straight line. In fact, this line represents one of the six MHD characteristics, the one for the right-moving fast wave.

As we have remarked, at each step the jumps in the quantities must satisfy 5 relations so that each step is a single mode propagating in the right direction (forward or backward). Integrating these conditions over all the steps, we arrive at 5 nonlinear relations between the quantities that must be satisfied to get a simple wave. If we label each of the steps by some quantity θ, then all quantities are functions of θ alone, just as for the unmagnetized sound wave. The relations are different for each mode and depend on ϕ. The corresponding relation for a right-moving sound mode is that equation 7 is a constant and this condition is nonlinear in ρ.

If these five MHD relations are all satisfied initially, then we do get a simple wave of one type. In general, the MHD simple wave will steepen and shock just as the sound wave does. We can usually estimate the distance to where the shock forms from the equation $\delta V_{\|} t = \lambda$, where $\delta V_{\|}$ is the variation of the component of velocity along the direction of propagation. If the wave is a small-amplitude, parallel-propagating Alfven wave, then this component will be second order in the wave amplitude. In this case the distance to shocking will be much larger than for the other modes but shocking will still take place.

The above considerations are for an ideal fluid. The waves will also damp due to nonideal effects, such as viscosity and thermal conductivity, but these effects are roughly independent of amplitude, while, except for parallel propagation, the shocking rate is proportional to amplitude. Thus, very small-amplitude waves will damp viscously, while larger amplitudes waves will shock. For the parallel propagating case and also for the intermediate mode, the damping rate is proportional to the wave amplitude, while shocking is proportional to the amplitude squared. So, again, even for this case, the larger amplitude waves end up as shocks while the smaller amplitudes damp by viscous processes.

We have discussed steepening and shocking for simple waves, which are seen to be quite special. Namely, $2c_s/(\gamma - 1) - V$ must be a constant initially and for all time. Do other waves also steepen and shock, i.e., those that are right propagating, but that do not satisfy this relation exactly? This question is answered by writing the equations of motion in terms of Riemann invariants (Lamb 1945; Landau and Lifshitz 1959; Courant and Friedricks 1948).

It can be directly verified from the fluid equations that the nonlinear equations

$$\left[\frac{\partial}{\partial t} + (V + c_s)\frac{\partial}{\partial x}\right] J_+ = 0$$

$$\left[\frac{\partial}{\partial t} + (V - c_s)\frac{\partial}{\partial x}\right] J_- = 0 \tag{12}$$

are valid where

$$J_+ = \frac{2}{\gamma - 1} c_s + V$$

$$J_- = -\frac{2}{\gamma - 1} c_s + V \tag{13}$$

Conversely, the fluid equations (1) follow from these equations, so they are completely equivalent to the fluid equations.

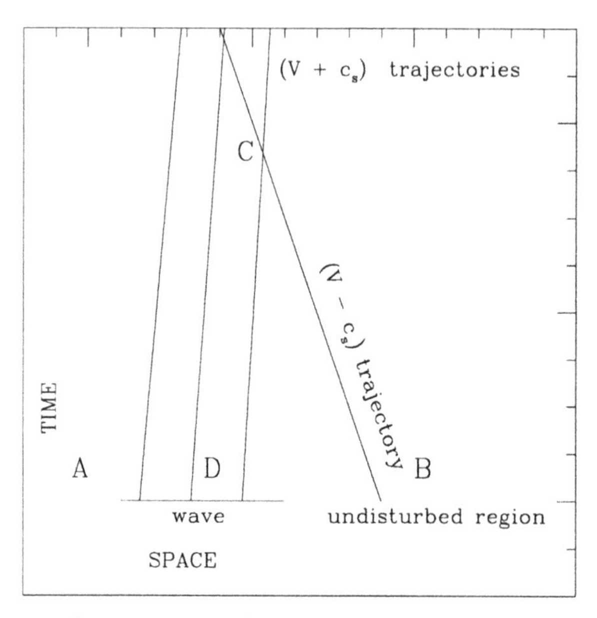

Figure 6.3. Lines of constant J_+ and constant J_- for a wave packet propagating into an undisturbed region

We can obtain the simple wave solution from them. Take J_- constant in space initially. Then we can see that J_- is constant for all time and space by equation 12. Now J_+ is constant along the curves $dx/dt = c_s + V$ and so is J_-, so that c_s and V are constants along these curves. Thus, these curves are straight lines, and we have recovered the simple right-moving solution.

Next, consider a wave packet localized initially to a region D, $x_1 < x < x_2$, and with J_- roughly constant (see figure 6.3). Let the fluid quantities in the regions, A, $x < x_1$, and B, $x > x_2$, be uniform and stationary so that $V = 0$, $c_s = c_s^0$ with c_s^0 a constant.

We expect that the initial disturbance in region D will generate a right-moving and a left-moving sound wave. The right-moving wave will lie between the right characteristics emerging from the ends of region D. Now, $J_- = 2c_s^0/(\gamma - 1)$ is a constant along the left-propagating characteristics $dx/dt = -c_s + V$ coming from region B. These characteristics cross the right-propagating characteristics emerging from the region $x_1 < x < x_2$, $t = 0$, which represent the right-moving disturbance. Let the left characteristic from $x = x_2$ cross the right-moving characteristics of the wave packet in a space–time region **C**.

For times later than **C**, $J_- = 2c_s/(\gamma - 1)$ is a constant. Therefore, the solution emerging from **C** as an initial condition is a simple wave. Thus, after a time $t \approx (x_2 - x_1)/c_s^0$ this wave will then shock with a shocking time of $\lambda/\delta V$, where δV is the variation of V along **C**. It is somewhat less than the variation of V at time $t = 0$.

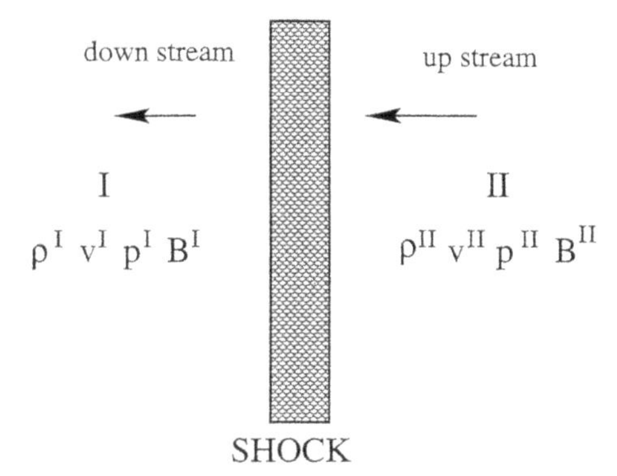

Figure 6.4. Upstream and downstream quantities about a shock

6.2 Shocks

In the previous section, we showed that a propagating wave solution of the ideal fluid equations leads to infinite gradients in a finite time, and beyond this time no solution of the ideal equations exist. This loss of a solution is not surprising when we remember that the ideal equations are really an approximation, valid only when scales of variation of all quantities are large compared to the mean free path. As the gradient increases the fluid equations are no longer valid, and it is not possible to use them to follow the fluid quantities passed the infinite gradient catastrophe inferred from the ideal equations.

However, the breakdown in the ideal equations occurs only in a very thin region, and the fluid equations remain valid everywhere else. It is only in this very thin region that it is difficult to describe the plasma in detail. The region is usually termed a *shock*. However, even though we have no simple means to handle the plasma in this region, we know this region cannot interfere with the general conservation laws for mass, momentum, energy, and magnetic flux, which must be true no matter what happens to the plasma in this region. This is because the plasma continues to be made of innumerable numbers of particles for which we know such conservation laws hold.

In figure 6.4, consider a narrow region or shock separating two regions I and II in which the fluid description holds. Let region II be the region into which the shock is propagating. For simplicity, we first consider the shock in a frame in which the shock is at rest.

In general, this region is so narrow that when considering the shock properties we can consider the region on either side to be uniform. We denote the density, velocity, pressure, and magnetic field in region I by ρ^{I}, $\mathbf{V}^{\mathrm{I}}$, p^{I}, $\mathbf{B}^{\mathrm{I}}$, respectively, and similarly for region II. Let us first apply mass conservation to regions I and II. The rate of flow of mass *out* of region II is $-\rho^{\mathrm{II}}\mathbf{V}^{\mathrm{II}} \cdot \mathbf{n}$ per unit area, where $\mathbf{n}$ is a unit vector perpendicular to the shock pointing from region I to region II. The flow *into* region I is $-\rho^{\mathrm{I}}\mathbf{V}^{\mathrm{I}}\cdot\mathbf{n}$ per unit area. Since the mass inside the shock is negligible and does not change, and since there is negligible flow along the shock, these two rates must be equal:

$$\rho^{\mathrm{I}}\mathbf{V}^{\mathrm{I}}\cdot\mathbf{n} = \rho^{\mathrm{II}}\mathbf{V}^{\mathrm{II}}\cdot\mathbf{n} \tag{14}$$

or more compactly

$$\langle \rho\mathbf{V}\cdot\mathbf{n}\rangle = 0 \tag{15}$$

where here the brackets around any quantity ϕ denote the jump of the quantity across the shock from region II to region I, $\langle\phi\rangle = \phi_{\mathrm{I}} - \phi_{\mathrm{II}}$.

From the conservation of momentum we get a similar relation. The rates of flow of momentum are given by the surface term in equation 20 in chapter 4. Equating these rates gives

$$\left\langle \rho\mathbf{V}\mathbf{V}\cdot\mathbf{n} + \left(\frac{p+B^2}{8\pi}\right)\mathbf{n} - \mathbf{B}\frac{\mathbf{B}\cdot\mathbf{n}}{4\pi}\right\rangle = 0 \tag{16}$$

Similarly, from energy conservation we get the rates of energy flow from the surface terms in equation 42 in chapter 4. Equating the two flows we must have

$$\left\langle \rho\mathbf{V}\frac{V^2}{2} + \frac{\gamma}{\gamma-1}p\mathbf{V}\cdot\mathbf{n} - \frac{(\mathbf{V}\times\mathbf{B})\times\mathbf{B}}{4\pi}\cdot\mathbf{n}\right\rangle = 0 \tag{17}$$

Finally, any magnetic flux leaving one region must end up in the other region. This conservation relation is most easily expressed by the equality of the tangential part of $\mathbf{E}\times\mathbf{n}$ on both sides. Since $\mathbf{E} = -\mathbf{V}\times\mathbf{B}/c$, this gives the relation

$$\langle(\mathbf{V}\times\mathbf{B})\times\mathbf{n}\rangle = 0 \tag{18}$$

There is one further relation we can derive that comes from the $\nabla\cdot\mathbf{B} = 0$ equation:

$$\langle\mathbf{B}\cdot\mathbf{n}\rangle = 0 \tag{19}$$

We now have eight relations between eight quantities ρ, $\mathbf{V}$, p, and $\mathbf{B}$ in region I (downstream) and the corresponding eight quantities in region II (upstream). (Equation 18 involves only the tangential part of $\mathbf{V}\times\mathbf{B}$ and so it gives just two relations.) If we know the eight upstream quantities in region II, we can use the eight relations to determine the eight downstream quantities. However, because we have written the quantities in the shock frame, the upstream velocity $\mathbf{n}\cdot V^{\mathrm{II}}$ (in this frame) involves the shock velocity normal to itself, V_S, which is not known a priori. The shock speed represents one more free parameter, so we need one further piece of information to determine it. This comes from whatever physics, in the downstream region, is responsible for the shock in the first place. For example, this could be

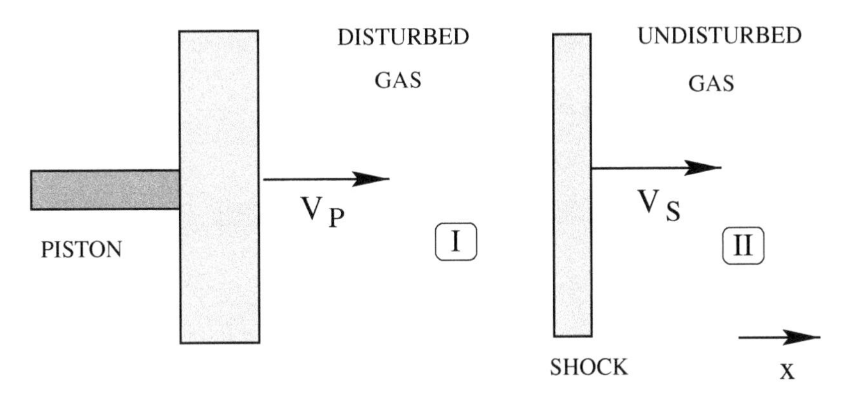

Figure 6.5. The piston problem

a piston moving into an undisturbed region at supersonic or superalfvenic speed. The velocity of the piston will then give the extra information.

If the shock arises from steepening of a nonlinear wave, then the behavior of the downstream characteristics provides the information. For simplicity, we will restrict ourselves to the fast piston as the source of the shock.

To see how all this works let us first consider a simple hydrodynamic, one-dimensional shock. Assume that we start with a uniform plasma at rest in which $\mathbf{B} = 0$. Let there be a piston whose face is perpendicular to the x axis and that moves into the undisturbed plasma with a positive velocity V_P (see figure 6.5). As the piston moves into the plasma it will drive a planar shock that is also perpendicular to the x axis and that has velocity V_S, which we want to determine. A moment after the piston starts, the plasma will be divided into two regions, I and II. Region II is the undisturbed upstream region and lies to the right of disturbed downstream region I, in contact with the piston. The two regions are separated by the shock. Both regions are uniform but have different values of $\rho, \mathbf{V} = V_x \hat{\mathbf{x}}$, and p. In region II, $\rho^{\mathrm{II}} = \rho_0$ and $p^{\mathrm{II}} = p_0$ are known, and $V_x^{\mathrm{II}} = 0$. In region I, $\rho^{\mathrm{I}} = \rho$ and $p^{\mathrm{I}} = p$ are unknown, but $V_x^{\mathrm{I}} = V_P$.

To find, ρ, p, and V_S we apply the jump conditions with $\mathbf{B} = 0$. (V_y and $V_z = 0$ by symmetry.) We must first transform to the moving shock frame and denote velocities in this frame by $U = V_P - V_S$ and $U_0 = -V_S$. Then the jump conditions become

$$\rho U = \rho_0 U_0 \tag{20}$$

$$\rho U^2 + p = \rho_0 U_0^2 + p_0 \tag{21}$$

$$\tfrac{1}{2}\rho U^3 + \tfrac{5}{2} p U = \tfrac{1}{2}\rho_0 U_0^3 + \tfrac{5}{2} p_0 U_0 \tag{22}$$

where we have taken $\gamma = \frac{5}{3}$ for simplicity. (Upstream quantities have the subscript 0 and downstream quantities are unsubscripted.) These equations are the Rankine–Hugoniot conditions for a hydrodynamic shock (Courant and Friedricks (1948).

We can eliminate ρ from equation 20 to obtain

$$p = p_0 + \rho_0 U_0 (U_0 - U) \tag{23}$$

Using equations 20 and 23 to eliminate ρ and p from equation 22, and multiplying it by two we have

$$\rho_0 U_0 (U^2 - U_0^2) + 5p_0(U - U_0) + 5\rho_0 U U_0 (U_0 - U) = 0 \tag{24}$$

This equation is quadratic in U. It obviously has the solution $U = U_0$, since the jump conditions apply to to a surface of continuity as well. This root $U = U_0$ can be excluded since it corresponds to $V_p = U - U_0 = 0$. Canceling the factor $U - U_0$ we have

$$4U = U_0 + \frac{3c_s^2}{U_0} \tag{25}$$

where $c_s^2 = 5p_0/3\rho_0$.

Substituting $U = V_P - V_S$ and $U_0 = -V_S$ in the second solution, equation 25 gives

$$4V_P = 3\left(V_S - \frac{c_s^2}{V_S}\right) \tag{26}$$

a quadratic equation for V_S. Regarding it as an equation for the piston velocity V_P in terms of V_S, we see that it is a monotonic function of V_s. When V_P is small, then V_S is close to the speed of sound c_s, and when V_P is large compared to the speed of sound, then

$$V_S \approx \tfrac{4}{3} V_P \tag{27}$$

If $p_0 = 0$ the latter is always the case. (The solution for general γ can be obtained by replacing the 4/3 by $(3\gamma - 1)/(3\gamma - 2)$.) If V_P is small compared to the speed of sound, the shock wave is weak and is just a linearized sound wave. The jump in entropy p/ρ^γ is third order in V_P and the shock speed minus the sound speed is linear in V_P. On the other hand, if V_p is large compared to the sound speed, the shock is strong. We easily find the downstream density, $\rho = 4\rho_0$. Further, the pressure behind the shock is

$$p \approx \tfrac{3}{4}\rho_0 V_S^2 \gg p_0 \tag{28}$$

The kinetic energy per unit volume is $\left(\frac{1}{2}\right)\rho V_P^2 = \left(\frac{9}{8}\right)\rho_0 V_S^2$. The increase in the downstream energy per unit time per unit area is

$$\left(\frac{1}{2}\rho V_S^2 + \frac{3}{2}p\right)(V_S - V_P) = \left(\frac{9}{8}\rho_0 V_S^2 + \frac{9}{8}\rho_0 V_S^2\right)\frac{V_s}{4} = \frac{9}{16}\rho_0 V_S^3 \tag{29}$$

since the region between the piston and the shock increases at the rate $V_S - V_P$. On the other hand, this is equal to the work done by the piston per unit area per unit time is

$$pV_P = \left(\tfrac{3}{4}\rho_0 V_S^2\right)\left(\tfrac{3}{4}V_S\right) = \tfrac{9}{16}\rho_0 V_S^3 \tag{30}$$

Half the work done by the piston goes into kinetic energy and the other half goes into thermal energy. The shock, of course, produces no energy, but it is the agent for converting the work done by the piston (which produces the shock in the first place) into thermal and kinetic energy.

Although conservation relations determine downstream properties behind the shock, this is only the case if we are certain that the shock exists. Not all the solutions of the jump conditions correspond to real shocks. For example, if $V_P > 0$, then $V_S > c_s$, and the shock does exist. However, if $V_P < 0$ so that the piston withdraws from the undisturbed region, then the solution is $V_S < c_s$, and it turns out that the narrow region comprising the shock expands. The conditions used in deriving the shock (namely, the mass, momentum, and energy inside the shock region are conserved) no longer hold, and the flow out of region II of these quantities is not equal to the flow of them into region I. A necessary condition for the shock thickness to stay narrow is that the entropy, $\ln(p/\rho^\gamma)$, must increase. This is because the steepening in the shock is balanced by the dissipative processes of viscosity and thermal conduction, which must increase the entropy.

Let us show that the entropy per unit mass, $S = C_V \ln p/\rho^\gamma$, is larger in region I than in region II, if and only if $V_S > c_s^0$. For simplicity, we take $\gamma = \frac{5}{3}$. Let $S = p/\rho^\gamma$ and note that

$$\frac{\rho}{\rho_0} = \frac{U_0}{U} = \left(\frac{1}{4} + \frac{3}{4}x\right)^{-1}$$

$$\frac{p}{p_0} = \frac{5}{4x} - \frac{1}{4} \tag{31}$$

where $x = c_s^2/U_0^2 = 1/M^2$, where M is the Mach number of the shock. Then

$$\frac{S}{S_0} = \left(\frac{5}{4x} - \frac{1}{4}\right)\left(\frac{1}{4} + \frac{3}{4}x\right)^{5/3} \tag{32}$$

and

$$\frac{d}{dx}\frac{S}{S_0} = -\frac{5}{16x^2}(x-1)^2\left(\frac{1}{4} + \frac{3}{4}x\right)^{2/3} \tag{33}$$

Now, for $x = 1$ (the weak shock or the sound wave limit), $\rho = \rho_0$, $p = p_0$, and $S = S_0$. Also, S/S_0 decreases monotonically with x and, therefore, increases monotonically with the Mach number, $M = V_S/c_s^{\mathrm{II}}$. Thus, for $M > 1$, $S > S_0$ and entropy increases from region II to I, as it should, while for $M < 1$ the entropy decreases. In fact, if we attempt to make a subsonic shock by withdrawing the piston we find a solution without discontinuities that is actually an expansion fan. This solution is easily constructed from the Riemann invariants (see problem 5). The point is that there is a nonlinear solution of this negative piston problem in which no steepening occurs and in which there is no breakdown in the fluid equations. We can also see from equation 33 that for weak shocks, small $x - 1$, the entropy jump is cubic in $x - 1$, and thus in the amplitude $\Delta\rho/\rho$ of the shock.

The other important property of the shock is that in the frame of the shock the flow is supersonic upstream and subsonic downstream. That is, $c_s^0 < V_S$ in region II, as we have just emphasized by the entropy discussion, and $c_s^{\rm I} > U$ in region I. To demonstrate the latter result at least for $\gamma = \frac{5}{3}$ note that

$$\frac{U^2}{c_s^{\rm I2}} = \left(\frac{U}{U_0}\right)^2 \frac{\rho U_0^2}{\gamma p} = \frac{U}{U_0}\left(\frac{\rho_0 U_0}{U_0}\right)\frac{U_0^2}{\gamma \rho_0}\frac{p_0}{p}$$

$$= \left(\frac{1}{4} + \frac{3x}{4}\right)\frac{1}{5/4 - x/4} \tag{34}$$

or

$$\frac{U^2}{c_s^2} = \frac{1+3x}{5-x} = -3 + \frac{16}{5-x} \tag{35}$$

which is monotonically increasing with x and so monotonically decreasing with M. Since it is equal to one for $M = x = 1$, it must be less than one for all acceptable shocks.

One of the most important properties of shocks is that they can take a supersonic flow and reduce it to a subsonic one. When a supersonic flow hits an object, a shock must develop that reduces the flow to a subsonic flow so that the pressure at the stagnation point on the object is able to deflect the flow to be tangential to the object and flow around it. An obvious example is the encounter of the solar wind with the earth's magnetic field. Since it cannot penetrate the field it is forced to flow around it, and for this to be accomplished there must be a shock in front of the earth's magnetosphere. This standoff shock is termed the earth's *bow shock*.

6.3 MHD Shocks

How are shocks modified if a magnetic field is present? The magnetic field in the upstream (undisturbed) region may make an arbitrary angle θ to the shock normal, the direction of propagation of the shock. For the moment, assume that we know the upstream quantities in the shock frame. (This, of course, implies that we know the shock speed.)

Choose the axes so that the shock propagates in the x direction and the upstream magnetic field is in the x–y plane, and choose the velocity frame so that the shock is at rest and so that $V_y^{\rm II} = V_z^{\rm II} = 0$. Let us simplify our notation as we did for the hydrodynamic piston problem. We represent quantities in the upstream (II) region by a 0 superscript and those in the downstream (I) region by no superscript. Then there are 5 known quantities B_x^0, B_y^0, ρ^0, p^0, and $V_x^0 = -V_S$. There are six unknown quantities in the downstream region, ρ, p, V_x, V_y, B_x, and B_y. We assume, for the moment, that the downstream z components of **V** and **B** are zero. Thus, six equations are needed to specify them. This is to be contrasted with the field-free case where there were only three unknowns, ρ, V_x, and p. So there must be three additional conditions in the magnetic case. Two additional conditions arise from the conservation of magnetic flux, and the continuity of the x component of **B**. The third condition is the conservation of the y component of the momentum or, equivalently, the balance of the y component of the stress on either side of the shock.

The six conditions are from equations 15–19:

$$\rho V_x = \rho^0 V_x^0$$

$$\rho V_x^2 + p + \frac{B_y^2}{8\pi} = \rho^0 V_x^{02} + p^0 + \frac{B_y^{02}}{8\pi}$$

$$\rho V_x V_y + \frac{B_x B_y}{4\pi} = \frac{B_x^0 B_y^0}{4\pi}$$

$$\frac{1}{2}\rho V_x (V_x^2 + V_y^2) + \frac{\gamma}{\gamma - 1} p V_x + \frac{1}{4\pi}\left(-V_y B_y B_x + V_x B_y^2\right) = \frac{1}{2}\rho^0 V_x^{03} + \frac{\gamma}{\gamma - 1} p V_x^0 + \frac{1}{4\pi} V_x^0 B_y^{02}$$

$$B_x = B_x^0$$

$$V_x B_y - V_y B_x = V_x^0 B_y^0 \tag{36}$$

These equations are the Rankine–Hugoniot conditions for an MHD shock. They are made clearer by the following remarks. The first equation says that the amount of mass flowing across a unit area of the shock is the same in both regions. The second equation says that the deceleration of the plasma as it crosses the shock is equal to the increase of the total pressure, including the magnetic pressure, as the plasma passes to the downstream region. To understand the third equation, note that $\langle B_x B_y \rangle / 4\pi = B_x \langle B_y / 4\pi \rangle$ is $(B_x \times j_z) \times \delta$, where δ is the shock thickness. This is δ times the Lorentz force in the shock layer that an element of plasma experiences as it passes through the shock. The time to pass the shock layer is δ / V_x, so the net sideways impulse is roughly the right-hand side divided by V_x. (Of course, V_x and ρ are not constant, so this argument is only roughly correct.) For the fourth or energy equations the magnetic terms added to the hydrodynamic jump condition are just the jump in the Poynting flux, i.e., $c\mathbf{E} \times \mathbf{B}/4\pi$. The fifth equation follows from $\nabla \cdot \mathbf{B} = 0$. The last equation actually represents the fact that the jump in the z component of the electric field must vanish. It

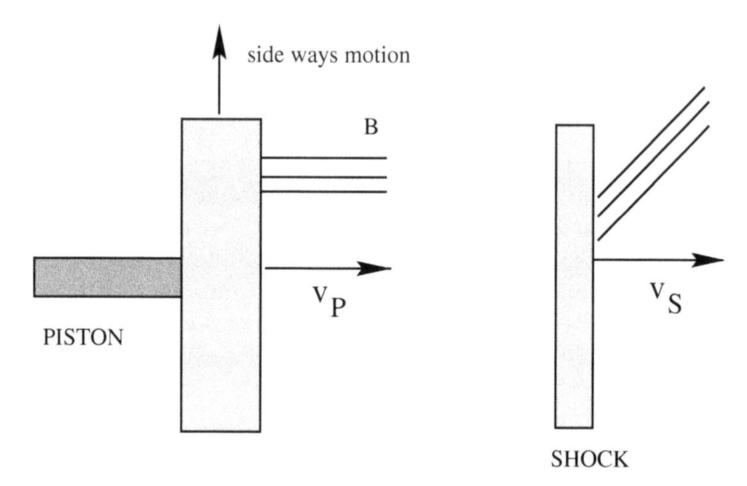

Figure 6.6. Two motions of a piston in a magnetic field, forward and sideways

also represents flux conservation. This is clear if $B_x = 0$, since in this case it reduces to $\langle V_x B_y \rangle = 0$. To see what happens in the general case of nonzero B_x, go to a transverse moving frame with velocity $V_y' = (B_y^0/B_x)V_x^0$. Then in this frame the upstream plasma is flowing parallel to **B**. By flux freezing the plasma continues to flow along **B** in the downstream region. This implies that in the rest frame $V_y + V_y' = V_x B_y / B_x$, which is the same as the last equation in equation 36.

Now, if the shock becomes weak, it should go over to a linearized wave, just as it went over to a linearized sound wave in the field-free case. But we know that there are three linearized MHD waves, so we expect there to be three different shocks: a fast shock, a slow shock, and an intermediate shock. In the above discussion we have excluded the latter because we have taken V_z and B_z to be zero. Thus, the above jump conditions should lead to only two shocks, the slow and fast shocks.

The general discussion of MHD shocks that we now present is really just a condensation of the lucid and comprehensive article of Kantrovitch and Petschek (1966). In general, when we follow the steepening of the simple wave solutions for the slow and fast simple waves we would expect them to lead to these two shocks. On the other hand, we can arrive at the two shocks by considering the motion of a piston into a uniform undisturbed MHD equilibrium. Such a piston can have two arbitrary motions: a forward motion along x and a sideways motion along y (see figure 6.6).

A single MHD shock can always be found so that its downstream value of V_x is equal to the V_x motion of the piston or so that its downstream value of V_y is equal to the V_y of the piston, but not both. Thus, two shocks are necessary to get both velocities correctly. The shock conditions of the second shock have to be chosen using the downstream quantities of the first shock as its upstream values. In practice it is not easy to find these two shocks, but it should be clear in principle that it can be done. To get a steady-state solution we must have two different shocks, since if there were only one type, then the second shock would catch up with the first shock. This is plausible when we recall that in the hydrodynamic case the downstream flow is always subsonic.

Let us now take up the intermediate shock. From considerations of the linearized jump solutions we can see that the intermediate shock would involve a B_z component in the downstream region. To investigate it more closely let us combine the third and fifth equations of equation 36, eliminate V_y, and obtain

$$\left(\frac{V_x^2}{b_x^2} - 1\right) B_y^2 = \left(\frac{V_x^{02}}{b_x^{02}} - 1\right) B_y B_y^0 \tag{37}$$

where $b_x = B_x/\sqrt{4\pi\rho}$ and similarly for b_x^0. Performing the same operation for the analogous jump equations for the z components of **B** and **V**, we get

$$\left(\frac{V_x^2}{b_x^2} - 1\right) B_z^2 = 0 \tag{38}$$

Since for the intermediate shock B_z is not zero, the last equation shows that the downstream velocity must be equal to b_x, i.e., the downstream intermediate velocity, and from equation 37 we see that if $B_y \neq 0$, then the same must be true for the upstream velocity. (If $B_y^0 = 0$, then we can rotate our coordinate system to eliminate the z component.) Thus, $\rho V_x^2 = \rho^0 V_x^{02}$, which with equation 36, implies that $\rho = \rho^0$, $V_x = v_x^0$, and $p = p^0$. But there actually is a discontinuity in which the only change is a rotation of the transverse **B** about the x axis. Thus, the intermediate discontinuity is really not a shock but simply a discontinuous limit of a large amplitude intermediate wave.

We now exclude the case of z components of **B** and **V**, and restrict ourselves to the slow and fast shocks. The six equations of equation 36 are still quite complicated and allow a complicated set of solutions for the downstream quantities for the fast and slow shocks. For a graphic tabulation of these solutions see Ericson and Bazer (1959). Normally, we think of shocks as starting out as weak shocks with the properties of the linearized jumps and gradually going faster as we increase their amplitudes, which are parameterized by the relative jump in the density across the shock. As this jump increases, the shocks can go faster and faster, reaching unlimited velocities as in the case of the field-free case of strong shocks. This could lead to the rather confusing situation that the slow shock can end up going faster than the fast wave speed and the regions of shock velocities could overlap.

However, as we discussed in the field-free case, the jump conditions are only necessary conditions for a discontinuity, if it is to exist as a persistent structure. There are actually solutions of the jump conditions that do not persist and in which the corresponding discontinuity will spread out. An example of such a situation was given in the field-free case of the last subsection in which the piston had a negative velocity and the discontinuity did not persist. For shocks we have to restrict ourselves to solutions in which the density increases, i.e., compressional shocks. For MHD shocks, it turns out that there is an additional condition that must be satisfied for the solution of the jump condition to lead to a persistent MHD shock. This condition leads to the convenient result that the slow shock speed must always be smaller than the fast wave speed. This makes the totality of solutions must simpler to grasp.

The condition can be stated as follows: for any shock, in its rest frame, we must have that either (1) the local intermediate velocity b_x must be less than V_x, the flow velocity normal to the shock, in both the upstream and downstream regions, or (2) it must be greater than the flow velocity in both the upstream and downstream regions. Such shocks are called admissible. (Note that because of the density jump, the intermediate velocity is different upstream and downstream.)

To reword this condition, note that

$$\frac{V_x}{b_x} = \sqrt{\frac{\rho^0}{\rho}} \frac{V_x^0}{b_x^0} < \frac{V_x^0}{b_x^0} \tag{39}$$

so the statement is this: both ratios of V_x to b_x are larger than one or smaller than one. If this condition is satisfied, then condition 1 holds for fast shocks and condition 2 for slow shocks. From equation 37 we see that this condition is equivalent to B_y having the same sign both upstream and downstream.

A solution that does not satisfy the two conditions—first that it is compressional and second that it satisfies the intermediate wave speed condition—is called an extraneous solution and leads to a discontinuity that does not persist but immediately splits into two shocks that are admissible (or into a rarefaction wave). Thus, we can exclude these extraneous solutions, which actually are the more complex solutions of the jump conditions of equation 36. The admissible solutions are much easier to handle.

How does such a condition as the above arise physically? Suppose it were violated and assume that the intermediate wave velocity is smaller than the flow velocity upstream but larger downstream. Let us suppose that the discontinuity arose in the first place from a piston pushing into an undisturbed plasma and also moving in the y direction. At the time the piston started and the shock was launched, let it also have a tiny motion in the z direction. Then we would have to have a weak intermediate wave to take care of the z velocity and magnetic field that arises from the z motion of the piston. Where would this intermediate wave reside? It cannot be ahead of the shock because the shock would catch up with it. On the other hand, it cannot be behind the shock in the downstream region because it would catch up with the shock. There is no place where it could be and lead to a persistent solution.

It turns out that the situation is generally resolved by splitting the shock into two discontinuities. We shall shortly consider one case where we get a pair of shocks, a switch-on shock followed by a switch-off shock. The only resolution would be for it to split the shock into two discontinuities, which would consist of a shock satisfying the condition and an intermediate or rotational discontinuity following it. As we shall see, the typical resolution of a single nonadmissible shock is a pair of shocks: a switch-on shock followed by a switch-off shock.

Now the linearized solutions obviously satisfy the second condition, with the fast wave satisfying condition 1 and the slow wave satisfying condition 2. As the shocks strengthen they should continue to satisfy these respective conditions and a direct calculation from the jump conditions shows this to be true. Thus, the slow shock velocity is always less than the intermediate speed, while the fast shock velocity is always greater, and they never can have the same speed.

Let us look at various limiting cases to get a rough idea of the properties of MHD shocks. The parameters of the upstream plasma are characterized by its pressure, its magnetic field strength, the angle θ that the magnetic field makes with the shock normal, and, of course, the ratio γ of specific heats, which we take as $\frac{5}{3}$.

The simplest case to treat is $\theta = 90°$, where the magnetic field lies in the shock plane and is perpendicular to the shock normal. This shock is called a perpendicular shock. Inspection of the jump conditions shows that with $B_x^0 = 0$ there are only three relevant jump conditions. They are closely analogous to the jump conditions for the field-free case except the upstream and downstream pressures are replaced by the total pressure $p + B^2/8\pi$ and γ is replaced by a value somewhere between $\frac{5}{3}$ and 2, the effective γ for the magnetic field pressure. Thus, the perpendicular shock is a fast shock whose properties are very close to an ordinary field-free shock. When the shock is strong it should be even closer to this case since both upstream pressures are small. The magnetic pressure increases by the compression squared and the remainder of the pressure estimated from the field-free case ends up in the plasma. There is no slow shock for strictly perpendicular propagation since the intermediate speed is zero. In point of fact, the behavior of the perpendicular fast shock is rather insensitive to angle and the behavior of the heating and compression is essentially the same as a function of shock speed down to an angle of 45°.

Let us next consider the case of parallel propagation of a fast shock $\theta = 0$. First, let the upstream plasma pressure be zero. Since the propagation is along **B**, we would expect that the shock would be the pure strong hydrodynamic shock discussed in the field-free case. If the upstream velocity (the shock speed) is smaller than the intermediate speed b_x^0, then the flow velocity downstream is $\frac{1}{4}$ of this, while, since the density is compressed by 4, the downstream intermediate speed is reduced by 2, and the hydrodynamic shock is admissible as a slow shock. But when the upstream velocity increases to larger than the intermediate speed, the downstream velocity is still smaller than the downstream intermediate speed and the hydrodynamic shock is not admissible. Finally, when the upstream velocity exceeds twice the intermediate speed the hydrodynamic shock becomes admissible again. Thus, in the low shock speed and high shock speed limits, the shock is a pure gas dynamic shock. In between, the shock splits into a fast switch-on shock and a slow switch-off shock.

In the fast switch-on shock propagating strictly along **B** the symmetry is broken and a B_y field develops downstream. This B_y field enables the shock to satisfy the required condition because it supplies the pressure to slow down the flow in the shock frame. As a result not so much pressure is required from the plasma and the compression is reduced to the point that the downstream intermediate speed is larger, and the condition is satisfied.

To connect the flows in a switch-on shock to the piston a second switch-off shock will be set up. From equation 37 for a switch-off shock, we see that its upstream velocity will be also equal to its upstream Alfven speed, so it will travel at the exactly the same speed as the switch-on shock. The two shocks will propagate with a constant distance between them, and the conservation conditions between the upstream flow of the switch-on shock and the downstream flow of the switch-off shock will be identical to the unmagnetized shock conditions. In fact, if the distance between the two shocks is small enough, the turning on and off of the transverse magnetic field may not be visible and the two shocks may appear as a single hydrodynamic shock. However, the small perhaps invisible region enables the two shocks to satisfy the admissibility condition, while the apparent hydrodynamic shock would not.

From equation 37 we see that the downstream flow is exactly equal to the intermediate speed. Then the slow shock can reduce B_y to zero (switch off B), and from the same equation with the downstream $B_y = 0$, we see that the upstream velocity of the slow shock is equal to the intermediate speed. Thus, the slow shock stays a fixed distance behind the fast shock. This pair of switch-on, switch-off shocks resolves the hydrodynamic shock in the parameter region where it is nonadmissible.

When the sound velocity is not zero the situation is very similar, there being a smaller gap in nonadmissible gas dynamic shock propagation along **B**. However, when the sound speed exceeds the Alfven velocity, the gap disappears entirely. The shock velocity being larger than the sound speed, which is larger than the Alfven speed upstream, the flow is automatically larger than the Alfven speed downstream, because one is reduced by the compression and the other by the square root of the compression.

There is nothing very dramatic in the behavior of the slow shock except that its speed is limited by the intermediate speed. Its behavior with respect to density and pressure is quite normal, but it has the ability to reduce the B_y field considerably. When the sound speed is large compared to the intermediate speed, the shock speed is correspondingly small and the wave has essentially the properties of a weak wave with respect to ρ, p, and V_x. However, it still has a large effect on the transverse field and velocity.

This latter property plays a very important role in the Petschek theory of reconnection, as we will see in chapter 14. For a more thorough discussion of admissible and nonadmissible shocks and some illustrative graphs of shock properties the reader is referred to the excellent aforementioned article of Kantrovitch and Petschek (1966).

6.4 The Shock Thickness and Collisionless Shock Waves

We have stated that a propagating sound wave will steepen until its rate of steepening is balanced by diffusion of the velocity produced by viscosity. The viscous diffusion is of order λv_T, where λ is the mean free path, and v_T is the thermal velocity. ΔV is the jump in the fluid velocity across the discontinuity that is formed. The rate of steepening due to the $\mathbf{V} \cdot \nabla \mathbf{V}$ term is of order $\Delta V/\delta$, where δ is the thickness of the discontinuity. The rate of diffusion is of order $v_T \lambda/\delta^2$. Equating these two rates gives $\delta = \lambda v_T/\Delta V$ for the thickness the shock. (Note that weak shocks have larger thicknesses.)

However, for perpendicularly propagating MHD shocks, the relevant viscosity is the perpendicular viscosity, which is discussed in chapter 8. It is $\nu_i \rho_i^2$, where ν is the collision rate. This is so, provided that $\lambda > \rho_i$, the usual situation. This is much smaller than the unmagnetized viscosity. The resultant shock thickness then should be $(\rho_i^2/\lambda)v_T/\Delta V$. If $\Delta V \sim v_T$ and $\rho_i \ll \lambda$, then the predicted thickness would be smaller than the ion gyroradius. It is hard to believe that the thickness would be smaller than ρ_i, for in this case the steepening process would fail because the ions would slip through the magnetic field.

A key question concerning shocks is the mechanism for increasing the entropy across the shock. For the gas dynamic shock the heating rate is $\rho\nu(\Delta V/\delta)^2$. The amount of heating is this multiplied by δ/v_T, since this is the time the fluid spends in the shock. The amount of heating is thus $\rho(\Delta V)^2$, as is expected. On the other hand, for a magnetized shock the same calculation with the modified viscosity gives a heating of $(\rho_i/\lambda)^2\rho(\Delta V)^2$, so if $\rho_i \ll \lambda$ the entropy does not change by the required amount and there cannot be a strong shock. However, this is true only for smooth flows with ordinary viscous behavior.

Actually, we expect that the counterflowing ions gyrating in the magnetic field and passing through the layer several times will give rise to strong plasma instabilities. These instabilities will convert the upstream kinetic energy into energetic small-scale waves. These waves, will propagate downstream and quickly dissipate to thermal energy that will provide the required mechanism to increase the entropy and support a shock.

If we compare the downstream quantities far enough away from the shock discontinuity that the waves have damped, we can invoke the same conservation equations used to derive equations 36 for the ordinary MHD shock. The main unknown not settled by these conservation equations is how the energy is split between the ions and the electrons. This is an important astrophysics question since it determines the amount of radiation in the shock heated material and thereby the information to be gained concerning the shock by observing the radiation.

6.5 Problems

1. (Simple Alfven wave) Derive a simple wave solution for a magnetosonic wave propagating perpendicular to a magnetic field in the x direction in a cold plasma. The approximate nonlinear equations are

$$\mathbf{B} = B_x(z,t)\hat{\mathbf{x}}$$

$$\frac{\partial \rho}{\partial t} + \frac{\partial}{\partial z}(\rho V) = 0$$

$$\rho\left(\frac{\partial V}{\partial t} + V\frac{\partial V}{\partial z}\right) = -\frac{\partial}{\partial z}\frac{B_x^2}{8\pi}$$

$$B_x^2 = B_0^2\frac{\rho^2}{\rho_0^2}$$

where B_0 and ρ_0 are reference values.

Show that we get the same solution as that for a sound wave if we set $c_s^2 = \tilde{B}_x^2/4\pi\rho$, where $\tilde{B}_x$ is the initial amplitude of the simple wave.

2. (Heating the chromosphere by sound waves) Consider a sound wave moving upward from the photosphere into the chromosphere, with a wavelength λ and a perturbed velocity δV at the base of the chromosphere. The density decreases exponentially with a constant scale height $H = 150$ km. Estimate at what height you expect it to shock. Take a constant temperature in the chromosphere.

To do this use equation 11 to show that it shocks when $\int \delta V dt = \lambda/2\pi$ and show that the shock occurs when

$$\frac{\delta V}{c_s}\left(e^{z/2H} - 1\right) = \frac{\lambda/2\pi}{2H}$$

As an example take typical values of $\lambda = 100$ km, $T = 10^4$K, $H = 150$ km, and $(\delta V)_0/c_s = 0.02$ and find the z at which the shock occurs. Is this lower than the height at which the corona starts, ≈ 2000 km? (Note that the wavelength remains constant as the wave rises) (see Osterbrock 1961).

3. (Can MHD waves heat the chromosphere?) Repeat problem 2 for magnetosonic wave propagating perpendicularly across a horizontal field $B_0\hat{\mathbf{x}}$. Treat the plasma as cold. Take into account that V_A as well as the density varies with height, so that the wavelength λ also varies. Show that the condition on z for shocking is now

$$\frac{(\delta B)_0}{B_0}\left(e^{-z/4H} - e^{-z/2H}\right) = \frac{\lambda_0/2\pi}{4H}$$

where $(\delta B)_0$ is the magnetic fluctuation and λ_0 the wavelength at the photosphere. Show that if $(\lambda_0/2\pi)/H$ is greater than one, then there is no height at which shocking occurs. This is because the wave becomes weaker with height.

Take the typical parameters, $(\delta B)_0/B_0 = 0.2$ and the other parameters as in problem 2, and find the height z at which shocking occurs.

4. (Dissipation of a shock wave) A strong hydrodynamic shock propagating into a cold uniform medium produces kinetic and thermal energy in its downstream region. The rate of production is given by equation 30. If the shock comes from a freely moving piston that has suffered a sudden velocity V_P and has a mass per unit area σM_P, estimate how far the piston will go before it starts to slow down. For your estimate assume that the piston continues at the initial velocity V_P until it has transferred a significant fraction of its energy to the plasma through the shock mechanism. Compare the mass that is processed by the shock with the mass of the piston. Repeat the problem using the pressure on the piston from equation 28.

5. (A rarefaction wave) Consider a one-dimensional piston problem in which at $t = 0$ a piston at $x = 0$ is suddenly given a negative velocity $V_P = -V_0$. Show that the Riemann invariant

$$J_- = -\frac{2c_s}{\gamma - 1} + V = -\frac{2c_s^0}{\gamma - 1}$$

is a constant everywhere, where c_s^0 is the undisturbed speed of sound. Show that the other invariant

$$J_- = \frac{2c_s}{\gamma - 1} + V$$

is constant along the "forward" characteristic,

$$\frac{dx}{dt} = V + c_s$$

Show that $V = 0$ and $c_s = c_s^0$ in the space–time region $x > c_s^0 t$, and that V decreases from zero to $V_P = -V_0$ in the region $\{c_s^0 - [(\gamma + 1)/2]V_0\}t < x < c_s^0 t$. Show that when $V_P > -(2/\gamma - 1)c_s^0$, c_s and V are constant and $V = V_p$ in the region $V_P t < x < \{c_s^0 - [(\gamma + 1)/2]V_0\}t$.

If V_P is smaller than this, we get cavitation, i.e., $\rho = 0$ in the region $V_P t < x < \{c_s^0 - [(\gamma + 1)/2]V_0\}t$.

References

Andersen, J. E. 1963. *Magnetohydrodynamic Shock Waves*, MIT Press, Cambridge, MA.

Courant, R., and K. O. Friedricks. 1949. *Supersonic Flow and Shock Waves*, p. 87, Interscience, New York.

Ericson, W. B., and J. Bazer. 1959. *Astrophysical Journal* **129**, 758.

Kantrovitch, A., and H. E. Petschek. 1966. (p. 158) In *Plasma Physics in Theory and Application*, p. 158, ed. W. B. Kunkel, McGraw Hill, New York.

Kulikovskii, A. G., and G. A. Lyubimov. 1965. *Magnetohydrodynamics*, trans. L. Oster, Addison-Wesley, Reading, MA.

Lamb, H. 1945. *Hydrodynamics*, 6th ed., p. 482, Dover, New York.

Landau, L. D., and E. M. Lifshitz. 1959. *Fluid Mechanics*, p. 382, trans. J. B. Sykes and W. H. Reid, Pergamon Press, Oxford, UK.

Osterbrock, D. E. 1961. *Astrophysical Journal* **134**, 347.

Shercliff, J. A. 1960. *Journal of Fluid Mechanics* **9**, 481.

Chapter 7

THE ENERGY PRINCIPLE AND INSTABILITIES

7.1 Stability

In our discussion of waves, we have sought to determine the frequency ω from a dispersion relation. In the situations we have discussed where the undisturbed medium varies slowly over the scale of the waves we have found that ω is always real and the perturbations excited about the undisturbed regions are purely oscillatory. When the scale of variation of the perturbations is comparable with the scale of variation of the unperturbed equilibrium, the situation becomes more complicated, and the perturbations can grow. The classic example is the Raleigh–Taylor instability, where a heavy fluid such as water is supported by a lighter fluid such as air. When the scales of variation are comparable, two alternatives for the perturbations are possible: an oscillating motion and an unstable motion.

The situation is somewhat similar to quantum mechanics, where a potential well is given and the wave function for the motion of a particle in it is analyzed. The normal modes with the negative eigenfrequencies correspond to bound modes, while the others with positive eigenstates correspond to unbound states.

The question of stability is easiest to analyze when we consider static situations where the velocity is zero and the equilibrium is bounded by rigid, infinitely conducting walls, so that no energy escapes from the system. For such a system an instability corresponds to a conversion of the potential forms of energy—magnetic energy, pressure energy, or gravitational energy—into kinetic energy. As the perturbation grows, the kinetic energy must grow, and this can happen only by such a conversion. If the equilibrium is nonstatic, then kinetic energy is already present and the instability can involve the conversion of one type of kinetic energy, such as rotation, into a different form, such as turbulence. The first case corresponds to a static atmosphere being unstable to convection, while the latter case is illustrated by a differentially rotating system, such as an accretion disk being unstable to turbulent radial and vertical motion, either by the Taylor instability or, if a magnetic field is present, the magnetic rotational instability.

We can ask how we could set up an equilibrium state that is unstable, since it would quickly change and would be difficult to set up in the first place. We have difficulty in exactly balancing a pencil or a top on its point, since when we get close to such an equilibrium, the instability would destroy it. But in the case of the top, we could first balance it by rotating it rapidly. Then as it slows down, it would reach a point where it first becomes unstable and starts to wobble.

An example of this occurs in the outer convection zone of a star where we could imagine gradually increasing the temperature gradient by increasing the luminosity until the critical temperature gradient is reached and convection becomes unstable. At this point the motion need not be large, and the convection can carry enough of the luminosity that the temperature gradient is very close to the marginal state, just large enough to drive the convection unstable. Thus, the equilibrium exists in a slightly unstable state.

Finally, a stability analysis can show which equilibria cannot exist, i.e., those with rapidly growing instabilities. For this analysis, we often take the point of view that somehow we start with a specific unstable equilibrium and then show that the growth rates are fast. However, we can argue that when we try to arrive at this equilibrium by varying some parameter, e.g., the luminosity in our last example, we first pass through a series of stable equilibria and then arrive at one that is slowly unstable with a growth rate comparable to the rate at which the parameter changes. Then, the equilibrium is destroyed at this point, or at any rate the perturbation becomes large enough to prevent any further change in the critical parameter for stability. Thus, we would never expect to see a temperature gradient much larger than the critical one. (This is not strictly true, but certainly it is if the heat capacity of the gas is such that slow motions can carry the required luminosity.) We have swept under the rug the question of saturation of the instabilities. In convective stars there is actually a thin region near the surface where the temperature gradient is considerably larger than the critical one.

7.2 The Energy Principle

Let us proceed to a formal treatment of MHD stability. This will lead us to a very useful energy principle for discussing stability in general. We first consider a static equilibrium satisfying

$$\mathbf{j}_0 \times \mathbf{B}_0 = \nabla p_0 \tag{1}$$

$$4\pi \mathbf{j}_0 = \nabla \times \mathbf{B}_0 \tag{2}$$

$$\nabla \cdot \mathbf{B}_0 = 0 \tag{3}$$

where $p_0(\mathbf{r}_0)$, $\rho_0(\mathbf{r}_0)$, $\mathbf{B}_0(\mathbf{r}_0)$, $\mathbf{j}_0(\mathbf{r}_0)$ are now functions of position and independent of time. Let us assume that the equilibrium is bounded by a rigid wall S at which the magnetic field is tangent, $\mathbf{B} \cdot \mathbf{n} = 0$ (see figure 7.1). This assumption will assure that the energy contained in the system is conserved.

We perturb the system from its static state by a small displacement $\boldsymbol{\xi}$, and during this initial displacement, we assume that all the MHD equations are

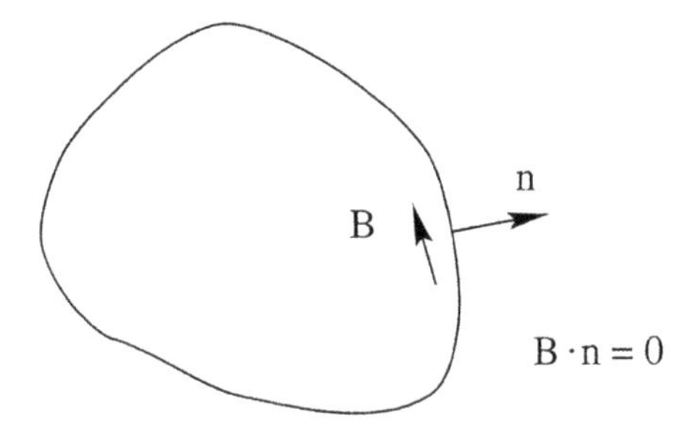

Figure 7.1. Volume V bounded by S on which $\mathbf{B} \cdot \mathbf{n} = 0$

satisfied except the equation of motion. Thus, at $t = 0$ we have

$$\begin{aligned} \boldsymbol{\xi}(\mathbf{r}) &= \boldsymbol{\xi}_0(\mathbf{r}) \\ \frac{\partial \boldsymbol{\xi}}{\partial t}(\mathbf{r}) &= \hat{\boldsymbol{\xi}}_0(\mathbf{r}) \end{aligned} \tag{4}$$

We keep only linear terms in $\boldsymbol{\xi}$. Thus, $\mathbf{r} = \mathbf{r}_0 + \boldsymbol{\xi}(\mathbf{r}_0, t)$, and, in the argument of perturbed qunatities, we may replace $\mathbf{r}_0$ by $\mathbf{r}$. Since the displacement $\boldsymbol{\xi}$ induces changes in the MHD forces, the plasma will accelerate from $t = 0$ onward, and we examine the subsequent motion.

During the premotion $t < 0$ and the subsequent motion $t > 0$, the plasma density, pressure, and magnetic field will satisfy their corresponding equations and we have for all t

$$\begin{aligned} \rho(t, \mathbf{r}) &= \rho_0 - \nabla \cdot (\rho_0 \boldsymbol{\xi}) \\ p(t, \mathbf{r}) &= p_0 - \gamma p_0 \nabla \cdot \boldsymbol{\xi} - \boldsymbol{\xi} \cdot \nabla p_0 \\ \mathbf{B}(t, \mathbf{r}) &= \mathbf{B}_0 + \nabla \times (\boldsymbol{\xi} \times \mathbf{B}_0) \\ 4\pi \mathbf{j}(t, \mathbf{r}) &= \mathbf{j}_0 + \nabla \times [\nabla \times (\boldsymbol{\xi} \times \mathbf{B}_0)] \end{aligned} \tag{5}$$

(see equations 66, 68, and 69 in chapter 4).

The equation of motion for $\boldsymbol{\xi}$ to first order in $\boldsymbol{\xi}$ for $t > 0$ is

$$\begin{aligned} \rho_0 \frac{\partial^2 \boldsymbol{\xi}}{\partial t^2} = \frac{1}{4\pi} \{\nabla \times [\nabla \times (\boldsymbol{\xi} \times \mathbf{B}_0)]\} \times \mathbf{B}_0 + \mathbf{j}_0 \times [\nabla \times (\boldsymbol{\xi} \times \mathbf{B}_0)] \\ + \nabla [\boldsymbol{\xi} \cdot \nabla p_0 + \gamma p_0 (\nabla \cdot \boldsymbol{\xi})] - \mathbf{g} \nabla \cdot \rho_0 \boldsymbol{\xi} \end{aligned} \tag{6}$$

Because the wall is rigid we must also satisfy $\boldsymbol{\xi} \cdot \mathbf{n} = 0$ on S. The boundary need not be assumed infinitely conducting if $\mathbf{B} \cdot \mathbf{n} = 0$ on S. The right-hand side is a second-order linear vector operator on $\boldsymbol{\xi}$, which depends on the equilibrium functions $p_0(\mathbf{r}), \mathbf{B}_0(\mathbf{r})$, and $\mathbf{j}_0(\mathbf{r})$. For brevity we denote it by $\mathbf{F}(\boldsymbol{\xi})$ so that the equation of motion can be written

$$\rho_0 \frac{\partial^2 \boldsymbol{\xi}}{\partial t^2} = \mathbf{F}(\boldsymbol{\xi}) \tag{7}$$

Clearly, to set up the initial perturbation $\boldsymbol{\xi}_0, \dot{\boldsymbol{\xi}}_0$ we need an additional force $\mathbf{F}_0(\mathbf{r})$, so for $t < 0$ we have

$$\rho_0 \frac{\partial^2 \boldsymbol{\xi}}{\partial t^2} = \mathbf{F}(\boldsymbol{\xi}) + \mathbf{F}_0(\mathbf{r}, t) \tag{8}$$

It is clear that $\mathbf{F}_0(\mathbf{r})$ can be chosen to produce any arbitrary initial perturbation $\boldsymbol{\xi}_0$ and $\dot{\boldsymbol{\xi}}_0$ of equation 4. For example, we can make $\boldsymbol{\xi} = \boldsymbol{\xi}(t, \mathbf{r}_0) = f(t)\boldsymbol{\xi}_0$ for $-t_0 < t < 0$, so that f and df/dt vanish at $-t_0$ and $f = 1$, and $df/dt = 0$ at $t = 0$. Then between $-t_0$ and 0, we impose the external force

$$\mathbf{F}_0(\mathbf{r}, \mathbf{t}) = -f(t)\mathbf{F}[\boldsymbol{\xi}_0(\mathbf{r})] + \rho_0 \frac{d^2 f}{dt^2} \boldsymbol{\xi}_0(\mathbf{r}) \tag{9}$$

This external force will lead to $\boldsymbol{\xi} = \boldsymbol{\xi}_0, \partial\boldsymbol{\xi}/\partial t = 0$ at $t = 0$. On the other hand, imposing $\mathbf{F}_0 = \rho_0 \delta(t) \dot{\boldsymbol{\xi}}_0$ leads to $\boldsymbol{\xi} = 0, \partial\boldsymbol{\xi}/\partial t = \dot{\boldsymbol{\xi}}_0$ at $t = 0$. An appropriate linear combination gives us any initial conditions we want.

A very important property of the linear operator $\mathbf{F}(\xi)$ is that it is self-adjoint. By self-adjointness we mean that for all vector functions $\boldsymbol{\xi}(\mathbf{r})$ and $\boldsymbol{\eta}(\mathbf{r})$ satisfying $\boldsymbol{\xi} \cdot \mathbf{n} = 0, \mathbf{n} \cdot \boldsymbol{\eta} = 0$ on the boundary, we have

$$\int_V \eta \cdot \mathbf{F}(\boldsymbol{\xi}) d^3x = \int_V \boldsymbol{\xi} \cdot \mathbf{F}(\boldsymbol{\eta}) d^3x \tag{10}$$

We will call any perturbation satisfying $\mathbf{n} \cdot \boldsymbol{\xi} = 0$ at the boundary an admissable perturbation. A direct proof of self-adjointness is given in problem 6. However, because self-adjointness is closely related to energy conservation, there is an indirect formal proof of self-adjointness that avoids the algebraic complexities of the direct proof. See Kulsrud (1964) for both proofs.

Before presenting this proof we remind the reader that the energy is

$$\mathcal{E} = \frac{1}{2} \int_V \rho V^2 d^3x + \int_V \left(\frac{p}{\gamma - 1} + \frac{B^2}{8\pi} + \rho\phi \right) d^3x \tag{11}$$

where ϕ is the gravitational potential ($\mathbf{g} = -\nabla\phi$). This energy is exactly conserved for the ideal motions such that ρ, p, $\mathbf{v}$, and $\mathbf{B}$ satisfy the ideal equations. This holds for arbitrary displacements to all orders in their magnitude. For our purposes, we need it only to second order.

Second, note that ρ, p, and $\mathbf{B}$ can be expressed exactly in terms of the displacement $\boldsymbol{\xi}(\mathbf{r}_0, t)$, where the perturbed quantities ρ, etc. are to be considered as a functions the initial position $\mathbf{r}_0$ from which of the displacement starts at $t = 0$. Thus, to second order we can write

$$\mathcal{E} = \frac{1}{2} \int_V \rho_0(\mathbf{r}) \left(\frac{\partial \boldsymbol{\xi}}{\partial t} \right)^2 d^3x + W(\boldsymbol{\xi}, \boldsymbol{\xi}) \tag{12}$$

where the functional W is quadratic in $\boldsymbol{\xi}$.

The actual direct calculation of W to second order is quite complicated, but we do not need to carry it out explicitly once we are convinced of its existence. W is to be considered an expansion in the $\boldsymbol{\xi}$.

Now write

$$W = W_0 + W_1(\boldsymbol{\xi}) + W_2(\boldsymbol{\xi}, \boldsymbol{\xi}) \tag{13}$$

The energy must be conserved for any motion $\boldsymbol{\xi}$, with the initial conditions $\boldsymbol{\xi}_0, \dot{\boldsymbol{\xi}}_0$ provided that the motions satisfy $\boldsymbol{\xi} \cdot \mathbf{n} = 0$ and $\dot{\boldsymbol{\xi}} \cdot \mathbf{n} = 0$. Now set $\dot{\boldsymbol{\xi}}_0 = \boldsymbol{\eta}$. Differentiating the energy given by equation 12 we have

$$\frac{d\mathcal{E}}{dt} = \int_V \rho_0 \frac{\partial \boldsymbol{\xi}}{\partial t} \cdot \frac{\partial^2 \boldsymbol{\xi}}{\partial t^2} d^3x + W_1\left(\frac{\partial \boldsymbol{\xi}}{\partial t}\right) + W_2\left(\frac{\partial \boldsymbol{\xi}}{\partial t}, \boldsymbol{\xi}\right) + W_2\left(\boldsymbol{\xi}, \frac{\partial \boldsymbol{\xi}}{\partial t}\right) = 0 \tag{14}$$

Since this is true for all times, it holds for $t = 0$, so

$$\int_V \boldsymbol{\eta} \cdot \mathbf{F}(\boldsymbol{\xi}) d^3x + W_1(\boldsymbol{\eta}) + W_2(\boldsymbol{\eta}, \boldsymbol{\xi}) + W_2(\boldsymbol{\xi}, \boldsymbol{\eta}) = 0 \tag{15}$$

where we have made use of equation 7. Now, in first order, $W_1(\boldsymbol{\eta}) = 0$. In second order we see that the sum of the two W_2 terms is symmetric under interchange of $\boldsymbol{\xi}$ and $\boldsymbol{\eta}$, so the first term is also symmetric under this interchange. But this is just equation 10, the self-adjointness property.

Further, by setting $\boldsymbol{\eta} = \boldsymbol{\xi}$ and dividing by 2 we obtain an alternative expression for W_2:

$$W_2(\boldsymbol{\xi}, \boldsymbol{\xi}) = -\frac{1}{2} \int_V \boldsymbol{\xi} \cdot \mathbf{F}(\boldsymbol{\xi}) d^3x \tag{16}$$

We see that the perturbed potential energy is just the work the external force $\mathbf{F}_0$ must do to set up the initial conditions $\boldsymbol{\xi}_0$, and $\dot{\boldsymbol{\xi}}_0$ during the time $-t_0 < t < 0$. For the first case, $\boldsymbol{\xi} = \boldsymbol{\xi}_0$ and $\dot{\boldsymbol{\xi}}_0 = 0$, we have from equation 9 that the work done is

$$\begin{aligned} W &= \int_{t_0}^{0} dt \int_V d^3x \frac{\partial \boldsymbol{\xi}}{\partial t} \cdot \mathbf{F}_0(t, \mathbf{r}) \\ &= -\int_{-t_0}^{0} \frac{\partial f}{\partial t} f dt \int_V d^3x \boldsymbol{\xi}_0 \cdot \mathbf{F}(\boldsymbol{\xi}_0) \\ &\quad + \int_{-t_0}^{0} \frac{\partial f}{\partial t} \frac{d^2 f}{dt^2} dt \int_V d^3x \rho \boldsymbol{\xi}_0 \cdot \boldsymbol{\xi}_0 \\ &= -\frac{1}{2} \int_V d^3x \boldsymbol{\xi}_0 \cdot \mathbf{F}(\boldsymbol{\xi}_0) \end{aligned} \tag{17}$$

since $\int_{t_0}^{0} f(\partial f/\partial t)dt = 1/2[f^2(0) - f^2(-t_0)] = \frac{1}{2}$ and the other time integral vanishes.

Now let us consider the implications that can be drawn from the self-adjointness of $\mathbf{F}$. In treating motions governed by equations such as equation 7, it is natural to look at normal mode solutions:

$$\boldsymbol{\xi} = \hat{\boldsymbol{\xi}} e^{i\omega t} \tag{18}$$

Of course, after the complex solution is obtained we take the real part as usual. In general, $\hat{\boldsymbol{\xi}}(\mathbf{r})$ is a complex vector function of $\mathbf{r}$. Substituting equation 18 in equation 7 we have

$$-\rho_0\omega^2\hat{\boldsymbol{\xi}}(\mathbf{r}) = \mathbf{F}\left[\hat{\boldsymbol{\xi}}(\mathbf{r})\right] \tag{19}$$

where we remind the reader that that $\mathbf{F}$ stands for the real second-order operator on the complex quantity $\boldsymbol{\xi}$ on the right-hand side of equation 19. The complex conjugate of equation 19 is

$$-\rho_0\omega^{*2}\hat{\boldsymbol{\xi}}^*(\mathbf{r}) = \mathbf{F}\left[\hat{\boldsymbol{\xi}}^*(\mathbf{r})\right] \tag{20}$$

$\mathbf{F}$ is real, so it is the same in both equations.

Dotting equation 19 with $\boldsymbol{\xi}^*$ and equation 20 with $\boldsymbol{\xi}$, subtracting, and then integrating the result over the volume V yields

$$(\omega^{*2} - \omega^2)\int_V \rho_0\boldsymbol{\xi}^* \cdot \boldsymbol{\xi} d^3x = \int_v \left[\boldsymbol{\xi}^* \cdot \mathbf{F}(\boldsymbol{\xi}) - \boldsymbol{\xi} \cdot \mathbf{F}(\boldsymbol{\xi}^*)\right] d^3x \tag{21}$$

The right-hand side vanishes by the self-adjointness of $\mathbf{F}$, as we can see by writing $\boldsymbol{\xi} = \boldsymbol{\xi}_1 + i\boldsymbol{\xi}_2, \boldsymbol{\xi}^* = \boldsymbol{\xi}_1 - i\boldsymbol{\xi}_2$ and writing out the four cross terms involving $\boldsymbol{\xi}_1$ and $\boldsymbol{\xi}_2$ (problem 1). (We have dropped the hat symbol over $\boldsymbol{\xi}$, which is notationally convenient and also correct, since $\hat{\boldsymbol{\xi}}$ differs from $\boldsymbol{\xi}$ by a constant factor.) The integral on the left-hand side is positive and nonzero so

$$\omega^{*2} = \omega^2 \tag{22}$$

and, therefore, ω^2 is real.

If $\omega^2 > 0$, then ω is purely real and the normal mode is stable, while if $\omega^2 < 0$, then $\omega = i\lambda$ is pure imaginary, and the normal mode is either purely growing or purely damped, or some linear combination of these. There is no mode that both grows and oscillates at the same time (no overstability).

Let us vary some parameter P and track this normal mode as it passes from stability to instability. At the transition point we must have $\omega = 0$. Chandrasekhar (1961) refers to this as the principle of exchange of stabilities. This principle greatly simplifies the task of finding for what range of parameters, P, Q, etc., the equilibrium is unstable. We simply set $\omega = 0$ in the normal mode equation and look for values of P, Q, etc. for which

$$\mathbf{F}(\boldsymbol{\xi}) = 0 \tag{23}$$

has an eigensolution. These values give the boundary of the unstable region in parameter space.

Once we know ω^2 is real it is trivial to conclude that $\boldsymbol{\xi}$ can always be taken as real. If we have a solution in which $\boldsymbol{\xi} = \boldsymbol{\xi}_1 + i\boldsymbol{\xi}_2$ is complex, we may simply take the real part of equation 19 and recognize that $\boldsymbol{\xi}_1$ satisfies it.

A second important consequence of the self-adjointness of $\mathbf{F}$ is the orthogonality of its eigenfunctions. Let $\boldsymbol{\xi}_n$ and $\boldsymbol{\xi}_m$ be two different eigenfunctions with unequal eigenvalues ω_n^2 and ω_m^2. Then starting from

$$-\rho_0 \omega_n^2 \boldsymbol{\xi}_n = \mathbf{F}(\boldsymbol{\xi}_n) \tag{24}$$

and

$$-\rho_0 \omega_m^2 \xi_m = \mathbf{F}(\boldsymbol{\xi}_m) \tag{25}$$

we dot the first relation with $\boldsymbol{\xi}_m$, the second with $\boldsymbol{\xi}_n$, subtract the two results, and integrate over the volume to get

$$(\omega_m^2 - \omega_n^2) \int_V \rho_0 \boldsymbol{\xi}_n \cdot \boldsymbol{\xi}_m d^3x = \int_V \left[\boldsymbol{\xi}_m \cdot \mathbf{F}(\boldsymbol{\xi}_n) - \boldsymbol{\xi}_n \cdot \mathbf{F}(\boldsymbol{\xi}_m)\right] = 0 \tag{26}$$

Thus, $\boldsymbol{\xi}_n$ and $\boldsymbol{\xi}_m$ are orthogonal with a weighting factor of ρ_0.

If $\omega_n^2 = \omega_m^2$, we must proceed a little differently. Assume that there are only a finite number p of independent eigenfunctions $\boldsymbol{\xi}_{n,s}$ belonging to ω_n^2. Then any linear combinations of these eigenfunctions are also eigenfunctions. We can select, in a known way, p such linear combinations such that they are orthogonal to each other.

Further, any constant times an eigenfunction is an eigenfunction, and we can find constants C_n for each eigenfunction such that the weighted integral of $\xi'_n = C_n \boldsymbol{\xi}_n$ is unity. That is,

$$\int_V \rho_0 C_n^2 \boldsymbol{\xi}_n^2 d^3x = \int_V \rho_0 \xi'^2_n d^3x = 1 \tag{27}$$

Combining all this we can choose the set of all eigenfunctions to be orthonormal:

$$\int_V \rho_0 \boldsymbol{\xi}_n \cdot \boldsymbol{\xi}_m d^3x = \delta_{nm} \tag{28}$$

How does all the bear on stability? Let us make the common assumption for such sets of eigenfunctions that they are complete. Then, we can assume that we start with a general initial perturbation $\boldsymbol{\xi}_0$ and expand it in eigenfunctions $\boldsymbol{\xi}_n$:

$$\boldsymbol{\xi}_0(\mathbf{r}) = \Sigma a_n \boldsymbol{\xi}_n(\mathbf{r}) \tag{29}$$

The time evolution of $\boldsymbol{\xi}$ for which $\boldsymbol{\xi} = \boldsymbol{\xi}_0, \dot{\boldsymbol{\xi}} = 0$ is

$$\boldsymbol{\xi} = \tfrac{1}{2} \Sigma a_n \hat{\boldsymbol{\xi}}_n(\mathbf{r}) \left(e^{i\omega_n t} + e^{-i\omega_n t}\right) \tag{30}$$

Now, if some ω_n has a negative imaginary part and if $a_n \neq 0$, then the $\boldsymbol{\xi}_n$ term will grow indefinitely. This means the whole perturbation must grow. In fact, the weighted mean square of $\boldsymbol{\xi}$

$$\begin{aligned} K = \frac{1}{2} \int_V d^3x \rho_0 \boldsymbol{\xi}^2 &= \frac{1}{8} \Sigma a_n b_m \int_V d^3x \rho_0 \hat{\boldsymbol{\xi}}_n \cdot \hat{\boldsymbol{\xi}}_m \left[e^{i(\omega_n + \omega_m)t} + \cdots\right] \\ &= \frac{1}{8} \Sigma a_n b_m \delta_{nm} \left[e^{i(\omega_n + \omega_m)t} + \cdots\right] \\ &= \frac{1}{8} \Sigma a_n^2 \left[e^{2i\omega_m t} + 2 + e^{-2i\omega_n t}\right] \end{aligned} \tag{31}$$

so K must grow. We can always choose an initial perturbation with $a_n \neq 0$ ($\boldsymbol{\xi}_0 = \boldsymbol{\xi}_n$ is one). Thus, we can conclude that if any ω_n^2 is negative, then there are initial perturbations that lead to growth away from the equilibrium, and the equilibrium is unstable.

We can reduce the above discussion to the following statement: if any ω_n^2 is negative, then the equilibrium is unstable, and if all ω_n^2 are positive, then it is stable. We can reword it further. Let the ω_n^2 be arranged in increasing order of magnitude, so that

$$\omega_1^2 < \omega_2^2 < \omega_3^2 < \cdots \tag{32}$$

Then, if ω_1^2 is negative we have instability, and if it is positive we have stability.

This is a simple rewording of the standard normal mode approach to stability that if any normal mode is unstable, i.e., growing, then the system is unstable; conversely, if *all* modes are stable, then the system is stable. We now extend this result to a different criteria for stability, which is easier to apply, but gives up some information. This is the MHD energy principle criteria (Bernstein et al. 1958).

Before proceeding, it is necessary to clear up the logic of the above discussion. First, we cannot simply give the initial value of $\boldsymbol{\xi}$, but must consider its initial time derivative as well. ω_n^2 is the eigenvalue but it gives two solutions, $\hat{\boldsymbol{\xi}}_n \exp i\omega_n t$ and $\hat{\boldsymbol{\xi}}_n \exp -i\omega_n t$. Finally, the real part must be taken. Although this causes complications, it does not change the conclusion.

In fact, by linearity, we can divide up a general initial perturbation into one that has a $\boldsymbol{\xi}_0$ displacement and zero initial time derivative, and one that has a finite initial velocity $\dot{\boldsymbol{\xi}}$, but zero initial displacement $\boldsymbol{\xi}_0$. For the first perturbation with $\boldsymbol{\xi}_0 = \Sigma a_n \hat{\boldsymbol{\xi}}_n$ we have, as above in equation 30,

$$\boldsymbol{\xi} = \Sigma \frac{a_n}{2} \boldsymbol{\xi}_n \left(e^{i\omega_n t} + e^{-i\omega_n t} \right) \tag{33}$$

For the second perturbation we have with $\dot{\boldsymbol{\xi}}_0 = \Sigma b_n \hat{\boldsymbol{\xi}}_n$

$$\boldsymbol{\xi} = \Sigma \frac{a_n}{2i\omega_n} \boldsymbol{\xi}_n \left(e^{i\omega_n t} - e^{-i\omega_n t} \right) \tag{34}$$

and the rest of the argument proceeds as before. If $\omega_n^2 < 0$, then ω_n is pure imaginary and, in either case, one of the terms grows.

We now return to the instability statement $\omega_1^2 < 0$, with ω_1^2 given by equation 32. How can we find ω_1^2? For a general *trial* perturbation $\boldsymbol{\xi}$, let us evaluate the ratio

$$\Lambda = \frac{W_2(\boldsymbol{\xi}, \boldsymbol{\xi})}{K(\boldsymbol{\xi}, \boldsymbol{\xi})} \tag{35}$$

where $W_2(\boldsymbol{\xi}, \boldsymbol{\xi})$, the perturbed potential energy is given by equation 16 as

$$W_2(\boldsymbol{\xi}, \boldsymbol{\xi}) = -\frac{1}{2} \int_V \boldsymbol{\xi} \cdot \mathbf{F}(\boldsymbol{\xi}) d^3x \tag{36}$$

while the weighted mean square of $\boldsymbol{\xi}$

$$K(\boldsymbol{\xi}, \boldsymbol{\xi}) = \frac{1}{2}\int_V \rho_0 \boldsymbol{\xi}^2 d^3x \tag{37}$$

is somewhat inappropriately referred to as the kinetic energy. Using the expansion of $\boldsymbol{\xi} = \Sigma a_n \boldsymbol{\xi}_n$ we find that

$$\begin{aligned}
\Lambda = \frac{-\int_V \boldsymbol{\xi} \cdot \mathbf{F}(\boldsymbol{\xi}) d^3x}{\int_V \rho_0 \boldsymbol{\xi}^2 d^3x} &= \frac{-\Sigma a_n a_m \int_V \rho_0 \boldsymbol{\xi}_n \cdot \mathbf{F}(\boldsymbol{\xi}_m) d^3x}{\Sigma a_n a_m \int_V \rho_0 \boldsymbol{\xi}_n \cdot \boldsymbol{\xi}_m d^3x} \\
&= \frac{-\Sigma a_n a_m \int_V -\rho_0 \omega_m^2 \boldsymbol{\xi}_n \cdot \boldsymbol{\xi}_m d^3x}{\Sigma a_n a_m \delta_{nm}} \\
&= \frac{\Sigma \omega_n^2 a_n^2}{\Sigma a_n^2} \\
&\geq \omega_1^2
\end{aligned} \tag{38}$$

where the normal mode expansion, equation 29, and the orthonormality condition, equation 28, have been used. The ratio Λ actually achieves the minimum ω_1^2 if we take $\boldsymbol{\xi} = \boldsymbol{\xi}_1$.

Thus, stability reduces to the following question: is Λ ever less than zero? If it is for some $\boldsymbol{\xi}$, then the situation is unstable, while if Λ is greater than zero for all $\boldsymbol{\xi}$, then the situation is stable. But because K is always positive we can state the following: A necessary and sufficient condition for stability is that the change in the perturbed potential energy

$$W_2(\boldsymbol{\xi}, \boldsymbol{\xi}) > 0, \qquad \text{for all } \boldsymbol{\xi} \tag{39}$$

If it is ever negative, then we have an unstable motion and instability, while if it is always positive the situation is stable.

Note that we can vary the expression for Λ with respect to $\boldsymbol{\xi}$. But equivalently we may also vary any one of the a's. Thus,

$$\frac{\partial \Lambda}{\partial a_q} = -\frac{W_2}{K^2} 2a_q + \frac{\omega_q^2 2a_q}{K} = \frac{2a_q}{K}(\omega_q^2 - \Lambda) \tag{40}$$

and this vanishes if and only if either $a_q = 0$ or $\omega_q^2 = \Lambda$. Thus, Λ at some value of $\boldsymbol{\xi}$ is stationary to all variations of $\boldsymbol{\xi}$ (or the a's) if and only if it equals ω_q^2 for some q, and $\boldsymbol{\xi}$ is an eigenmode (or equivalently all the a's but a_q are zero). In fact, a direct variational calculation with the $\boldsymbol{\xi}$'s and making use of the self-adjointness shows that Λ is stationary if and only if

$$\mathbf{F}(\boldsymbol{\xi}) = -\Lambda \rho \boldsymbol{\xi} \tag{41}$$

everywhere. Thus, the normal mode problem can be replaced by a variational problem and vice versa. The *minimization* of Λ over $\boldsymbol{\xi}$ leads to the lowest eigensolution with $\omega_1^2 = \Lambda$.

Although the initial calculation of $W_2(\boldsymbol{\xi}, \boldsymbol{\xi})$ that we outlined was very complicated, once we are assured of its existence, there turns out to be a much simpler way to find it by using equation 16 and substituting **F** from (6). For this reason it was not necessary to carry out the direct complicated calculation of W_2.

The original discussion, which showed how in principle it was possible to explicitly evaluate the potential energy, was needed to show that the change in the potential energy depends only on $\boldsymbol{\xi}$ and not its time derivative and further that it is a quadratic expression in $\boldsymbol{\xi}$, $W_2(\boldsymbol{\xi}, \boldsymbol{\xi})$. Its existence could actually be inferred by considering the work done when we apply an external force to the equilibrium that produces the initial displacement $\boldsymbol{\xi}$ but no velocity. We can see that the work done by such a force, which is equal to the potential energy of the perturbation, must depend only on the final value of the displacement and not how it is set up (or how the external force depends on time). In a sense, this is what we accomplished by our example of how to set up an initial perturbation in equation 9. We prefer the first argument because it seems logically more direct and easier to follow.

Let us now derive the expression for the perturbed potential energy W_2 making use of equations 6 and 16 and integrating by parts. We separate **F** into four parts as follows:

$$\begin{aligned} \mathbf{F} = \mathbf{F}_a + \mathbf{F}_b + \mathbf{F}_c + \mathbf{F}_d = & \frac{1}{4\pi} \{\nabla \times [\nabla \times (\boldsymbol{\xi} \times \mathbf{B}_0)]\} \times \mathbf{B}_0 \\ & + \mathbf{j}_0 \times [\nabla \times (\boldsymbol{\xi} \times \mathbf{B}_0)] \\ & + \nabla [p_0 \nabla \cdot \boldsymbol{\xi} + (\boldsymbol{\xi} \cdot \nabla p_0)] \\ & + \nabla \cdot (\rho_0 \boldsymbol{\xi}) \nabla \phi \end{aligned} \tag{42}$$

where the expressions for the four different forces are on the four different lines. The potential energy will have four corresponding contributions from each of the the four forces. The first one, W_a, is

$$\begin{aligned} W_a &= -\frac{1}{2} \int_V \boldsymbol{\xi} \cdot \mathbf{F}_a d^3x \\ &= \frac{1}{8\pi} \int_V d^3x \, \{\nabla \times [\nabla \times (\boldsymbol{\xi} \times \mathbf{B}_0)]) \cdot (\boldsymbol{\xi} \times \mathbf{B}_0\} \\ &= \frac{1}{8\pi} \int_V d^3x \nabla \cdot [\nabla \times (\boldsymbol{\xi} \times \mathbf{B}_0) \times (\boldsymbol{\xi} \times \mathbf{B}_0)] \\ &\quad + \frac{1}{8\pi} \int_V d^3x [\nabla \times (\boldsymbol{\xi} \times \mathbf{B}_0)] \cdot [\nabla \times (\boldsymbol{\xi} \times \mathbf{B}_0)] \end{aligned} \tag{43}$$

Let

$$\mathbf{Q} = \nabla \times (\boldsymbol{\xi} \times \mathbf{B}_0) \tag{44}$$

the surface term from Gauss's theorem is

$$\frac{1}{8\pi}\int_S d\mathbf{S}\cdot\mathbf{Q}\times(\boldsymbol{\xi}\times\mathbf{B}_0) = -\frac{1}{8\pi}\int_S dS[(\mathbf{n}\cdot\mathbf{B}_0)(\boldsymbol{\xi}\cdot\mathbf{Q}) - (\mathbf{n}\cdot\boldsymbol{\xi})(\mathbf{B}_0\cdot\mathbf{Q}] \\ = 0 \tag{45}$$

since $\boldsymbol{\xi}\cdot\mathbf{n}$ and $\mathbf{B}_0\cdot\mathbf{n}$ are zero by the boundary conditions we assumed. The other terms are quite straightforward, the surface terms appearing in W_c and W_d vanishing from Gauss's theorem and the boundary conditions.

It is conventional to call W_2, δW. This is because this expression is frequently used for testing stability, and it is useful to have a familiar and characteristic name for it. The result of our calculation for δW is

$$\delta W = \frac{1}{2}\int_V d^3x\left[\frac{Q^2}{4\pi} + \mathbf{j}_0\cdot(\boldsymbol{\xi}\times\mathbf{Q}) + \gamma p_0(\nabla\cdot\boldsymbol{\xi})^2 + \boldsymbol{\xi}\cdot\nabla p_0)(\nabla\cdot\boldsymbol{\xi}) \\ - \boldsymbol{\xi}\cdot\nabla\phi\nabla\cdot(\rho_0\boldsymbol{\xi})\right] \tag{46}$$

From now on we will drop the zero subscript on the equilibrium quantities.

How can we use this energy principle, $\delta W > 0$ for all $\boldsymbol{\xi}$, as the necessary and sufficient test for stability? To determine whether a given equilibrium is unstable we need to examine δW for all $\boldsymbol{\xi}$ to see if any one of them makes δW negative. In general, to do this it would seem necessary to minimize δW over every possible $\boldsymbol{\xi}$ and find out if the minimum is negative.

Clearly, a direct minimization of δW gets us nowhere. If it is ever negative, then the minimum is minus infinity, while if it is always positive the minimum is zero. To effectively examine δW we must normalize $\boldsymbol{\xi}$ somehow, or divide δW by a normalizing integral before taking the minimum, so that the expression to be minimized is homogeneous in $\boldsymbol{\xi}$.

A natural normalizing integral is $K = \left(\frac{1}{2}\right)\int \rho\boldsymbol{\xi}^2 d^3x$, but if we use this we get back to the normal mode equations, and we get no gain from the energy principle. Moreover, using K, the full kinetic energy factor, often leads to great complications, although if we wish to know the exact growth rate, this is the necessary normalization.

If we are willing to give up information about the exact value of the growth rate, we may choose a different normalization integral, which leads to a simpler minimization problem. This minimization is more tractable and still gives a reliable decision about stability, but it leads to a growth rate that is not accurate. As an example, if we have a cylindrical equilibrium we could use

$$K' = \frac{1}{2}\int_V \rho\xi_r^2 d^3x \tag{47}$$

instead of

$$K = \frac{1}{2}\int_V \rho\left(\xi_r^2 + \xi_\theta^2 + \xi_z^2\right) d^3x \tag{48}$$

and indeed this does give a much simpler minimization problem. The minimizing $\boldsymbol{\xi}, \boldsymbol{\xi}_0$, gives

$$\Lambda' = \frac{\delta W(\boldsymbol{\xi}_0, \boldsymbol{\xi}_0)}{\int_V \rho \xi_{0r}^2 d^3x} > \frac{\delta W(\boldsymbol{\xi}_0, \boldsymbol{\xi}_0)}{\int_V \rho \left(\xi_{0r}^2 + \xi_{0\theta}^2 + \xi_{0z}^2\right) d^3x} > \Lambda_{\min} \tag{49}$$

Thus, for any $\boldsymbol{\xi}$ the resulting $\Lambda'(\boldsymbol{\xi}, \boldsymbol{\xi}) > \Lambda(\boldsymbol{\xi}, \boldsymbol{\xi}) > \omega_1^2$. So, by choosing a different normalization we get a lower bound for the growth rate, provided the equilibrium is unstable. It is probably still correct in order of magnitude. On the other hand, if Λ' is always positive, then the numerator must always be positive and therefore we have stability.

In our example, we have chosen a normalization integral closely related to the true K. But if we choose a more arbitrary normalization, we can still substitute the minimizing $\boldsymbol{\xi}$ into $\delta W/K$ and, if we find that δW can be negative, get a lower bound for the growth rate. (If it is stable we get an upper limit of the lowest frequency.)

We can even go a little further. By making use of some physical idea as to the general nature of the perturbation that is likely to lead to instability, we can simply substitute this $\boldsymbol{\xi}$ into δW and see if we get a negative value. If we do, we can conclude (rigorously) that the system is unstable. If δW is positive for our guess, then we cannot conclude stability or instability. However, if the equilibrium depends on some parameter, say the pressure p, then from this $\boldsymbol{\xi}$ alone we can find a range of p for which the equilibria are unstable. Simply examine δW with this $\boldsymbol{\xi}$ and from the range in p for which it is negative we have a range of p for which we have instability. We even get an approximate idea of where the stability boundary in p lies.

In summary, the energy principle is useful because by giving up some information, e.g., the exact growth rate, we can simplify the normal mode equation through modifying the normalization integral. If we give up more, i.e., the exact determination of stability, we can find some of the unstable equilibria by choosing a physically appropriate trial function $\boldsymbol{\xi}$ and testing for which equilibria δW is negative.

Before proceeding to the discussion of more general forms of the energy principle, let us illustrate it by applying it to the Schwarzschild criteria for convective instability in stars (Schwarzschild 1958). Let us take a one-dimensional equilibrium with quantities depending only on z, and with g in the negative z direction. Let us also take $\mathbf{B} = 0$. Then the equilibrium satisfies

$$\rho g_z = \frac{\partial p}{\partial z} \tag{50}$$

Since $g_z = -g < 0$, p decreases with increasing z. Let us also ignore the boundary condition for the moment. Then

$$2\delta W = \int_V \left[\gamma p(\nabla \cdot \boldsymbol{\xi})^2 + \boldsymbol{\xi} \cdot \nabla p(\nabla \cdot \boldsymbol{\xi}) + \boldsymbol{\xi} \cdot \mathbf{g}(\rho \nabla \cdot \boldsymbol{\xi} + \boldsymbol{\xi} \cdot \nabla \rho)\right] d^3x \tag{51}$$

We assume that $\boldsymbol{\xi}$ is sinusoidal in the x direction and independent of y:

$$\boldsymbol{\xi} = \mathrm{Re}\left[\hat{\boldsymbol{\xi}}(z)e^{ikx}\right] \tag{52}$$

$$\nabla \cdot \boldsymbol{\xi} = \mathrm{Re}\left(\frac{\partial \hat{\xi}_z}{\partial z} + ik\hat{\xi}_x e^{ikx}\right) \tag{53}$$

(When we minimize δW over a complex $\hat{\boldsymbol{\xi}}$ we are really minimizing over the real and imaginary parts of $\hat{\boldsymbol{\xi}}$. The real part of $\hat{\boldsymbol{\xi}}$ times the exponential must always be taken before substitution in δW.)

Now, ξ_z and ξ_x are the functions over which we wish to test δW for stability. But from equation 53 we see that we can equally well test the stability over all functions ξ_z and $(\nabla \cdot \boldsymbol{\xi})$. Using $\rho \mathbf{g} = \nabla p$ in δW as given in equation 50, we see that the second and third terms are equal. Then completing the square in the integrand we have

$$2\delta W = \int_V \left[\gamma p\left(\nabla \cdot \boldsymbol{\xi} + \frac{\boldsymbol{\xi} \cdot \nabla p}{\gamma p}\right)^2 - \frac{(\boldsymbol{\xi} \cdot \nabla p)^2}{\gamma p} + \boldsymbol{\xi} \cdot \nabla p \frac{\boldsymbol{\xi} \cdot \nabla \rho}{\rho}\right] d^3x$$

$$= \int_V \gamma p\left(\nabla \cdot \boldsymbol{\xi} + \frac{\boldsymbol{\xi} \cdot \nabla p}{\gamma p}\right)^2 - \frac{1}{\gamma}\left[\boldsymbol{\xi} \cdot \nabla \ln\left(\frac{p}{\rho^\gamma}\right)\right](\boldsymbol{\xi} \cdot \nabla p) d^3x \tag{54}$$

The first term is positive and can be minimized to zero by choosing $\nabla \cdot \boldsymbol{\xi}$. The second term depends on the relative direction of the variation of the entropy $s = C_v \ln(p/\rho^\gamma)$ and p. Since $\partial p/\partial z < 0$, the sign of δW depends on the sign of ds/dz, and when the entropy decreases upward this term is negative. Now suppose this is the case for some local range of z. Then choose ξ_z to be nonzero only in this range, and choose $\nabla \cdot \boldsymbol{\xi}$ to make the first term zero everywhere. For this choice of $\boldsymbol{\xi}$, $\delta W < 0$ and we have instability.

On the other hand, if the entropy s increases everywhere with z, then the integrand is positive for all $\boldsymbol{\xi}$, $\delta W > 0$ for all $\boldsymbol{\xi}$, and we have stability. Thus, for our problem, the necessary and sufficient condition for instability is that s decrease upward somewhere or, equivalently, that p/ρ^γ decrease upward somewhere. Note that the first term is proportional to the square of the perturbed pressure, at a fixed point, and so our perturbation has been chosen so that the rising blob of plasma expands at just the rate to stay in local pressure equilibrium with the ambient pressure. For such a motion, the last two terms give ξ_z time the perturbed force, which is upward where ξ_z is positive if the plasma is unstable. This relates the energy principle δW calculation to the usual physical argument (see Schwarzschild 1958). However, while the physical argument is approximate, the δW argument gives an exact result for stability and instability.

What about the boundary conditions that we imposed in deriving the δW criterion, but ignored in our stability test? We should suppose that walls are imposed at two layers, z_0 and z_1, and demand that $\xi_z = 0$ on these walls. But since $\xi_z \neq 0$ only locally, these conditions are automatically satisfied. Since the walls can be imagined to be placed anywhere, they can be supposed far enough away that they do not physically affect the perturbation. Thus, the boundary conditions play little if any role in the stability condition.

What about the growth rate? Let us suppose that ξ_z is nonzero in some region and let us substitute the trial perturbation into $2\delta W/K$ to get a lower bound for the growth rate. To estimate K note that ξ_x is obtained from ξ_z by equating the square brackets in the second line of equation 54 to zero. If the scale height of the pressure is H and the extent of the instability region is δ, then we have

$$\xi_x \approx \xi_z \left(\frac{1}{\gamma k H} + \frac{1}{k\delta} \right) \tag{55}$$

(We ignore signs and factors of i.) Take $\delta \ll H$, then take $k\delta \gg 1$. Then we have for the growth rate Γ,

$$\Gamma^2 >= \frac{2\delta W}{2K} \approx \frac{\int_V \boldsymbol{\xi} \cdot \nabla \ln(p/\rho^\gamma)(\boldsymbol{\xi} \cdot \nabla p) d^3x}{\gamma \int_V \rho \boldsymbol{\xi}^2 d^3x}$$

$$\approx \frac{1}{\gamma} \frac{p}{\rho H^2} \Delta \approx \frac{c_S^2}{\gamma^2 H^2} \Delta \tag{56}$$

where $\Delta = d\ln(p/\rho^\gamma)/d\ln p$ is the relative entropy gradient, which measures the extent to which the entropy gradient is above the stability gradient. If we take the entropy content ΔH of the plasma in the unstable region and multiply it by T, and by the growth rate $\Gamma = \Delta^{1/2} c_s/H$, we arrive at the usual mixing length estimate for the convective flux. (The assumption is that the perturbation grows to a finite value and turns an element of fluid of size H over in a time approximately equal to a reciprocal growth, $1/\Gamma$ (see Schwarzschild 1958).

We can go a little farther and ask how a magnetic field might affect convection. Let us consider a uniform magnetic field $\mathbf{B}_0$ in the y direction. Then if we take the same $\boldsymbol{\xi}$ as above, δW will have an additional term

$$2\delta W = \int_V \frac{Q^2}{4\pi} d^3x = \frac{1}{4\pi} \int_V [\nabla \times (\boldsymbol{\xi} \times \mathbf{B})]^2 \, d^3x$$

$$= \frac{B_0^2}{4\pi} \int_V d^3x (\nabla \cdot \boldsymbol{\xi})^2 = \frac{B_0^2}{4\pi} \int_V d^3x \frac{1}{\gamma^2 p^2} \xi_z^2 \left(\frac{\partial p}{\partial z} \right)^2$$

$$= \frac{v_A^2}{4\pi \gamma^2 H^2} \int_V d^3x \rho \xi_z^2 \tag{57}$$

and adding this positive term to equation 56 we get

$$\frac{\delta W}{K} = \frac{v_A^2}{4\pi\gamma^2 H^2} - \frac{c_s^2}{\gamma^2 H^2} \Delta \tag{58}$$

Since the second term represents the unstable energy, we find that our trial perturbation gives a negative δW if

$$v_A^2 < \Delta c_s^2 \tag{59}$$

If this condition is violated, we cannot say anything about stability since we looked only at a specific trial function $\boldsymbol{\xi}$. (Another trial function could increase the pressure term more than it might raise the magnetic term.) However, if it is satisfied, we definitely have instability, even in the presence of the magnetic field.

This example should make clear the possibilities and limitations of stability analysis based on the energy principle. It does enable us to establish simple correct but imprecise criteria for instability, avoiding complex normal mode analysis.

The derivation of the energy principle does invoke the completeness of the eigenmodes, which, in general, is not provable. (Sometimes the normal modes themselves do not exist.) What happens if they should happen to be in incomplete? It is certainly still the case that $\delta W > 0$ for all $\boldsymbol{\xi}$ is sufficient for stability. To see this we need merely check that the energy

$$\mathcal{E} = \frac{1}{2}\int_V d^3x\rho\dot{\boldsymbol{\xi}}^2 + \delta W(\boldsymbol{\xi}, \boldsymbol{\xi}) \tag{60}$$

is a constant. This is easily checked (problem 1). But then, if we start with a weak perturbation, $\mathcal{E}$ is small, and because the second term is positive and $\mathcal{E}$ remains small, the first term on the right-hand side cannot grow. With regard to necessity for stability: if δW is negative for some $\boldsymbol{\xi}_0$ and $\boldsymbol{\xi}_0$ is not expandable in normal modes, then we cannot conclude from our proof that ω_1^2 is negative, since the connection between the eigenmode $\boldsymbol{\xi}_1$ and the trial $\boldsymbol{\xi}$ is no longer present. (If the modes were not complete but the trial function that made δW negative was expandable in normal modes, then of course we could conclude that ω_1^2 is negative, but there is no convenient way to restrict our trial perturbations to those that are so expandable.) The possible non completeness of the normal modes does not affect the self-adjointness of $\mathrm{F}(\boldsymbol{\xi})$, which is still self-adjoint for nonexpandable $\boldsymbol{\xi}$ and $\boldsymbol{\eta}$.

However, it turns out that we can avoid the whole question of normal modes and their possible noncompleteness altogether by the following theorem due to Laval, Mercier, and Pellat (1965).

Suppose that for some $\boldsymbol{\xi}_0$, $\delta W < 0$. Set

$$p_0^2 = -\frac{\delta W}{K} = -\frac{\delta W(\boldsymbol{\xi}_0, \boldsymbol{\xi}_0)}{\frac{1}{2}\int_V d^3x\rho\boldsymbol{\xi}_0^2} \tag{61}$$

(p_0 is our approximation for the growth rate associated with the trial perturbation $\boldsymbol{\xi}_0$.) Set

$$I(\boldsymbol{\xi}, \boldsymbol{\xi}) = \frac{1}{2}\int_V d^3x\rho\boldsymbol{\xi}^2 \tag{62}$$

The theorem says: if at time $t = 0, \boldsymbol{\xi} = \boldsymbol{\xi}_0$ and $\partial\boldsymbol{\xi}/\partial t = p_0\boldsymbol{\xi}_0$, then for all $t > 0$

$$I(\boldsymbol{\xi}, \boldsymbol{\xi}) > I_0 e^{2p_0 t} \tag{63}$$

where $I_0 = I(\boldsymbol{\xi}_0, \boldsymbol{\xi}_0)$.

The proof of this theorem, (Laval, et al. 1965) is as follows. First, the energy $\mathcal{E}$ is zero, since at time $t = 0$,

$$\frac{1}{2}\int_V d^3x\rho\left(\frac{\partial\boldsymbol{\xi}}{\partial t}\right)^2 + \delta W(\boldsymbol{\xi}, \boldsymbol{\xi}) = \frac{p_0^2}{2}\int_V d^3x\rho\boldsymbol{\xi}_0^2 + \delta W(\boldsymbol{\xi}_0, \boldsymbol{\xi}_0) = 0 \tag{64}$$

by the definition of p_0. The time derivative of $I(\boldsymbol{\xi}, \boldsymbol{\xi})$ is

$$\frac{dI}{dt} = \int_V d^3x\rho\boldsymbol{\xi} \cdot \dot{\boldsymbol{\xi}} \tag{65}$$

The second time derivative is

$$\begin{aligned}\frac{d^2I}{dt^2} &= \int_V d^3x\rho\frac{\partial^2\boldsymbol{\xi}}{\partial t^2} \cdot \boldsymbol{\xi} + \int_V d^3x\rho\dot{\boldsymbol{\xi}}^2 \\ &= \int_V d^3x\boldsymbol{\xi} \cdot \mathbf{F}(\boldsymbol{\xi}) + \int_V d^3x\rho\dot{\boldsymbol{\xi}}^2 \\ &= -2\delta W + 2\mathcal{E}_V \\ &= 4\mathcal{E}_V\end{aligned} \tag{66}$$

since $2\mathcal{E}_V + 2\delta W = 2\mathcal{E} = 0$, where $\mathcal{E}_V = \frac{1}{2}\int_V d^3x\rho\dot{\boldsymbol{\xi}}^2$ is the actual kinetic energy. Now, by the Schwarz inequality (Courant and Hilbert 1953),

$$\dot{I}^2 = \left(\int_V d^3x\rho\boldsymbol{\xi} \cdot \dot{\boldsymbol{\xi}}\right)^2 < \int_V d^3x\rho\boldsymbol{\xi}^2 \int_V d^3x\rho\dot{\boldsymbol{\xi}}^2 = \ddot{I}I \tag{67}$$

Divide this equation by I^2 and rearrange the inequality to get

$$0 < \frac{\ddot{I}}{I} - \frac{\dot{I}^2}{I^2} = \frac{d}{dt}\left(\frac{\dot{I}}{I}\right) \tag{68}$$

or

$$\frac{d(\ln I)}{dt} = \frac{\dot{I}}{I} > \frac{\dot{I}}{I}(\text{at } t = 0) = 2p_0 \tag{69}$$

Integrating this with respect to time we get

$$\begin{aligned}&\ln I > \ln I_0 + 2p_0 t \\ &\text{or } I > I_0 e^{2p_0 t}\end{aligned} \tag{70}$$

which is the stated result. With this theorem the energy principle is freed from any discussion of the completeness of normal modes, or even the normal modes themselves.

What actually went into the establishment of the energy principle? First, there was an operator, $\mathbf{F}(\boldsymbol{\xi})$, which represents to first order in $\boldsymbol{\xi}$ the force per unit volume on each fluid particle in terms of its displacement, Second, the equation for $\boldsymbol{\xi}$ has no $\dot{\boldsymbol{\xi}}$ term,

$$\rho\ddot{\boldsymbol{\xi}} = \mathbf{F}(\boldsymbol{\xi}) \tag{71}$$

Third, the energy, which is constant, can be expressed as the sum of a kinetic term and a quadratic expression in $\boldsymbol{\xi}$,

$$\mathcal{E} = \frac{1}{2}\int_V d^3x\rho\dot{\boldsymbol{\xi}}^2 + \delta W(\boldsymbol{\xi},\boldsymbol{\xi}) \tag{72}$$

Fourth, the boundary conditions on all $\boldsymbol{\xi}$ are chosen so that the energy is conserved in the volume of interest. That is, either $\mathbf{B}_0 \cdot \mathbf{n} = 0, \boldsymbol{\xi} \cdot \mathbf{n} = 0$, or $\mathbf{B} \cdot \mathbf{n} \neq 0, \boldsymbol{\xi} = 0$ on S. These conditions give the self-adjointness of $\mathbf{F}$.

If some acceptable $\boldsymbol{\xi}$ can be found such that $\delta W < 0$, then there exists an initial perturbation that grows exponentially, faster than $\exp 2p_0t$, where $p_0^2 = -\delta W/K$ for this perturbation, and this continues till nonlinear terms become important.

So far, we have considered only equilibria with $\mathbf{B}_0$ satisfying $\mathbf{B}_0 \cdot \mathbf{n} = 0$, and $\boldsymbol{\xi} \cdot \mathbf{n} = 0$ on the boundary S. But everything we have said holds if the equilibria satisfy $\mathbf{B} \cdot \mathbf{n} \neq 0$ and $\boldsymbol{\xi} = 0$ on S. This is true for the calculation of δW as well, which has exactly the same form with this boundary condition. As an important example, the energy principle applies to the solar atmosphere whose field lines enter the solar surface, provided that only perturbations that vanish on the solar surface are considered.

A more general case for which the energy principle is still valid is the case of a self-gravitating system. Here $\mathbf{F}$ must be an integral operator to properly include self-gravitational forces, and the boundary S must be taken to infinity. The energy principle still holds, but the perturbed gravitational energy term in δW is now a double integral of $\boldsymbol{\xi}$ over space.

Finally, it is sometimes the case that two different plasmas are separated by a contact surface, which moves with the perturbation. For example, the chemical composition and the related molecular weight could change abruptly. In this case, it is necessary to restrict ourselves to the requirement that $\mathbf{B}_0 \cdot \mathbf{n} = 0$ on this contact surface. In our example, the contact surface could be the photosphere, and the second plasma the solar atmosphere. The boundary condition between two plasmas when they are in dynamic motion is as follows: first, $\mathbf{V} \cdot \mathbf{n}$ is the same on both sides of the contact surface and is equal to the velocity of the this surface. Second, the jump in the total plasma pressure across the surface must be zero.

$$\mathbf{V}_a \cdot \mathbf{n} = \mathbf{V}_b \cdot \mathbf{n}$$

$$\left(p + \frac{B^2}{8\pi}\right)_a = \left(p + \frac{B^2}{8\pi}\right)_b \tag{73}$$

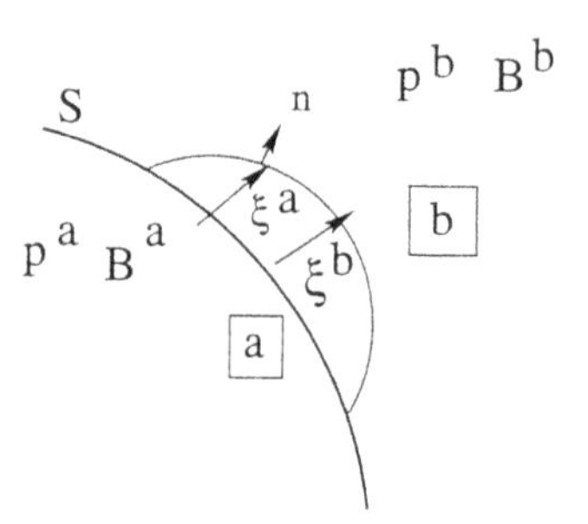

Figure 7.2. Boundary conditions on a perturbed surface

The second equation arises because volume forces acting on the infinitely thin layer between the plasmas do not enter. The pressure forces must balance, since otherwise there is a finite force on a zero amount of matter in the surface. In general, such a free surface problem is very difficult to handle, but when we are looking for small displacements about a static equilibrium the procedure is more tractable.

The perturbed boundary conditions projected back to the unperturbed surface (as in figure 7.2) are as follows. The velocity condition is

$$\boldsymbol{\xi}_a \cdot \mathbf{n} = \boldsymbol{\xi}_b \cdot \mathbf{n} \tag{74}$$

The pressure condition is more complicated. It reduces to

$$\left[\left(\gamma p + \frac{B^2}{4\pi}\right)\nabla\cdot\boldsymbol{\xi} - \frac{\mathbf{B}_0\cdot(\nabla\boldsymbol{\xi})\cdot\mathbf{B}_0}{4\pi}\right]_a = \left[\left(\gamma p + \frac{B^2}{4\pi}\right)\nabla\cdot\boldsymbol{\xi} - \frac{\mathbf{B}_0\cdot(\nabla\boldsymbol{\xi})\cdot\mathbf{B}_0}{4\pi}\right]_b \tag{75}$$

The previous calculations when applied to this case are all formal except for the evaluation of δW, which is essentially the sum of the δW's from the two regions plus a surface term that arises from the surface terms that no longer vanish. Thus,

$$\delta W = \delta W_a + \delta W_b + \delta W_S \tag{76}$$

where

$$\delta W_S = \frac{1}{2}\int_S dS(\mathbf{n}\cdot\boldsymbol{\xi})\mathbf{n}\cdot\left\langle\nabla\left(p + \frac{B^2}{8\pi}\right)\right\rangle \tag{77}$$

where the angle brackets denote the jump from region a to region b along $\mathbf{n}$, which points from a to b. In this case the energy principle should read as follows: the combined system is stable if and only if all $\boldsymbol{\xi}$'s satisfying equations 74 and 75 make δW (as given by equation 76) positive. If one such $\boldsymbol{\xi}$ makes δW negative, then the system is unstable.

Equation 75 is rather complex. Thus, it is fortunate that we can dismiss it. In fact, if all $\boldsymbol{\xi}$ in the enlarged set that satisfy only equation 74 make δW positive, then the smaller set satisfying both equation 74 and equation 75 also must make δW positive. On the other hand, if it turns out that some $\boldsymbol{\xi}$ satisfying only equation 74 makes $\delta W < 0$, then it is always possible to modify this $\boldsymbol{\xi}$ slightly near the boundary, by working with $\nabla \cdot \boldsymbol{\xi}$, in such a way that δW is still negative for the modified $\boldsymbol{\xi}$, but such that both equation 74 and equation 75 are satisfied.

Taking this result into account we can now state the energy principle for an equilibrium with a discontinuous surface as follows: If $\delta W > 0$ for all $\boldsymbol{\xi}$ satisfying only the first boundary condition, equation 74, then we have stability. Otherwise, we have instability.

Finally, it may be the case that one region, say, b, is either a vacuum or has such a low conductivity that $\mathbf{j} = 0$. In this case, the energy principle is based on still a further expression for δW. This expression is

$$\delta W = \delta W_a + \delta W_S + \delta W_V \tag{78}$$

where δW_V is the vacuum contribution to δW given by

$$\delta W_V = \frac{1}{8\pi} \int_V d^3x \left(\nabla \times \mathbf{A}_1\right)^2 \tag{79}$$

where $\mathbf{A}_1$ is the perturbed vector potential, considered to be the free independent function in region b, while $\boldsymbol{\xi}$ is a free function in region a. The boundary condition in this case is simply

$$\mathbf{n} \times \mathbf{A}_1 = -(\mathbf{n} \cdot \boldsymbol{\xi})\mathbf{B}_b \tag{80}$$

where $\mathbf{B}_b$ is the equilibrium field in the vacuum. The expression for δW_S is the same as that given in equation 77 but with the pressure gradient term dropped in region b. The proof of all these facts is given by Bernstein et al. (1958).

The variety of forms we have given for the energy principle makes it possible to discuss a large variety of stability problems in a highly physical manner (provided that we abandon an attempt to get exact results for the growth rate or the structure of the eigenmode). This is more normally the astrophysical situation, where the equilibria are not very well known in the first place.

The energy principle is patterned on the simple problem of a ball on a hill, or in a valley, or at a saddle point (see figure 7.3).

When the ball is slightly displaced a horizontal distance $\boldsymbol{\xi}$ from the top of the hill, the potential energy is changed by $\delta W = g\delta h = g(a\xi_x^2 + b\xi_y^2)$, where the height of the hill is Taylor expanded in x and y. For a hill, a and b are both negative so the displacement, either in the x or y direction, results in a negative change in potential energy δW. For a saddle point at least one of the coefficients, a or b, is negative, so the corresponding ξ_x or ξ_y gives a negative value to δW. However, in this case there is a range in directions for which $\boldsymbol{\xi}$ makes δW positive. Thus, not all, but only a limited range of

Figure 7.3. Ball on a hill or in a valley

$\boldsymbol{\xi}$'s makes δW negative. This is sufficient to make the ball at a saddle point unstable, the expected result. For a valley, both a and b are positive and δW is positive for all $\boldsymbol{\xi}$'s. This is the only stable case.

In this example of a ball on the hill we readily see that the perturbation grows, with potential energy being converted to kinetic energy. The potential energy drops and kinetic energy grows. We might think that when the potential energy can decrease for some displacement $\boldsymbol{\xi}$, the equilibrium should be unstable, but this is not always the case. We can modify the problem of the ball on the hill by giving it a charge Q and imposing a vertical magnetic field B. Then as the ball picks up speed rolling down the hill, the Lorentz force bends its orbit around and, if B is large enough, forces it back up the hill. Indeed, a normal mode analysis shows that if $\Omega = qB/Mc > 2\gamma$, where γ is the growth rate for $B = 0$, then the ball is stabilized. But imposing this field does not change either its potential or it kinetic energy. In fact, when we write out the equation of motion of the ball, we have a $\dot{\boldsymbol{\xi}}$ term,

$$\ddot{\boldsymbol{\xi}} = \mathbf{M}(\dot{\boldsymbol{\xi}}) + \mathbf{F}(\boldsymbol{\xi}) \tag{81}$$

where $\mathbf{M}$ is the operator representing the Lorentz force. (The magnetic field would be even more effective in the saddle point case.) Thus, it is not correct to infer that merely lowering the potential energy of a system necessarily leads to instability. In our formal derivation of the energy principle for static MHD equilibria we did not have the $\mathbf{M}(\dot{\boldsymbol{\xi}})$ term, so its presence did not interfere with our conclusion that lowering the potential energy would lead to instability (see problem 2). Thus, our formal proof of the energy principle is really necessary.

If the MHD equilibrium has flows, it is even less likely that appealing to the potential energy $\delta W(\boldsymbol{\xi}, \boldsymbol{\xi})$ will give a stability criterion. There are additional contributions to the energy from the perturbed kinetic energy of the zero-order equilibrium flows. There are actually three possibilities.

The first possibility is that some of the equilibrium kinetic energy can be converted to the kinetic energy of the perturbation, so that even if δW is always positive, we can still have instability. The second possibility is that the opposite can happen, in a way analogous to stabilizing the charged ball on the hill by Lorentz forces. The MHD system could start off in an unstable perturbation (or, anyway, one that makes δW negative) and then the perturbation could be rotated by the flow into a different direction (in function space) that makes δW positive. The third possibility is that if the velocity has shear, the shear could tear the unstable coherent mode apart, stabilizing it. Examples of all three of these possibilities exist and their existence should warn people off from any direct employment of the energy principle when flows are present.

We could argue that the idea of a displacement $\boldsymbol{\xi}$ is not useful when we have equilibrium flows because the flows would stretch $\boldsymbol{\xi}$ into a finite displacement in a finite time. This is actually not a major difficulty. Frieman and Rotenberg (1960) have redefined $\boldsymbol{\xi}$ at $\mathbf{r}$ and t as the displacement of the fluid particle from $\mathbf{r}$ that would have been there in the absence of the perturbation. This enabled them to still find the perturbed force in terms of a self-adjoint operator, $\mathbf{F}(\boldsymbol{\xi})$. The perturbed potential energy δW has the same form as before with some extra inertial terms. However, in the equation of motion for this $\boldsymbol{\xi}$ there are terms involving $\dot{\boldsymbol{\xi}}$ and this spoils the validity of the energy principle. From now on we will mainly restrict ourselves to instabilities of static equilibria, with the one exception of the magnetorotational instability (MRI) that plays such a significant role in accretion disks.

7.3 Instabilities

Now let us examine some important MHD instabilities using the energy principle to obtain a simple physical picture of their character. However, there is one very important instability, the magnetorotational instability (the MRI), that is not treatable by the energy principle, and we will discuss it by a direct normal mode analysis.

7.3.1 The Interchange Instability

The simplest example of the interchange instability is the one-dimensional equilibria in which a cold plasma ($p = 0$) is supported against gravity in the negative z direction by a magnetic field in the x direction, where all unperturbed quantities depend only on z (see figure 7.4). The equilibrium is given by

$$\rho \mathbf{g} = \nabla \left(\frac{B^2}{8\pi} \right) = \frac{\partial}{\partial z} \left(\frac{B^2}{8\pi} \right) \hat{\mathbf{z}} \tag{82}$$

The Lorentz force is represented by the gradient of the the magnetic pressure. $\mathbf{g}$ is in the negative z direction, so B^2 decreases upward. Since $p = 0$, δW is given by

$$2\delta W = \int_V d^3x \left[\frac{Q^2}{4pi} + \mathbf{j} \cdot (\boldsymbol{\xi} \times \mathbf{Q}) + \nabla \cdot (\rho \boldsymbol{\xi}) \boldsymbol{\xi} \cdot \mathbf{g} \right] \tag{83}$$

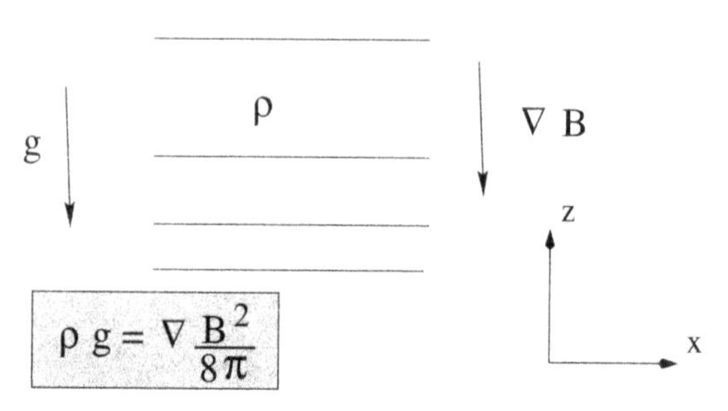

Figure 7.4. Gravitational interchange instability

where

$$\mathbf{Q} = \nabla \times (\boldsymbol{\xi} \times \mathbf{B}) = \mathbf{B} \cdot \nabla \boldsymbol{\xi} - \boldsymbol{\xi} \cdot \nabla \mathbf{B} - \mathbf{B}(\nabla \cdot \boldsymbol{\xi}) \tag{84}$$

Now, instead of trying to obtain an exact criterion for instability by minimizing δW with some normalization, let us simply take a perturbation that will simplify δW and see what we get. An obvious perturbation to take is a $\boldsymbol{\xi}$ that makes $\mathbf{Q} = 0$, since this gets rid of the first two terms. For this we must take $\boldsymbol{\xi}$ independent of x so that we do not bend the lines of force. Then equation 84 for $\mathbf{Q}$ gives us a relation between $\boldsymbol{\xi}$ and $\nabla \cdot \boldsymbol{\xi}$,

$$\nabla \cdot \boldsymbol{\xi} = \frac{-\boldsymbol{\xi} \cdot \nabla B}{B} \tag{85}$$

But we may substitute this into $\nabla \cdot \rho\boldsymbol{\xi} = \boldsymbol{\xi} \cdot \nabla\rho + \rho(\nabla \cdot \boldsymbol{\xi})$ in the last term of equation 83 to get

$$\begin{aligned} 2\delta W &= \int_V d^3x\, (\rho\nabla \cdot \boldsymbol{\xi} + \boldsymbol{\xi} \cdot \nabla\rho)\, \boldsymbol{\xi} \cdot \mathbf{g} \\ &= \int_V d^3x \left(-\frac{\rho\boldsymbol{\xi} \cdot \nabla B}{B} + \boldsymbol{\xi} \cdot \nabla\rho \right) \boldsymbol{\xi} \cdot \mathbf{g} \\ &= \int_V d^3x \rho\boldsymbol{\xi} \cdot \nabla \ln \left(\frac{\rho}{B} \right) \boldsymbol{\xi} \cdot \mathbf{g} \end{aligned} \tag{86}$$

Thus, if there is any point where $\nabla \ln(\rho/B)$ increases upward, then we have instability. This, of course, need not be a necessary condition for instability, since we have examined δW for only a subclass of $\boldsymbol{\xi}$'s, those that make $\mathbf{Q} = 0$, but it is certainly sufficient.

Let us examine the nature of the perturbation we chose to test for instability. Clearly, $\mathbf{Q} = 0$ corresponds to no first-order perturbation in $\mathbf{B}$ at any point. This is accomplished physically by imagining the convection pattern as seen in figure 7.5, where there are a number of tubes around the convection cell. These tubes are chosen so that they each have the same amount of flux. Then, because B decreases upward, the tubes on top have a bigger cross-sectional area.

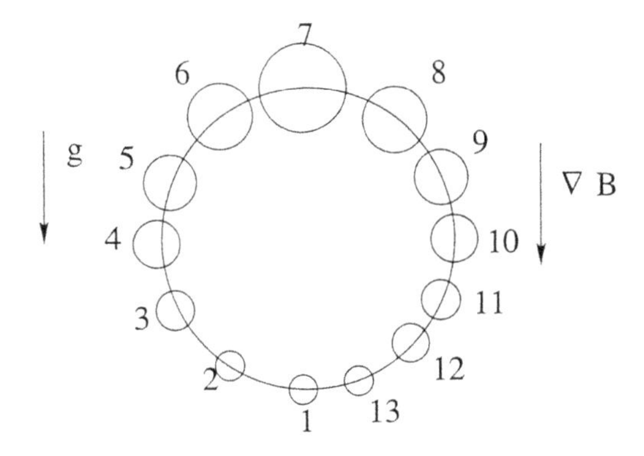

Figure 7.5. The nature of the interchange instability

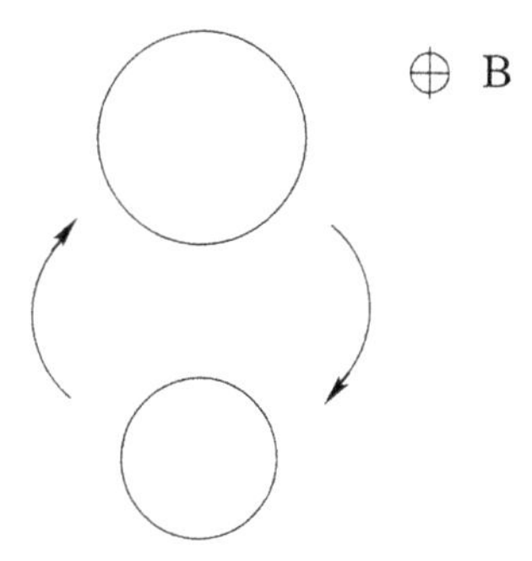

Figure 7.6. The interchange instability as the interchange of only two tubes

Tube 1 is imagined to move into tube 2. This motion does not change **B** in tube 2, since the plasma in tube 2 is replaced by the plasma of tube 1, but the number of lines in tube 1 is the same as in tube 2. We continue these motions in the same way, with tube 2 going into tube 3, tube 3 into tube 4, etc., until finally tube 13 goes into tube 1. The amount of plasma in tube 1 is raised into tube 2 etc., up to tube 7. The plasma in tube 7 is lowered into tube 9 etc., down to tube 13. If the mass per flux is linear in height, then the convection leads to a change in energy proportional to that from simply moving tube 1 into tube 7 and tube 7 into tube 1 with no other change. The proportionality factor is $\delta h/H$, where δh is the change in height of each tube in the convection, and H is the difference in heights between tube 1 and tube 7. Thus, we can think of the change in energy of the combined motions as proportional to the effect of interchanging the plasmas in tube 1 and tube 7 with no change in magnetic energy (see figure 7.6).

But the amount of mass in each tube per unit length is equal to ρ/B times the flux ψ in any one of the tubes, so that if the mass per unit flux in tube 1 is less than the mass per unit flux in tube 7, then we have actually lowered the gravitational potential energy without changing the magnetic energy. If there is a region in z where ρ/B increases upward, then taking our convection

Figure 7.7. Plasma slipping downward along the magnetic lines of force

pattern in this region we find that the motion between all the tubes produces a net decrease in δW in agreement with the direct evaluation of δW given above.

Even this simple picture can be short-circuited by considering only two tubes and directly interchanging their plasmas, as in figure 7.6, rather than dealing with the chain of interchanges of the convection cells of figure 7.5. It was actually this picture that led to the name "interchange instability."

We have picked a special displacement for a test of whether δW can be negative. Is this the worst possible $\boldsymbol{\xi}$? If ρ/B decreases upward, would the plasma be stable. Note that B decreases upward, so our tentative condition for instability is that ρ does not decrease as fast. Actually, if we restrict our displacements to the $\boldsymbol{\xi}$ that do not vary along $\mathbf{B}$, i.e., $\mathbf{B} \cdot \nabla\boldsymbol{\xi} = 0$, then it turns out (as can be shown by an explicit calculation) that this $\boldsymbol{\xi}$ is as bad for stability as any other for the test. Thus, the mass per flux ρ/B increasing upward is the necessary and sufficient condition for instability to such perturbations.

7.3.2 The Parker Instability

However, if we allow the lines of force to bend, then the zero pressure system is unstable no matter how ρ varies with z. This is the well-known Parker instability, which plays such an important role in molecular cloud formation in the interstellar medium and is important in many other contexts as well. Parker (1966, 1967).

Parker's rough idea is to imagine that the line is buoyed so that it forms a sinusoidal structure but B_x still does not change. Then let the matter slip down the line from the top to the bottom, as in figure 7.7. The mass holding the line down against magnetic pressure now is reduced at the top and increased at the bottom, allowing the line to distort further. It is clear that a magnetic tension force arises from the buoying, but if the distance over which the line buoys is made sufficiently long, then the downward tension force on the top is smaller than the imbalance in the magnetic pressure force and the weight of the plasma. The same holds in the valley of the line. A little thought will show that the next line in the y direction will buoy oppositely for the B_x component of the field to remain undisturbed at a fixed point.

How do we directly show that the Parker instability is actually unstable from the energy principle? To do this, take the y and z components of the displacement, ξ_y and ξ_z, to be the same as for the pure interchange given above but let them vary sinusoidally in x. Then, as before, Q_x is zero and

$$\nabla \cdot \boldsymbol{\xi}_\perp = -\frac{\boldsymbol{\xi} \cdot \nabla B}{B} \tag{87}$$

where now $\boldsymbol{\xi}_\perp$ is the perpendicular part of $\boldsymbol{\xi}$. However, now $\mathbf{Q}_\perp = \mathbf{B} \cdot \nabla \boldsymbol{\xi}$. Next add an additional displacement ξ_x to the original displacement, so that $\rho \nabla \cdot \boldsymbol{\xi}$ has an additional term $\rho \partial \xi_x / \partial x$, which must be added to equation 86. The resulting expression for δW is now

$$2\delta W = \int_V d^3x \left[\frac{1}{4\pi} \left(\mathbf{B} \cdot \nabla \boldsymbol{\xi}_\perp \right)^2 + \rho \boldsymbol{\xi} \cdot \nabla \ln \left(\frac{\rho}{B} \right) (g_z \xi_z) + 2\rho \frac{\partial \xi_x}{\partial x} (\xi_z g_z) \right] \tag{88}$$

(g_z is negative. The factor two in the last term comes from the $\mathbf{j} \cdot \boldsymbol{\xi} \cdot \mathbf{Q}$ term as can be shown by an elementary calculation.)

The last term is the only one involving the parallel displacement ξ_x. By taking ξ_x out of phase with ξ_z, we can make it nonzero and as negative as we please. Even if both of the first two terms are positive, we can choose ξ_x large enough to make the last term larger in magnitude than sum of the other terms, and, thus, make $\delta W < 0$. Note that ξ_x corresponds to the sliding of the plasma down the bent line. (Since the perturbation is linear, this sliding displacement along $\mathbf{B}$ appears as a displacement in the x direction.) Its derivative corresponds to the additional change in density during the unloading of the plasma from the top of the line and the loading of the plasma at the bottom. Since $\nabla \cdot \boldsymbol{\xi}_\perp$ is not zero, we must have the perturbation vary in the y direction, so the different lines buoy up and down as we look in the y direction. Thus, with the Parker type of displacement we see that there is no stable equilibrium in which a *cold* plasma is supported against gravity by a magnetic field.

If ρ/B increases upward, then the pure interchange instability is unstable, without the Parker instability. However, if ρ/B decreases upward, then the Parker buoying of the lines is necessary for an instability. The unloading of the field by the slipping down of the matter generally requires a large parallel displacement. Can we estimate the growth rate in this case? Let the wave number of the variation in x be k, and make $\xi_z \gg \xi_y$ by choosing a large wave number in the y direction. Set $g_z = g$. Then roughly

$$-\gamma^2 \approx \frac{\delta W}{K} \approx \frac{k^2 v_A^2 \xi_z^2 + 2kg\xi_x\xi_z - kg\xi_z^2 \Delta'}{\xi_x^2 + \xi_z^2}$$

where $\Delta' = d(\ln \rho/B)/dz$ and we treat quantities in the two integrands as constant.

Choose a case where $\Delta' = -1$, and take k so that $k^2 v_A^2 = kg$. Then

$$\gamma^2 = \frac{-\delta W}{K} = 2\frac{g^2}{v_A^2} \frac{\xi_x^2 + \xi_x \xi_z}{\xi_x^2 + \xi_z^2} \tag{89}$$

and the minimum for γ is greater than $0.6g/v_A$ since second fraction in equation 89 takes the value 0.2 for $\xi_x = -3\xi_z$, and $\sqrt{2 \times 0.2}$ is greater than 0.6. Again, we have chosen our $\boldsymbol{\xi}$ for ease of calculation. However, we have found the complete answer to the stability question once we include the buoying of the Parker instability. The conclusion we can draw, which is rigorous, is that there is no stable one-dimensional slab equilibrium in which a cold plasma is supported against gravity by a magnetic field.

Let us return to the simpler criterion for instability ρ/B increasing upward. If we consider $B^2/8\pi \to p_B$ to be a form of pressure, we can reword the criterion as ρ^2/B^2 or ρ^2/p_0 increasing upward for instability. If we consider the magnetic field as a $\gamma_B = 2$ gas, then the criteria for instability is p_B/ρ^{γ_B} decreasing upward. This is essentially the Schwarzschild criterion for convective instability. That is, if we assume only displacements with $\mathbf{B} \cdot \nabla \boldsymbol{\xi} = 0$ we get the Scwarzschild criterion.

If we have support from both pressure and magnetic field, then the criteria should be $\rho^{\gamma'}/P$ increasing upward, where $P = p + B^2/8\pi$ is the total pressure and γ' is a number somewhere between γ_{gas}, and γ_B. γ' would be nearer γ_{gas} if $p \gg B^2/8\pi$, and would be nearer 2 in the opposite limit. (A simple algebraic calculation shows that $\gamma' = 2(\gamma p + B^2/8\pi)/(2p + B^2/8\pi)$, where in the Schwarzschild criterion γ is to be replaced by γ'.

But replacing the supporting pressure entirely by a magnetic pressure shows that the system is always unstable when we take into account the Parker instability buoying. We can get a necessary criterion as above for the Parker instability when pressure is included by taking $\boldsymbol{\xi}_\perp$ as before and again adding an out of phase ξ_x. The resulting δW has additional terms involving ξ_x. But now there is a quadratic term arising from the $\gamma p(\nabla \cdot \boldsymbol{\xi})^2$ term in δW so the system is not automatically unstable. Physically, this arises because the warm plasma sliding along the field compresses and increases its pressure adiabatically. This increase in pressure resists the tendency for the warm plasma to slip down along the buoyed field lines, unlike the cold plasma case. We do not carry out the derivation of the new stability criteria. The calculation is straightforward, but does not lead to a particularly simple or informative result. Instead, we note that the stability condition can be expressed in terms of the adiabatic index, γ. For $\gamma = 0$, the $\gamma p(\nabla \cdot \boldsymbol{\xi})^2$ is absent and δW is again linear in ξ_x, so the system is always unstable to the Parker instability. Further, for any $\boldsymbol{\xi}$, δW is monotonic and continuous in γ. Thus, for any choice of $\boldsymbol{\xi}$ that makes δW negative when $\gamma = 0$ there is a critical γ, $\gamma_c(\boldsymbol{\xi})$, below which the system is unstable. Maximizing this critical $\gamma_c(\boldsymbol{\xi})$ over all $\boldsymbol{\xi}$ that are unstable for $\gamma = 0$, we see that there is a constant, γ_c, below which the Parker instability is unstable. The calculation of this γ_c is some what complex. It is carried out as a function of the various equilibrium parameters, ρ, p, B, g, etc., in the papers of Parker (1966, 1967), and directly from the energy principle in a paper of Zweibel and Kulsrud (1975).

7.3.3 The Interchange Without Gravity

The simplest instability is that driven by gravity, where the interchange of two flux tubes with equal flux lowers the gravitational energy, leaving the magnetic energy unchanged. If gravity is absent but the lines are curved, then

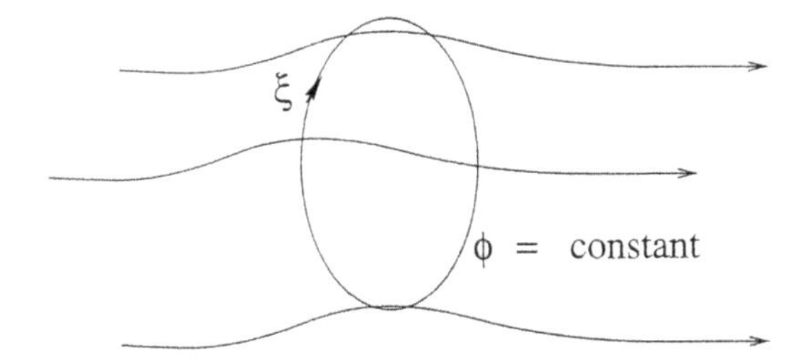

Figure 7.8. $\phi =$ constant convective pattern in the interchange instability

the centrifugal acceleration of the particles moving along the curved lines simulates gravity. However, it turns out that the extra degree of instability associated with the Parker buoying of the lines no longer works the same way since the effective gravity comes from the same pressure that resists the parallel displacement of the plasma necessary for the Parker instability.

In this section we drop gravity and imagine that there is curvature of the lines of force. An interesting example of this is that part of the earth's dipole field that faces the oncoming solar wind and in which there is pressure confined by this field. The stability of such a system is difficult to treat exactly, but using the energy principle we can extract some simple results that give an approximate idea of when the system is unstable.

Again (without specifying the equilibrium in detail), let us imagine we do have an equilibrium and consider a specific trial perturbation $\boldsymbol{\xi}$ that makes $\mathbf{Q} = 0$. Then for this $\boldsymbol{\xi}$,

$$2\delta W = \int_V d^3x \left[\gamma p(\nabla \cdot \boldsymbol{\xi})^2 + \boldsymbol{\xi} \cdot \nabla p(\nabla \cdot \boldsymbol{\xi})\right] \tag{90}$$

The actual expression for such a $\boldsymbol{\xi}$ (that makes $\mathbf{Q} = 0$) is as follows:

$$\mathbf{Q} = \nabla \times (\boldsymbol{\xi} \times \mathbf{B}) = 0 \tag{91}$$

so $\boldsymbol{\xi} \times \mathbf{B}$ must be the gradient of some scalar ϕ, say,

$$\xi \times \mathbf{B} = \nabla\phi \tag{92}$$

ϕ cannot be arbitrary since dotting equation 92 with $\mathbf{B}$ gives

$$\mathbf{B} \cdot \nabla\phi = 0 \tag{93}$$

That is to say, ϕ is constant along the lines of force.

Each line of force can have a different value for ϕ. The convection of the lines of force into each other in a circular pattern must carry lines of force of equal ϕ into each other since $\boldsymbol{\xi}$ is perpendicular to $\nabla\phi$ (figure 7.8). Any ϕ that is a function only of the lines, or a line label, gives such an interchange. It is easy to see that a line with flux ψ is displaced by such a $\boldsymbol{\xi}$ into another line that occupies the position of another flux tube of the equilibrium with the same flux. In this way the magnetic field at a fixed point is not changed under the displacement. We can express $\boldsymbol{\xi}$ in terms of ϕ and

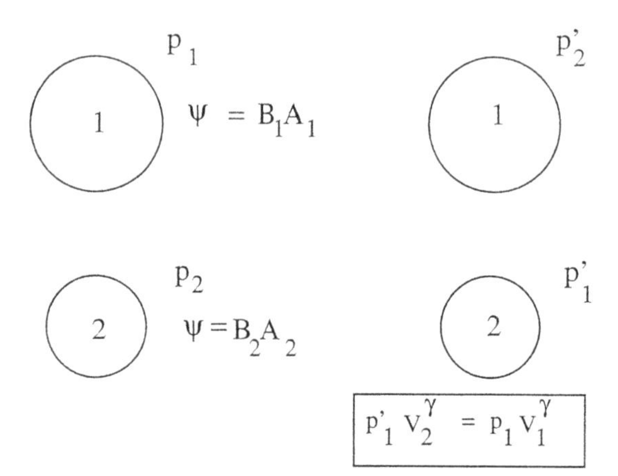

Figure 7.9. Change in pressure during an interchange of two tubes of equal flux

substitute into our expression for δW to find the analytic result for δW in terms of ϕ. We can carry this out formally and by minimizing the result over the parallel component of $\boldsymbol{\xi}$ to show that $\nabla \cdot \boldsymbol{\xi}$ must itself be a function of ϕ only. Carrying this out we can show that δW depends only on the pressure gradient and how the volume per flux changes along it (Kulsrud 1967). This is closely analogous to the gravitational problem where the mass per flux has been replaced by the volume per flux.

We can arrive at essentially the correct result for the interchange stability in a more physical fashion. We simply reduce the change in energy to that produced by an exchange of the plasma in just two different flux tubes with the same fluxes as we did in our physical discussion of the gravitational interchange instability (see Frieman and Kulsrud 1958). Denote the flux tubes, as in the figure 7.9, by 1 and 2. For a given cross section of the tubes we have

$$\begin{aligned} \psi &= B_1 A_1 \\ \psi &= B_2 A_2 \end{aligned} \tag{94}$$

If $B_1 < B_2$, then the cross-section area A_1 is bigger than A_2.

Let p_1 and p_2 be the pressures in tubes 1 and 2. (The pressures are constant in each flux tube because in equilibrium $\mathbf{B} \cdot \nabla p = 0$, which follows from $\mathbf{j} \times \mathbf{B} = \nabla p$.) Let p_1' be the pressure of plasma 1 after it has been placed in tube 2, and p_2' be the changed pressure of the plasma originally in tube 2 after it is placed in tube 1. Let V_1 and V_2 be the volumes of tubes 1 and 2. Then the change in the plasma energy is

$$\frac{p_1' V_2 - p_1 V_1}{\gamma - 1} + \frac{p_2' V_1 - p_2 V_2}{\gamma - 1} \tag{95}$$

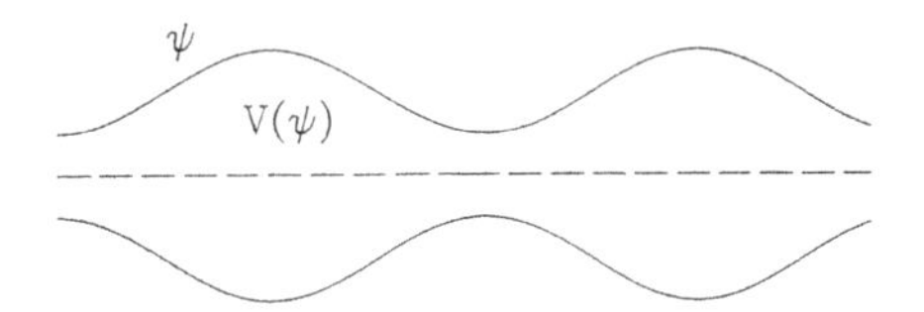

Figure 7.10. A magnetic field periodic in z

Moreover, the plasmas preserve their entropies $\ln(p/\rho^\gamma)$ during the displacements so that

$$p_1' V_2^\gamma = p_1 V_1^\gamma$$
$$p_2' V_1^\gamma = p_2 V_2^\gamma \tag{96}$$

so the change in the sum of the energies becomes

$$(\gamma - 1)\Delta W = p_1 \left[\left(\frac{V_1}{V_2} \right)^\gamma V_2 - V_1 \right] + p_2 \left[\left(\frac{V_2}{V_1} \right)^\gamma V_1 - V_2 \right] \tag{97}$$

We have carried out the calculation for a finite change in the plasma energy so we use ΔW instead of δW. Let the tubes be close to each other and let

$$V_2 - V_1 = \delta V$$
$$p_2 - p_1 = \delta p \tag{98}$$

By expanding the result for ΔW to second order in δV and δp we get

$$\delta W = p_1 \delta V \left(\frac{\gamma \delta V}{V_1} + \frac{\delta p}{p_1} \right)$$
$$= p_1 \delta V \left[\delta \ln \left(p V^\gamma \right) \right] \tag{99}$$

The resulting criteria for instability, that this be less than zero, is reminiscent of the Schwarzschild criteria if we identify V with $1/\rho$, the specific volume.

V_1 is the volume of tube 1 with flux ψ, and $V_2 = V_1 + \delta V$ is the volume of the tube with the same flux. If we write

$$V = U\psi \tag{100}$$

then U_1 and U_2 are the volume per flux of the two tubes, and the instability criteria can be written

$$\frac{\delta W}{\psi} = p \delta U \frac{\delta \left(p U^\gamma \right)}{p U^\gamma} < 0 \tag{101}$$

Let us now apply this to the specific equilibrium of figure 7.10, an axially symmetric equilibrium periodic in the axial coordinate z. Let ψ now be the flux out to a given surface, so ψ is a flux label, and let $V(\psi)$ be the volume enclosed by the surface in one period. Let $p(\psi)$ be the pressure on the flux surface ψ. Then the criteria for instability for our specific pure interchange displacement ξ, equation 99, reduces to

$$V''\left(\frac{p'}{p} + \frac{\gamma V''}{V'}\right) < 0 \tag{102}$$

where primes denote derivatives with respect to ψ. If V'' increases outward from the axis, which is usually the case, then

$$\frac{p'}{p} + \frac{\gamma V''}{V'} < 0 \tag{103}$$

is the sufficient condition for instability. The direct calculation sketched above before the physical argument gives exactly this criterion. If p increases outward, then the criterion fails and we cannot conclude either stability or instability since we have tested δW only for a subclass of all particular displacements.

Let p decrease outward, i.e., let the pressure be "confined" toward the axis. Then we get instability if

$$pV^{\gamma} \tag{104}$$

decreases outward. If p goes to zero at some value of ψ, then near this ψ, pV^{γ} must decrease and if we localize our interchange to the neighborhood of this ψ we have δW negative and the equilibrium must be unstable. This can also be seen from equation 102, since the first term goes to negative infinity while the second term is only finite. This result can be reworded as follows: Systems that confine pressure are unstable if V'' is greater than zero.

When is $V'' > 0$? Or, more generally, when does the volume per flux increase, when we choose a given direction of variation? A very simple example is provided by the earth's dipole field. The volume of a thin tube of force of flux $d\psi$ is

$$V'd\psi = \int dl dA = \int \frac{dl}{B} B dA = d\psi \int \frac{dl}{B} \tag{105}$$

since $BdA = d\psi$, so

$$V' = \int \frac{dl}{B} \tag{106}$$

the length of a tube of flux is proportional to L the distance from the center of the earth at which the tube crosses the equatorial plane. The mean field of a dipole is $\overline{B} = 1/L^3$, so

$$V' \sim L^4 \tag{107}$$

and a sufficient condition for instability is that

$$pV'^{\gamma} \sim pL^{4\gamma} \tag{108}$$

increase outward with L. (V'' is clearly positive.) Since $4\gamma = 4 \times \frac{5}{3} = \frac{20}{3}$ we would not expect the pressure to drop faster than $L^{6.67}$. If it did, the interchange instability would set in automatically and resultant convection would alter the pressure profile until it dropped at the slower rate $L^{-6.67}$.

The presumed situation is that plasma is supplied at the outer edge of the magnetosphere where it is in contact with the solar wind, and that the plasma diffuses inward by random magnetic fluctuations induced by irregularities in the solar wind. If the diffusion is rapid, then the critical gradient is exceeded, and convection induced by the interchange instability sets in and restricts the plasma to the stable pressure profile.

For small values of L the plasma is precipitated to the earth by microinstabilities at the same rate that it is supplied at the magnetosphere. As a result, a steady state is achieved that is sustained by a balance of diffusion and convective turbulence driven by the interchange.

This picture, which is only a crude approximation to what goes on, is closely analogous to the Schwarzschild convection in outer regions of solar-type stars. In both the stars and the magnetosphere the pressure profile is held close to a marginal state that is slightly on the unstable side of the marginal profile for the convective instability. The pressure profile is just unstable enough to maintain the low level of turbulence needed to balance the flow of plasma in the magnetosphere and the flow of luminosity above the radiative flux in stars.

It is striking that the observed pressure profile in the magnetosphere is close to the marginal profile that we derived by our seemingly arbitrary choice of a displacement $\boldsymbol{\xi}$ that makes $\mathbf{Q} = 0$. However, this choice, which was made to simplify the evaluation of δW, is not as arbitrary as it appears. The plasma pressure energy is small compared to the vacuum magnetic energy so any other choice for $\boldsymbol{\xi}$ would lead to a large positive value for Q^2, large compared to the contributions from the pressure term and even compared to the $\mathbf{j} \cdot \boldsymbol{\xi} \times \mathbf{Q}$ term, so that for these choices $\delta W > 0$ and we would have stability. Thus, for a strong magnetic field, the only trial $\boldsymbol{\xi}$ that we need to look at are actually the ones we chose to make $\mathbf{Q} = 0$. That is to say, our sufficient condition for instability should also be close to the necessary condition of instability.

Let us consider the volume per flux for another simple example shown in figure 7.11, that of a cylindrical field B_θ produced by a current at the origin $r = 0$, i.e., $B_\theta = \alpha/r$. If the pressure is small, we can neglect plasma currents and this is close to actual field. The result for V' is

$$V' = \int \frac{dl}{B} = \int \frac{r d\theta}{\alpha/r} = \frac{2\pi}{\alpha} r^2 \tag{109}$$

The volume per flux increases with r. Thus, when a line is concave inward, the volume per flux increases outward, while the opposite happens when the line is concave outward (provided that the pressure decreases outward). The first case, concave inward, is the unstable case of volume per flux increasing outward. But, if we remember the centrifugal force argument we see that the effective gravity is outward, which for a confined plasma with more pressure

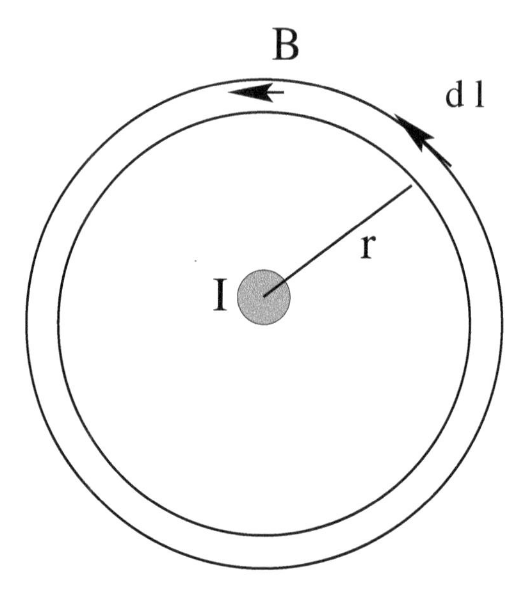

Figure 7.11. Magnetic field due to a current at the origin

on the inside is also the unstable direction. Conversely, if the line is concave in the direction of decreasing plasma the situation tends toward stability. Often a given line has both signs of curvature, and it is necessary to carry out the specific integral of equation 106 to determine whether the net volume per flux increases outward or not.

The direction and rate of increase of the volume per flux in general gives the sufficient condition for the pure interchange instability, $Q = 0$. If the pressure energy is also small compared to the magnetic energy, this gives the necessary condition. However, it turns out that if the pressure energy is not negligible, then a displacement $\boldsymbol{\xi}$ that is not a pure interchange displacement can lead to a lower value for δW. Such a perturbation could be one that is larger than the interchange $\boldsymbol{\xi}$ in the region of bad curvature and smaller in the region of good curvature, as in figure 7.12. This alteration in $\boldsymbol{\xi}$ clearly increases the amount of negative change in the pressure energy. However, in addition, we must take into account the amount of positive energy that must develop because $\mathbf{Q}$ is no longer zero. We can get a crude idea of stability by taking a $\boldsymbol{\xi}$ that is zero in the good curvature region and nonzero in the region where the curvature is bad.

If the magnetic field were uniform, we could derive a frequency for the positive energy of the magnetic field, due to the ballooning of the perturbation, and this would satisfy the dispersion relation for the wave we expect to develop, namely a toroidal Alfven wave,

$$k^2 v_A^2 = \frac{\delta W_{\text{magnetic}}}{K} \tag{110}$$

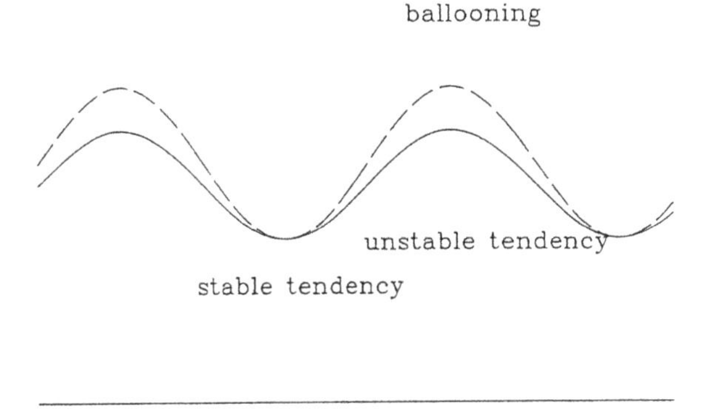

Figure 7.12. Ballooning mode

or

$$\delta W_{\text{magnetic}} = k^2 v_A^2 K \tag{111}$$

This is still approximately correct for the nonuniform field with good and bad curvature, if the perturbation is nonzero in the bad curvature region and zero in the good curvature region.

Let us estimate the pressure contribution to δW:

$$\delta W_{\text{pressure}} = \gamma^2 K \tag{112}$$

where γ is the growth rate we would expect if the bad curvature filled the equilibrium, i.e., the interchange growth rate if the curvature were everywhere bad. K is the same in both equations, so

$$\delta W = \delta W_{\text{magnetic}} + \delta W_{\text{pressure}} = \left(k^2 v_A^2 - \gamma^2\right) K \tag{113}$$

This means we have instability if

$$\gamma > k v_A \tag{114}$$

This type of perturbation is called a ballooning instability since it balloons out in the bad curvature region, much more than in the good region, just as an inner tube will balloon out at a weak spot when the inner tube is filled too full. The resulting normal mode that emerges from this situation when detailed stability calculations are carried out is called a ballooning mode and is very important.

The message is that if the interchange criterion for instability involving the volume per flux is satisfied, then the plasma is unstable to the pure interchange. This instability occurs no matter how small the plasma pressure is compared to the magnetic pressure, since the interchange makes the change in the magnetic energy zero.

On the other hand, if the interchange criterion is not satisfied, then altering the displacement from the pure interchange to emphasize the bad curvature regions of the magnetic field can lead to a negative change in the pressure energy. If this is large enough to overcome the change in the magnetic energy associated with this altered magnetic field, then the plasma is unstable to the ballooning mode. But since the change in the pressure energy is directly proportional to the pressure, the ballooning mode requires a certain nonzero amount of plasma pressure to be unstable.

7.3.4 Line Tying and Shear

A closely related situation to the above is line tying (see Kulsrud 1967). The simplest example is that of the solar atmosphere. In the solar atmosphere the lines emerge from the photosphere into the atmosphere and, by and large, return to the photosphere. Because the photosphere is so much denser than the atmosphere, the motion in the photosphere can be much smaller than the motion in the atmosphere. This is because the motions we are interested in are driven by forces in the atmosphere. A good approximation is that the lines are anchored in the photosphere as though it were a rigid wall. Thus, we would take the trial $\boldsymbol{\xi}$ to vanish at the solar surface.

Let us consider a prominence of heavy material supported against gravity by magnetic field lines (see figure 4.4). If the field lines extended indefinitely in the horizontal direction, then the prominence would be subject to the gravitational driven interchange instability and the growth rate would be given by $\delta W/K$, where

$$\delta W \approx -\int \frac{\rho g}{H} \xi^2 dl \tag{115}$$

and the integral is taken along a line if force. But the lines are anchored by the photosphere over a length L, giving rise, as in the ballooning mode, to an additional positive energy

$$\left(\frac{\pi}{L}\right)^2 \int \rho v_A^2 \xi^2 dl \tag{116}$$

so the criteria for instability is roughly

$$\gamma_0^2 = \frac{g}{H} > v_A^2 \left(\frac{\pi}{L}\right)^2 \tag{117}$$

where γ_0 is the growth rate in the absence of line tying (see problem 5). We are thus led to a criteria reminiscent of the ballooning mode

$$\gamma_0 > \pi \frac{v_A}{L} \tag{118}$$

which has a clear interpretation. If the mode can locally double its amplitude in a time comparable to the time it takes for Alfven wave to travel the length L and back to find out that it is actually line tied, then the larger amplitude is not modified significantly by the line tied message, which only knows about the smaller amplitude that the Alfven wave carried down and back. Thus, under the condition of equation 118, the mode will continue to grow. If the

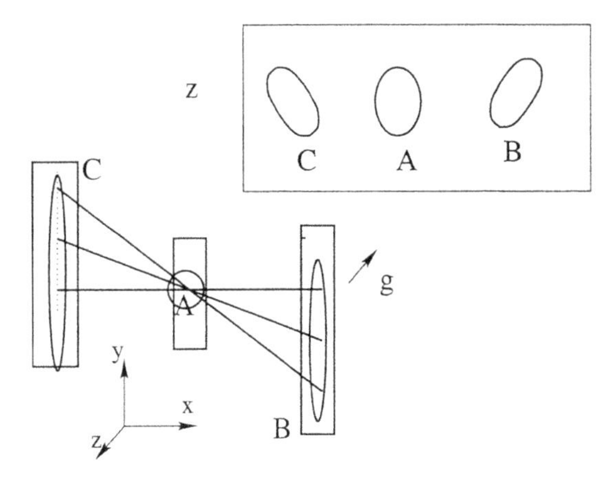

Figure 7.13. Shearing of the pure interchange modes

lines supporting the prominence are not too long, then the mode is stabilized and prominence can be stably supported by the magnetic field.

A further related stabilization occurs if the lines of force in the system are not parallel but are sheared relative to each other (see figure 7.13). Let us model this by a field

$$\mathbf{B} = B_0\hat{x} + \frac{z}{L_S}B_0\hat{y} \tag{119}$$

with g in the negative z direction. Now let us examine what happens with a pure interchange convective cell at $x = 0$. As we move along the line $x = z = 0$ the interchange is undisturbed. But if we follow another line at $z = \Delta$, the line moves sideways at the rate $dy/dx = \Delta/L_S$. When $x = L_S$ the line has been displaced in y a distance Δ, and as x increases still further, the line displacement increases indefinitely. To have a finite perturbation, $\boldsymbol{\xi}$ must be made to vanish eventually and the maximum growth rate occurs when this length is approximately L_S. Thus, the shear seems to provide a wall at $x = L_S$, at which the effective line tying occurs. The corresponding stability criterion is

$$\gamma_0 \approx \sqrt{\frac{g}{H}} > \frac{\pi v_A}{L_S} \tag{120}$$

This can be verified qualitatively by a model $\boldsymbol{\xi}$ and an estimate of δW. In fact, we can prove from the energy principle that equation 120 gives a sufficient condition for instability within a factor of order unity.

Roughly, the physical argument is as follows: Let the extent of the mode be some value L. When L is longer than L_S the kinetic factor K starts to increase as L^2, rather than L. This increase in the denominator reduces the growth rate γ_0. On the other hand, unstable energy in the numerator increases as L, so for $L > L_S$ the estimated growth rate (without line tying, $\sqrt{-\delta W/K}$), decreases as $1/L$. But the reciprocal Alfven wave time grows as L, so if line tying stabilizes the mode when $L = L_S$, it will continue to stabilize it for larger L. On the other hand, if it doesn't stabilize the mode at $L = L_S$ the mode is unstable. Thus, this choice for L results in the approximately correct stability criteria.

7.4 The Magnetorotational Instability (MRI)

An important instability that involves rotation is the MRI instability. It occurs primarily in accretion disks. In this instability the action is opposite to those we have discussed before. Here, kinetic energy of rotation is converted to magnetic energy rather than the reverse direction, as in the interchange in which the instability converts magnetic and pressure energy into kinetic energy.

This instability is considered very important because it seems to underlie the greatly enhanced viscous dissipation inferred to be present when matter falls onto a compact star from a companion. Because the companion is circling about the compact star, the infalling matter has considerable angular momentum, which produces a rotational barrier to direct infall of the matter onto the compact object. Instead, the infalling matter goes into a Keplerian orbit about the compact object. When a sufficient amount of the infalling matter accumulates, it forms a rotating disk about the object. Because the circular velocity of the fluid is Keplerian, there is differential rotation and any viscosity leads to an evolution of the density distribution of the disk, the inner regions of which rotate faster, but have lower angular momentum compared to the slower rotating outer regions.

As a result, the inner regions would lose angular momentum and move inward, while the outer regions if not perturbed by more infalling matter, would gain angular momentum and move outward. However, this outer motion is suppressed by the continuing flow of new matter with new angular momentum from the companion star. The net consequence is that, due to viscosity, matter everywhere flows in toward the compact object. As it settles into the deeper gravitational well of the compact object it lowers its gravitational energy. This lowering of energy is actually due to the viscous dissipation resulting from the differential rotation. Presumably this heat of dissipation coming from the changed gravitational energy is converted to radiation, usually X-rays, which is how the object is detected.

This scenario is due to Shakura and Sunyaev (1973), and it is generally referred to as the Shakura–Sunyaev model of accretion disks. Their picture explains many of the observations. Unfortunately, there is one serious drawback to their picture, namely, that the actual molecular viscosity is much too small to lead to the speed of inflow inferred from the observations of the emitted energy from the majority of the observed objects. Shakura and Sunyaev get around this objection by postulating that there is some process, due to macroscopic turbulence, that leads to a large enough value of the effective viscosity to produce the observed large energy generation. They parameterize this assumed value for the viscosity ν as

$$\nu = \alpha H c_s \tag{121}$$

where H is the thickness of the disk and c_s is the speed of sound in the disk, while the coefficient α is a parameter of order unity, which characterizes the unknown process that produces the anomalous viscosity. The process has become known as the α effect, and the inferred viscosity is called the α viscosity. It turns out that to fit the observations α must be somewhere between 0.1 and 1.

Many attempts were made to discover a process that could produce such a large viscosity. These attempts were unsuccessful until 1991 when Balbus and Hawley (1991) rediscovered an instability that had originally been found by Velikov (1959) and Chandrasekhar (1961) in the late 1950s. Before it was appreciated that Balbus and Hawley's discovery was actually a rediscovery of the Chandrasekhar–Velikov instability, the instability was generally called the Balbus–Hawley instability. During the 1990s it was known by that name. It is now called the magnetorotational, or MRI, instability in deference to its many discoverers—Chandrasekhar and Velikov for their original discovery and Balbus and Hawley for their emphasis on its importance for accretion disks and compact objects.

How does this instability work? It was immediately appreciated that a magnetic field with a sizable radial and toroidal component in the disk would exert a torque, "slowing down" the inner disk and "accelerating" the outer disk. Of course, "slowing down" really means decreasing the angular momentum and "accelerating" means increasing the angular momentum, so "slowing down" results in an increased angular velocity, and, correspondingly, "acceleration" results in a decreased angular velocity!

A basic problem with the mechanism of magnetic tension producing the required viscosity is that to make α of order unity, $B_r B_\theta / 4\pi$ must be comparable with ρc_s^2. On the other hand, line stretching makes B_θ increase linearly with time, so that it tends to be much larger than B_r, so for α to be of order of unity we must have $B_\theta^2/4\pi \gg \rho c_s^2$. Such a magnetic field is too strong to be confined to the disk. This problem is resolved by the MRI instability, which leads to the production of a radial field from the θ field at the same rate as the toroidal field is produced by the radial field, so that the two fields can be comparable in magnitude. Presumably, magnetic reconnection of the resulting (somewhat turbulent) field continually destroys the fields at the rate at which they are generated, so the combination of the MRI instability and magnetic reconnection can lead to a field with $B_r^2 \approx B_\theta^2 \approx 4\pi\rho c_s^2$. A number of detailed numerical simulations of the MRI have led to such fields. The simulations also yield $\alpha \approx 0.1$, so this does seem to provide a mechanism to generate the desired effective viscosity proposed by Shakura and Sunyaev. Note that the actual heating mechanism is the destruction of magnetic energy by resistive magnetic reconnection.

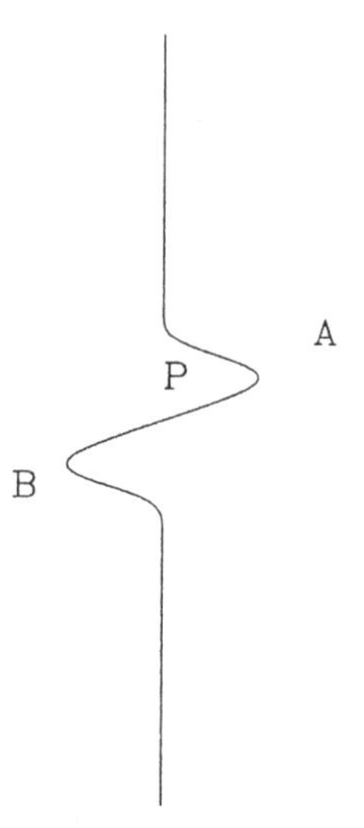

Figure 7.14. The distorted field line in the MRI

Let us now turn to a description of the MRI itself. The physics of the linearized MRI is similar for the case of an equilibrium field that is a pure θ field and the case where it is a pure z field. The latter case, however, is much easier to understand intuitively so we restrict ourselves to this case.

Imagine that we start with a differentially rotating disk, which for definiteness we assume is in nearly Keplerian rotation:

$$\Omega(r) = \frac{\beta}{r^{3/2}} \tag{122}$$

(where β is a constant), and which possesses a magnetic field in the z direction. Let us displace a fluid element in the radial direction by $\delta r = \epsilon \cos kz$, as in figure 7.14. Conditions adjust so that the displaced fluid element immediately adapts its angular velocity to be the Keplerian velocity at the displaced position. Then, since the resulting angular velocities vary along the disturbed line of force, it is stretched and a perturbed B_θ field is produced.

The resulting picture of the line of force at A and at B is shown in figure 7.15. The tension of the the distorted magnetic field line is such as to transfer angular momentum outward to A from B. If the fluid element A remained at the same radius, it could not increase its angular momentum, which is fixed by the Keplerian motion. The only way it can increase its angular momentum is to increase its radius further. Similarly, the loss of angular momentum by B causes it to move inward.

But these two motions enhance the original perturbation of figure 7.14, which leads to the instability. Since the mechanism involves magnetic tension we expect the growth to involve the magnetic field and a simple estimate shows that $\gamma \approx kv_A$, with $v_A = B_z/\sqrt{4\pi\rho}$. In this simplified picture we have neglected the the radial magnetic tension that would disturb the Keplerian

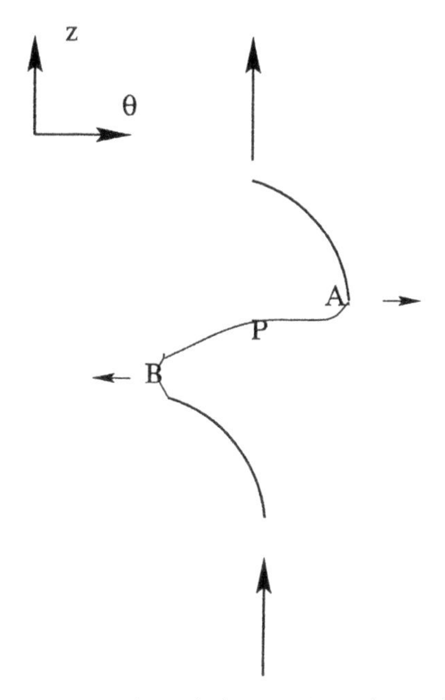

Figure 7.15. The radial separation of A and B

motion. The actual critical criterion for instability turns out to be

$$\omega_A = kv_A < \sqrt{3}\Omega \tag{123}$$

ω_A depends on the vertical wave number of the perturbation k. For small k, as it increases the growth rate increases until $\omega_A \sim \Omega$. Thus, the growth rate maximized over k is of order of the angular rotational velocity. Therefore, the growth is quite fast, since Ω is large. The instability is indeed a powerful one.

Let us examine the mathematical derivation in some detail to see how it reflects the physical picture just given. The Euler equation can be written

$$\rho\left(\frac{\partial \mathbf{V}}{\partial t} + \mathbf{V}\cdot\nabla\mathbf{V}\right) = -\nabla\left(p + \frac{B^2}{8\pi}\right) + \frac{\mathbf{B}\cdot\nabla\mathbf{B}}{4\pi} \tag{124}$$

Linearizing we write $\mathbf{V} = r\Omega\hat{\theta} + \mathbf{u}$ and $\mathbf{B} = B_z^0\hat{z} + \mathbf{b}$. We assume incompressibility and look for axisymmetric modes, $\partial/\partial\theta = 0$, so

$$\nabla\cdot\mathbf{u} = \frac{1}{r}\frac{\partial(ru_r)}{\partial r} + \frac{\partial u_z}{\partial z} = 0 \tag{125}$$

We assume further that the perturbation is localized in r to a region small compared to r, so that we can take a pure Fourier mode

$$\mathbf{u} = \hat{\mathbf{u}}e^{\lambda t + i(k_r r + k_z z)} \tag{126}$$

and we further assume that $k_z \gg k_r$.

Before writing out the equations, let us consider the linearized value of $\mathbf{V} \cdot \nabla \mathbf{V}$. Taking the cylindrical symmetry into account we find by a straight forward computation that the r and θ components are

$$
\begin{aligned}
(\mathbf{V} \cdot \nabla \mathbf{V})_1 &= (r\Omega)\hat{\theta} \cdot \nabla \mathbf{u} + \mathbf{u} \cdot \nabla (r\Omega\hat{\theta}) \\
&= \left(r\frac{d\Omega}{dr} + 2\Omega \right) u_r \hat{\theta} - 2u_\theta \Omega \hat{r} \\
&= \tfrac{1}{2}\Omega u_r \hat{\theta} - 2u_\theta \Omega \hat{r}
\end{aligned}
\tag{127}
$$

where we use the Keplerian relation, equation 122, for Ω to obtain the last line. The $\hat{\theta}$ component of this equation comes from the Coriolis force, while the $\hat{r}$ component is the perturbed centrifugal force. Substituting this into the equation of motion, equation 124, we find its r, θ, and z components are

$$
\begin{aligned}
\lambda u_r - 2\Omega u_\theta &= -ik_r \frac{P_1}{\rho} + \frac{ik_z B_0 b_r}{4\pi\rho} \\
\lambda u_\theta + \frac{\Omega}{2} u_r &= \frac{ik_z B_0 b_\theta}{4\pi\rho} \\
\lambda u_z &= -ik_z \frac{P_1}{\rho}
\end{aligned}
\tag{128}
$$

where $P = p + B^2/8\pi$ is the total pressure and P_1 is the linear part.

We can eliminate P_1 from the r and z equations and also eliminate u_z from the incompressibility condition, equation 125. If $k_z \gg k_r$, then the u_r equation becomes

$$
\lambda u_r - 2\Omega u_\theta = ik_z \frac{B_0 b_r}{4\pi\rho}
\tag{129}
$$

or, in other words, with incompressibility and $k_z \gg k_r$, the perturbed pressure drops out of the r equation. We find $\mathbf{b}$ from the induction equations

$$
\begin{aligned}
\lambda b_r &= ikB_0 u_r \\
\lambda b_\theta &= -\tfrac{3}{2}\Omega b_r + ikB_0 u_\theta
\end{aligned}
\tag{130}
$$

where from now on we replace k_z with k since they are essentially equal. The first term on the right of the second equation arises from the shearing of the b_r field. The other terms represent the shearing of the B_0 field by the u_r and u_z motions. Eliminating b_r from the b_θ equation, we can write

$$
\lambda b_\theta = -\tfrac{3}{2}\frac{ikB_0\Omega}{\lambda} u_r + ikB_0 u_\theta
\tag{131}
$$

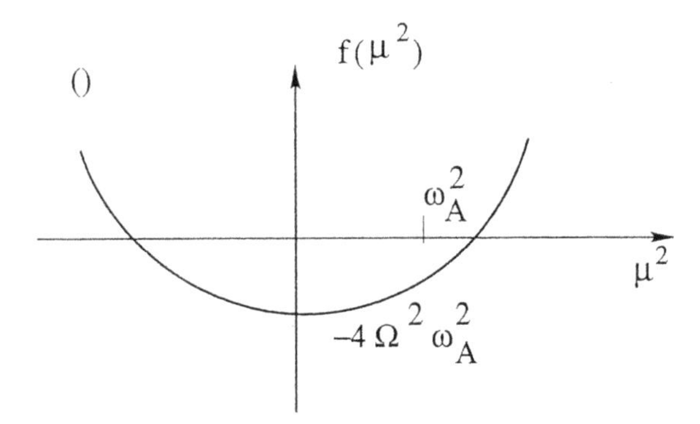

Figure 7.16. $f(\omega^2)$

Then substituting the expression for b_θ and b_r into the u_r and u_θ equations we get

$$(\lambda^2 + \omega_A^2)u_r = 2\lambda\Omega u_\theta$$

$$(\lambda^2 + \omega_A^2)u_\theta = \left(-\frac{\lambda}{2}\Omega + \frac{3}{2}\frac{\omega_A^2\Omega}{\lambda}\right)u_r \tag{132}$$

where $\omega_A^2 = k^2B_0^2/4\pi\rho$. Setting the determinant of these equations to zero, we get the dispersion relation for λ^2. Letting

$$\mu^2 = \lambda^2 + \omega_A^2 \tag{133}$$

the dispersion relation can be written

$$f(\mu^2) = \mu^4 + \Omega^2\mu^2 - 4\Omega^2\omega_A^2 = 0 \tag{134}$$

$f = 0$ is a quadratic equation in μ^2 with real roots. $f(\mu^2)$ is sketched in figure 7.16. We have drawn it so that the positive roots for μ^2 are greater than ω_A^2, so that λ is real. For the negative roots of λ^2, λ is imaginary and these roots represent stable modes.

The condition that the positive root of μ^2 is greater than ω_A^2 is found by evaluating $f(\omega_A^2)$

$$f(\omega_A^2) = \omega_A^4 - 3\Omega^2\omega_A^2 \tag{135}$$

If $f(\omega_A^2)$ is negative, then we see from the figure that the root μ^2 is greater than ω_A^2, and so $\lambda^2 > 0$. Hence, one of the λ is real and positive, corresponding to instability. Thus, the instability condition is

$$\frac{k^2B^2}{4\pi\rho} = \omega_A^2 < 3\Omega^2 \tag{136}$$

The larger k modes are stable because the restoring tension dominates the rotational terms. The smaller k modes are unstable. Note that very weak B_0 fields are unstable, but if B_0 is large, then k has to be small so that the modes will not fit into the disk. Thus, there is a limit to how large B_0 can be and still trigger the MRI. (B has to be such that $v_A/H < 3\Omega$.) However, B_0 fields within this limit can still lead to a significant value for α.

Although the above calculation is not too complicated to follow in detail it is still sufficiently complicated that the simple intuitive picture that we have given is lost in the details. We can recover the simple picture by going to the limiting case $\omega_A \ll \Omega$.

In this limit we find, from the dispersion relation, that λ is of order ω_A for the unstable root and of order Ω (actually of order $\pm i\Omega$) for the stable roots. Let us try both approximations in our examination of the equations for the MRI instability. First, assume $\lambda = O(\omega_A)$. The important equations we concentrate on are equations 129 for u_r, the second of equations 128 for u_θ, and equations 130 for b_r and b_θ. From the b_r equation

$$b_r \approx \frac{kB_0u_r}{\lambda} \tag{137}$$

so that in the u_r equation the b_r term is of order $(k^2B_0^2/4\pi\rho)u_r/\lambda = (\omega_A^2/\lambda)u_r$, and is the same size as λu_r. But Ωu_θ cannot be larger than either of these, so

$$u_\theta \leq \frac{\lambda u_r}{\Omega} \ll u_r \tag{138}$$

This is the condition, in the intuitive picture, that the rotation at r must always be close to the Keplerian rotation. (The u_r equation does not yield any further information, since the smaller value of u_θ adds to the b_r term to give λu_r, and we do not care what its value is.)

Now, in the b_θ equation 130 we see from equation 138 that the u_θ term is small, so the b_θ time derivative comes only from the finite differential rotation $-\frac{3}{2}\Omega$ acting on the perturbed b_r. Finally, in the λu_θ equation, the u_r and the b_θ terms must balance, since they are both greater than the u_θ term ($\lambda \ll \Omega, u_\theta \ll u_r$). Thus, the b_θ term, which is the tension term that transfers angular momentum outward, "drives" the Coriolis term, and forces the plasma to move further outward in the r direction.

Putting all this together we have

$$b_\theta = -\frac{3}{2}\frac{\Omega}{\lambda}b_r \tag{139}$$

from the b_θ equation, and

$$b_r = \frac{ikB_0u_r}{\lambda} \tag{140}$$

from the b_r equation. Also, we have

$$u_r = \frac{ikB_0}{4\pi\rho}\left(\frac{2}{\Omega}\right)b_\theta \tag{141}$$

from the u_θ equation. Eliminating b_r and u_r from these equations, we get

$$b_\theta = -\frac{3\Omega}{2\lambda}\frac{ikB_0}{\lambda}\frac{ikB_0}{4\pi\rho}\frac{2b_\theta}{\Omega} = \frac{3\omega_A^2}{\lambda^2}b_\theta \tag{142}$$

and $\lambda = \pm\sqrt{3}\omega_A$. The upper sign gives the instability. This result agrees with the exact equation for λ in the assumed limiting case. It is clear that these equations correspond exactly to the intuitive picture of the instability given at the beginning of this section.

For λ of order Ω, it is easy to see, by similar reasoning, that the magnetic terms in the u_r and the u_θ equations are negligible and $u_r \sim u_\theta$. In this case we get

$$\lambda^2 = -\Omega^2 \tag{143}$$

or, in other words, the epicyclic motion, which is equal to the Keplerian motion. The fluid elements are all on elliptical orbits near circular orbits, but at all toroidal angles θ they are in phase, so that the fluid velocity is still axisymmetric.

7.5 Problems

1. (Self-adjointness) If it is known that the operator $\mathbf{F}(\boldsymbol{\xi})$ is self-adjoint when $\boldsymbol{\xi}$ is a real vector field, what can be said about self-adjointness when $\boldsymbol{\xi}$ and $\boldsymbol{\eta}$ are complex? Is $\int \boldsymbol{\eta}^* \cdot \mathbf{F}(\boldsymbol{\xi}) = \int \boldsymbol{\xi}^* \cdot \mathbf{F}(\boldsymbol{\eta})$? Or is $\int \boldsymbol{\eta} \cdot \mathbf{F}(\boldsymbol{\xi}) = \int \boldsymbol{\xi} \cdot \mathbf{F}(\boldsymbol{\eta})$?

Assume that $\mathbf{F}(\boldsymbol{\xi})$ is self-adjoint, where $\mathbf{F}(\boldsymbol{\xi})$ is the force such that any displacement $\boldsymbol{\xi}(\mathbf{r})$ satisfies

$$\rho \frac{\partial^2 \boldsymbol{\xi}}{\partial t^2} = \mathbf{F}(\boldsymbol{\xi})$$

Show directly that the energy

$$\mathcal{E} = \frac{1}{2} \int \left[\rho \left(\frac{\partial \boldsymbol{\xi}}{\partial t} \right)^2 - \boldsymbol{\xi} \cdot \mathbf{F}(\boldsymbol{\xi}) \right] d^3x$$

is a constant in time.

2. (The ball on the hill) Consider a charged ball (charge q and mass M) near the top of a round hill in a vertical magnetic field B. Show that the linearized equations of motion can be written

$$\ddot{x} + \nu\dot{x} - \lambda_0^2 x = \Omega\dot{y}$$

$$\ddot{y} + \nu\dot{y} - \lambda_0^2 y = -\Omega\dot{x}$$

where ν is a small frictional coefficient, $\Omega = qBMc$, and λ_0 is the growth rate in the absence of the field and viscosity. Show that if $\nu = 0$ and we take the perturbations proportional to $e^{\lambda t}$, then λ has four roots that are given by

$$2\lambda = \pm i\Omega \pm \sqrt{4\lambda_0^2 - \Omega^2}$$

and that if $\Omega > 2\lambda_0$, then all the roots are imaginary and the top of the hill is a stable position for the ball.

Finally, when the ball is thus stabilized, calculate the change λ_1 in each of the eigenvalues λ if ν is nonzero but very small and show that they are given by

$$\frac{\lambda_1}{\nu} = \frac{\sqrt{\Omega^2 - 4\lambda_0^2} \pm i\Omega}{\sqrt{\Omega^2 - 4\lambda_0^2}}$$

Which roots have become unstable? This instability is called a secular instability. Note that the perturbed potential energy, which is all gravitational energy, is always negative, so when $\nu = 0$ the energy principle fails. However, when $\nu \neq 0$ we again have instability but with a slow growth rate.

3. (Pinch instability) Show that in cylindrical coordinates,

$$p = p_0, \quad B = 0; \qquad r < a$$

$$p = 0, \quad B = B_0 \frac{a}{r}; \qquad a < r < b$$

where p_0 is a constant, is an equilibrium solution if $p_o = B_0^2/8\pi$. (Note that even though there is a discontinuity, the pressure in the inner region balances the magnetic pressure in the outer region across $r = a$.)

Show that if we choose $\boldsymbol{\xi}_\theta = 0$, $\xi_r = \hat{\xi}_r(r)\cos(kz)$, and $\xi_z = \hat{\xi}_z(r)\sin(kz)$ with $\hat{\xi}_r$ chosen to make $\nabla \cdot \boldsymbol{\xi} = 0$ inside and to make

$$r\frac{\partial}{\partial r}\left(\frac{\hat{\xi}_r}{r}\right) + \frac{\partial \hat{\xi}_z}{\partial z} = 0$$

outside, then $\mathbf{Q} = \nabla \times (\boldsymbol{\xi} \times \mathbf{B}) = 0$ outside $r = a$. Then show that only the surface term of δW, equation 77, remains, and

$$2\delta W = -\frac{\pi}{2}\int adz\hat{\boldsymbol{\xi}}_r^2(a)\frac{B_0^2}{8\pi a}$$

so that the surface current pinch is unstable.

4. (MHD waves) The theory of MHD waves is really a special case of the energy principle theory. We can write the force $\mathbf{F}(\boldsymbol{\xi})$ for sinusodial waves as ρ times a constant, three-dimensional matrix M times the three-dimensional vector amplitude (ξ_x, ξ_y, ξ_z). This is already done for two components in equation 20 in chapter 5. What does the full matrix look like? What is δW in terms of M? Show that self-adjointness is equivalent to the symmetry of M. Demonstrate orthogonality of the three MHD modes. Show that the lowest eigenvalue for a fixed k (i.e., the frequency of the slow mode) is gotten by minimizing

$$\omega^2 = \frac{\boldsymbol{\xi} \cdot M \cdot \boldsymbol{\xi}}{\boldsymbol{\xi} \cdot \boldsymbol{\xi}}$$

By taking $\xi_x = \cos\theta$, $\xi_y = 0$, $\xi_z = -\sin\theta$ show that

$$\omega_s^2 < k^2 v_A^2 \cos^2\theta$$

and thus show that the slow mode is slower than the intermediate mode.

5. (Stability of a prominence) Suppose that a prominence is supported 10^9 cm above the photosphere by a horizontal magnetic field of 100 G. Take the vertical half-thickness of the prominence to be 10^9 cm and its density 10^{10}cm^{-3}. Take the solar gravity as 2×10^4 cm/ sec^2. Are the lines greatly compressed by the weight of the prominence?

If the prominence density is a smooth function of z (say a parabolic function), what is the approximate timescale for the growth of the gravitational interchange instability? If the lines reach the photosphere, where they are anchored 10^{10} cm from the center of the prominence, do you think the instability is stabilized by line tying? Calculate v_A using the prominence density.

6. (Self-adjointness) Prove the self-adjointess of the operator $\mathbf{F}(\boldsymbol{\xi})$, for the case $\mathbf{g} = 0$. Break $\mathbf{F}$ into $\mathbf{F} = \mathbf{F}_1 + \mathbf{F}_2$ and show $\mathbf{F}_1$ and $\mathbf{F}_2$ are each separately self-adjoint. First, for

$$\mathbf{F}_1(\boldsymbol{\xi}) = \frac{1}{4\pi}\{\nabla \times [\nabla \times (\boldsymbol{\xi} \times \mathbf{B})]\} \times \mathbf{B} + \nabla[\gamma p(\nabla \cdot \boldsymbol{\xi})]$$

show that

$$\begin{aligned}
\boldsymbol{\eta} \cdot \{\nabla \times [\nabla \times (\boldsymbol{\xi} \times \mathbf{B})]\} \times \mathbf{B} &= -(\boldsymbol{\eta} \times \mathbf{B}) \cdot \nabla \times [\nabla \times (\boldsymbol{\xi} \times \mathbf{B})] \\
&= \nabla \cdot \{(\boldsymbol{\eta} \times \mathbf{B}) \times [\nabla \times (\boldsymbol{\xi} \times \mathbf{B})]\} \\
&\quad - \nabla \times (\boldsymbol{\eta} \times \mathbf{B}) \cdot \nabla \cdot (\boldsymbol{\xi} \times \mathbf{B})
\end{aligned}$$

The divergence term integrates to zero and the last term is symmetric under the interchange of $\boldsymbol{\xi}$ and $\boldsymbol{\eta}$ from which the self-adjointness of the first operator in $\mathbf{F}_1$ follows. Show in a similar way that the second term is also self-adjoint.

The expression for $\mathbf{F}_2$ is

$$\begin{aligned}
F_2 &= \mathbf{j} \times [\nabla \times (\boldsymbol{\xi} \times \mathbf{B})] + \nabla(\boldsymbol{\xi} \cdot \nabla p) \\
&= \mathbf{j} \times (\mathbf{B} \cdot \nabla\boldsymbol{\xi}) - \mathbf{j} \times (\boldsymbol{\xi} \cdot \nabla\mathbf{B}) - \mathbf{j} \times \mathbf{B}(\nabla \cdot \boldsymbol{\xi}) + \nabla(\boldsymbol{\xi} \cdot \nabla p)
\end{aligned}$$

We wish to show that $\int \boldsymbol{\eta} \cdot \mathbf{F}_2(\boldsymbol{\xi}) = \int \boldsymbol{\xi} \cdot \mathbf{F}_2(\boldsymbol{\eta})$. Consider the third term in $\boldsymbol{\eta} \cdot \mathbf{F}_2(\boldsymbol{\xi})$ minus the fourth term in $\boldsymbol{\xi} \cdot \mathbf{F}_2(\boldsymbol{\eta})$

$$-\boldsymbol{\eta} \cdot (\mathbf{j} \times \mathbf{B})(\nabla \cdot \boldsymbol{\xi}) - \boldsymbol{\xi} \cdot \nabla(\boldsymbol{\eta} \cdot \nabla p)$$

and use $\mathbf{j} \times \mathbf{B} = \nabla p$ to show that this reduces to

$$-\nabla \cdot [\boldsymbol{\xi}(\boldsymbol{\eta} \cdot \nabla p)]$$

which integrates to zero, and similarly for the third term of $\boldsymbol{\xi} \cdot \mathbf{F}_2(\boldsymbol{\eta})$ and the fourth term of $\boldsymbol{\eta} \cdot \mathbf{F}_2(\boldsymbol{\xi})$.

The first term in $\boldsymbol{\eta} \cdot \mathbf{F}_2(\boldsymbol{\xi})$ can be written

$$\begin{aligned}
\boldsymbol{\eta} \cdot \mathbf{j} \times (\mathbf{B} \cdot \nabla)\boldsymbol{\xi} &= \nabla \cdot [\mathbf{B}\boldsymbol{\eta} \cdot (\mathbf{j} \times \boldsymbol{\xi})] \\
&\quad - (\nabla \cdot \mathbf{B})\boldsymbol{\eta} \cdot \mathbf{j} \times \boldsymbol{\xi} - (\mathbf{B} \cdot \nabla)\boldsymbol{\eta} \times \mathbf{j} \cdot \boldsymbol{\xi} \\
&\quad - \boldsymbol{\eta} \times [(\mathbf{B} \cdot \nabla)\mathbf{j}] \cdot \boldsymbol{\xi}
\end{aligned}$$

The first term on the right-hand side integrates to zero, and the second term vanishes. Combine the remaining terms with the rest of terms in

$\eta \cdot \mathbf{F}_2(\xi) - \xi \cdot \mathbf{F}_2(\eta)$ and show that the sum is

$$\eta \cdot [-(\mathbf{j} \cdot \nabla)\mathbf{B} \times \xi + (\xi \cdot \nabla)\mathbf{B} \times \mathbf{j} + (\nabla\mathbf{B}) \cdot (\xi \times \mathbf{j})]$$

For this you need to show that $\mathbf{B} \cdot \nabla\mathbf{j} = \mathbf{j} \cdot \nabla\mathbf{B}$, which comes from the curl of $\mathbf{j} \times \mathbf{B} = \nabla p$.

Finally, show that the bracket vanishes for any ξ, $\mathbf{B}$, and $\mathbf{j}$ either directly by a subscript calculation or vectorially by expanding

$$[(\mathbf{j} \times \xi) \times \nabla] \times \mathbf{B}$$

in two different ways: first, by expanding the bracket as a triple product, and, second, by treating $\mathbf{j} \times \xi$ as a single vector and expanding out the resulting triple product. As a result conclude the self-adjointness of $\mathbf{F}$.

7. (Parker instability) What is the timescale for the maximum growth of the Parker instability in the interstellar medium. Take $g = 10^{-9}$cm /sec ^{2}, $v_A = 10^6$ cm/sec.

References

Balbus, S. A., and J. F. Hawley. 1991. *Astrophysical Journal* **376**, 214.

Bernstein, I. B., E. Frieman, M. Kruskal, and R. Kulsrud. 1958. *Proceedings of the Royal Society* **244**, 17.

Chandrasekhar, S. 1961. *Hydrodynamic and Hydromagnetic Stability*, p. 384, Oxford University Press, London.

Courant, R. and D. Hilbert. 1953. *Methods of Mathematical Physics*, Vol. 1, p. 49, Interscience, New York.

Frieman, E., and R. M. Kulsrud. 1958. *Problems in Hydromagnetics*, in *Advances in Applied Mechanics*, Vol. 5, p. 195, Academic Press, New York.

Frieman, E., and M. Rotenberg. 1960. *Reviews of Modern Physics*, **32**, 898.

Kulsrud, R. M. 1964. *General Stability Theory in Plasma Physics*, in Proceedings of the International School of Physics, Enrico Fermi, Course XXV, Advanced Plasma Theory, p. 54, Academic Press, New York.

Kulsrud, R. M. 1967. *Plasma Instabilities*, in Proceedings of the International School of Physics, Enrico Fermi, Course XXXIX, Plasma Astrophysics, p. 46, Academic Press, New York.

Laval, G., C. Mercier, and R. Pellat. 1965. *Journal of Nuclear Fusion*, **5**, 156.

Parker, E. N. 1966. *Astrophysics Journal* **145**, 811.

Parker, E. N. 1967. *Astrophysics Journal* **149**, 535.

Shakura, N. I., and R. A. Sunyaev. 1973. *Astronomy and Astrophysics* **24**, 337.

Schwarzschild, M. 1958. *Structure and Evolution of Stars*, p. 25, Princeton University Press, Princeton, NJ.

Velikov, E. P. 1959. *Soviet Physics, JETP* **9**, 995.

Zweibel, E. G., and R. M. Kulsrud. 1975. *Astrophysics Journal* **201**, 63.

Chapter 8

COLLISIONS AND THE BRAGINSKI EQUATIONS

8.1 Introduction

In the last few chapters we have treated a plasma as a fluid even though we are aware that it consists of many individual electrons and ions. The justification for this has been that the collision rate is large and the mean free path is small. As a consequence, if we look at a small piece of plasma for a time long compared to a collision time but short compared to an MHD evolution time (and similarly for its scale size), then each particle distribution function would relax to a local Maxwellian distribution function

$$f = \frac{n(\mathbf{r}, t)}{\sqrt{[2\pi T(\mathbf{r}, t)/m]^3}} \exp - \frac{m[\mathbf{v} - \mathbf{U}(\mathbf{r}, \mathbf{t})]^2}{2T(\mathbf{r}, t)} \tag{1}$$

characterized by the three parameters: density n, temperature T, and mean velocity $\mathbf{U}$. These, as far as collisions go, can be arbitrary functions of space and time. However, we expect the next correction to evolve these parameters and, indeed, this leads to the fluid equations that determine the time dependence of n, $\mathbf{U}$, and T. If the fluid settles down to a steady state, evolution then occurs on a longer timescale due to the processes of viscosity, thermal condition, and resistivity.

Formally, we arrive at these results by deriving a general equation for the distribution function f, the Fokker–Planck equation. Then the fluid equations arise from a formal expansion of this Fokker–Planck equation in the limit of strong collisions. The method of expansion is an extension of a method first developed by Chapman and Enskogg (Chapman and Cowling 1960) to solve the Boltzmann equation for collisional evolution of a neutral gas, now applied to a plasma. Although this expansion is formal, it leads to a useful set of equations that represent the complexities of the dissipative processes in a complete and useful form. In the presence of a magnetic field, there are too many complexities—perpendicular and parallel heat conductivities, viscous processes, and resistivity—for it to be possible to write them down ab initio from purely intuitive reasoning. However, once they are derived formally, the various terms in them lend themselves to simple physical interpretation.

In this chapter we proceed first with a discussion of simple two-body collisions. We next treat the effect of these collisions statistically to arrive at the Fokker–Planck equation for the distribution function of electrons and ions. Then we expand the solution of the Fokker–Planck equation in the strong collision limit of this equation to arrive at the well known Braginski equations. We finally apply these equations to a simple problem to illustrate their use. The formal expansion processes parallel the description in the first paragraph.

8.2 Binary Collisions

Consider the path of a single electron in a collisional plasma. As it moves it passes many ions and other electrons. The Coulomb force of these charged particles changes its velocity and therefore its path. If it passes another particle within a distance b of order

$$h = \frac{e^2}{\mathcal{E}} \tag{2}$$

where $\mathcal{E}$ is its energy, then its velocity $\mathbf{v}$ is changed by a finite amount and its path is finitely bent. On the other hand, if it passes a distance $b \gg h$, its velocity is changed by a much smaller value. If b is larger than the mean interparticle distance $n^{-1/3}$, then during the time it takes to pass a single particle, b/v, it will pass many others. Thus, simultaneous "collisions" are superimposed on the electron. However, since the changes in velocity and the deviations from its undisturbed straight line motion are small, we can simply add up the effect of these collisions independently. Since it is logical that the particles it encounters are randomly distributed in location relative to our electron, these collisions are random and can be added statistically.

Now, the actual time for the particle to change its velocity by an appreciable amount is dominated by the weak small-angle collisions because there are so many more of them than the large-angle collision ones. The picture we should have is that between two close encounters with other charged particles the electron has finitely changed its velocity many times, in direction and magnitude. Thus, even if we ignore large-angle collisions entirely (in a manner that will be made clear), we still have a reliable result for the statistics of the electron's motion. Therefore, we restrict our attention to weak collisions.

An electron encounters other electrons and much more massive ions. Similarly, we also wish to determine the motion of the ions due to collisions with other ions and even with other electrons. It turns out that by introducing the reduced mass for a collision between particles of mass m_A and mass m_B,

$$\mu = \frac{m_A m_B}{m_A + m_B} \tag{3}$$

we can treat all four types of collisions simultaneously (in a well-known way). Namely, the equations of motion for two particles, A and B, are

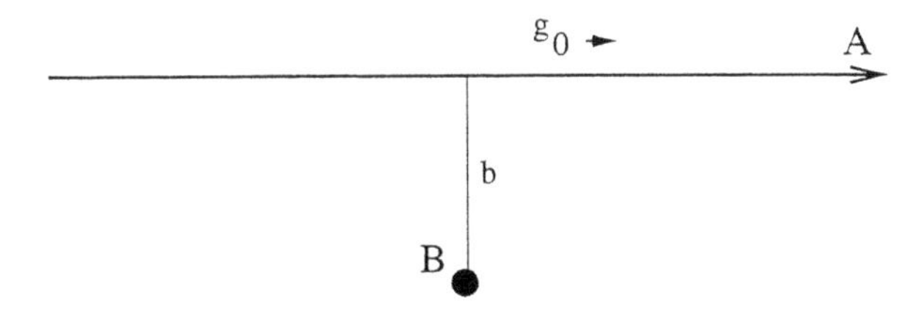

Figure 8.1. Particle *A* passing particle *B*

$$m_A \frac{d\mathbf{v}_A}{dt} = \frac{q_A q_B}{r_{AB}^3} \mathbf{r}_{AB}$$

$$m_B \frac{d\mathbf{v}_B}{dt} = -\frac{q_A q_B}{r_{AB}^3} \mathbf{r}_{AB} \tag{4}$$

where q_A and q_B are the charges, $\mathbf{r}_A$ and $\mathbf{r}_B$ are the positions, $\mathbf{v}_A$ and $\mathbf{v}_B$ are the velocities, and $\mathbf{r}_{AB} = \mathbf{r}_A - \mathbf{r}_B$.

Adding these equations we have

$$(m_A + m_B)\mathbf{V} = m_A \mathbf{v}_A + m_B \mathbf{v}_B = \text{constant} \tag{5}$$

where $\mathbf{V}$ is the center of mass velocity. Dividing the equations by m_A and m_B, respectively, and subtracting we have

$$\frac{d}{dt}\mathbf{v}_{AB} = \left(\frac{1}{m_A} + \frac{1}{m_B}\right) \frac{q_A q_B}{r_{AB}^3} \mathbf{r}_{AB} = \frac{1}{\mu} \frac{q_A q_B}{r_{AB}^3} \mathbf{r}_{AB} \tag{6}$$

where $\mathbf{v}_{AB}$ is the relative velocity. For $m_A = m_B$, $\mu = m_A/2$, the case for like–like collisions. To avoid too many subscripts, set $\mathbf{r} = \mathbf{r}_{AB}$, $\mathbf{g} = \mathbf{v}_A - \mathbf{v}_B$.

We label our electron as particle *A*, and the particle it passes during a collision as particle *B* (see figure 8.1). During each collision we evaluate the change in the relative velocity $\mathbf{g}$. From the change in $\mathbf{g}$ we get the change in $\mathbf{v}_A$ due to this collision. We sum up the changes $\mathbf{v}_A$ of particle *A* due to its many collisional encounters with other particles, assuming that each collision is independent of the others. We do this by first summing only those collisions with particles *B* of the same type and having the same velocity $\mathbf{v}_B$ but randomly disposed in space. Then we sum over the different velocities $\mathbf{v}_B$ and finally over different species. Because we are considering only weak collisions, there are many encounters before the electron, particle *A*, changes its velocity appreciably, so it is possible to assume that there are enough collisions of each category to randomly sum them.

Now consider a single collision in which particle *A* passes particle *B*. In the reduced frame particle *B* is at rest. We solve equation 6 for $\mathbf{v}_{AB} = \mathbf{g}$ as an expansion in h/b, where b is the impact parameter. To lowest order $\mathbf{g} = \mathbf{g}_0$ is constant, so particle *A* pursues a linear straight line motion.

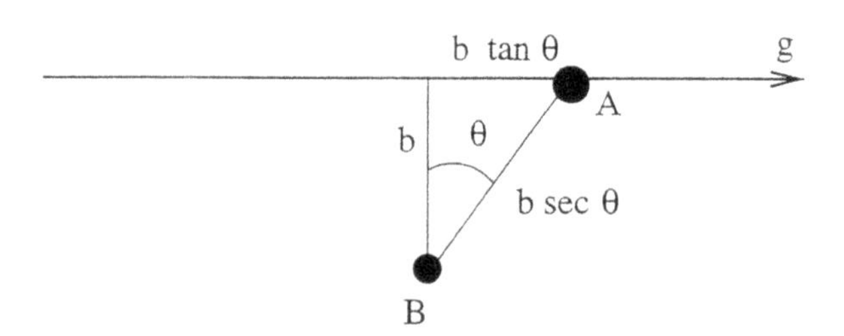

Figure 8.2. Geometry of the collision

The change in $\mathbf{g}$ to first order is given by

$$\frac{d\mathbf{g}_1}{dt} = \frac{q_A q_B}{r_{AB}^3}\mathbf{r}_{AB} \tag{7}$$

or, since $\mathbf{g}_0$ is constant,

$$\frac{d\mathbf{g}_1}{ds} = \frac{q_A q_B}{g_0 r_{AB}^3}\mathbf{r}_{AB} \tag{8}$$

where s is distance measured along the unperturbed orbit. The total change in $\mathbf{g}_1$ is in the $\hat{\mathbf{b}}$ direction, the direction perpendicular to $\mathbf{g}_0$ and in the plane of the orbit (see figure 8.2). The integral

$$\int_{-\pi/2}^{\pi/2} \frac{b d(\tan\theta)/d\theta}{b\sec^3\theta} d\theta = 2 \tag{9}$$

and, thus, the first-order change in $\mathbf{g}$ at the end of the collision is

$$\Delta\mathbf{g} = \mathbf{g}_1 = 2\frac{q_A q_B}{\mu b g_0}\hat{\mathbf{b}} \tag{10}$$

We can find the component of $\mathbf{g}_2$ along $\mathbf{g}_0$ from energy conservation, since g^2 is constant during the collision. Thus,

$$2\mathbf{g}_0 \cdot \mathbf{g}_2 = -\mathbf{g}_1^2 \tag{11}$$

But the average value of $\mathbf{g}_2$ is along $\mathbf{g}_0$, so that

$$\langle \mathbf{g}_2 \rangle = \mathbf{g}_{2\parallel} = \frac{-g_1^2 \hat{\mathbf{g}_0}}{2g_0} = -2\frac{q_A^2 q_B^2}{\mu^2 b^2 g_0^3}\hat{\mathbf{g}_0} \tag{12}$$

We have found the first- and second-order changes in the relative velocity $\Delta\mathbf{g} = \Delta\mathbf{v}_A - \Delta\mathbf{v}_B$. Combining these with a zero change in the undisturbed center of mass velocity $m_A\Delta\mathbf{v}_A + m_B\Delta\mathbf{v}_B = 0$ we find the change in $\mathbf{v}_A$

$$\Delta\mathbf{v}_A = \frac{m_B}{m_A + m_B}\Delta\mathbf{g} = \frac{\mu}{m_A}\Delta\mathbf{g} \tag{13}$$

Next, we sum the changes in $\mathbf{g}$ over many encounters with particles B with the same $\mathbf{v}_B$, but encountered with different values of $\mathbf{b}$. First, for a fixed $\mathbf{g}_0$, we average over all the different orientations of $\hat{\mathbf{b}}$ (which is always perpendicular to $\mathbf{g}_0$). The factor multiplying $\hat{\mathbf{b}}$ in equation 10 is constant, so we find that the probability distribution for $\Delta\mathbf{g}_1$ is axisymmetric about $\mathbf{g}_0$.

We introduce two unit vectors $\mathbf{e}_1$ and $\mathbf{e}_2$ perpendicular to $\mathbf{g}_0$ and perpendicular to each other. Then, averaging the components of the tensor $\hat{\mathbf{b}}\hat{\mathbf{b}}$ about this axis, we have

$$\left\langle \hat{\mathbf{b}}\hat{\mathbf{b}} \right\rangle = \tfrac{1}{2}\left(\mathbf{I} - \hat{\mathbf{g}}_0\hat{\mathbf{g}}_0\right) \tag{14}$$

Let the density of particles B with velocity in $d^3\mathbf{v}_B$ about $\mathbf{v}_B$ be Δn. Then in a time t the number of collisional encounters with particles of type B in this velocity range and with impact parameters between b and $b + db$ is

$$g_0 t(\Delta n) 2\pi b\, db \tag{15}$$

The average second-order change in $\mathbf{v}_A$ per unit time due to these encounters summed over all impact parameters is

$$\langle \Delta \mathbf{v}_A \rangle = \frac{\mu}{m_A}\langle \mathbf{g}_2 \rangle = -4\pi \frac{q_A^2 q_B^2}{m_A \mu g_0^2}(\Delta n)\hat{\mathbf{g}}_0 \int \frac{db}{b} \tag{16}$$

The average change per unit time of the tensor $\Delta\mathbf{v}\Delta\mathbf{v}$ is

$$\langle \Delta \mathbf{v}_A \Delta \mathbf{v}_A \rangle = 4\pi \frac{q_A^2 q_B^2}{m_A^2 g_0}(\Delta n)\left(\mathbf{I} - \hat{\mathbf{g}}_0\hat{\mathbf{g}}_0\right) \int \frac{db}{b} \tag{17}$$

The b integral is logarithmically divergent at both small b and large b. The small b divergence is related to our neglect of large-angle collisions, so the lower limit should be of order h. The divergence at large b occurs because beyond a certain $b_{\max}$ the collisions are so numerous they begin to interfere with each other and no longer can be summed up independently. This length, $b_{\max}$, is the Debye length λ_D where $\lambda_D \approx T/(4\pi n e^2$. This is the length beyond which the plasma shields out electrostatic fields (Cohen et al. 1950; Spitzer 1962).

We discuss this more fully below. Thus, we replace the divergent b integral by $\ln(\lambda_D/h)$. This logarithm is conventionally denoted as $\ln \Lambda$.

Finally, we want the average of the changes of $\Delta\mathbf{v}_A$ and $\Delta\mathbf{v}_A\Delta\mathbf{v}_A$ summed over all velocities $\mathbf{v}_B$. Introducing the distribution function $f_B(\mathbf{v})$ for the B particles by

$$\Delta n = f(\mathbf{v}_B) d^3\mathbf{v}_B \tag{18}$$

we find

$$\langle \Delta \mathbf{v}_A \Delta \mathbf{v}_A \rangle = 4\pi \frac{e^4}{m_A^2} \ln \Lambda \int \frac{\mathbf{I} - \hat{\mathbf{g}}_0\hat{\mathbf{g}}_0}{g} f(\mathbf{v}_B) d^3\mathbf{v}_B \tag{19}$$

and

$$\langle \Delta \mathbf{v}_A \rangle = -4\pi \frac{e^4}{m_A \mu} \ln \Lambda \int \frac{\mathbf{g}}{g^3} f(\mathbf{v}_B) d^3\mathbf{v}_b \tag{20}$$

where we have set $q_A^2 = q_B^2 = e^2$ (Rosenbluth et al. 1957; Trubnikov 1965).

The usual explanation for the upper cutoff in the b integral is that the Coulomb interaction is cut off at the Debye length. But where in our picture does this cutoff actually emerge when we consider particles randomly going by each other? Let us return to the first-order equation (8) and find the change at minimum impact in position of the particle away from the zero-order straight line. We assume like–like collisions, for definiteness, and assume that an electron is encountering other electrons. The direction of $\mathbf{g}_1$ is the same as $\hat{\mathbf{b}}$. The time to pass the center of force (in the reduced picture) is of order b/g_0. Thus, the electron is displaced outward. When it is near the other electron this displacement is of order

$$\Delta \mathbf{r} \approx g_1 \frac{\mathbf{b}}{g_0} = \frac{2e^2}{\mu {g_0}^2} \hat{\mathbf{b}} \approx h \hat{\mathbf{b}} \tag{21}$$

where h is the distance of closest approach, equation 2. Similarly, when the electron encounters an ion it is displaced inward by h. Note that this displacement is independent of b (see problem 2). Now, while most aspects of the collision are random in their directional effects, the spatial shift in the particle motion is coherent and always of the same sign: for like particle collision it is always outward, and for unlike particles it is always inward. This shift in position of the particles A passing B thus represents a polarization of the plasma. As b gets larger the amount of charge shifted into or out of a sphere with radius b becomes larger until eventually the shifted charge is just equal and opposite to the charge on B. To estimate the radius at which this occurs consider a large sphere of radius $b_{\max}$ about a positively charged particle. The negative passing particles are shifted inward across the sphere by $\approx h$, while the positive charges are shifted outward by $\approx h$. Thus, the resulting change in charge in the sphere is of order

$$-4\pi b_{\max}^2 hne \tag{22}$$

and this equals e when

$$b_{\max}^2 = \frac{1}{\sum 4\pi hn} = \frac{1}{\sum 4\pi ne^2/mv^2} \approx \frac{T}{8\pi ne^2} \approx \lambda_D^2 \tag{23}$$

where the sum is over particles of different species. The last quantity is the square of the Debye length. Thus, the Coulomb force is that due to charged particle minus that due to the polarized charge. (In our calculation we have assumed that the Coulomb force is that due to e alone, ignoring the polarized charge in estimating $b_{\max}$. Actually, when b approaches λ_D in order of magnitude, the value of h is reduced and the charge is not totally shielded. In fact, the Coulomb force falls off exponentially but the cutoff distance is essentially correct.)

The logarithm, generally referred to as $\log \Lambda$, has a value usually between 10 and 30. Its value represents the extent to which small-angle collisions dominate large-angle collisions. This is because the b integrals in equations 16 and 17 can be thought of as the sum of many logarithmic intervals all contributing equally to the integral and with only the first one being in order of magnitude equal to the large-angle collision contribution to the collision. The value of $\log \Lambda$ is tabulated in Spitzer (1962).

We have calculated only the two changes, $< \Delta\mathbf{v}_A >$ and $< \Delta\mathbf{v}_A\Delta\mathbf{v}_A >$, for collision with particles of only one species B. To get the general result we must sum over all species B including species A. The result is

$$\langle \Delta\mathbf{v}\Delta\mathbf{v}\rangle_A = 4\pi\,\Sigma_B \frac{e^4 \ln\Lambda}{m_A^2} \int d^3\mathbf{v}_B \frac{(\mathbf{I} - \hat{\mathbf{g}}\hat{\mathbf{g}})_{AB}}{g_{AB}} f_B(\mathbf{v}_B) \tag{24}$$

and

$$\langle \Delta\mathbf{v}\rangle_A = -4\pi\,\Sigma_B \frac{e^4 \ln\Lambda}{m_A\mu} \int d^3\mathbf{v}_B \frac{\mathbf{g}_{AB}}{g_{AB}^3} f_B(\mathbf{v}_B) \tag{25}$$

where in the integral $\mathbf{g} = \mathbf{v} - \mathbf{v}'$, and $\hat{\mathbf{g}}$ is a unit vector in the $\mathbf{g}$ direction. (In the sum over B on the right-hand side B can equal A, i.e., like–like collisions.)

The significance of the collision formulas is as follows: After a time dt the particle A undergoes a number of collision with particles of its type and particles of the other type. The change in the particle velocity is not definite, but after a time t, has a Gaussian distribution about the displaced velocity $\mathbf{v} - \langle\Delta\mathbf{v}\rangle t$ and has a spread in different directions characterized by the dyadic moments, $\langle\Delta\mathbf{v}\Delta\mathbf{v}\rangle t$. As time increases the particle spreads out in velocity space by an amount proportional to $\sqrt{dt}$ and its mean velocity changes by an amount proportional to dt. From now on we will take $dt = 1$, and $\Delta\mathbf{v}$ and $\Delta\mathbf{v}\Delta\mathbf{v}$ will always be the changes in a unit time.

The distribution in changes in velocities is symmetrical about the initial $\mathbf{v}$ and its drag is parallel to the initial $\mathbf{v}$ only if the velocity distribution of the scattering particles is isotropic. We have discussed only a single particle, but all the particles have Gaussian probability distribution for their velocities with $\langle\Delta v\rangle$'s and $\langle\Delta\mathbf{v}\Delta\mathbf{v}\rangle$ that depend on the initial velocity. To actually find these changes we have to know f and then carry out the integrals. If f is a Maxwellian, then these integrations can be carried out explicitly, and this is done in the last section of this chapter.

As time changes, the distribution function f will also change and, therefore, so do the mean rates of change of velocities $\langle\Delta\mathbf{v}\rangle$ and $\langle\Delta\mathbf{v}\Delta\mathbf{v}\rangle$. The processes of collision seem almost infinitely complex, but the situation is saved when there are many collisions, because the distribution rapidly approaches a Maxwellian distribution. Although it is not necessary to go into this in complete detail, let us follow the logic in a little more detail.

8.3 The Fokker–Planck Equation

We define the probability function $P(\mathbf{v}, \Delta\mathbf{v})$ (Chandrasekhar 1943b), so that

$$P(\mathbf{v}, \Delta\mathbf{v}, \Delta t)d^3\Delta\mathbf{v} \tag{26}$$

is the probability that a particle with velocity $\mathbf{v}$ at time t has a velocity $\mathbf{v} + \Delta\mathbf{v}$ with $\Delta\mathbf{v}$ in $d^3\Delta\mathbf{v}$ at the time $t + \Delta t$. We define the distribution function $f(\mathbf{v}, t)$, where

$$f(\mathbf{v}, t)d^3\mathbf{v} \tag{27}$$

as the number of particles with $\mathbf{v}$ in $d^3\mathbf{v}$. It will change in time. This change is governed by the Fokker–Planck equation that arises as follows.

Let us assume that we know $f(\mathbf{v}, t)$ at t. Then, at time $t + \Delta t$, a particle will end up in $d^3\Delta\mathbf{v}$ at $\mathbf{v}$ if it was in $d^3\mathbf{v}$ at $\mathbf{v} - \Delta\mathbf{v}$ at time t, and during the interval dt changed its velocity by $\Delta\mathbf{v}$. The probability of this change $\Delta\mathbf{v}$ is

$$P(\mathbf{v} - \Delta\mathbf{v}, \Delta\mathbf{v}, \Delta t)d^3\mathbf{v} \tag{28}$$

It is important to remember that the probability of a change in a particle's velocity depends on its initial velocity $\mathbf{v} - \Delta\mathbf{v}$, so that the first argument in P must be $\mathbf{v} - \Delta\mathbf{v}$. Summing the distribution function at time t times the probability that it ends up in $d^3\mathbf{v}$ at $t + dt$ over all velocities, we get the distribution function at $\mathbf{v}$ and $t + dt$:

$$\mathrm{f}(\mathbf{v}, t + dt)d^3\mathbf{v} = d^3\mathbf{v} \int f(\mathbf{v} - \Delta\mathbf{v}, t)P(\mathbf{v} - \Delta\mathbf{v}, \Delta\mathbf{v}, \Delta t)d^3\Delta\mathbf{v} \tag{29}$$

Now, if dt is small enough P falls off exponentially before $\Delta\mathbf{v}$ becomes too large. Thus, for short times, we can expand the factors under the integral as a Taylor expansion in $\Delta\mathbf{v}$. That is, we expand f in its argument and P in its first argument. Thus,

$$\begin{aligned} f(\mathbf{v}, t + dt) = \int (fP)(v)d^3\Delta\mathbf{v} - \int \Delta\mathbf{v} \cdot \frac{\partial}{\partial\mathbf{v}}(fP)d^3\Delta\mathbf{v} \\ + \frac{1}{2}\int \Delta\mathbf{v} \cdot \left[\Delta\mathbf{v} \cdot \frac{\partial}{\partial\mathbf{v}}\frac{\partial}{\partial\mathbf{v}}(fP)\right] d^3\Delta\mathbf{v} \end{aligned} \tag{30}$$

where the derivatives of fP are with respect to their first arguments.

But in these three integrals only P depends on $\Delta\mathbf{v}$ and by definition

$$\int P(\mathbf{v}', \Delta\mathbf{v})d^3\Delta\mathbf{v} = 1 \tag{31}$$

$$\int \Delta\mathbf{v}P(\mathbf{v}', \Delta\mathbf{v}, dt)d^3\Delta\mathbf{v} = \langle\Delta\mathbf{v}\rangle\, dt \tag{32}$$

$$\int P(\mathbf{v}', \Delta\mathbf{v}, dt)\Delta\mathbf{v}\Delta\mathbf{v}d^3\Delta\mathbf{v} = \langle\Delta\mathbf{v}\Delta\mathbf{v}\rangle dt \tag{33}$$

according to our notation in equations 24 and 25. Thus, collecting terms and dividing by dt, we find that

$$\frac{\partial f_j}{\partial t} = -\frac{\partial}{\partial\mathbf{v}} \cdot \left(\langle\Delta\mathbf{v}\rangle f_j\right) + \frac{1}{2}\frac{\partial}{\partial\mathbf{v}} \cdot \left[\frac{\partial}{\partial\mathbf{v}} \cdot \left(\langle\Delta\mathbf{v}\Delta\mathbf{v}\rangle f_j\right)\right] \tag{34}$$

where $\langle\Delta\mathbf{v}\rangle$ and $\langle\Delta\mathbf{v}\Delta\mathbf{v}\rangle$ on $\mathbf{v}$ are given in equations 24 and 25. Their dependence on $\mathbf{v}$ is evident in these equations.

To apply the Fokker–Planck equation to a plasma, we have inserted the subscript j on f, where j gives the species. $j = e$ corresponds to electrons and $j = i$ corresponds to ions. The coefficients $\langle \Delta \mathbf{v} \rangle$ and $\langle \Delta \mathbf{v} \Delta \mathbf{v} \rangle$ themselves depend on the distribution functions f_e and f_i. It turns out that after substituting the expressions for $\langle \Delta \mathbf{v} \rangle$ and $\langle \Delta \mathbf{v} \Delta \mathbf{v} V \rangle$ into the Fokker–Planck equation and employing several simple vector operations and several integration by parts over the velocity space, we can write this equation in a simple and convenient form called the Landau form. This is the form everybody generally starts from (Landau 1937; Trubnikov 1965).

Consider the evolution of the electron distribution due to collisions with electrons alone. The Fokker–Planck equation in the Landau form is then

$$\left(\frac{\partial f}{\partial t}\right)_{c,ee} = \frac{2\pi e^4 \ln \Lambda}{m^2} \frac{\partial}{\partial \mathbf{v}} \cdot \int \frac{I - \hat{\mathbf{g}}\hat{\mathbf{g}}}{g} \cdot \left[\frac{\partial f(\mathbf{v})}{\partial \mathbf{v}} f(\mathbf{v}') - \frac{\partial f(\mathbf{v}')}{\partial \mathbf{v}'} f(\mathbf{v})\right] d^3\mathbf{v}' \tag{35}$$

where, as before, $\mathbf{g}$ stands for $\mathbf{v} - \mathbf{v}'$ and f is the electron distribution. From now on f will denote the electron distribution function and F will denote the ion distribution function.

The right-hand side has the pleasing property that it vanishes when f is a Maxwellian distribution, f_M. Indeed, for any Maxwellian, $\partial f_M/\partial \mathbf{v} = -[m(\mathbf{v} - \mathbf{U})/T] f_M$ and $\partial f_M/\partial \mathbf{v}' = -[m(\mathbf{v}' - U)/T] f_M$. Thus, the integrand reduces to

$$\left[\frac{m(\mathbf{v} - \mathbf{U})}{T} - \frac{m(\mathbf{v}' - \mathbf{U})}{T}\right] \cdot \frac{\mathbf{I} - \hat{\mathbf{g}}\hat{\mathbf{g}}}{g^2} f(\mathbf{v}) f(\mathbf{v}') = \frac{m\mathbf{g}}{T} \cdot \frac{\mathbf{I} - \hat{\mathbf{g}}\hat{\mathbf{g}}}{g^2} f(\mathbf{v}) f(\mathbf{v}') = 0 \tag{36}$$

It is possible to show that a Maxwellian distribution is the only distribution function that makes the electron–electron collision integral vanish. Thus, as long as f differs from a Maxwellian, then collisions will continue to alter it.

The collision integral also conserves the fundamental quantities of number, momentum, and energy. Let us consider the energy,

$$\mathcal{E} = \int \frac{mv^2}{2} f d^3\mathbf{v} \tag{37}$$

Its rate of change due to collisions is

$$\begin{aligned} \frac{d\mathcal{E}}{dt} &= \int \frac{mv^2}{2} \left(\frac{\partial f}{\partial t}\right)_{c,ee} d^3\mathbf{v} \\ &= \frac{2\pi e^4 \ln \Lambda}{m^2} \cdot \int d^3\mathbf{v} \frac{mv^2}{2} \frac{\partial}{\partial \mathbf{v}} \cdot \int d^3\mathbf{v}' (\mathbf{I} - \hat{\mathbf{g}}\hat{\mathbf{g}}) \cdot \left(\frac{\partial f}{\partial \mathbf{v}} f' - f \frac{\partial f'}{\partial \mathbf{v}'}\right) \\ &= -\frac{2\pi e^4 \ln \Lambda}{m^2} \int d^3\mathbf{v} d^3\mathbf{v}' \quad m\mathbf{v} \cdot (\mathbf{I} - \hat{\mathbf{g}}\hat{\mathbf{g}}) \cdot \left(\frac{\partial f}{\partial \mathbf{v}} f' - f \frac{\partial f'}{\partial \mathbf{v}'}\right) \end{aligned} \tag{38}$$

where we have integrated by parts in $\mathbf{v}$. But the expression in the last parentheses is antisymmetric in $\mathbf{v}$ and $\mathbf{v}'$. Symmetrizing the right-hand side by interchanging $\mathbf{v}$ and $\mathbf{v}'$ and adding and then dividing by 2 simply replaces $m\mathbf{v}$ by $\frac{1}{2}m\mathbf{g}$ and so, because $\mathbf{g}\cdot(\mathbf{I}-\hat{\mathbf{g}}\,\hat{\mathbf{g}})$ vanishes, the entire expression vanishes and the electron energy is conserved under electron–electron collisions. A similar symmetry argument shows that momentum is conserved. Number conservation is even easier since $(\partial f/\partial t)_{c,ee}$ is a divergence. We get identical results for ion–ion collisions that also conserve ion number, momentum, and energy.

These conservation relations are just what is to be expected, since, after all, the collisional expression was derived by starting with individual binary electron–electron collisions with each binary collision conserving the three fundamental quantities: number, momentum, and energy. However, the Landau form of the collision integral was derived after a number of operations, so it is a gratifying check on this equation that the fundamental quantities are still conserved.

The electron distribution function is, of course, also changed by collision with ions. If the ions are singly charged, the electron–ion collision integral is

$$\left(\frac{\partial f}{\partial t}\right)_{c,ei} = \frac{2\pi n e^4 \ln\Lambda}{m}\frac{\partial}{\partial\mathbf{v}}\cdot\int\frac{\mathbf{I}-\hat{\mathbf{g}}\hat{\mathbf{g}}}{g}\cdot\left[\frac{1}{m}\frac{\partial f}{\partial\mathbf{v}}F(\mathbf{v}')-\frac{1}{M}\frac{\partial F}{\partial\mathbf{v}'}F(\mathbf{v})\right]d^3\mathbf{v} \tag{39}$$

where

$$\mathbf{g}=\mathbf{v}-\mathbf{v}' \tag{40}$$

The positioning of the m and M factors guarantees that if f and F are both Maxwellians with the same temperature T and mean velocity U, then the integral vanishes. The number of electrons is still conserved by collision but electron momentum is not. However, the ion–electron collisions are obtained by replacing the m in front by M and interchanging f and F in the integrand. When we consider the rate of change of electron momentum plus ion momentum or electron energy plus ion energy, essentially the same symmetry argument shows that these sums of momentum and energy are preserved.

Let us now consider the effect of just electron–electron collisions on the electron distribution function. Assume that we start off with a non-Maxwellian f_e. It will change, but in such a way so as to preserve number, momentum, and energy, and it will continue to change until it turns into a Maxwellian distribution function whose parameters have to be such that the number, momentum, and energy are still the same.

Do we expect it to reach such a Maxwellian distribution? The answer is yes, because it turns out that for fixed quantities n, U, and $\mathcal{E}$ the Maxwellian is the distribution function that occupies the greatest volume in phase space. Thus, a generalized entropy argument shows that f should evolve toward a Maxwellian. This generalized entropy argument can be given explicitly in terms of Boltzmann's famous H theorem. Let

$$H=\int f\ln f d^3v \tag{41}$$

whose negative is essentially the entropy. The rate of change of H due to electron–electron collisions is

$$\begin{aligned}\frac{dH}{dt} &= \int \frac{\partial}{\partial f}(f \ln f)\left(\frac{\partial f}{\partial t}\right)_c d^3v \\ &= \int \ln f \left(\frac{\partial f}{\partial t}\right)_c d^3\mathbf{v} + \int \left(\frac{\partial f}{\partial t}\right)_c d^3\mathbf{v}\end{aligned} \tag{42}$$

The second integral is the rate of change of the number density, and it vanishes. Let us write the electron–electron collision integral as

$$\frac{\partial f}{\partial t} = \Gamma \frac{\partial}{\partial \mathbf{v}} \cdot \int d^3\mathbf{v}'\mathbf{J} \tag{43}$$

where $\mathbf{J}$ is the usual integrand and $\Gamma = 2\pi e^4 \ln \Lambda / m^2$. Then, integrating by parts, we find

$$\begin{aligned}\frac{dH}{dt} &= \Gamma \int d^3\mathbf{v} \ln f \frac{\partial}{\partial \mathbf{v}} \cdot \int d^3\mathbf{v}'\mathbf{J} \\ &= -\Gamma \int \frac{1}{f}\frac{\partial f}{\partial \mathbf{v}} \cdot \mathbf{J} d^3\mathbf{v} d^3\mathbf{v}'\end{aligned} \tag{44}$$

Interchange $\mathbf{v}$ and $\mathbf{v}'$ in this integral and add, noting that $\mathbf{J}$ is odd in $\mathbf{v}$ and $\mathbf{v}'$. Thus,

$$\frac{dH}{dt} = -\frac{\Gamma}{2} \int \frac{1}{ff'}\left(f'\frac{\partial f}{\partial \mathbf{v}} - f\frac{\partial f'}{\partial \mathbf{v}'}\right) \cdot \mathbf{J} d^3\mathbf{v} d^3\mathbf{v}' \tag{45}$$

But $\mathbf{J}$ has exactly the same factor as the expression in the parentheses, so

$$\frac{dH}{dt} = -\frac{\Gamma}{2} \int \frac{d^3\mathbf{v} d^3\mathbf{v}'}{ff'}\left(f'\frac{\partial f}{\partial \mathbf{v}} - f\frac{\partial f'}{\partial \mathbf{v}'}\right) \cdot \frac{I - \hat{\mathbf{g}}\hat{\mathbf{g}}}{g} \cdot \left(f'\frac{\partial f}{\partial \mathbf{v}} - f\frac{\partial f'}{\partial \mathbf{v}'}\right) < 0 \tag{46}$$

This is negative since the factor under the integral is

$$\left(\frac{\partial f}{\partial \mathbf{v}} f' - \frac{\partial f}{\partial v'} f\right)_\perp^2$$

where $\perp$ means the component perpendicular to $\mathbf{g}$. Thus, H decreases and its minimum value under the conservation constraints is easily shown to be given by the Maxwellian with the same conservation constraints. The result of all this discussion is that electron–electron collisions relax any electron distribution function to a Maxwellian. It is not difficult to see that the time of relaxation is the electron–electron collision time.

We have ignored collisions of electrons with ions, which are almost as fast except for the reduced mass factor. Electron–ion collisions rapidly remove the electron momentum, but change its energy by only a small amount. Thus, in the electron–electron collision time, the electron–ion collisions will reduce the mean velocity of the electrons to that of the ions. It turns out that the loss of energy is very small during this time. Thus, the electrons first relax to a Maxwellian with mean velocity equal to the mean velocity of the ions. Then after a time about a factor $\sqrt{M/m}$ longer the ions relax to a Maxwellian with the same mean velocity and with the same initial energy. At this stage the electron and ion temperatures can still be different. Finally, the electron and ion temperatures equalize in a time longer than the electron–electron collision time by a factor M/m.

8.4 Collision Rates

It is important to have better idea as to the order of magnitude of the collision rates than was presented in chapter 3. First, consider electron–electron collisions. We see from equation 16 that the effect of small-angle collisions is roughly $\ln \Lambda$ times that of large-angle collisions. But the cross section for large-angle collisions is

$$\sigma_{\text{large}} = \pi h^2 \tag{47}$$

where h is the minimum impact parameter (equation 2). For a large-angle collision to happen, electrons have to approach close enough for a fair fraction of their kinetic energy to be converted to potential energy. But a typical relative energy for a pair of electrons is T, so

$$h \simeq \frac{e^2}{T} = \frac{1.5 \times 10^{-7}}{T_{\text{eV}}} \tag{48}$$

and taking $\ln \Lambda = 15$ we have a cross section of

$$\pi h^2 \ln \Lambda = \frac{10^{-12}}{T_{\text{eV}}^2} \text{cm}^2 \tag{49}$$

When the relative energy of the electrons is larger than T, the cross section is reduced by a factor $(\mathcal{E}/T)^{-2}$. For a smaller relative energy it is higher by the same factor.

The collisional cross section of electron–ion collisions is about the same as for electron–electron collisions. However, the effect of the collisions is quite different because the ions are so much more massive than the electrons. After an electron–ion collision the electron changes the direction of its velocity. However, it hardly changes its energy at all. In fact, let the electron change its velocity by $\Delta v \approx v$. We can see from equation 13 that the change in the ion velocity is smaller than the change in the electron velocity by m/M, so the change in the ion energy is $(M/m) \times (m/M)^2 \times m(\Delta v)^2 \approx (m/M)mv^2$, which is very small. Since the sum of the energies is constant, the electrons also hardly change their energy during an electron–ion collision.

The above estimate for like–like collision cross sections does not depend on mass, so collisions of ions with ions have the same cross sections as collisions of electrons with electrons. Consider an electron with energy very small compared to T. Its relative energy for collisions with the other electrons is still of order T, so its cross section with other electrons is only slightly greater than equation 49. However, these collisions do lead to an increase in its energy. It should be noted that collisions of slow electrons with ions have a much larger cross section than this because the ions have small thermal motions so that the relative energy for collisions with ions is very small.

Collisions of ions with electrons are a little confusing. Because the ions are heavy they are hardly disturbed by such encounters to lowest order. However, there is a small energy exchange and this process gradually leads to an equalization of the electron and ion temperatures.

On the other hand, if we have an energetic ion with energy E, its ion–ion cross section decreases with energy as E^{-2}, and the collision rate as $E^{-3/2}$. However, because its relative velocity with respect to the electrons does not change very much, the rate of transfer of energy to electrons does not decrease with E, and there is actually an energy E_c, above which the drag on the energetic ion by electrons dominates the drag by ions. At $E = T$ the ion drag is larger than the electron drag by $\sqrt{M/m}$, so the two drag rates are equal at when $(E/T)^{3/2} \approx \sqrt{M/m}$ or at

$$E_c = \left(\frac{M}{m}\right)^{1/3} T \approx 12T \tag{50}$$

(Actually the electron drag is a little more complicated. Because of a cancellation, only electrons with energy below the fast ion velocity $V_i = \sqrt{2E/M}$ yield a net drag. The fraction of electrons below V_i is proportional to $(V_i/v_e)^3$, where v_e is the electron thermal speed, but the collision rate for such slow electrons is larger by $(v_e/V_i)^3$, so the electron drag rate is indeed independent of E.)

Let us consider several examples. On the surface of the sun the neutral density is about 10^{16}cm^{-3}, but the electron and ion density, being due only to the ionization of metals, is smaller by a factor of 10^4 and is 10^{+12}cm^{-3}. The electron–electron cross section is 10^{-12} cm^2 since $T_{\text{eV}} \approx 1$. The mean free path λ for electron–electron collision is $(10^{-12})(10^{12}) = 1$ cm. Notice that at 1eV, $\sigma_{ee} \approx 10^4\sigma_{\text{atom}}$, so $\lambda_{e\,\text{atom}} \approx 1$ cm. At the center of the sun $T_{\text{eV}} \approx 10^3$, $\sigma = 10^{-18}$ cm^2, and $n \approx= 10^{26}$ cm^{-3}, so $\lambda \approx 10^{-8}$ cm.

In the interstellar medium let us take $T_{\text{eV}} \approx 1$ again and $n = 1$ cm^{-3}, so $\lambda = 10^{12}$ cm. In a molecular cloud $T = 10\text{K} \approx 10^{-3}$ eV, so $\sigma = 10^{-6}$ cm^2. The density is of order $n \approx 10^3$ cm^{-3} and if the fractional ionization $\chi = 10^{-5}$, then $\lambda \approx 10^8$ cm. On the other hand, $\sigma_{e-\text{atom}} = 10^{-16}$ cm^2, so $\lambda \approx 10^{13}$ cm. Thus, even with this very low degree of ionization, Coulomb collisions still dominate. This is because the Coulomb cross section becomes very large at low temperatures. At high temperatures the Coulomb cross section is small but there are no neutrals, so, again, Coulomb collisions dominate. In fact, in general, it can be taken as a rule of thumb that Coulomb collisions dominate neutral collisions nearly everywhere except for plasmas in which the degree of ionization is extremely low.

Thus, if electrons and neutrals are present, then the electrons and ions are bound together tightly by their collisions. They are more loosely tied to neutrals, and so the charged fluid can stay together and will flow as a unit through the neutrals if the forces on them are different; for example, if there are magnetic forces acting on the ions and electrons.

In the solar surface, v_e is order 10^8 cm/sec and $v_i \approx 10^6$ cm/sec, so $\nu_{ii} = v_i/\lambda \approx 10^6$/sec, and $\nu_{ee} \approx 10^8$/sec. Near the center of the sum $v_e = 10^9$cm/sec and $v_i = 10^7$ cm/sec, so $\nu_{ee} \approx 10^{17}$/sec, and $\nu_{ii} \approx 10^{15}$/sec. In the interstellar medium, $\nu_e = 10^8/10^{12} = 10^{-4}$ sec, $\nu_{ii} \approx 10^{-6}$/sec, so $\sigma = 10^{-6}\text{cm}^2$. In molecular clouds if the fractional ionization $\chi = 10^{-5}$, then $\nu_{ii} \approx 3 \times 10^{-4}$/sec.

From these examples it may be seen that in almost all cases λ is enormously smaller than the scale of any phenomenon in which we might be interested, while ν is enormously faster than dynamical rates. For this reason, we can generally treat the plasma as a fluid. Let us see how by making use of this small parameter λ/L, where L is the macroscopic scale size, the fluid equations for a collisional plasma actually arise from the complicated Fokker–Planck equation we have derived in the last section.

8.5 The Space-Dependent Fokker–Planck Equation

Our discussions so far have been for homogeneous systems for which the distribution function f is independent of position. This is not a bad approximation even in inhomogeneous systems for sufficiently short times, times short compared to the dynamical evolution time of f due to particle flows over the macroscopic scale L of the inhomogeneities. This latter time is usually the sonic time L/c_s or the Alfven time L/v_A. But we need to follow the evolution of the plasma over these longer times and for this we need to consider that f is a function of position $\mathbf{r}$ and time t as well as of velocity $\mathbf{v}$. We need a more general evolution equation.

To derive this equation we start at some time t, consider a small six-dimensional "box" in phase space, $x_0 < x < x_0 + dx, \ldots, v_{0z} < v_x < v_{0z} + dv_z$, and find the rate of change of the number of particles in this box. The change occurs not only by collisions, as given by the Fokker–Planck collision operator of the previous section, but also by "flows" in phase space (see figure 8.3).

We first ignore collisions and calculate how fast the number of particles changes in the box due to these flows. The only difficulty with this calculation is the number of dimensions of phase space, six. If we were to ask the same question in ordinary space of three dimensions, the answer would be well known and simple, that is, it would be given by the ordinary continuity equation:

$$\frac{\partial n}{dt} + \nabla \cdot (n\mathbf{U}) = 0 \tag{51}$$

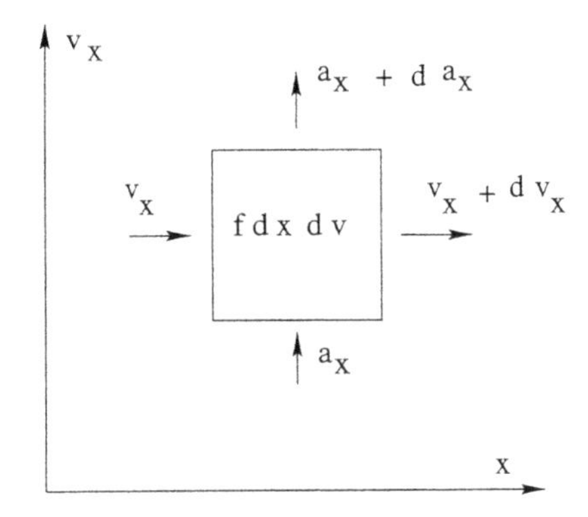

Figure 8.3. Particles entering and leaving a box in phase space

where **U** is the fluid velocity $\mathbf{U}(\mathbf{x}, t)$, which is a function of position and gives the rate of flow of particles across each surface of the box. (In this chapter we use capital **U** for the fluid velocity, since it is easier to distinguish from the particle velocity **v** than **V** is.)

Now, with a little thought the continuity equation can be extended to six-dimensional phase space. At any point in phase space $\mathbf{v}, \mathbf{x}, t$, the particles change their position in a unique way. The flow in the x direction is given by v_x, and since this is itself a coordinate in phase space, all the particles in the box move in the x direction with the same unique velocity v_x. If there is an electric field **E**, a magnetic field **B**, and a gravitational field **g**, then all the particles in the box at $\mathbf{x}, \mathbf{v}$ accelerate according to Newton's law:

$$m\frac{d\mathbf{v}}{dt} = q\left[\mathbf{E}(\mathbf{x}) + \frac{\mathbf{v} \times \mathbf{B}(\mathbf{x})}{c}\right] + \mathbf{g}(\mathbf{x}) \tag{52}$$

and for given **x** and **v** in phase space this acceleration is the same for all the particles in the box. They move in the velocity direction with this "velocity." Thus, in equation 51 we replace $n(\mathbf{x})$ by $f(\mathbf{x}, \mathbf{v})$, introduce the flows **v** and $d\mathbf{v}/dt$ of the continuous "fluid," and write down the six-dimensional fluid equation (Chandrasekhar 1943b):

$$\frac{\partial f}{\partial t} + \nabla \cdot (\mathbf{v} f) + \nabla_{\mathbf{v}} \cdot \left[\frac{q}{m}\left(\mathbf{E} + \frac{\mathbf{v} \times \mathbf{B}}{c}\right) f\right] + \nabla_{\mathbf{v}} \cdot (\mathbf{g} f) = 0 \tag{53}$$

where ∇ stands for the x derivative,

$$\nabla\hat{\mathbf{x}} = \frac{\partial}{\partial x} + \hat{\mathbf{y}}\frac{\partial}{\partial y} + \hat{\mathbf{z}}\frac{\partial}{\partial z}$$

as before, and $\nabla_{\mathbf{v}}$ stands for velocity derivative,

$$\nabla_{\mathbf{v}} = \hat{\mathbf{x}}\frac{\partial}{\partial v_x} + \hat{\mathbf{y}}\frac{\partial}{\partial v_y} + \hat{\mathbf{z}}\frac{\partial}{\partial v_z}$$

Now, $\nabla \cdot (\mathbf{v} f) = \mathbf{v} \cdot \nabla f$, since **v** is an independent variable and is held fixed when ∇ operates. Similarly, $\nabla_{\mathbf{v}} \cdot [\mathbf{E}(\mathbf{x}) f] = \mathbf{E} \cdot \nabla_{\mathbf{v}} f$ and $\nabla_{\mathbf{v}}(\mathbf{g} f) = \mathbf{g} \cdot \nabla_{\mathbf{v}} f$, since

$\mathbf{x}$ is held fixed when $\nabla_{\mathbf{v}}$ operates. Moreover, $\nabla_{\mathbf{v}} \cdot (\mathbf{v} \times \mathbf{B}) = (\nabla_{\mathbf{v}} \times \mathbf{v}) \cdot \mathbf{B} = 0$. Thus, equation 51 can be written

$$\frac{\partial f}{\partial t} + \mathbf{v} \cdot \nabla f + \frac{q}{m}\left(\mathbf{E} + \frac{\mathbf{v} \times \mathbf{B}}{\mathbf{c}}\right) \cdot \nabla_{\mathbf{v}} f + \mathbf{g} \cdot \nabla_{\mathbf{v}} f = 0 \tag{54}$$

This is called the Vlasov equation. It is valid only when collisions are negligible. In general, the opposite is true and collisions are very important.

The phase velocity fluid is carried smoothly through a phase space box by phase fluid flow. But at the same time particles stochastically diffuse into and out of the phase box by the random collisions at a rate determined by the Fokker–Planck operator. The full space-dependent Fokker–Planck equation is obtained by adding the Fokker–Planck operator to the Vlasov equation:

$$\begin{aligned} &\frac{\partial f}{\partial t} + \mathbf{v} \cdot \nabla f + \frac{q}{m}\left(\mathbf{E} + \frac{\mathbf{v} \times \mathbf{B}}{c}\right) \cdot \nabla_{v} f \\ &\qquad + \mathbf{g} \cdot \nabla_{v} f = \sum \left(\frac{\partial f}{\partial t}\right)_{c} \end{aligned} \tag{55}$$

where the sum on the right is over the two types of particles (ions and electrons) with which collisions occur. This is the Fokker–Planck operator of equation 35. This space-dependent Fokker–Planck equation is conventionally referred to as simply the Fokker–Planck equation. When collisions are absent it is called the Vlasov equation.

There are actually two Fokker–Planck equations: one for the electron distribution function f with $q = -e$, $m = m$ and one for the ions distribution function F with $q = e$, $m = M$. If we drop the collision term and reduce the full equation to the pure Vlasov equation, we notice a remarkable result. The Vlasov equation can be written

$$\frac{df}{dt} = 0 \tag{56}$$

where the total time derivative of f is taken along a particle orbit as it changes its $\mathbf{x}$ position in phase space due to $\mathbf{v}$ and its $\mathbf{v}$ position in phase space due to acceleration by the electric and magnetic fields. Since this is true no matter how complicated the fields, this equation, which is sometimes known as Liouville's equation, has strong implications.

For example, when cosmic rays enter the solar system they pass through the interplanetary system and the earth's magnetosphere before striking the surface of the earth. If we measure the flux in a unit solid angle and in a unit velocity (energy) range, we obtain f at the earth. In a steady state this f must be the same as the f of the same cosmic ray taken along its orbit, just before it enters the solar system.

However, at the earth f, at the same energy, is constant in angle and time. If the magnetic fields of the solar system and magnetosphere are constant in time (as they are supposed to be to the lowest order), then the cosmic rays have same energy when they enter the solar system as when they arrive at the earth. Thus, because of the observed isotropy of f at the earth, it must be nearly isotropic in the interstellar medium outside the solar system. The only reasonable explanation is that the cosmic rays are nearly isotropic (to about one part is 10^4) in the interstellar medium. The explanation for this near isotropy is a plasma effect, which is discussed in chapter 12.

8.6 The Fluid Equations

We now show how the ideal fluid equations arise from the solution of the electron and ion Fokker–Planck equations in the limit when collisions are strong. First note that the collision rates are assumed large compared to the dynamic terms v/L in the Fokker–Planck equation but not compared to the cyclotron frequency. For example, the electron collision rate in the photosphere is $v/\ell \approx 10^8/\,\text{sec}$. If the field is of order 10 G, the electron cyclotron frequency Ω is $\approx 2 \times 10^8$/sec. The dynamic rate for scales of $L = 10^3$ km is $v/L < 10^6/10^8 \approx 10^{-2}$/sec, so there is a large separation of rates between the dynamic rates and the cyclotron and collision rates.

We first derive the ion fluid equations. To lowest order, the ion Fokker–Planck equation for the zero-order ion distribution function F_0 is

$$\frac{e}{M}\left(\mathbf{E} + \frac{\mathbf{v}\times\mathbf{B}}{c}\right)\cdot\nabla_v F_0 = \left(\frac{\partial F_0}{\partial t}\right)_{c,ii} + \left(\frac{\partial F_0}{\partial t}\right)_{c,ie} \tag{57}$$

since the other terms such as $\mathbf{v}\cdot\nabla f \sim (v/L)f$ are of first order. The two terms on the right are the Fokker–Planck integrals for the ion–ion collisions and the ion–electron collisions, respectively. The ion–electron collisions are small and we will treat them as first order. Also, as we will see from Maxwell's equations, $E_\parallel$ must also be of first order. If it were zero order, then the electrons would all be accelerated rapidly along B until $E_\parallel$ is shorted out.

We can eliminate $\mathbf{E}_\perp$ by transforming to the $\mathbf{E}\times\mathbf{B}$ velocity frame

$$\mathbf{v} = \mathbf{v}' + \mathbf{U}_E \tag{58}$$

where

$$\mathbf{U}_E = \frac{c\mathbf{E}\times\mathbf{B}}{B^2} \tag{59}$$

so that $\mathbf{E} + \mathbf{U}_E \times \mathbf{B}/c = 0$. That is, we transform to the frame in which the electric field is zero. The zero-order Fokker–Planck equation in this frame is

$$-\Omega\frac{\partial F_0}{\partial \phi'} = \left(\frac{\partial F_0}{\partial t}\right)_{c,ii} \tag{60}$$

where we have introduced cylindrical coordinates $v'_\perp$, ϕ', $v'_\parallel$ with the cylindrical axis along $\mathbf{B}(\mathbf{x}, \mathbf{t})$. (Since there are no spatial derivatives in equation 60, we treat $\mathbf{B}$ as a constant.) The steady time-independent solution of this equation is the Maxwellian distribution,

$$F(v, x, t) = \frac{n_i(\mathbf{x}, t)}{\left(\sqrt{2\pi T_i(x.t)/M}\right)^3} \exp - \frac{M[\mathbf{v}' - \mathbf{U}'_i(\mathbf{x}, \mathbf{t})]^2}{2T_i(x, t)} \tag{61}$$

where $\mathbf{U}'_i$ must be parallel to $\mathbf{B}$. F_0 is axisymmetric in $\mathbf{v}'$, so the left-hand side of equation 60 vanishes. F_0 is of the Maxwellian form, so the right-hand side vanishes also.

Equation 61 can be shown to be the most general solution of equation 60. Simply apply the H theorem. Use equation 60 to calculate the time derivative of $H = \int F_0 \ln F_0 d^3 v$. The contribution from the cyclotron terms vanishes and therefore the contribution from the collisions must also vanish. For the collisional contribution to vanish F_0 must be a Maxwellian with parameters $n_i, \mathbf{U}'_i, T_i$. Now substitute this Maxwellian into equation 60. The right-hand side vanishes. For the left-hand side to vanish, F_0 must be cylindrically symmetric about $\mathbf{B}$, and so $\mathbf{U}'_i$ must be parallel to $\mathbf{B}$. Since equation 60 has no space or time derivatives, the parameters n_i, $U_{i\parallel}$ and T_i can be arbitrary functions of $\mathbf{x}$ and t. In the laboratory frame, the solution is

$$F_0 = \frac{n_i(x, t)}{\left(\sqrt{2\pi T_i(x, t)/M}\right)^3} \exp - \frac{M[\mathbf{v} - \mathbf{U}_i(x, t)]^2}{2T_i(x, t)} \tag{62}$$

where $\mathbf{U}_i = \mathbf{U}_E + U_{i\parallel}\hat{\mathbf{b}}$. $\mathbf{U}_E$ is specified by $\mathbf{E}(\mathbf{x}, \mathbf{t})$ and $\mathbf{B}(\mathbf{x}, \mathbf{t})$.

We regard $\mathbf{E}$ and $\mathbf{B}$ as given for this discussion, but they are actually to be found self-consistently from Maxwells's equations in terms of the currents and charge densities.

The lowest-order Fokker–Planck equation for the electron distribution function f_0 is, in analogy to equation 60,

$$-\frac{e}{m}\left(\mathbf{E} + \frac{\mathbf{v} \times \mathbf{B}}{c}\right) \cdot \nabla_v f_0 = \left(\frac{\partial f_0}{\partial t}\right)_{c,ee} + \left(\frac{\partial f_0}{\partial t}\right)_{c,ei} \tag{63}$$

The electron–ion collision term is zero order and must be kept. Again transforming to the $\mathbf{E} \times \mathbf{B}$ frame, we find that the solution is the same as equation 61 and in the laboratory frame the same as equation 62. Substituting this into the lowest-order electron Fokker–Planck equation we find that all the terms vanish except the electron–ion collision term. This term will vanish to zero order if and only if $U_e = U_i$, because the rate of exchange of momentum between electrons and ions is of order of the electron–ion collision rate, which is large. However, T_e can be different from T_i, because the rate of exchange of energy between electron and ions by collisions is small. The electron distribution function to lowest order in the laboratory frame is, thus,

$$f_0 = \frac{n}{(2\pi T_e/m)^{3/2}} \exp - \frac{m(\mathbf{v} - \mathbf{U})^2}{2T_e} \tag{64}$$

where, since $\mathbf{U}_i = \mathbf{U}_e$ we can drop the subscripts on $\mathbf{U}$.

Now if $n_i \neq n_e$, there will be a zero-order charge density. But this charge density will be much larger than is allowed by the Poisson equation. We can get an estimate of the required charge density as follows:

$$\nabla \cdot E = 4\pi e(n_i - n_e) \tag{65}$$

or

$$\frac{E}{L} = 4\pi e(n_i - n_e) \tag{66}$$

But, $E \approx UB/c$, so the relative difference in the electron and ion densities is

$$\frac{n_i - n_e}{n} \approx \frac{UB}{4\pi necL} = \frac{U}{v_i}\frac{\rho_i}{L}\frac{v_A^2}{c^2} \tag{67}$$

This is generally very small so we consider it as first order and take $n_i = n_e$ to zero order.

Now because in zero order $\mathbf{U}_i - \mathbf{U}_e = 0$, there is also no zero-order current. We can see that the current required by Ampere's law is also small, compared to $n_0 eU$, and can be considered to come from a first-order difference in $\mathbf{U}_i$ and $\mathbf{U}_e$. We can estimate this $\Delta\mathbf{U} = \mathbf{U}_i - \mathbf{U}_e$ as follows. From Ampere's law

$$4\pi \mathbf{j} = 4\pi(\mathbf{j}_i - \mathbf{j}_e) = \frac{4\pi ne}{c}(\mathbf{U}_i - \mathbf{U}_e) = \nabla \times \mathbf{B} \approx \frac{B}{L} \tag{68}$$

so

$$\Delta U = \frac{cB}{4\pi neL} \tag{69}$$

If we assume $B^2/8\pi \approx p = nT$ and compare ΔU with v_i, we get

$$\frac{(\Delta U)^2}{v_i^2} = \frac{c^2 B^2}{(4\pi)^2 n^2 e^2 v_i^2 L^2} \approx \frac{c^2 M}{4\pi ne^2 L^2} = \frac{c^2}{\omega_{pi}^1 L^2} \tag{70}$$

c/ω_{pi} is generally called the ion skin depth. The difference in U's divided by v_i is of order the ratio of the ion skin depth to the scale size. This is usually small, so we consider ΔU to be of first order.

We have shown that in the strong collision limit (with large cyclotron frequencies) the distributions must be Maxwellians to lowest order. However, the Maxwellians are different at different places and times. Now, in a dynamic time, $\sim L/v$, particles will move from one region where n, $\mathbf{U}$, and T have certain values to another region where they are substantially different. The Maxwellians from these two regions should collisionally relax to a different Maxwellian with different parameters. Thus, this mixing will change n, $\mathbf{U}$, and T in time. The rate of evolution is given by the fluid equations, which we derive from the next order in the Fokker–Planck equation. We know physically that the way this happens is governed by the fluid equations, which act on the large dynamic timescale to change $n, U_\parallel, T$.

To see how this happens from the systematic expansion we go to first order. First, consider the ions and write their distribution function F as

$$F = F_0 + F_1$$

Substitute this expansion into the Fokker–Planck equation. To first order

$$\frac{\partial F_0}{\partial t} + \mathbf{v} \cdot \nabla F_0 - C_{ie} = -\frac{e}{m}\left(\mathbf{E} + \frac{\mathbf{v} \times \mathbf{B}}{\mathbf{c}}\right) \cdot \nabla_v F_1 + C_{ii}(F_0, F_1) + C_{ii}(F_1, F_0) \tag{71}$$

where C_{ii} denotes the the ion–ion collision integral $(\partial F/\partial t)_{cii}$ and C_{ie} denotes the ion–electron collision integral $(\partial F/\partial t)_{cie}$, which is already first order when F_0 and f_0 are substituted in it.

From the Landau form of the ion–ion collision integral, equation 35 with f replaced by F, we see that C_{ii} is quadratic in F. Thus, its expansion involves two different sets of terms, one from perturbing $F(\mathbf{v})$ and the other from perturbing $F(\mathbf{v}')$.

Equation 71 is a linear integral equation for F_1 that is difficult to solve. In fact, it can be solved for F_1 only if the inhomogeneous terms on the left-hand side satisfy certain consistency conditions. These conditions arise from the conservation relations that are satisfied by the operators on the right-hand side. We know that ion–ion collisions cannot change the number of ions, their total momentum, or their total energy.

Since the right-hand side is the linearization of the collision integral, this operator should vanish when multiplied by 1, $M\mathbf{v}$, or $Mv^2/2$ and integrated over $\mathbf{v}$. On the other hand, the Lorentz term vanishes when multiplied by $1, M\mathbf{v}_\parallel$, or $M(\mathbf{v} - \mathbf{U}_E)^2$ and integrated over $\mathbf{v}$. Let $\mathbf{v} = \mathbf{U} + \mathbf{v}'$, where $\mathbf{v}'$ is the peculiar velocity. From this we can see that the right-hand side also vanishes when multiplied by 1, $M\mathbf{v}_\parallel$ and $Mv'^2/2$ and integrated. Therefore, the same must be true for the left-hand side, and the three resulting conditions are just what we need to determine the evolution of the three quantities, n_i, $\mathbf{U}_{i\parallel}$, and T_i. (These, in fact, can be shown to be exactly the required consistency conditions for solving the integral equation.)

First, multiply equation 71 by 1 and integrate over $\mathbf{v}$. We get

$$\frac{\partial}{\partial t}\int F_0 d^3\mathbf{v} + \nabla \cdot \int (F_0 \mathbf{v}) d^3\mathbf{v} = 0 \tag{72}$$

Now, from equation 62 we get

$$\int F_0 d^3\mathbf{v} = n \tag{73}$$

and

$$\int F_0 \mathbf{v} d^3 v = n\mathbf{U} \tag{74}$$

so

$$\frac{\partial n}{\partial t} + \nabla \cdot (n\mathbf{U}) = 0 \tag{75}$$

the fluid continuity equation.

Next, instead of multiplying equation 71 by $M\mathbf{v}_{\parallel}$ we multiply by it by $M\mathbf{v}$ and then take the parallel component after the integrations. Thus, we get

$$\frac{\partial}{\partial t}\int F_0(\mathbf{v})M\mathbf{v}d^3\mathbf{v} + \nabla\cdot\int MF_0\mathbf{v}\mathbf{v}d^3\mathbf{v}$$
$$= -e\int \mathbf{v}\frac{\partial}{\partial \mathbf{v}}\cdot\left[\left(\mathbf{E}+\frac{\mathbf{v}\times\mathbf{B}}{\mathbf{c}}\right)F_1 + \mathbf{E}_1F_0\right]d^3\mathbf{v} \tag{76}$$

This moment of C_{ie} vanishes because the ions and electrons have the same velocity. However, the electric and magnetic terms vanish only to lowest order, and because they are large we must keep F to first order in them. (We also keep the small first-order $\mathbf{E}_1$.)

Now, again from equation 62,

$$M\int F_0\mathbf{v}d^3\mathbf{v} = \rho\mathbf{U} \tag{77}$$

and

$$M\int F_0\mathbf{v}\mathbf{v}d^3\mathbf{v} = nT_i\mathbf{I} + \rho\mathbf{U}\mathbf{U} \tag{78}$$

On integration by parts the $\mathbf{E}$ and $\mathbf{B}$ terms become

$$M\frac{e}{M}\int \mathbf{v}\frac{\partial}{\partial \mathbf{v}}\cdot\left[\left(\mathbf{E}+\frac{\mathbf{v}\times\mathbf{B}}{c}\right)F\right]d^3\mathbf{v} = -e\int F\left(\mathbf{E}+\frac{\mathbf{v}\times\mathbf{B}}{c}\right)$$
$$= n_{i1}e\mathbf{E} + \mathbf{j}_{i1}\times\mathbf{B} + neE_1 \tag{79}$$

where

$$n_{i1} = \int F_1 d^3\mathbf{v} \tag{80}$$

is the first-order ion density and

$$\mathbf{j}_{1i} = e\int F_1\frac{\mathbf{v}}{c}d^3\mathbf{v} \tag{81}$$

is the first-order ion current, in the laboratory frame. Collecting these results we have

$$\frac{\partial}{\partial t}(\rho\mathbf{U}) + \nabla\cdot(\rho\mathbf{U}\,\mathbf{U}) + \nabla p_i = n_{i1}e\mathbf{E} + \mathbf{j}_{i1}\times\mathbf{B} + ne\mathbf{E}_1 \tag{82}$$

This is a form of the ion momentum equation.

The parallel component of this can be written

$$\rho\left(\frac{\partial \mathbf{U}_{\parallel}}{\partial t}\right) + \rho(\mathbf{U}\cdot\nabla\mathbf{U})_{\parallel} + \nabla_{\parallel}p_i = ne\mathbf{E}_{1\parallel} \tag{83}$$

where we have used the equation of continuity. This reduced equation gives us what we need to determine the evolution of the $U_{\parallel}$ in the Maxwellian, equation 62.

Finally, let us multiply equation 71 by $Mv'^2/2$, and integrate:

$$M \int \frac{v'^2}{2} \frac{\partial F_0}{\partial t} d^3\mathbf{v} + M \int \frac{v'^2}{2} \mathbf{v} \cdot \frac{\partial F_0}{\partial \mathbf{x}} d^3\mathbf{v} - M \int \frac{v'^2}{2} C_{ie} d^3\mathbf{v} = 0 \tag{84}$$

Now the first integral on the left is

$$\frac{\partial}{\partial t} M \int F_0 \frac{v'^2}{2} d^3\mathbf{v} = \frac{\partial}{\partial t}\left(\frac{3}{2} n_i T_i\right) = \frac{3}{2}\frac{\partial p_i}{\partial t} \tag{85}$$

(The term from differentiating the $v'^2/2$ at fixed t vanishes.) The $\partial/\partial\mathbf{x}$ term gives

$$M\nabla \cdot \int \left(\frac{v'^2}{2} \mathbf{v} F_0 d^3\mathbf{v}\right) - M \int \mathbf{v} \cdot \nabla \frac{v'^2}{2} F_0 d^3\mathbf{v} \tag{86}$$

But $\mathbf{v} \cdot \nabla v'^2/2 = -\mathbf{v} \cdot \nabla \mathbf{U} \mathbf{v}'$, so by evaluating the velocity integrals for the Maxwellian F_0 we get

$$\nabla \cdot (\tfrac{3}{2} p_i \mathbf{U}) + p_i \nabla \cdot \mathbf{U} \tag{87}$$

Therefore, the $Mv'^2/2$ moment of equation 71 is

$$\frac{3}{2}\frac{\partial p_i}{\partial t} + \frac{3}{2}\nabla \cdot (p_i \mathbf{U}) + p_i \nabla \cdot \mathbf{U} = Q_i \tag{88}$$

where $Q_i = \int Mv'^2/2 C_{ie} d^3\mathbf{v}$ is this moment of the ion–electron collision integral. A direct integration (with Maxwellian distribution functions F_0 and f_0) shows that

$$Q_i = \frac{3m}{M}\frac{n}{\tau_e}(T_e - T_i) \tag{89}$$

where the electron collision time τ_e is defined in equation 150 below.

Because of the mass ratio in Q_i it can be smaller than the other terms in the equation and we drop it in the zero-order fluid equations, where Q_{ie} represents the rate of energy exchange by collisions between the ions and electron zero-order Maxwellians. Thus, we have

$$\frac{\partial p_i}{\partial t} + \mathbf{U} \cdot \nabla p_i + \frac{5}{3} p_i \nabla \cdot \mathbf{U} = 0 \tag{90}$$

the fluid equation for the ion pressure. Notice that the forces do not enter directly into this equation for p_i. Also, $p_i/\rho^{5/3}$ is a constant following the motion, so this equation states that the ion entropy is constant following the fluid motion.

The electron fluid equations come from their Fokker–Planck equation to first order. Write the electron distribution function f as

$$f_e = f_0 + f_1 \tag{91}$$

Substitute this expansion into the electron Fokker–Planck equation. To first order we get

$$\frac{\partial f_0}{\partial t} + \mathbf{v} \cdot \nabla f_0 = \frac{e}{m}\left(\mathbf{E} + \frac{\mathbf{v} \times \mathbf{B}}{c}\right) \cdot \nabla_v f + C_{ee}(f_1, f_0)$$

$$+ C_{ei}(f_1, F_0) + C_{ei}(f_0, F_1) \tag{92}$$

where $C_{ee} = (\partial f/\partial t)_{1,ee}$ is the first-order electron–electron collision integral and $C_{ei} = (\partial f/\partial t)_{1,ei}$ is the first-order electron–ion collision integral.

Now, this is the integral equation for f_1 and the condition for it to be satisfied is that the moments of 1 and $mv'^2/2$ of the equation vanish. Since U is regarded as known, there are only two parameters in f_0 whose time evolution needs to be determined. (The $mv'^2/2$ moments of the C_{ei} terms do not vanish, but, as we have shown, they are of higher order.)

In a strict analogy to the calculation of the ion fluid equations we get

$$\frac{\partial n}{\partial t} + \nabla \cdot (n\mathbf{U}) = 0 \tag{93}$$

$$\frac{\partial p_e}{\partial t} + \mathbf{U} \cdot \nabla p_e + \frac{5}{3} p_e \nabla \cdot \mathbf{U} = 0 \tag{94}$$

Although we do not need it for the evolution of f_0 we also take the $m\mathbf{v}$ moment of the electron Fokker–Planck equation, which is

$$m\frac{\partial (n\mathbf{U})}{\partial t} + m\nabla \cdot (n_e\mathbf{U}\mathbf{U}) + \nabla p_e = -n_{e1}e\mathbf{E} + \mathbf{j}_{e1} \times \mathbf{B} - ne\mathbf{E}_1 + \mathbf{R}_{ei} \tag{95}$$

where n_{e1} and j_{e1} are the first-order electron density and current. These involve f_1. $R_{ei} = \int m\mathbf{v}C_{ei}d^3\mathbf{v}$ is the velocity moment of the electron-ion collision integral to first order, which also involves f_1. It is the frictional term on the electrons from electron–ion collisions and is the term responsible for resistivity. In fact, equation 95 is a generalized form of Ohm's law.

The inertial terms are very small because of the small electron mass. Neglecting them we find that the parallel component of this equation is simply

$$\nabla p_{e\parallel} = -neE_\parallel + R_{ei} \tag{96}$$

By estimating R_{ie} physically in terms of j we can show that, indeed, $E_\parallel$ is small compared to $E_\perp$.

Let us add the electron fluid equations 93–95 to the ion fluid equations (75), (82), and (88):

$$\frac{\partial \rho}{\partial t} + \nabla \cdot (\rho\mathbf{U}) = 0 \tag{97}$$

$$\frac{\partial}{\partial t} p + p\nabla \cdot \mathbf{U} + \frac{5}{3} p(\nabla \cdot \mathbf{U}) = 0 \tag{98}$$

$$\rho\frac{\partial \mathbf{U}}{\partial t} + \rho\mathbf{U} \cdot \nabla\mathbf{U} + \nabla p = q\mathbf{E} + \mathbf{j} \times \mathbf{B} \tag{99}$$

These equations give the evolution of ρ, p, and $U_\parallel$ in time. Because $\mathbf{j}$ is determined by F_1 and f_1 we cannot use equation 99 to advance $\mathbf{U}_\perp$. However if the evolution of $\mathbf{E}$ and $\mathbf{B}$ are known, then so is $\mathbf{U}_\perp$ and we do not need equation 99 to find it. In the last equation we drop the electron inertial terms. They are small because of the small electron mass. The $\mathbf{R}_{ei}$ and the Q_{ei} terms in the electron equations cancel the $\mathbf{R}_{ie}$ and Q_{ie} terms in the ion equations. The $\mathbf{E}_1$ also terms cancel. $p = p_i + p_e$ is the total pressure and $\mathbf{j} = \mathbf{j}_{i1} + \mathbf{j}_{e1}$ is the total current. The zero-order currents cancel. q is the total charge $q = e(n_{1i} - n_{1e})$. The zero-order charges also cancel.

To complete this fluid equation, we might think that we would have to solve for F_1 and f_1 and then take their first-order moments. But it is actually not necessary to completely solve for the first-order $f's$ to find their moments. In fact, the solution for their moments leads right back to this equation.

Up to this point in our discussion we have assumed that $\mathbf{E}$ and $\mathbf{B}$ were specified and we solve for F_0 and f_0 as functions of space and time in terms of them. $\mathbf{U}_\perp = \mathbf{U}_E$ is given by equation 59, F_0 and f_0 are Maxwellians with parameters n, $\mathbf{U}$, T_i, and T_e with $\mathbf{U}_i = \mathbf{U}_e = \mathbf{U}$ and $n_i = n_e = n$. Thus, only the functions $U_\parallel, n, T_i$, and T_e are needed to specify the zero-order distribution functions. But $\mathbf{U}_\perp$ is given by equation 99 as well as directly from $c\mathbf{E} \times \mathbf{B}/B^2$ and these two determinations must agree, Thus, we may take $\mathbf{U}_\perp$ and $\mathbf{B}$ as the fundamental variables instead of $\mathbf{E}$ and $\mathbf{B}$, so $\mathbf{E}$ is expressed in terms of $\mathbf{U}$ by

$$\mathbf{E} + \frac{\mathbf{U} \times \mathbf{B}}{c} = 0 \tag{100}$$

The magnetic field is evolved by Maxwell's equation

$$\frac{\partial \mathbf{B}}{\partial t} = -c\nabla \times \mathbf{E} = \nabla \times (\mathbf{U} \times \mathbf{B}) \tag{101}$$

To evolve $\mathbf{E}$ (and $\mathbf{U}_\perp$) we now need equation 99, which requires $\mathbf{j}$. But $\mathbf{j}$ is given by Ampere's law

$$\nabla \times \mathbf{B} = 4\pi \mathbf{j} \tag{102}$$

so we are able to advance $\mathbf{U}_\perp$ (and $\mathbf{E}$). (The charge density q in the $q\mathbf{E}$ term can be obtained from Poisson's equation, but, of course, as we saw in chapter 4 this term is relativistically small.) Our zero-order expansion of the Fokker–Planck equations thus leads to the MHD equations of chapter 3 as well as giving the zero-order distribution functions. Finally, we may incorporate the resistivity into equation 100 by making use of equation 96 since $\mathbf{R}_{ei}$ is proportional to $\mathbf{j}$. In fact, $\mathbf{R}_{ei} = ne\eta\mathbf{j}$.

Why did we go to all this trouble to derive the collision integral, the Fokker–Planck equation, and its solution if the only result is the fluid equations? There are basically two reasons for this. One is to show that the fluid equations are essentially correct in the appropriate limit, which emerges from this discussion. The second is to set the stage to determine the next corrections to the fluid equations, by going to higher order in the Fokker–Planck equations. These corrections comprise the various dissipation terms or transport effects, such as viscosity and thermal conductivity.

8.7 Transport Effects

The transport effects are found by going to the second order in the solution of the Fokker–Planck equations for the electrons and the ions. To do this we first solve the integral equations (71) and (92) for F_1 and f_1, the first-order ion and electron distribution functions. Then in the second-order Fokker–Planck equations we derive the conditions that we can solve for F_2 and f_2. These conditions are obtained by taking the 1, $\mathbf{v}_\parallel$, and $Mv'^2/2$ moments of these equations. These moments involve F_1 and f_1 but not F_2 nor f_2.

We illustrate the procedure by considering just the ions. Their first-order Fokker–Planck equation 71 is

$$\frac{\partial F_0}{\partial t} + \mathbf{v} \cdot \nabla F_0 = -\frac{e}{M}\left(\mathbf{E} + \frac{\mathbf{v} \times \mathbf{B}}{c}\right) \cdot \nabla_v F_1 + C_{ii} \tag{103}$$

where C_{ii} is the first-order part of the ion–ion collision integral. (We drop the ion–electron collision integral C_{ie} because it is small and involves the electron distribution function f_1, which complicates the procedure.)

Now, this is a linear integral equation for F_1. It has homogeneous solutions as well as particular ones. These homogeneous solutions are obtained by solving the homogeneous equation, obtained by dropping the left-hand side of the Fokker–Planck equation,

$$0 = -\frac{e}{M}\left(\mathbf{E} + \frac{\mathbf{v} \times \mathbf{B}}{c}\right) \cdot \nabla_v F_1 + C_{ii}(F_1, F_0) + C_{ii}(F_0, F_1) \tag{104}$$

The right-hand side is the linearized part of the exact Lorentz and ion–ion collision terms. Now, we know the exact solution of the nonexpanded form. Namely, it is a Maxwellian function involving n, $U_\parallel$, and T:

$$F = F_{\max}(n, \mathbf{U}, T : \mathbf{v}) = \frac{n}{(2\pi T/M)^{3/2}} \exp -\frac{M(\mathbf{v} - \mathbf{U})^2}{2T} \tag{105}$$

Thus, the homogeneous solution of the linearized form can be obtained by simply perturbing these three quantities by any amount we wish. We can choose these perturbations so that in the breakup of F_1 into the sum of the homogeneous and particular solutions

$$F_1 = F_h + F_p, \tag{106}$$

the particular solution F_p is chosen to have zero moments of 1, $Mv_\parallel$, and $Mv'^2/2$. This choice makes the particular solution F_p unique and it is obtained by solving equation 103. The homogeneous solution is

$$F_h = n_1 \frac{\partial F_0}{\partial n} + U_{1\|}\mathbf{b} \cdot \frac{\partial F_0}{\partial \mathbf{U}} + T_1 \frac{\partial F_0}{\partial T}$$

$$= \left[\frac{n_1}{n} + MU_{1\|}\mathbf{b} \cdot \frac{\mathbf{v}'}{T} + \frac{T_1}{T} \left(\frac{Mv'^2}{2T} - \frac{3}{2} \right) \right] F_0 \qquad (107)$$

where $\mathbf{v}' = \mathbf{v} - \mathbf{U}$ is the peculiar velocity, and $n_1, U_{1\|}$, and T_1 are arbitrary constants ($F_0 = F_{\max}(n_0, \mathbf{U}_0, T_0; \mathbf{v})$). We use the freedom provided by these constants to make the corresponding moments of F_p vanish and render F_p unique.

The homogeneous solution still has these arbitrary constants. They are determined by the condition mentioned above that F_2 can be solved for. If we carry through this procedure, we find that $n_1 \mathbf{U}_{1\|}$ and T_1 will grow until F_1 becomes comparable to F_0 and the expansion breaks down. This is physically plausible since

$$F_0(t, \mathbf{x}, \mathbf{v}) = F_{\max}(n_0, U_0, T_0 : \mathbf{v}) \qquad (108)$$

is governed by the fluid equations with no dissipation. But over a sufficiently long time, dissipative terms will change the density n, the velocity $\mathbf{U}$, and the temperature T finitely away from n_0, $\mathbf{U}_0$, and T_0. (If they did not, we would never be interested in them.) Thus, to avoid this we need to follow $n, \mathbf{U}$, and T given by the full equations with dissipation rather than $n_0, \mathbf{U}_0$, and T_0 given by the fluid equations.

In this connection it is useful to note that

$$F_0 + F_h \approx F_{\max}(n_0 + n_1, \mathbf{U}_0 + \mathbf{U}_{1\|}, T_0 + T_1 : \mathbf{v}) \qquad (109)$$

so that the effect of these first-order terms is to keep the distribution near a Maxwellian distribution different from the fluid Maxwellian. (As a result of this F_p is always small.) The transverse velocity $\mathbf{U}_\perp$ is closely tied to the $\mathbf{E} \times \mathbf{B}$ velocity $\mathbf{U}_E$. Thus, we expect the solution to be better represented by

$$F = F_{\max}(n, \mathbf{U}, T : \mathbf{v}) \qquad (110)$$

where we follow $n, \mathbf{U}$, and T correctly. The $\mathbf{U}_1$ in this ion distribution function is different from the corresponding elecron $\mathbf{U}_1$ that produces $\mathbf{j}$ in Ampere's law.

To get the modified equations for n, U, and T we take the moments of the full Fokker–Planck equation for the ions, equation 55. (We will suppress the subscript i until we get to the final ion equations.) These calculations are closely analogous to those we carried out to obtain the ion fluid equations. However, there are some extra terms from the F_p deviation from a local Maxwellian.

Let us pick out these terms. Our notation will be $\mathbf{v} = \mathbf{U} + \mathbf{v}'$, with $\mathbf{U}$ the true mean velocity given by

$$\int F\mathbf{v}d^3\mathbf{v} = \int (F_{\max} + F_p)\mathbf{v}d^3\mathbf{v} = n\mathbf{U} \qquad (111)$$

(In $F_{\max}$, $\mathbf{U}_\perp$ now differs from the $\mathbf{U}_E$ by a small contribution from F_p to the perpendicular velocity.) But F_p produces no change in $n, U_\parallel$, and T, since the corresponding moments of F_p are zero. Therefore, we have

$$n = \int F d^3\mathbf{v} = \int F_{\max} d^3\mathbf{v}$$

$$nU_\parallel = \int F v_\parallel d^3\mathbf{v} = \int F_{\max} v_\parallel d^3\mathbf{v}$$

$$nT = \int F \frac{Mv'^2}{2} d^3\mathbf{v} = \int F_{\max} \frac{Mv'^2}{2} d^3\mathbf{v} \tag{112}$$

The zeroth moment obtained from multiplying the Fokker–Planck equation by 1 and integrating gives no correction to the continuity equation. For the first moment, gotten from multiplying by $M\mathbf{v}$, the $\mathbf{v}'\mathbf{v}'$ term in equation 78 is changed to

$$M \int F \mathbf{v}'\mathbf{v}' d^3\mathbf{v} = nT\mathbf{I} + \int F_p \left(\mathbf{vv} - \frac{1}{3}\mathbf{v}'^2\mathbf{I} \right) d^3\mathbf{v}$$

$$= nT\mathbf{I} + \boldsymbol{\pi} \tag{113}$$

where $\boldsymbol{\pi}$ is the viscous tensor responsible for viscosity. For the second moment, gotten by multiplying by $Mv'^2/2$ and integrating, we find that the first term on the right-hand side of equation 86 under the divergence becomes

$$M \int F \left(\frac{v'^2}{2} \mathbf{v} \right) d^3\mathbf{v} = M \int F_{\max} \left(\frac{v'^2}{2} \mathbf{U} \right) d^3\mathbf{v} + M \int F_p \left(\frac{v'^2}{2} \mathbf{v}' \right) d^3\mathbf{v}$$

$$= \frac{3}{2} p\mathbf{U} + \mathbf{q} \tag{114}$$

where $\mathbf{q}$ is the last term and is the heat flow vector. Finally, in equation 86 the term

$$M \int F(\mathbf{v} \cdot \nabla\mathbf{U} \cdot \mathbf{v}') d^3\mathbf{v} = p_i \nabla \cdot \mathbf{U} + \boldsymbol{\pi} : \nabla\mathbf{U} \tag{115}$$

With these changes we get the more accurate dissipative equations for the fluid quantities. The continuity equation becomes

$$\frac{\partial n}{\partial t} + \nabla(n\mathbf{U}_i) = 0 \tag{116}$$

The equation of motion becomes

$$\frac{\partial}{\partial t}(\rho\mathbf{U}) + \nabla \cdot (\rho\mathbf{U}_i\mathbf{U}_i) + \nabla p_i + \nabla \cdot \boldsymbol{\pi}_i = q_i\mathbf{E} + \mathbf{j}_i \times \mathbf{B} \tag{117}$$

The entropy equation for the temperature or rather the pressure $p = nT$ becomes

$$\frac{3}{2}\frac{\partial p_i}{\partial t} + \frac{3}{2}\mathbf{U}_i \cdot \nabla p_i + p\nabla \cdot \mathbf{U}_i + \boldsymbol{\pi}_i : \nabla\mathbf{U}_i = -\nabla \cdot \mathbf{q}_i \tag{118}$$

In these equations the stress tensor $\boldsymbol{\pi}_i$ is

$$\pi_i = \int F\left(\mathbf{v}'\mathbf{v}' - \frac{1}{3}\mathbf{v}'^2\mathbf{I}\right) d^3\mathbf{v} = \int F_p\left(\mathbf{v}'\mathbf{v}' - \frac{1}{3}\mathbf{v}'^2\mathbf{I}\right) d^3\mathbf{v} \tag{119}$$

The heat flow vector $\mathbf{q}$ is

$$\mathbf{q}_i = M \int F\frac{v'^2}{2}\mathbf{v}' d^3\mathbf{v} = M\int F_p\frac{v'^2}{2}\mathbf{v}' d^3\mathbf{v} \tag{120}$$

To complete these equations we need to solve the first-order Fokker–Planck equation for F_p, and from it calculate $\boldsymbol{\pi}_i$ and $\mathbf{q}_i$. Making use of the fluid equations to lowest order we can write this as

$$C_{1ii}(F_p) - \frac{e}{M}\Omega_i\frac{\partial F_p}{\partial \phi} = \left(\frac{Mv'^2}{2T} - \frac{5}{2}\right)\mathbf{v}\cdot\nabla T + \frac{M}{2T}\left(\mathbf{v}'\mathbf{v}' - \frac{v'^2}{3}I\right) : \mathbf{W} \tag{121}$$

where $\mathbf{W} = \nabla\mathbf{U} + (\nabla U)^{\text{tr}} - (\frac{2}{3})\nabla\cdot UI$, and C_{1ii} is the first-order expansion of the collision operator. (The superscript tr indicates that the transpose of the tensor $\nabla\mathbf{U}$ should be taken.)

As we have stated, we get rid of the ambiguity of the particular solution by demanding that the appropriate moments of F_p vanish. We then find $\boldsymbol{\pi}$ and $\mathbf{q}$ by substituting the solution into equations 119 and 120 for them.

The right-hand side of equation 121 can be broken up into separate components, involving $\partial T/\partial \ell_\parallel$, $\partial T/\partial \ell_\perp$, and the tensor components of $\mathbf{W}$, and for each such term the corresponding contribution to F_p can be found. F_p is the sum of these contributions. It turns out that only ∇T terms contribute to the heat flow vector $\mathbf{q}$ and only the $\mathbf{W}$ terms contribute to the viscous tensor $\boldsymbol{\pi}$.

If $\Omega \gg \nu$, then the results for $\boldsymbol{\pi}$ and $\mathbf{q}$ are very dependent on the directions of ∇T and $\nabla\mathbf{U}$ relative to $\mathbf{B}$. If B is strong. the explicit results for the ion heat flow are

$$\mathbf{q} = \kappa_\perp \nabla_\perp T_i + \kappa_\parallel \nabla_\parallel T_i + \frac{5}{2}\frac{nT_i}{eB^2}\mathbf{B}\times\nabla T_i$$

$$\kappa_\parallel = 3.9\frac{nT_i\tau_i}{M}$$

$$\kappa_\perp = 2\frac{nT_i}{M\Omega_i^2\tau_i} \tag{122}$$

where

$$\tau_i = \frac{3}{4\sqrt{\pi}}\frac{\sqrt{M}T_i^{3/2}}{e^4 n_i \ln\Lambda} \tag{123}$$

is Braginski's ion–ion collision time. The last term in $\mathbf{q}$ is called the gyrotropic heat flow. It is somewhat artificial, since it arises because the guiding centers move with the $\mathbf{E} \times \mathbf{B}$ flow with $\mathbf{E}$ averaged over the circumference of of the cyclotron orbit. Since particles of different energies have different size orbits, the motion of the faster particles is slightly different than that of the slower ones, so the $\mathbf{E} \times \mathbf{B}$ motions are slightly different. Since the heat flow is a different velocity moment than the mass flow, this accounts for this term.

The parallel conductivity $\kappa_{i\parallel}$ does not depend on B since it involves only particle motion along $\mathbf{B}$. If $\Omega \gg \nu$, then the perpendicular thermal conductivity is much smaller than the parallel conductivity. This is the usual situation in astrophysics. We see that $\kappa_{i\parallel}$ is of order $n\lambda v_i$ and $\kappa_{i\perp}$ is of order $n\rho_i^2\nu$, as expected from simple physical estimates. The result of the solution of the integral equation is simply to establish the numerical coefficients, 3.16 and 4.66. (The expression for the ion stress tensor π is given at the end of full Braginski equations in the next section.)

The treatment of the electrons is very similar except for the complications introduced by C_{ei}, the electron–ion collision integral. Without this term we get expressions for the electron thermal conductivity and viscosity very like the ion thermal conduction and viscosity. The electron fluid equations would be the same, with the ion properties T_i and U_i replaced by T_e and U_e and the ion–ion collision time replaced by the electron–electron collision time. (However, as we saw in the zero-order fluid treatment the ion and electron densities and velocities are forced to be nearly equal by Maxwell's equations, so we need only the electron entropy equation.)

The C_{ei} integral is of the same order of magnitude as the C_{ii} integral, while the C_{ie} is much smaller than the C_{ii} integral because of the large mass ratio. This introduces other transport coefficients, such as resistivity, thermal–electric forces, Ohmic heating, and electron–ion energy exchange. While it would be straightforward to present the electron transport corrections in the same manner as we did the ion transport corrections, the complications are such that little insight would be gained. The treatment of the ions illustrates how we arrive at a complete set of equations correct to first order.

There was considerable confusion during the 1950s in trying to arrive at this correct set of equations, since the transport coefficients were each treated separately although they are physically linked. It was only when Braginski (1965) and others (e.g., Robinson and Bernstein (1962)) carried out the systematic solution of the Fokker–Planck equation by a correct expansion to first order that the correct set of equations was derived.

The equations derived by Braginski (1965) have now been universely accepted, and because of their correctness and completeness they are considered standard and are universely referred to as the Braginski equations. It is now the case that we can treat almost any transport problem by starting with these equations and keeping just the terms of interest. This is the most satisfactory way to proceed, since the assumptions used in solving transport problems are clear and definite. For this reason we proceed to write down the complete set of Braginski equations even though they are somewhat long. It turns out that for any special case they are reasonably simple.

8.8 The Braginski Equations

The complete set of fluid equations including first-order terms are the six Braginski equations, three for electrons and three for ions. They are listed in Braginski (1965) together with a discussion of their physical significance and an outline of their derivation. This paper is well worth reading since it is valuable from both a physical and a practical standpoint. These equations are extremely useful, and they are thus presented here in full, with the exception that only the leading terms in an ν/Ω expansion are kept.

$$\frac{\partial n_e}{\partial t} + \nabla \cdot (n_e \mathbf{U}_e) = 0 \tag{124}$$

$$\frac{\partial n_i}{\partial t} + \nabla \cdot (n_i \mathbf{U}_i) = 0 \tag{125}$$

$$m n_e \frac{d_e \mathbf{U}_e}{dt} = -\nabla p_e - \nabla \cdot \boldsymbol{\pi}^e - e n_e \left(\mathbf{E} + \frac{\mathbf{U}_e \times \mathbf{B}}{c} \right) + \mathbf{R} \tag{126}$$

$$M n_i \frac{d_i \mathbf{U}_i}{dt} = -\nabla p_i - \nabla \cdot \boldsymbol{\pi}^i + e n_i \left(\mathbf{E} + \frac{\mathbf{U}_i \times \mathbf{B}}{c} \right) - \mathbf{R} \tag{127}$$

$$\frac{3}{2} n_e \frac{d_e T_e}{dt} + p_e \nabla \cdot \mathbf{U}_e = -\nabla \cdot \mathbf{q}_e - \boldsymbol{\pi}_e : \nabla \mathbf{U}_e + Q_e \tag{128}$$

$$\frac{3}{2} n_i \frac{d_i T_i}{dt} + p_i \nabla \cdot \mathbf{U}_i = -\nabla \cdot \mathbf{q}_i - \boldsymbol{\pi}_i : \nabla \mathbf{U}_i + Q_i \tag{129}$$

These equations are completed by the definitions

$$p_e = n_e T_e, \quad p_i = n_i T_i \tag{130}$$

$$\frac{d_e}{dt} = \frac{\partial}{\partial t} + \mathbf{U}_e \cdot \nabla \tag{131}$$

$$\frac{d_i}{dt} = \frac{\partial}{\partial t} + \mathbf{U}_i \cdot \nabla \tag{132}$$

and the expressions for the transport coefficients

$$\mathbf{q}_e = \mathbf{q}_{eU} + \mathbf{q}_{eT} \tag{133}$$

$$\mathbf{q}_{eT} = -\kappa_{\parallel}^e \nabla_{\parallel} T_e \tag{134}$$

$$\mathbf{q}_{eU} = 0.71 n_e T_e \mathbf{U} \tag{135}$$

$$\kappa_\parallel^e = 3.16 n_e T_e \frac{\tau_e}{m_e} \tag{136}$$

$$\mathbf{q}_{iT} = -\kappa_\parallel^i \nabla_\parallel T_i - \kappa_\perp^i \nabla_\perp T_i \tag{137}$$

$$\kappa_\parallel^i = \frac{3.9 n_i T_i \tau_i}{M} \tag{138}$$

$$\kappa_\perp^i = 2\frac{n_i T_i}{M\Omega_i^2 \tau_i} \tag{139}$$

$$\mathbf{R} = \mathbf{R}_U + \mathbf{R}_T \tag{140}$$

$$\mathbf{R}_U = \frac{m n_e}{\tau_e}(0.51\mathbf{U}_\parallel + \mathbf{U}_\perp) = ne\left(\frac{\mathbf{j}_\parallel}{\sigma_\parallel} + \frac{\mathbf{j}_\perp}{\sigma_\perp}\right) \tag{141}$$

$$\mathbf{U} = \mathbf{U}_i - \mathbf{U}_e \tag{142}$$

$$\sigma_\perp = \frac{e^2 n \tau_e}{m} \tag{143}$$

$$\sigma_\parallel = 1.96\sigma_\perp \tag{144}$$

$$R_T = -0.71 n_e \nabla_\parallel T_e \tag{145}$$

$$Q_i = Q_\Delta = \frac{3m}{M}\frac{n_e}{\tau_e}(T_e - T_i) \tag{146}$$

$$Q_e = \mathbf{R}\cdot\mathbf{U} - Q_\Delta = \frac{j_\parallel^2}{\sigma_\parallel} + \frac{j_\perp^2}{\sigma_\perp} + \frac{\mathbf{j}\cdot\mathbf{R}_T}{n_e e} - Q_\Delta \tag{147}$$

$$\boldsymbol{\pi}_i = -\frac{\eta_0}{3}(\mathbf{I} - 3\mathbf{bb})[(\mathbf{I} - 3\mathbf{bb}) : \nabla\mathbf{U}] \tag{148}$$

$$\eta_0 = 0.96 n_i T_i \tau_i \tag{149}$$

These transport coefficients involve the definition of the collision terms τ_i and τ_e given by Braginski:

$$\tau_e = \frac{3}{4\sqrt{2\pi}}\frac{\sqrt{m}T_e^{3/2}}{n_1 e^4 \ln\Lambda} = \frac{3.5\times 10^4 T_e^{3/2}}{n}\frac{1}{\ln\Lambda/10} \tag{150}$$

$$\tau_i = \frac{3}{4\sqrt{\pi}}\frac{\sqrt{M}T_i^{3/2}}{n_i e^4 \ln\Lambda} = 2.1\times 10^6 \frac{T_i^{3/2}}{n}\frac{1}{\ln\Lambda/10} \tag{151}$$

(Note that the definition of τ_e has $\sqrt{2\pi}$ denominator, while that of τ_i has only $\sqrt{\pi_i}$. It is not clear why Braginski made this choice.)

These equations give the evolution of n_e, $\mathbf{U}_e$, T_e, n_i, $\mathbf{U}_i$, T_i once the evolution of **E** and **B** is given. But the evolution of **E** and **B** is coupled to the fluid quantities through Maxwell's equations, as we have discussed in the last two sections.

For hydrodynamic motions most of the viscous terms are small (except perhaps the electron thermal conductivity along the lines if the parallel scale is not sufficiently large) and for hydrodynamic times they can be neglected. These hydrodynamic motions are usually oscillatory and after a large number of periods the waves are damped by viscous terms and the situation approaches equilibrium. Over longer times the equilibrium evolves further, due to the viscous and thermal terms.

We note that the plasma energy flow in the Braginski equations is more complex than in the ideal equations (equation 42 in chapter 4). For example, kinetic energy can be dissipated by viscosity and the energy goes to heat the ions directly. Also, there is heat flow from one region to another.

The generalization of the total energy equation is as follows. Let

$$\mathcal{E}_{\text{tot}} = \int \left[\frac{\rho U^2}{2} + \frac{3}{2} n(T_i + T_e) + \frac{B^2}{8\pi} \right] d^3V \tag{152}$$

Then from Braginski's equations we can show that

$$\frac{d\mathcal{E}_{\text{tot}}}{dt} = - \int dS \cdot q_{\text{tot}} \tag{153}$$

where

$$q_{\text{tot}} = \left(\frac{1}{2} \rho U^2 + \frac{5}{2} p \right) \mathbf{U} + \boldsymbol{\pi}^i \cdot \mathbf{U}_i + \mathbf{q}_i + \mathbf{q}_e$$

$$+ \frac{c}{4\pi} (\mathbf{E} \times \mathbf{B}) \tag{154}$$

and where **E** is given by Ohm's law

$$\mathbf{E} + \frac{\mathbf{U} \times \mathbf{B}}{c} = \eta_\perp \mathbf{j}_\perp + \eta_\parallel \mathbf{j}_\parallel \tag{155}$$

8.9 Properties of the Transport Coefficients

We now discuss some of the properties of the physical quantities in the Braginski equations. The heat flows are predominantly along **B**. If $\nabla_\perp T$ and $\nabla_\parallel T$ are the same order of magnitude, then $q_\perp / q_\parallel \approx 1/\Omega^2 \tau^2$ (ignoring the gyro heat flows). This is usually a very small number for ions and a much smaller number for electrons.

Also, $\kappa_{\parallel}^{e} \sim T_e^{5/2}$ and $q_{\parallel}^{e} \approx T_e^{7/2}/L_{\parallel}$, so both quantities increase rapidly with T_e. The order of magnitude of $\kappa_{\parallel}$ is $n\lambda v_T$ for both ions and electrons, where λ is the mean free path and v_T is the thermal velocity. Because $v_{Te} \gg v_{Ti}$ and the λ's are the same, the electron heat flow along the lines is larger than ion heat flow by $\sqrt{M/m}$. On the other hand, perpendicular ion heat flow is larger than perpendicular electron flow because $\kappa_{\perp} \approx n\rho^2/\tau$. However, the perpendicular heat conductivity of both ions and electrons is, generally, very small. The heat flows can become comparable if the perpendicular temperature gradient is much larger than the parallel one. This is the case for coronal loops, which are much thinner than they are long, corresponding to very large perpendicular gradients. The magnetic field acts as a very good heat insulator.

There is an additional parallel heat flow of the electrons if there is a parallel electric field. This arises roughly as follows: The electric field induces a parallel velocity of the electrons relative to the ions. The mean parallel velocity is different for electrons with different velocities v because the Coulomb cross section for electron–ion collisions depends on velocity (as $1/v^3$). Thus, the induced velocity of the faster electrons is larger. The mean parallel velocity of the electrons is the weighted mean of the velocities

$$n\mathbf{U} = \int f_1 \mathbf{v} d^3\mathbf{v} \tag{156}$$

while the heat flow is a different mean

$$\mathbf{q} = \frac{1}{2}\int m(\mathbf{v} - \mathbf{U})^2(\mathbf{v} - \mathbf{U}) f_1 d^3\mathbf{v} \tag{157}$$

so it does not average out. The faster electrons have a smaller cross section and a larger induced velocity than $\mathbf{U}$, while the slower electrons with larger cross section have a smaller mean velocity than $\mathbf{U}$. The result is a heat flow due to $E_{\parallel}$, which, when expressed in terms of $\mathbf{U}$, is

$$\mathbf{q}_{eU} = 0.71 n_e T_e \mathbf{U} \tag{158}$$

In equation 121 the ion velocity gradients produce an F_p, which contributes a nonzero moment to the viscous stress tensor $\boldsymbol{\pi}^i$. The $-\nabla \cdot \boldsymbol{\pi}^i$ term in equation 117 represents the viscous force. The dominant components in $\boldsymbol{\pi}^i$ are

$$\pi_{zz} = -\eta_0 W_{zz} \tag{159}$$

$$\pi_{xx} = \frac{\eta_0}{2} W_{zz} \tag{160}$$

$$\pi_{yy} = \frac{\eta_0}{2} W_{zz} \tag{161}$$

where

$$\eta_0 = 0.96 n T_i \tau_i \tag{162}$$

and x, y, and z are local coordinates at a point P with the z axis along $\mathbf{B}$. Since the divergence of $\boldsymbol{\pi}$ enters into the equations of motion and $\mathbf{B}$ is space dependent, we have to take into account that the coordinate system rotates as P moves. Thus, it is best to write $\boldsymbol{\pi}$ vectorially

$$\boldsymbol{\pi} = -\frac{\eta_0}{2}\left(\mathbf{I} - 3\mathbf{bb}\right)\mathbf{b}\cdot\mathbf{W}\cdot\mathbf{b} \tag{163}$$

or

$$\boldsymbol{\pi} = \frac{\eta_0}{3}(\mathbf{I} - 3\mathbf{bb})(\mathbf{I} - 3\mathbf{bb}) : \nabla\mathbf{U} \tag{164}$$

where $\mathbf{b} = \mathbf{B}/B$ is a unit vector in the $\mathbf{B}$ direction. The viscosity due to electrons is always smaller because of their smaller mass (and their shorter collision time).

It may seem strange that there is a zero-order perpendicular viscosity, since, in analogy with the thermal flows, the ions are restricted from moving across lines. However, a perpendicular viscous stress actually should exist, as can be seen from the following argument: If the perpendicular velocity compresses the magnetic field, as can happen through the $\nabla\cdot\mathbf{U}$ term, then because of the conservation of the magnetic moment, we expect their perpendicular energy to increase relative to their parallel energy. The increase happens only for a collision time τ_i, since after that time, perpendicular and parallel pressures equalize.

Thus, during a time τ, B increases by $\delta B = -(\nabla_\perp\cdot U_\perp)\tau$ and $p_\perp$ increases by $(\delta B/B)nT = -(\nabla\cdot U_\perp)\tau p \approx \eta_0\nabla\cdot U_\perp$, in agreement with the above results. This increase gives rise to a mechanism for heating a plasma when the magnetic field is oscillated on a timescale comparable to τ. When one compresses, $p_\perp$ is larger than when one expands so the amount of work done during compression is larger than is recovered during expansion and net work is done. This is called "magnetic pumping" and is a useful way to heat a plasma.

Let us return to the electrons and let there again be a parallel electric field $E_\parallel$ giving rise to a mean velocity, U, of the electrons relative to the ions. The ratio of E to U is related to the resistivity of the plasma. In fact,

$$j_\parallel = ne\frac{U_\parallel}{c} = \sigma_\parallel E_\parallel \tag{165}$$

On the other hand, when we have a perpendicular $\mathbf{E}$ field the most direct effect is an $\mathbf{E}\times\mathbf{B}$ motion of both the electrons and the ions and there is actually very little current produced. Thus, there is no unique relation between the perpendicular electric field and the perpendicular current. We saw in chapter 2 that the current arises due to inertial effects, pressure gradients, gravity, and magnetization currents. Thus, the definition of $\sigma_\perp$ loses its normal meaning. On the other hand, for the case of a parallel electric field $E_\parallel$ we can find the difference in parallel velocities δU between the electrons and ions such that the mean collisional force between the electrons and the ions balances the mean force due to the electric field,

$$neE_{\parallel} = R_{ei} \tag{166}$$

(We can ignore the collision of electrons on each other since this leads to no mean force.) This frictional force is

$$R_{\parallel} = \frac{mn_e}{\tau} 0.51 U_{\parallel} \tag{167}$$

But if we wish $\eta_{\parallel}$ to be defined from $E = \eta j$ we have

$$\eta_{\parallel} j_{\parallel} = \frac{R_{\parallel}}{ne} = 0.51 \frac{m}{\tau e} U_{\parallel} = \frac{0.51 mc}{\tau n e^2} j \tag{168}$$

or

$$\eta_{\parallel} = \frac{0.51 mc}{\tau n e^2} \tag{169}$$

This suggests that we define $\eta_{\perp}$ in terms of the perpendicular frictional force, i.e., by

$$\eta_{\perp} j_{\perp} = \frac{R_{\perp}}{ne} = \frac{m}{\tau e} U_{\perp} = \frac{mc}{\tau n e^2} j \tag{170}$$

or

$$\eta_{\perp} = \frac{mc}{\tau n e^2} \tag{171}$$

Thus, $\eta_{\perp}$ and $\eta_{\parallel}$ are defined so that the frictional force on the electrons per unit volume is

$$ne \left(\eta_{\parallel} \mathbf{j}_{\parallel} + \eta_{\perp} \mathbf{j}_{\perp} \right) \tag{172}$$

and this frictional force should be added into the equation of motion of the electrons:

$$\mathbf{E} + \frac{\mathbf{U}_e \times \mathbf{B}}{c} = ne \left(\eta_{\parallel} \mathbf{j}_{\parallel} + \eta_{\perp} \mathbf{j}_{\perp} \right) - \frac{\nabla p_e}{ne} \tag{173}$$

$\mathbf{j}_{\parallel}$ can be found from this equation since the large $\mathbf{U}_e \times \mathbf{B}$ has no parallel component. However, $\mathbf{j}_{\perp}$ is found from the sum of the ion and electron equations. The role of perpendicular resistivity in Ohm's law is to cause a very small slippage between the field lines and the plasma.

Several things are worth noting: $\eta_{\parallel}$ and $\eta_{\perp}$ are essentially independent of density (except for a very weak dependence in the $\ln \Lambda$ factor in the definition of τ). This is easily seen for the parallel resistivity. If n is low, δU must be high for the same j, since there are fewer current carriers. The frictional force per electron is the number density of ions n_i times $\langle \sigma v \rangle$ times $m \delta \mathbf{U}$, so it is proportional to $n \delta \mathbf{U}$, which is thus independent of particle density for the same current density. A similar argument applies to $\eta_{\perp}$.

Second, $\eta_{\perp}$ is larger than $\eta_{\parallel}$. This is because for $j_{\parallel}$ the current-carrying electrons have a velocity larger than the thermal velocity, while for a perpendicular current the current-carrying electrons have about the thermal speed. Thus, the friction of parallel current-carrying electrons with ions is smaller than that of the perpendicular current-carrying electrons. It would be smaller by almost a factor of four, but the electron collisions of the faster current-carrying electrons with cold electrons becomes effective in removing mean electron momentum. We would think that electron–electron collisions would not produce any net force on the electrons. However, because of the large Coulomb cross section for the very slow electrons, the electron–ion collisions effectively transfer the momentum to the ions instantaneously. Therefore, while it is true that a collision between a fast electron and a slow electron does not immediately get rid of the total electron momentum, this momentum is very rapidly lost to ions and produces an effective friction in addition to the direct collision of current-carrying electrons with ions. Thus, the factor four is reduced to two.

Finally, it is worth remarking that there is a modification of the force on the electrons due to a parallel gradient of the temperature:

$$R_T = -0.71 n \nabla_{\|} T_e \tag{174}$$

This effectively cancels most of the force on the electrons due to the part of the parallel electron pressure gradient produced by a temperature gradient.

8.10 Summary

Let us now summarize. The basic kinetic equations for the electron and ion distribution functions are their kinetic Fokker–Planck equations. When collisions are strong we can expand their solutions in the reciprocal collision rates. The zero order of the expansion tells us that the two lowest-order distribution functions, F_0 and f_0, are both Maxwellians whose parameters are the two densities n_0, two mean velocities $\mathbf{U}$, and two temperatures T_0. Charge neutrality tells us that the two densities must be equal. Due to current neutrality and collisional interaction between the electrons and ions, the mean velocities are also equal. Further, the zero-order perpendicular velocities are the $\mathbf{E} \times \mathbf{B}$ velocities. The electron and ion temperatures need not be equal because the time to equilibrate these temperatures by electron–ion collision is very long.

The first-order expansion brings in the first-order distribution functions, f_1's. The condition that we can solve these first-order equations for the f_1's imposes constraints on the time evolution of the f_0s, which are actually the fluid equations. These give the zero-order evolution rates for the parameters in the Maxwellians. (The solutions for the F_1 and f_1 are not unique because there is a homogeneous solution for them as well. The first-order distribution functions give rise to small first-order velocity differences between the ions and electrons. These differences when multiplied by the large plasma charges produce finite current densities. It is these current densities that determine $\mathbf{B}$ and $\mathbf{E}$ through Maxwell's equations. These then couple back through the Fokker–Planck equations to give the finite $\mathbf{E} \times \mathbf{B}$ velocities.

Finally, the next order in the expansion brings in corrections to the fluid equations that are the transport coefficients. The fluid equations with these corrections constitute the Braginski equations, which we have listed.

8.11 An Example

We can illustrate the application of Braginski's equations by using them to calculate the damping of MHD waves. The dispersion relation of MHD waves with transport effects included is slightly different from the ideal fluid one. The real part of the wave frequency ω is not essentially different from the one obtained from the ideal equations. However, the dissipation terms in the Braginski equations lead to a small damping of the waves, which is not present in the ideal dispersion relation. If $\delta\omega$ is the change in ω due to dissipative effects, then its imaginary part gives the damping rate. The amplitude of a wave has a time dependence proportional to

$$e^{-i(\omega+\delta\omega)t} \tag{175}$$

Any MHD wave has an energy equal to twice its mean kinetic energy

$$\mathcal{E}_{\text{wave}} = \frac{1}{2}\int \rho\tilde{v}^2 d^3x \tag{176}$$

where $\tilde{v}$ is the velocity amplitude of the wave. Because energy is conserved, this wave energy must go somewhere when the wave disappears, and, in fact, it goes to raising the temperature of the background plasma. Such a temperature rise after the wave damps is associated with an increase $\Delta S = \mathcal{E}_{\text{wave}}/T_0 = \int \frac{3}{2}n\Delta T/T_0$ in the entropy of the plasma relative to its entropy before the wave damps. Now, entropy is a particularly convenient quantity to follow, because in calculating the change in the mean temperature we can ignore adiabatic changes, which are oscillatory, and simply concentrate on changes in the entropy, which increases steadily, due to the dissipative terms.

Now, the entropy per ion is

$$s_i = \ln\left(\frac{T_i^{3/2}}{n}\right) \tag{177}$$

with a similar expression for the electron entropy. Then a direct calculation from equations 125, 127, and 129 shows that

$$nT\left(\frac{\partial s_i}{\partial t} + \mathbf{U}_i \cdot \nabla s_i\right) = -\nabla\cdot\mathbf{q}_i - \boldsymbol{\pi} : \nabla\mathbf{U}^i + Q_i \tag{178}$$

$$Q_i = Q_\Delta \tag{179}$$

with a similar expression for the electron entropy. The Q's are given by equations 146 and 147. From equations 125 and 178, the rate of change in the total entropy per unit volume $S = n(s_e + s_i)$ is given (for $T_e = T_i$) by

$$\frac{\partial S}{\partial t} + \nabla \cdot \left[SU_i + S_e(U_e - U_i) + \frac{q_i + q_e}{T} \right] = \theta_i + \theta_e \tag{180}$$

where "the entropy production" rates θ_i and θ_e are given by

$$T_i\theta_i = -q_i \cdot \nabla \ln T_i - \pi : \nabla U_i + Q_\Delta \tag{181}$$

$$T_e\theta_e = -q_e \cdot \nabla \ln T_e + Q_\Delta + \eta_\perp j_\perp^2 + \eta_\parallel j_\parallel^2 + 0.71 T_{e\parallel} j_\parallel / ne \tag{182}$$

The electron viscous heating is negligible.

For our example, let us consider the damping rate of a perpendicular propagating magnetosonic wave in a medium with equal ion and electron temperatures. The rise in entropy per unit volume must be

$$\Delta S = \frac{1}{2}\rho \frac{\tilde{v}^2}{T_0} \tag{183}$$

where $\tilde{v}$ is the wave amplitude. But

$$\Delta S = \int \theta dt \tag{184}$$

In evaluating θ we can use the zero-order perturbed quantities from the ideal MHD wave, ignoring the modification due to viscosity. Because $\nabla_\parallel \delta T = 0$ there is no thermal contribution to the entropy production.

The averaged value of the viscous term in equation 181, with $\mathbf{U} = \tilde{\mathbf{v}}$ the wave amplitude, is

$$\frac{1}{3} < \eta_0[(\mathbf{I} - \mathbf{bb}) : \nabla \mathbf{U}][(\mathbf{I} - \mathbf{bb}) : \nabla \mathbf{U}] >= \frac{\eta_0}{3}\left\langle (\nabla \cdot U)^2 \right\rangle = \frac{\eta_0 k_\perp^2}{6}\tilde{v}^2 \tag{185}$$

The electron viscous heating is very small and we ignore it. The perturbed ion and electron temperatures are equal. Thus,

$$\bar{\theta} = \frac{\eta_0 \, (k_\perp \tilde{v})^2}{6T} \tag{186}$$

where $\tilde{v}_0$ is the initial value of $\tilde{v}$. From equation 181 and equation 183 we have

$$\frac{\rho \tilde{v}_0^2}{2T_0} = \Delta S = \int dt \eta_0 \frac{(k_\perp \tilde{v}_0)^2}{6T_0} e^{-2i\delta\omega t} = \eta_0 \frac{(k_\perp \tilde{v}_0)^2}{12 i \delta\omega T_0} \tag{187}$$

so

$$\delta\omega = -i \frac{\eta_0 k_\perp^2}{6\rho} \sim 0.1 k_\perp^2 v_i \lambda \tag{188}$$

(The resistive damping rate is actually $k^2\eta c/4\pi$, which is generally much smaller than the viscous damping.)

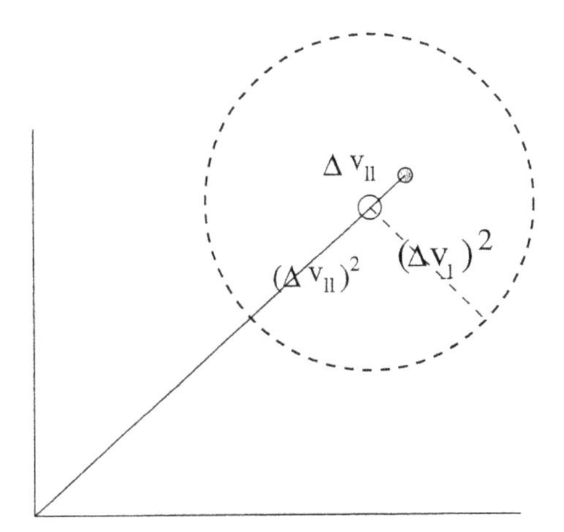

Figure 8.4. Spreading of the probability distribution of a particle in velocity space due to collisions

8.12 Maxwellian Collisions

We conclude this chapter with some expressions for the various collision rates discussed at the beginning of this chapter for the special case when particles are scattered by a Maxwell distribution of field particles. We are interested in four cases: electrons scattered by electrons and by ions, and ions scattered by ions and by electrons. The results presented here are identical with those in Spitzer (1962).

Suppose we have a Maxwell distribution of field particles with velocities w, mass m_f, charge $z_f e$, density n_f, temperature T_f, and no mean velocity:

$$f_f(w) = \frac{n_f \ell_f^3}{\pi^{3/2}} e^{-\ell_f^2 w^2} \tag{189}$$

where

$$\ell_f^2 = \frac{m_f}{2T_f} \tag{190}$$

Then let there be a test particle of mass m, charge ze, and initial velocity w in the presence of these field particles. After a time t, the probability distribution of these particles will be a Gaussian distribution centered at $w - \langle \Delta w \rangle t$, and with mean square transverse velocity of $\langle (\Delta w_\perp)^2 \rangle t$ and mean square changes in the longitudinal velocity of $\langle (\Delta w_\parallel)^2 \rangle t$, as in figure 8.4. Then, using the formulas for these diffusion coefficients given for a general $f(\mathbf{v}')$ in equations 19 and 20, taking this f to be the Maxwellian $f_f(w_f)$, equation 189, and carrying out the velocity integrations, we find

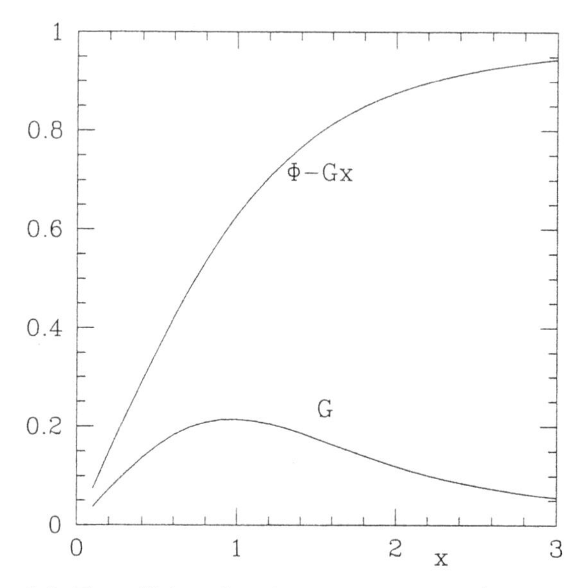

Figure 8.5. The collisions functions $\Phi(x) - G(x)$ and $G(x)$ versus x

that

$$\langle \Delta w_{\parallel} \rangle = -A_D \ell_f^2 \left(1 + \frac{m}{m_f}\right) G(\ell_f w) \tag{191}$$

$$\langle (\Delta w_{\parallel})^2 \rangle = \frac{A_D}{w} G(\ell_f w) \tag{192}$$

$$\langle (\Delta w_{\perp})^2 \rangle = \frac{A_D}{w} \left[\Phi(\ell_f w) - G(\ell_f w)\right] \tag{193}$$

where

$$A_D = \frac{8\pi e^4 n_f z_f^2 z^2}{m^2} \ln \Lambda \tag{194}$$

Φ is the error function

$$\Phi(x) = \frac{2}{\sqrt{\pi}} \int_0^x e^{-x^2} dx \tag{195}$$

and $G(x)$ is a function defined by Chandrasekhar (1943a):

$$G(x) = \frac{\Phi(x) - x\Phi'(x)}{2x^2} \tag{196}$$

$\Phi - G$ and G are given in the tables in Spitzer (1962), and are shown in figure 8.5.

For large x, $G \approx 1/(2x^2)$ and $\Phi \approx 1$, This is accurate for $x > 2$. For small x, $G \approx 0.37x$ and $\Phi - G \approx 0.75x$. Finally, for $x \approx 1$ we can use the values

$\Phi(1) - G(1) = 0.629$ and $G(1) = 0.214$. These approximate formulas give sufficiently accurate values for most purposes. The formulas are derived by Chandrasekhar in his book (1943a).

As an example, if we want electron–electron collisions rates, we take $m = m_f = m_e$, and $z = z_f = 1$. Let us define an effective collision cross section σ for $\ell_f w = 1$ by

$$\frac{(\Delta w_\perp)^2}{w^2} = n\sigma w = A_D \frac{\Phi - G}{w^3} \tag{197}$$

so

$$\begin{aligned} \sigma &= \frac{A_D}{w^4}(\Phi - G) \\ &= 0.629\ell_f^4 A_D \\ &= \frac{0.8 \times 10^{-12}}{T_{\text{eV}}^2}\ \text{cm}^2 \end{aligned} \tag{198}$$

in agreement with equation 49. (Here we take $\log\Lambda = 10$.) This is also the cross section for ion–ion collisions.

Now consider electron–ion collisions. Because of the large ion–electron mass ratio, the argument $\ell_f w$ is very large, so the 0.629 is replaced by 1. Then the ion mass does not enter into the expression for $(\Delta w_\perp)^2$. The angular diffusion rate for the electron velocity is larger by 1/0.629 than that for electron–electron collisions. Thus, for $\ell_e w = 1$ (the electron ℓ_f) and $T_e = T_i$, the electron–ion collision rate is faster than electron–electron collision rate by 1.6 and the total electron scattering rate due to electron scattering plus ion scattering is larger than electron scattering by $1 + 1.6 = 2.6$.

Next, consider ion–electron collisions The angular diffusion now involves the argument $\ell_f w \approx \sqrt{m/M} = \frac{1}{40}$ (this is the electron ℓ_f). Thus, the ion–electron angular scattering rate is smaller than ion–ion collision rate at $lw = 1$ by $1.19\sqrt{m/M}$. All these facts confirm our intuition.

On the other hand, from equation 193 we see that the electron–electron angular collision rate for slow electrons varies as $1/w^2$ and is much faster. ($< (\Delta w)^2 >$ is independent of w but the denominator is small. For electron–ion collisions this varies as $1/w^3$.) However, since an angular collision of such a slow electron does not usually change things much, it is not very important, except with respect to the role it plays is $\eta_\parallel$.

Now let us consider the drag term $\Delta w_\parallel$, equation 191. Again for electron–electron collisions, the factor multiplying $A_D l_f^2$ in equation 191 is $2 \times 0.214 \approx 0.43$ for $\ell_f w = 1$ and we see that the drag rate at this velocity

$$\frac{\Delta w_\parallel}{w} \approx 0.43 A_D l_f^3 \tag{199}$$

is comparable to the angular scattering rate $(\Delta w_\perp)^2/w^2 \approx 0.63 A_D/w^3$. For fast electrons $l_f w \gg 1$ these two rates are still equal and both are proportional to $1/w^3$.

For electrons slower than the thermal velocity ℓ_f^{-1}, the drag rate is a constant independent of w:

$$\frac{< \Delta w_\parallel >}{w_\parallel} = 2 \times 0.37 A_D \ell_f^3 \tag{200}$$

In fact, it turns out that only field electrons with velocity slower than w exert a net drag. Integration of the drag due to faster field electrons in a spherical shell of thickness Δw cancels out (in analogy to the gravitational force inside of a spherical shell of particles). The number of electrons producing drag is $\ell_f^3 w^3/\pi^{3/2}$ and the drag rate per each of these field electrons is proportional to $1/w^3$.

For the drag of electrons on fast ions the same result holds. Up to an energy $(M/m)T$ the drag rate is a constant. On the other hand, the drag rate on fast ions by ion–ion collisions decreases as $1/w^3$, so for fast enough ions, the drag on electrons is more important. In fact, the ratio of the electron to ion drag rate on the fast ion is

$$0.37 \left(\frac{E}{T_e}\right)^{3/2} \sqrt{\frac{m}{M}} \tag{201}$$

where $E = Mw^2/2$ and this is equal to one when $E = 24T_e$ for protons.

The fact that for $E \gg T$ the drag for ion–electron collisions is larger than that for ion–ion collisions may seem surprising since during an electron–ion collision the field electron velocity change is M/m times larger than the test ion velocity change. This is represented by the m/M factor in the parentheses of equation 191 for ion–electron collisions. However, the puzzle is resolved when we appreciate that the cross section for collisions is proportional to the square of the center of mass energy, $\mu g^2/2$. μ is $M/2$ for ion–ion collisions, while it is m for ion–electron (or electron–ion collisions). g is roughly the same, so, due to the much larger collisional cross section, ion–electron collisions are more frequent by $(M/m)^2$. Thus, for the drag on a fast test ion we can expect field electrons to dominate. However, not all electrons participate. The number is proportional to w^3 so only for E above $24T_e$ do electron collisions actually dominate. For $E > (M/m)T$ the electron drag rate decreases as $E^{-3/2}$, but it is larger than the ion drag rate by M/m.

One further result worth noting concerns the energy loss of a test electron due to collisions with field ions. We see from equation 191 that the rate of change of the mean velocity is

$$\frac{\langle w_\parallel \rangle}{2w} = -\frac{A_D}{2w^3}\left(1 + \frac{m}{M}\right) \approx -\frac{A_D}{2w^3} \tag{202}$$

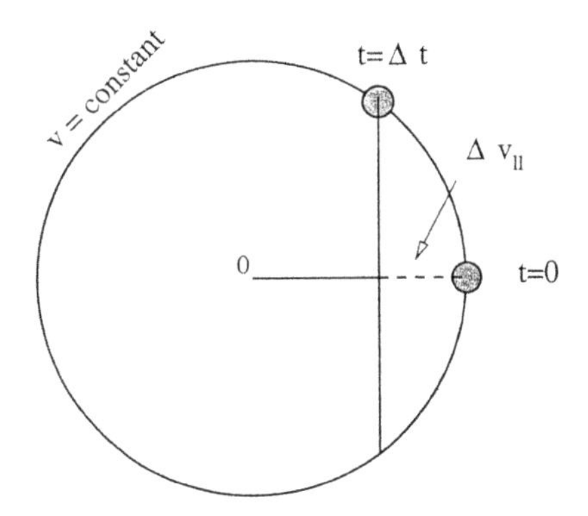

Figure 8.6. Diffusion of an electron on a constant velocity surface due to collisions with ions

which is just the electron–ion collision rate. But we do not expect the energy of electrons to be changed by collisions between electrons and ions. The resolution of this paradox can be seen from figure 8.6.

Due to electron–ion collisions the electrons spread at the electron–ion collision rate on a sphere of radius w. But the mean value of $w_\parallel$ is on the chord, so $w_\parallel$ can thus decrease rapidly without any energy change. The energy change rate is given by

$$\begin{aligned}\Delta E =& mw\langle\Delta w_\parallel\rangle + m\frac{(\Delta w_\perp)^2}{2} + m\frac{(\Delta w_\parallel)^2}{2}\\ =& -mA_D\ell_S^2\left(1+\frac{m}{M}\right)G(\ell_f w)w\\ &+ m\frac{A_D}{2w}\Phi(\ell_f w)\end{aligned} \tag{203}$$

In the natural limit, $\ell_f w \gg 1$, $G(\ell_f w) \approx 1/(2\ell^2 w^2)$, so we get

$$-mA_D\left(1+\frac{m}{M}\right)\frac{1}{2w} + \frac{m}{2w}A_D \approx -\frac{m}{M}\frac{A_D}{2w}E \tag{204}$$

so the rate of energy loss is m/M times the electron–ion collision rate, as we expect.

To conclude this discussion of collisions let us return to the logarithmic cutoff factor $\ln\Lambda$ (see Sivukhin 1966). Classically this is $\ln(D/h)$, where $h = e^2/E$ is the minimum impact parameter and D is the Debye length. We might at first expect the quantum mechanical cutoff parameter to be the same, since it is known that for a Coulomb force (both repulsive and attractive) the differential collision cross sections are the same. However, the actually interactive force is cut off at the Debye length and the potential is, in fact,

$$\frac{e^2}{r}e^{-r/\lambda_D} \tag{205}$$

For this potential the classical and quantum mechanical cross sections differ slightly. The quantum mechanical differential cross section for scattering through an angle θ from the Born approximation is

$$\sigma(\theta) = \left(\frac{e^2}{\mu g^2}\right)^2 \frac{1}{(\theta^2 + 4\theta_{\rm db}^2)^2} \tag{206}$$

where

$$\theta_{\rm db} = \frac{\lambda}{4\pi D} = \frac{\hbar}{2\mu g D} \tag{207}$$

$\lambda = h/\mu g$ is the de Broglie wavelength, $\theta_{\rm db}$ is the quantum mechanical diffraction angle from a slit of width D, and $\sigma d\Omega$ is the probability of a particle being scattered into a solid angle $d\Omega$ an angular distance θ from the unperturbed $\mathbf{g}$.

For $\theta \gg \theta_{\rm db}$ this agrees with the Rutherford cross section. Classically, θ is related to the impact parameter by equation 10, $\theta \approx h/b$, where h is the distance of closest approach in equation 2. Thus, if $\theta_{\rm db} \ll h/D$, then the cross section cuts off at the Debye length before the difference between the quantum mechanical and classical value is reached. On the other hand, if $\theta_{\rm db} < e^2/\mu g D$, then the cross section given by equation 206 leads to a different cutoff than the classical value.

Thus, we can distinguish between two cases leading to two values for the logarithmic factor. Case 1 is the classical limit:

$$\theta_{\rm db} < \frac{e^2}{\mu g D} \quad \text{or} \quad g < \frac{e^2}{\hbar} = \alpha c \tag{208}$$

where $\alpha = e^2/\hbar c$ is the fine structure constant. In this case

$$(\log \Lambda)_{\rm classical} = \ln\left(\frac{D\mu g^2}{e^2}\right) \tag{209}$$

Case 2 is the quantum mechanical limit:

$$\theta_{db} > \frac{e^2}{\mu g D} \quad \text{or} \quad g > \frac{e^2}{\hbar} = \alpha c \tag{210}$$

In this case

$$(\log \Lambda)_{\rm QM} = \ln\left(\frac{4\pi D}{\lambda}\right) = \ln\left(\frac{2D\mu g}{\hbar}\right) \tag{211}$$

The above discussion is not quite rigorous. The Born approximation is valid only if $g > \alpha c$, which is the limit of case 2 and for this limit equation 211 gives the correct logarithmic cutoff. However, fortuitously, in the opposite limit $g < \alpha c$, it turns out that the classical limit of quantum theory is valid and $\sigma = (e^2/\mu g^2)^2/\theta^2$ (Landau and Lifshitz 1977) and this leads to equation 209 as the correct logarithmic cutoff.

By matching the two formulas, equation 207 and equation 210, and taking $\mu g^2 = 3T/2$, we find that the transition between the two cases occurs at $g/c = e^2/\hbar c$. For electron–electron and electron–ion collisions this corresponds to a temperature of 9 eV . At lower temperatures the classical limit is correct and at higher temperatures the quantum mechanical limit must be used. For ion–ion collisions the transition temperature is 16 keV . In general, the distinction is not very important because $\log \Lambda$ can always be reasonably approximated by a value between 10 and 20.

8.13 Problems

1. (Numerical values for transport coefficients) To have convenient numerical values for thermal conductivity, viscosity, and resistivity, introduce the approximate values

$$v_{e0} = 5 \times 10^7 T_e^{1/2} \text{cm/sec}$$

$$v_{i0} = 10^6 T_i^{1/2} \text{cm/sec}$$

$$\ell_0 = \frac{10^{12}}{n} T^2 \text{ cm}$$

$$\tau_{e0} = \frac{\ell_0}{v_{e0}} \text{ sec}$$

$$\tau_{i0} = \frac{\ell_0}{v_{i0}} \text{ sec}$$

where the temperatures T_e and T_i are in electron volts. (Also $\sqrt{T/m} = 0.84 v_e$, $\sqrt{T/M} = 0.98 v_i$.)

Show that the Braginski collision times are

$$\tau_e = 1.75 \tau_{e0}/\lambda_{10}$$

$$\tau_i = 2.12 \tau_{i0}/\lambda_{10}$$

where $\lambda_{10} = \log \Lambda/10$.

Show that the parallel Braginski thermal conductivities are

$$\kappa_{e\parallel} = 3.93 n v_{e0} \ell_0/\lambda_{10}$$

$$\kappa_{i\parallel} = 7.9 n v_{i0} \ell_0/\lambda_{10}$$

Show that the ion viscosity is

$$\eta_0 = 1.95 \rho \ell_0 v_{i0}/\lambda_{10}$$

Show that the parallel resistivity in units of cm^2/sec used in the text is

$$\frac{\eta_{\|}c}{4\pi} = \frac{0.80 \times 10^7}{T_e^{3/2}} \lambda_{10} \text{ cm}^2/\text{ sec}$$

2. (Orbit shift) Show that the inward or outward displacement of a charged particle in a Coulomb orbit, $|b - r_{\text{min}}|$, tends to $e^2/\mu v^2 = k/v^2$ as the impact parameter b goes to infinity, as stated in the text, $k = e^2/\mu$. You can take the orbit as

$$\frac{1}{r} = \frac{\pm k}{b^2 v^2}\left(1 + \frac{\cos\theta}{\cos\theta_0}\right)$$

where θ_0 and $-\theta_0$ are the asymptotic directions of the hyperbolic orbit, b is the impact parameter, and the two signs correspond to attraction and repulsion, respectively, $\tan\theta_0 = -bv^2/|k|$.

Show that

$$|b - r_{\text{min}}| = \left|\frac{k\tan\theta_0}{v^2}\left(\frac{\sin\theta_0}{1 = \cos\theta_0} - 1\right)\right|$$

and take the limit as $\theta_0 \to \pi/2$.

3. (Drag on an iron nucleus cosmic ray) Consider a fully stripped iron nucleus cosmic ray passing through the interstellar medium whose density we take as 1 cm^{-3}. At what initial energy does the cosmic ray lose a good fraction of its energy in 10^6 years. Note that the cosmic ray moves faster than the free electrons at 1 eV or the bound electrons at 13 eV if it has an energy greater than 25 MeV per nucleon. *Hint:* The answer should be in the neighborhood of 100 MeV per nucleon so you may treat it as nonrelativistic and all the interstellar medium electrons as cold.

Is the mean loss due to collisions with electrons or ions? Compare the rate of loss of velocity, $\langle\Delta w\rangle/w$, to the rate of spread of velocity, $\langle(\Delta w)^2\rangle/w^2$. How does the loss term vary with energy? Due to drag what is the energy as a function of time? What is the loss time for a proton with the same energy per nucleon?

4. (Thermal relaxation of ions) Consider an unmagnetized plasma with a small sinusoidal electron and ion temperature variation in the z direction,

$$T_e = T_0 + T_{e1}(t)\sin kz T_i = T_0 + T_{i1}(t)\sin kz$$

where initially $T_{e1}(Z) = T_{i1}(Z)$. Write down the equations for the evolution of the two temperatures, including thermal conduction of each and the electron–ion energy exchange term. Discuss how the two temperatures vary in time. Assume that the electron temperature relaxes by thermal conduction faster than by energy exchange so that T_{e1} quickly goes to zero. Consider long wavelengths and short wavelengths and show that when the wavelength is long the ions relax by energy exchange and when the wavelength is short the relaxation is by ion thermal conduction. What is the critical wavelength expressed in terms of the ion or electron mean free path.

5. (Electron thermal conduction in the solar transition region) The temperature in the top of the chromosphere changes rapidly from 10^4 to 10^6K, and the region where this happens is termed the transition region. Take the density in this region to be 5×10^9 cm^{-3}. Because of the temperature gradient there is large electron thermal conduction of energy downward comparable with the energy emission of the corona, $\approx 2 \times 10^{-6}$ solar luminosities.

Calculate the temperature gradient of the electrons, at different temperatures, needed to carry this energy. Show that your result is independent of the density. If the thermal conduction continues to be equal to this luminosity what is the electron temperature variation with height? What is the scale of variation when a temperature of 10^5 K is reached? The actual scale height is only a few kilometers. Can you suggest why this might be?

6. (Magnetic pumping) Consider a uniform magnetized plasma where **B** is in the z direction. Let **B** undergo a small variation in magnitude

$$B = B_0 + B_1 \sin(\omega t)$$

due to a compressional motion in the x direction,

$$V = V_1 \frac{x}{L} \cos(\omega t)$$

From the viscous term in the ion pressure equation calculate the average rate of increase of the ion pressure and temperature expressed in terms of the ion–ion collision rate ν, ω, and B_1/B_0. Neglect any electron–ion interchange of energy. For fixed B_1 and collision rate ν, which value of ω maximizes this rate? Can you apply these results if $\nu < \omega$?

7. (Mirror trapping) Consider particles trapped in a magnetic mirror with a large mirror ratio $M = B_{\max}/B_0$, where B_0 is the field strength in the trapped region and $B_{\max}$ is the field strength in the mirror. Particles with $\sin^2\theta < 1/M$ are assumed to leave the trapped region instantaneously where θ is the polar or pitch angle in velocity space. If a particle is injected at $\mu = 0$, i.e., a 90-degree pitch angle, calculate its average length of time before it is lost. To do this take a source S of particles at $\mu = \cos\theta = 0$ and look for the steady-state solution of the velocity diffusion equation

$$\frac{\partial f}{\partial t} = \nu \frac{\partial}{\partial \mu}\left[(1-\mu^2)\frac{\partial f}{\partial \mu}\right] + S\delta(\mu)$$

Solve this equation for the time-independent f with the boundary condition $f = 0$ at $\mu = 1 - \mu_c$ and $\mu = -1 + \mu_c$, where $1 - \mu_c^2 = 1/M$. Integrate f over μ and take the mean time as this integral divided by S. For large M the answer should be a constant of order unity times $1/\nu \times \log M$, where the collision time is $1/\nu$. This is a little strange since we would think it would be proportional to the solid angle area of the loss cone, $\sim 1/M$. This is a direct result of the collisions being made up of many small-angle collisions rather than large-angle collisions.

References

Braginski, S. I. 1965. In *Reviews of Plasma Physics*, Vol I, p. 205, ed. M. A. Leontovitch, trans. H. Lashinsky, Consultants Bureau, New York.

Chandrasekhar, S. 1943a. *Principles of Stellar Dynamics*, p. 63, Dover, New York.

Chandrasekhar, S. 1943b. *Reviews of Modern Physics* **15**, 1. Reprinted in *Noise and Stochastic Processes*, ed. N. Wax, Dover, New York.

Chapman, S., and T. G. Cowling. 1960. *The Mathematical Theory of Nonuniform Gases*, Cambridge University Press, Cambridge, UK.

Cohen, R. S., L. Spitzer, and R. Routley. 1950. *Physical Review* **80**, 230.

Landau, L. D. 1937. *ZhETF (Journal Experimental and Theoretical Physics USSR)* **7**, 203.

Landau, L. D., and E. M. Lifshitz. 1977. *Quantum Mechanics*, 3rd ed., p. 317, trans. J. B. Sykes and J. S. Bell, Pergamon Press, Oxford, UK.

Robinson, B., and I. B. Bernstein. 1962. *Annals of Physics* **18**, 110.

Rosenbluth, M. N., W. Macdonald, and D. Judd. 1957. *Physical Review* **107**, 1.

Sivukhin, D. V. 1966. In *Review of Plasma Physics*, Vol. IV, p. 109, ed. M. A. Leontovitch, trans. H. Lashinsky, Consultants Bureau, New York.

Spitzer, Jr., L. 1962. *Physics of Fully Ionized Gases*, Interscience, New York.

Trubnikov, B. A. 1965. In *Reviews of Plasma Physics*, Vol. I, p. 105, ed. M. A. Leontovitch, trans. H. Lashinsky, Consultants Bureau, New York.

Chapter 9

COLLISIONLESS PLASMAS

9.1 Introduction

So far in this book we have confined ourselves to relatively slow, large-scale phenomenon, which are mostly dominated by collisions. When we look on smaller scales, or higher frequencies, collisions become less important.

Let us first consider the ideal situation of a uniform plasma in a uniform magnetic field that is collisionless. This system is governed by the Vlasov and Maxwell equations. It is important to clarify all the possible degrees of motion, or plasma waves, that such a system is capable of. If the waves are small amplitude, we can write down an exact solution of the linearized Vlasov equation in terms of the perturbed electric and magnetic fields. This solution involves an infinite sum over Bessel functions and at first sight is not very transparent. However, if the phase velocity of the waves is large compared to the thermal motions, and the frequency of the wave is not close to a harmonic of the cyclotron frequency, we can capture the bulk of the waves by taking the electrons and ions as cold and using fluid equations with zero pressure for both of them. Let us do this.

9.2 Dispersion Relation for Cold Plasma Waves

First consider the ions. Let us take the unperturbed magnetic field to be in the z direction. Since the plasma and the field are uniform we can Fourier analyze the perturbed quantities. For the $\mathbf{E}$ field we take

$$\mathbf{E} = \hat{\mathbf{E}} e^{i\mathbf{k}\cdot\mathbf{x} - i\omega t} \tag{1}$$

Also, we choose our coordinate system as in figure 9.1, where $\mathbf{k}$ is the x–z plane and makes an angle θ with $\mathbf{B}$. Let us first take $\mathbf{E}$ in the x direction. The x component of the equation of motion for a cold ion is

$$-i\omega M v_x = eE_x + \frac{ev_y B}{c}$$

or

$$-i\omega v_x - \Omega_i v_y = \frac{eE_x}{M} \tag{2}$$

and the y component is

$$\Omega_i v_x - i\omega v_y = 0 \tag{3}$$

where $\Omega_i = eB/Mc$.

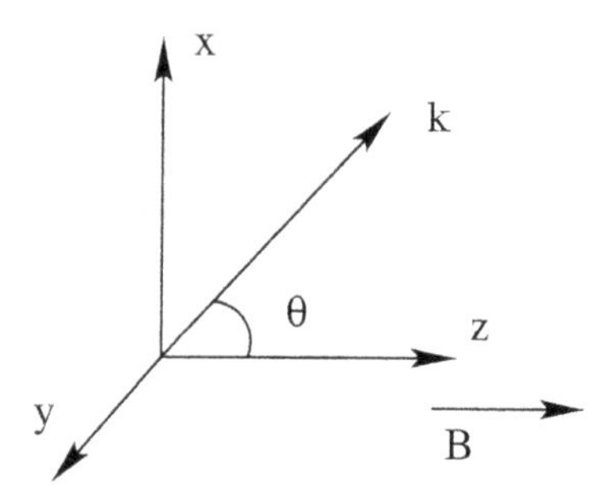

Figure 9.1. The coordinate system for k

The solution to these two equations in the two unknowns v_x and v_y is

$$v_x^i = \frac{-i\omega}{\Omega_i^2 - \omega^2}\frac{e}{M}E_x \tag{4}$$

$$v_y^i = -\frac{\Omega_i}{\Omega_i^2 - \omega^2}\frac{e}{M}E_x \tag{5}$$

Similar solutions apply when we have E_y alone, with v_x and v_y interchanged and the sign of v_x changed, as we can see by rotating the coordinate system. For **E** in the z direction we have

$$v_z = \frac{e}{-i\omega M}E_z \tag{6}$$

Note that if $\Omega_i = 0$, the v_x equation reduces to the v_z equations.

For electrons, the solution is identical, with Ω_i replaced by $-\Omega_e = -eB/mc$ and e/M replaced by $-e/m$. The electrical current is the sum of the electron current and the ion current:

$$\mathbf{j} = ne\mathbf{v}^i - ne\mathbf{v}^e \tag{7}$$

This is to be substituted into Maxwell's equation for $\mathbf{E}'$ and $\mathbf{B}'$, where the primes denote perturbed quantities,

$$\nabla \times \mathbf{B}' = i\mathbf{k} \times \mathbf{B}' = 4\pi\frac{\mathbf{j}'}{c} - i\frac{\omega}{c}\mathbf{E}' = -i\frac{\omega}{c}\boldsymbol{\epsilon}\cdot\mathbf{E}' \tag{8}$$

and

$$i\mathbf{k} \times \mathbf{E}' = i\frac{\omega}{c}\mathbf{B}' \tag{9}$$

The tensor $\boldsymbol{\epsilon}$ is defined implicitly by equation 10:

$$4\pi\mathbf{j}' - i\omega\mathbf{E}' = -i\omega\boldsymbol{\epsilon}\cdot\mathbf{E}' \tag{10}$$

Making use of our equations for $\mathbf{v}^i$ and $\mathbf{v}^e$, equations 4–6, we find from 10 that

$$\epsilon = \begin{pmatrix} S & -iD & 0 \\ iD & S & 0 \\ 0 & 0 & P \end{pmatrix} \tag{11}$$

where

$$S = 1 - \frac{\omega_{pi}^2}{\omega^2 - \Omega_i^2} - \frac{\omega_{pe}^2}{\omega^2 - \Omega_e^2} \tag{12}$$

$$D = \frac{\Omega_i \omega_{pi}^2}{\omega(\omega^2 - \Omega_i^2)} - \frac{\Omega_e \omega_{pe}^2}{\omega(\omega^2 - \Omega_e^2)} \tag{13}$$

$$P = 1 - \frac{\omega_{pi}^2}{\omega^2} - \frac{\omega_{pe}^2}{\omega^2} \tag{14}$$

Combining equations 9 and 10, we have

$$\mathbf{k} \times (\mathbf{k} \times \mathbf{E}) + \frac{\omega^2}{c^2}\epsilon \cdot \mathbf{E} = 0 \tag{15}$$

Together with equation 11, equation 15 represents three equations in the three unknowns E_x, E_y, and E_z. In matrix form

$$\begin{pmatrix} (S - n^2\cos^2\theta) & -iD & n^2\sin\theta\cos\theta \\ iD & S - n^2 & 0 \\ n^2\sin\theta\cos\theta & 0 & P - n^2\sin^2\theta \end{pmatrix} \begin{pmatrix} E_x \\ E_y \\ E_z \end{pmatrix} = 0 \tag{16}$$

where $\mathbf{n} = \mathbf{k}c/\omega$ is the vector index of refraction of magnitude $n = c/v_\phi$, where v_ϕ is the phase velocity. The condition that the three homogeneous equations (16) have a solution is that the determinant of their coefficients vanish. This condition reduces to the equation (Stix 1962)

$$(S\sin^2\theta + P\cos^2\theta)n^4 - [RL\sin^2\theta + PS(1 + \cos^2\theta)]n^2 + PRL = 0 \tag{17}$$

where

$$R = S + D = 1 - \frac{\omega_{pi}^2}{\omega(\omega + \Omega_i)} - \frac{\omega_{pe}^2}{\omega(\omega - \Omega_e)} \tag{18}$$

$$L = S - D = 1 - \frac{\omega_{pi}^2}{\omega(\omega - \Omega_i)} - \frac{\omega_{pe}^2}{\omega(\omega + \Omega_e)} \tag{19}$$

This equation is quadratic in n^2. For any ω all the coefficients can be found and there are two values for n^2 satisfying it. It can be shown that these values are always real, but sometimes negative. (The roots depend on the parameters ω_{pi}, ω_{pe}, Ω_i and Ω_e as well as ω). When n^2 is negative, k must be imaginary and there is no propagation. When n^2 is positive, the two values for n (and for k) are equal and of opposite signs, which corresponds to propagation in two opposite directions.

9.3 Parallel Propagation

To get an idea of what kind of waves there are let us first consider propagation parallel to **B**, $\theta = 0$. In this case, equation 17 factors and the roots are

$$n^2 = R \tag{20}$$

$$n^2 = L \tag{21}$$

The equation is also satisfied when $\theta = 0$ if $P = 0$, which yields the plasma oscillation.

The solutions R and L correspond to right and left polarized waves. In fact, from the second equation in (16),

$$iDE_x + (S - n^2)E_y = 0 \tag{22}$$

If we set $n^2 = R$ and note that $S = (R + L)/2$ and $D = (R - L)/2$, we have

$$i(R - L)E_x + (L - R)E_y = 0 \tag{23}$$

Thus, $E_x = -iE_y$. Let E_y be real, and take real parts

$$E_x = \hat{E}_y \cos\left(kz - \omega t - \frac{\pi}{2}\right) = \hat{E}_y \sin(kz - \omega t) \tag{24}$$

and

$$E_y = \hat{E}_y \cos(kz - \omega t) \tag{25}$$

At fixed z, **E** rotates in the right-hand sense about $\hat{\mathbf{z}}$. We can check this physically. Notice that the denominator of the electron term in R vanishes when $\omega = \Omega_e$. If $\omega = \Omega_e$, the electron rotates at the same rate and with the same sense as **E**, so that the interaction is resonant and leads to an infinite response. If $\omega = \Omega_i$, the denominator of the ion term is finite and there is no resonance since the ions rotate in the left-hand sense. For the L wave the ions resonate at $\omega = \Omega_i$ but the electrons do not resonate at $\omega = \Omega_e$. This agrees with the L wave being a left circularly polarized wave.

To plot R versus ω, first consider low frequencies, $\omega \ll \Omega_e$. We can neglect ω with respect to Ω_e in the $\omega + \Omega_e$ factors in the electron denominator and further $\omega_{pe}^2/\Omega_e = \omega_{pi}^2/\Omega_i$. Thus, if $\omega \ll \Omega_e$

$$R = 1 + \frac{\omega_{pi}^2}{\Omega_i(\omega + \Omega_i)} \tag{26}$$

For ω large compared to Ω_i the ion term is negligible and we have for high frequencies

$$R = 1 - \frac{\omega_{pe}^2}{\omega(\omega - \Omega_e)} \tag{27}$$

At very low frequencies below the ion cyclotron frequency, equation 18 reduces to $R = 1 + \omega_{pi}^2/\Omega_i^2 = 1 + c^2/v_A^2$ and

$$n^2 = \frac{k^2c^2}{\omega^2} = R = 1 + \frac{c^2}{v_A^2}$$

That is,

$$\omega^2 = \frac{k^2 v_A^2}{1 + v_A^2/c^2} \approx v_A^2 \tag{28}$$

if $v_A \ll c$. Thus, at low frequency the right circularly polarized wave is a right circularly polarized Alfven wave. If $c \gg v_A$, which is possible at very low densities, that is, if $nMc^2 \ll B^2/8\pi$, we have, essentially, a light wave:

$$\omega^2 = k^2c^2 \tag{29}$$

although it would not propagate at these low frequencies in the absence of the magnetic field.

For $v_A \ll c$ we have dropped the 1 in the denominator of equation 28 and obtained the regular Alfven wave. For all frequencies small compared to Ω_e we get the equation

$$\omega^2 = k^2 v_A^2 \left(1 + \frac{\omega}{\Omega_i}\right) \tag{30}$$

so above the ion cyclotron frequency the wave is faster than the Alfven speed by the factor $1 + \omega/\Omega_i$. Finally, if $\Omega_i \ll \omega \ll \Omega_e$, we get

$$\omega = \frac{k^2 v_A^2}{\Omega_i} \tag{31}$$

which is the dispersion relation for the whistler wave (Barkhausen 1930; Eckersley 1935).

As $\omega \to \Omega_e$, equation 27 shows that $R \to \infty$ just below Ω_e and the phase velocity of the whistler goes to zero. Just above Ω_e, R is negative and there is no propagation. For $\omega \gg \Omega_e$ the one in the denominator dominates, and we have

$$\omega^2 = k^2c^2 \tag{32}$$

which is the light wave. Thus, there is a frequency ω_R at which R becomes positive. This is called the right cutoff frequency

$$\omega_R = \frac{\Omega_e + \sqrt{\Omega_e^2 + 4\omega_{pe}^2}}{2} \tag{33}$$

It is the frequency at which reflection may occur. The phase velocity of the R wave as a function of ω is plotted in figure 9.2. For a fixed frequency it determines the values of the parameters ω_{pe}^2 and Ω_e^2 (or the density and B) at which reflection occurs for normal incidence (when ∇n and ∇B are parallel to $\mathbf{B}$).

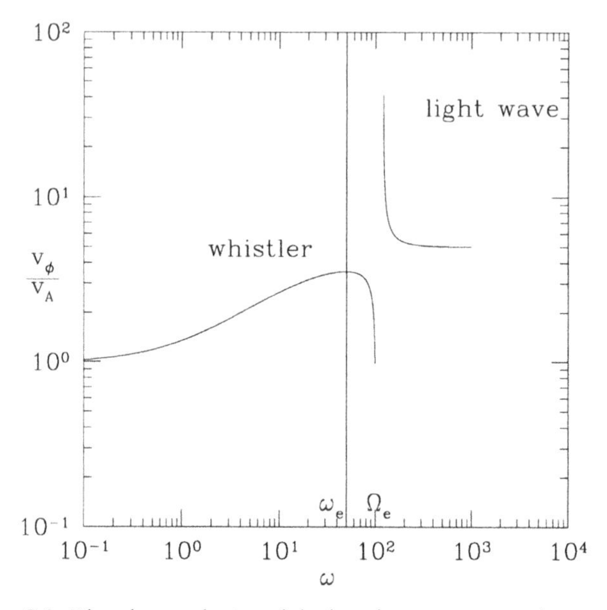

Figure 9.2. The phase velocity of the low frequency R mode versus ω

Consider a parallel propagating wave with frequency ω. Suppose that at some point ω_{pe} and Ω_e are such that $\omega > \omega_R$. Let Ω_e be fixed and let the density increase in the direction of propagation $\hat{z}$. Then from equation 33 we see that ω_R increases, and when a density is reached such that $\omega_R = \omega$ the wave can no longer propagate in the $+\hat{z}$ direction and it must reflect. If the magnetic field also increases with z, part of the wave may tunnel through to the region $\omega < \Omega_e$ and this part of the wave can continue to propagate in the z direction, the remaining part of the wave being reflected into the $-z$ direction. Thus, there is a gap in frequency $\Omega_e < \omega < \omega_R$ in which the R mode does not propagate.

The L wave has similar behavior. For $\omega \ll \Omega_e$, its phase velocity is given by

$$\omega^2 = k^2 v_A^2 \left(1 - \frac{\omega}{\Omega_i}\right) \tag{34}$$

For $\omega = \Omega_i$, L goes to an infinity and the phase velocity goes to zero. The L wave does not propagate above Ω_i ($n^2 < 0$) until the electron cyclotron frequency is approached. For $\omega \gg \Omega_i$,

$$L = 1 - \frac{\omega_{pe}^2}{\omega(\omega + \Omega_e)} = \frac{(\omega - \omega_L)(\omega + \omega_R)}{\omega(\omega + \Omega_e)} \tag{35}$$

where

$$\omega_L = \frac{-\Omega_e + \sqrt{\Omega_e^2 + 4\omega_{pe}^2}}{2} \tag{36}$$

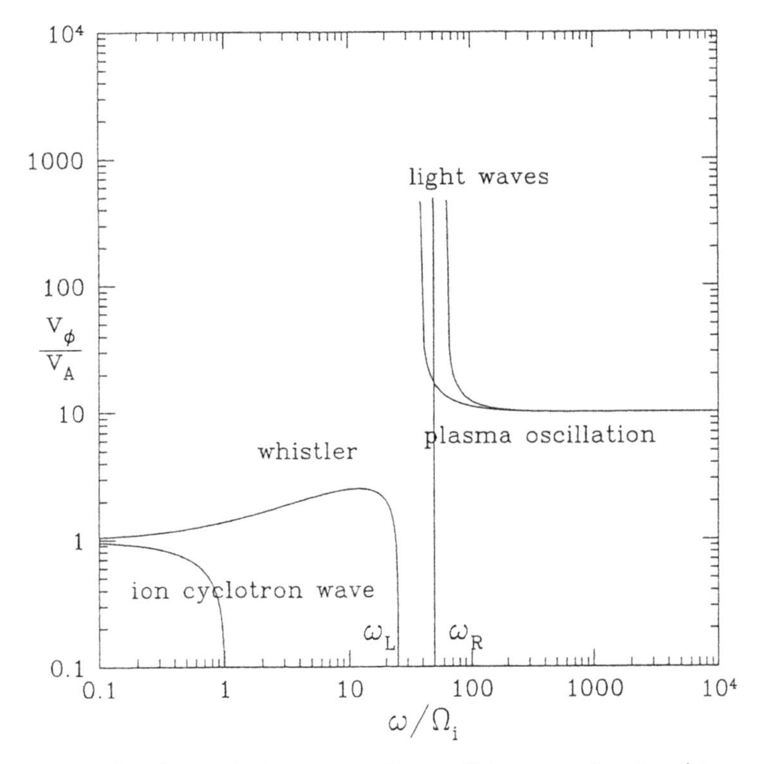

Figure 9.3. The phase velocity versus ω for parallel propagating R and L modes

This is the left cutoff frequency. The L wave does not propagate below ω_L. When $\omega \gg \Omega_e$,

$$\frac{\omega^2}{k^2} = \frac{c^2}{L} = c^2 \tag{37}$$

and we obtain a left circular polarized light wave.

The last parallel propagating wave is the P wave

$$P = 1 - \frac{\omega_{pe}^2}{\omega^2} = 0$$

or

$$\omega^2 = \omega_{pe}^2 \tag{38}$$

The P wave is the familiar plasma oscillation discovered by Langmuir. Its dispersion relation is independent of k and its phase velocity is

$$\frac{\omega}{k} = \frac{\omega_{pe}}{k} \tag{39}$$

while its group velocity is zero. **E** is in the z direction.

Superimposing, the L waves on the phase velocity diagram for the R waves gives figure 9.3. This plot corresponds to the limit $\omega_{pe} \gg \Omega_e$, which is the

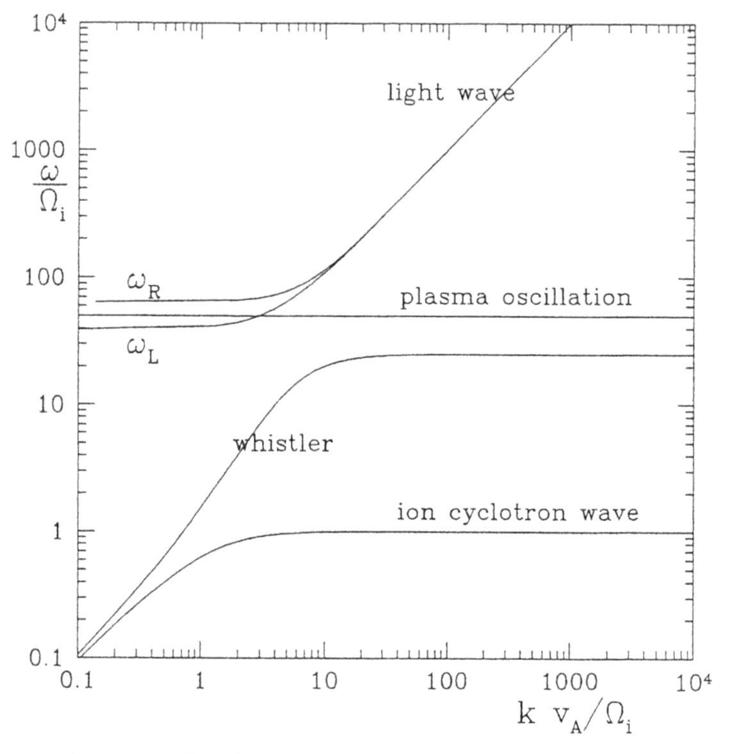

Figure 9.4. Plot of ω versus k for parallel propagating waves

case for the interstellar medium, as can be seen from

$$\frac{\omega_{pe}}{\Omega_e} = \sqrt{\frac{m}{M}} \frac{c}{v_A} \tag{40}$$

Typically, $v_A = 10$ km/sec, so $\omega_{pe}/\Omega_e \approx 1/40 \times 3 \times 10^4 = 10^3$. In this limit, ω_R and ω_L differ from ω_{pe} by $\Omega_e/2$

$$\omega_L = \omega_{pe} - \tfrac{1}{2}\Omega_e, \tag{41}$$

$$\omega_R = \omega_{pe} + \tfrac{1}{2}\Omega_e \tag{42}$$

It is interesting to turn these curves around and to plot ω versus k, which we do in figure 9.4 (see Krall and Trivelpiece 1973). We see that in this plot there are five different continuous wave branches with no gaps in k. There are still gaps in frequency. The L wave has a wide gap between Ω_i and $\omega_L \approx \omega_{pe}$, and the R wave has a narrow gap between ω_e and $\omega_R \approx \omega_{pe} + \Omega_e/2$.

However, no branch has a gap in k. To show this more analytically, expand out the quadratic equation in n^2, where equation 17 is now considered as a function ω. By multiplying it by $\omega^6(\omega^2 - \Omega_i^2)(\omega^2 - \Omega_e^2)$ we get a tenth-degree equation in ω or a fifth degree equation in ω^2. The coefficient of ω^{10} is 1 and is thus nonzero for all θ. Further, for fixed k and θ the roots for ω are all real.

We can see this clearly if we suppose that one of the roots is imaginary. Since the tenth-degree equation has real coefficients, the complex roots must appear in pairs and one of the complex roots must have a positive imaginary part corresponding to a growing solution. But the energy of any perturbation is quadratic and positive, $\mathcal{E} = \int(\rho v^2/2 + B^2/8\pi + E^2/8\pi)d^3x$ and is a constant, so it cannot grow, as would have to be the case for a growing mode, and this is a contradiction. Hence, all the roots must be real.

The constant coefficient is $-k^4c^4\omega_{pe}^2\Omega_i^2\Omega_e^2\cos^2\theta$, and it vanishes only for $\theta = 90^o$. Thus, for all other θ's there are five roots for ω^2.

9.4 The Number of Waves

Each one of the five roots for ω^2 corresponds to two equal roots for ω of opposite signs and, thus, two identical waves propagating in two different directions. However, they have opposite senses of circular polarization. Can we give an argument for why there should be five pairs of waves?

Earlier we showed that there should be $2 \times 3 = 6$ ideal MHD waves (and one $\omega = 0$ trivial entropy mode). This number of waves is necessary to provide the evolution from an arbitrary initial choice of ρ_1, $\mathbf{v}_1$, p_1, and $\mathbf{B}_1$ as independent quantities ($\mathbf{B}_1$ is constrained by $\nabla \cdot \mathbf{B} = 0$). To find the evolution we Fourier analyze in $\mathbf{r}$ and for each Fourier $\mathbf{k}$ harmonic we need the seven independent waves.

In a similar way, for the cold plasma waves, the independent initial quantities that can be chosen are n_{1i}, n_{1e}, $\mathbf{v}_{1i}$, $\mathbf{v}_{1e}$, $\mathbf{E}_1$, $\mathbf{B}_1$, 14 quantities. However, these are restricted by the time independent equations $\nabla \cdot \mathbf{B}_1 = 0$, and Poisson's equations $\nabla \cdot \mathbf{E}_1 = 4\pi(n_{1i} - n_{1e})e$, so there are actually 12 independent quantities. Again Fourier analyzing, we should have $2 \times 6 = 12$ different waves, that is, six types, each having two directions of propagation.

The five waves we have found so far, all assume that $\mathbf{E}_1 \neq 0$, see equation 16. There are, in addition, two degenerate, $\omega = 0$, modes that have $\mathbf{E}_1 = 0$. One mode has $n_{1i} = n_{1e}$, and everything else zero. The second mode has the parallel parts of $\mathbf{v}_{1i} = \mathbf{v}_{1e}$ nonzero, and everything else equals zero. These two modes clearly satisfy all the equations. Neither one has any perturbed current or charge, so $\mathbf{E}_1 = \mathbf{B}_1 = 0$ for both modes. The equation of motion is satisfied for both since $\omega = 0$ and since $\mathbf{v}$ is parallel to $\mathbf{B}_0$, so there is no force. None of these waves is damped, since there is no source of dissipation in the cold plasma equation, so ω must be real. This explains figure 9.4, which applies for $\theta = 0$. For $\theta \neq 0$ the waves are changed but there are still exactly five pairs of nondegenerate waves and two trivial waves.

9.5 Perpendicular Propagation

Let us now survey the perpendicularly propagating waves $\theta = \pi/2$. Equation 17 becomes

$$Sn^4 - (RL + PS)n^2 + PRL = 0 \tag{43}$$

and the two solutions are

$$n^2 = P$$
$$n^2 = \frac{RL}{S} \tag{44}$$

The first root for n^2, the P wave, is called the ordinary wave since, from equation 10, $E_x = E_y = 0$ and E is parallel to B. Its dispersion relation can be written as

$$\frac{k^2c^2}{\omega^2} = 1 - \frac{\omega_{pe}^2}{\omega^2} - \frac{\omega_{pi}^2}{\omega^2} \approx 1 - \frac{\omega_{pe}^2}{\omega^2}$$

or

$$\omega^2 = k^2c^2 + \omega_{pe}^2 \tag{45}$$

and it is independent of B. The dispersion relation is the same as the dispersion relation when the magnetic field is absent and for this reason the wave is called the ordinary wave.

However, it is necessary that $\omega > \omega_{pe}$ for the wave to propagate. If $\omega \ll \omega_{pe}$, then $k = \pm ic/\omega_{pe}$ and the wave damps in space in a distance c/ω_{pe} that is called the electron skin depth, since a wave can penetrate only this distance into a plasma.

For $\omega \ll \Omega_e$ the dispersion relation for the second perpendicularly propagating wave is

$$\frac{k^2c^2}{\omega^2} = RL/S = \left[-\frac{\omega_{pi}^2}{\Omega_i(\omega + \Omega_i)}\frac{\omega_{pi}^2}{\Omega_i(\omega - \Omega_i)}\right]\left(1 - \frac{\omega_{pi}^2}{\omega^2 - \Omega_i^2} + \frac{\omega_{pe}^2}{\Omega_e^2}\right)^{-1} \tag{46}$$

(We drop the 1's in R, L, and S since they are small.) The frequency at which the denominator vanishes and at which the phase velocity goes to zero is called the lower hybrid frequency, ω_{LH} (see problem 4). If $\Omega_e \ll \omega_{pe}$, then $\omega_{\text{LH}} \approx \sqrt{\Omega_e\Omega_i}$, which is the justification for the term hybrid. In this limit equation 46 is approximately equivalent to

$$\frac{k^2c^2}{\omega^2} = \frac{\omega_{pi}^2}{\Omega_i^2}\frac{1}{1 - (m/M)(\omega^2 - \Omega_i^2)/\Omega_i^2}$$

or since $\omega_{pi}^2/\Omega_i^2 = c^2/v_A^2$

$$\omega^2 = k^2v_A^2\left(1 - \frac{m}{M}\frac{\omega^2 - \Omega_i^2}{\Omega_i^2}\right) \tag{47}$$

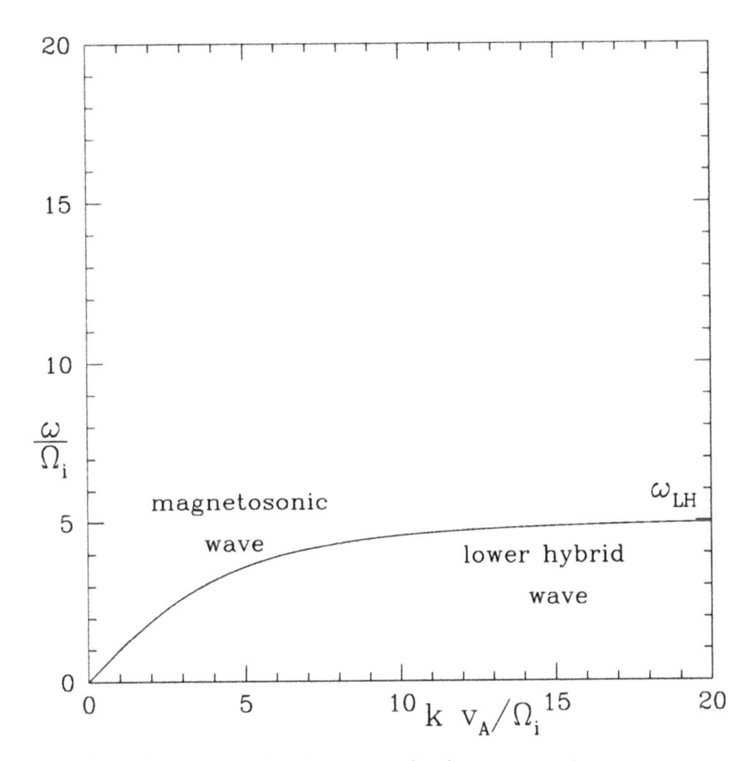

Figure 9.5. Plot of ω versus k for the perpendicular propagating magnetoacoustic-lower hybrid mode

For $\omega \ll \omega_{\mathrm{LH}}$ the wave propagates at the Alfven speed. As $\omega \to \omega_{\mathrm{LH}}$, $k \to \infty$ and the phase velocity approaches zero. In figure 9.5, we give a plot for $\theta = \pi/2$ of ω versus k for the low-frequency modes.

If $\omega \gg \omega_{\mathrm{LH}}$ the ion contribution to R, L, and S becomes small compared to the electron contribution. Keeping only the electron contribution we get

$$\frac{k^2c^2}{\omega^2} = \frac{RL}{S} = \frac{(\omega^2 - \omega_R^2)(\omega^2 - \omega_L^2)}{\omega^2(\omega^2 - \omega_{\mathrm{UH}}^2)} \tag{48}$$

where ω_{UH} is called the upper hybrid frequency and is given by $\omega_{\mathrm{UH}}^2 = \omega_{pe}^2 + \Omega_e^2$, and ω_R, where ω_L have been defined already. We cancel the ω^2's. In figure 9.6 we sketch kc/Ω_e versus ω/Ω_e. Rotating this figure and combining it with figure 9.5 for the low-frequency modes we get figure 9.7 for ω versus k for all the perpendicular propagating modes (see Krall and Trivelpiece 1973). The upper curve is called the extraordinary mode because $\mathbf{E}$ is perpendicular to $\mathbf{B}_0$ and for it the propagation is affected by the magnetic field, while the ordinary mode, the lower curve, is not.

We might ask what happened to the other root for ω^2 at $\theta = 90°$. For perpendicular propagation we can arrive at two more degenerate modes by

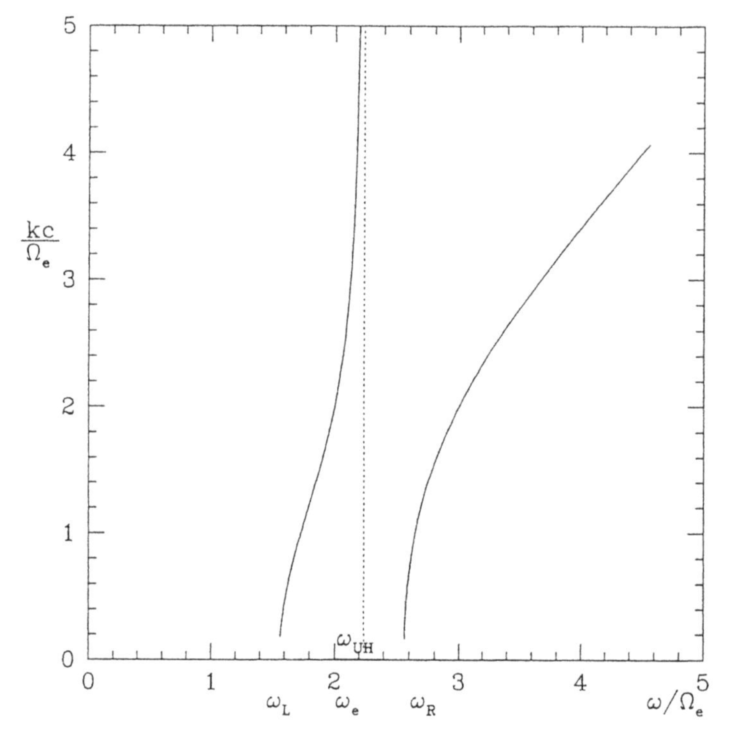

Figure 9.6. Plot of k versus ω for high frequency parallel propagating modes

combining two oppositely propagating Alfven modes, adding or subtracting them as θ goes to 90°. We find that one mode consists of $\mathbf{B}_1$ in the y direction with the ion and electron velocities in the z direction and $\mathbf{E} = 0$, so that

$$i\mathbf{k} \times \mathbf{B}_1 = 4\pi \mathbf{j}_1 = 4\pi n_i (\mathbf{v}_{1i} - \mathbf{v}_{e1})_z \hat{\mathbf{z}} \tag{49}$$

Since the velocities are parallel to $\mathbf{B}$ there is no force and for this mode $\omega = 0$.

In the other mode there is no perturbed $\mathbf{B}_1$, and $\mathbf{E}$ is in the x direction, while the ions and electrons have an $\mathbf{E} \times \mathbf{B}_0$ velocity in the y direction. Again, $\omega = 0$ and we can imagine that the lines are being transported rigidly without bending, in the y direction. In addition, there is a net charge, i.e., $n_i \neq n_e$, satisfying the Poisson equation

$$\nabla \cdot \mathbf{E} = ikE_x = 4\pi e(n_{1i} - n_{1e}) \tag{50}$$

These two additional degenerate modes exist only at $\theta = 90°$ where the other root is missing. It can be seen that for initial perturbations at $\theta = 90°$, with the two constraints, Poisson's law and $\nabla \cdot \mathbf{B} = 0$, there are again 12 independent quantities whose time evolution can be resolved by the eight = (two × four) nontrivial waves plus the two previous standard degenerate

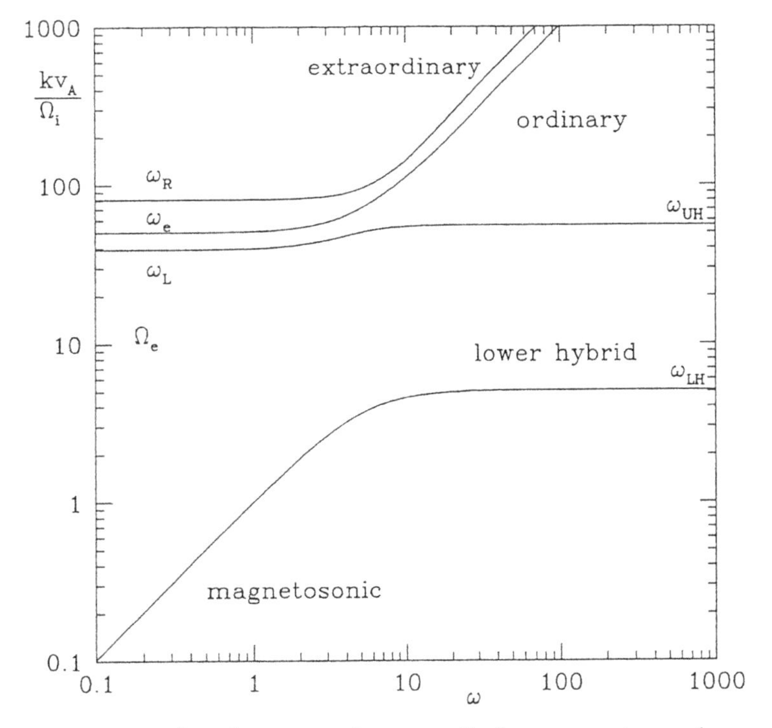

Figure 9.7. Plot of ω versus k for perpendicular propagating modes

modes (that exist at all θ) and plus these two special degenerate modes that exist only at $\theta = 90°$.

9.6 Propagation in a General Direction

In general, the capture of all the cold plasma modes for all θ is quite complicated. However, from equation 17 we can find the large k modes by setting

$$A = P\cos^2\theta + S\sin^2\theta = 0 \tag{51}$$

These modes are actually electrostatic modes in which $\mathbf{k} \times \mathbf{E} \ll kE$. For, in the limit $n \to \infty$, we have, from the first line of equation 16, $-E_x\cos\theta + E_z\sin\theta \to 0$ and so $\mathbf{E}$ is parallel to $\mathbf{k}$. For electrostatic modes, the dispersion relation can be derived from Poisson's equation. Indeed, if $\mathbf{E}$ is parallel to $\mathbf{k}$, then writing $\mathbf{E} = E_{ES}\hat{\mathbf{k}}$ and dotting equation 15 with $\mathbf{k}$ yields

$$\hat{\mathbf{k}} \cdot \boldsymbol{\epsilon} \cdot \hat{\mathbf{k}} = 0 \tag{52}$$

which from equation 11 yields equation 51.

For strictly parallel propagation (see figure 9.4), the only electrostatic mode obtained from equation 51 is the plasma oscillation $P = 0$, with

$\omega = \omega_{pe}$. For strictly perpendicular propagation (see figure 9.7), the only electrostatic modes are the upper and lower hybrid waves at $S = 0, \omega = \omega_{\text{LH}}$, and $\omega = \omega_{\text{UH}}$. However, at $\theta \to 0$ we find that there are additional electrostatic modes at Ω_i and Ω_e, as well as ω_{pe}. There is actually an exchange of limits when the electrostatic modes of equation 5 approach the electromagnetic modes as $\theta \to 0$.

The $k = \infty$ parts of the curves at $\theta = 0$, and $\theta = \pi/2$ are connected as follows. The Ω_i mode at $\theta = 0$ passes continuously to 0 at $\theta = \pi/2$. The Ω_e mode at $\theta = 0$ passes continuously to ω_{LH} at $\theta = \pi/2$ and the ω_{pe} mode at $\theta = 0$ passes continuously to ω_{UP} at $\theta = \pi/2$. The cutoff frequencies where $k \to 0$ for nonzero frequencies are at ω_L, ω_R, and ω_{pe}. They occur when $R = 0$, $L = 0$, and $P = 0$. They are the same for all θ.

With these remarks in mind we can compare the figures for propagation at $\theta = 0$ and $\theta = 90°$ and imagine how these curves are connected by surfaces for intermediate angles, one surface for each wave. We take the case appropriate for the interstellar medium that $\Omega_e \ll \omega_{pe}$. The lowest curve in figure 9.4 ($\theta = 0$) shrinks to zero as θ increases to 90°, just as is the case for the shear Alfven wave in MHD waves. The next curve in figure 9.4, the magnetosonic wave, goes to the whistler at higher wave numbers, which goes to the electron cyclotron frequency at $\theta = 0$. It connects to the lowest curve at $\theta = 90°$. As θ increases it maintains its character at lower frequencies but the whistler gradually goes over to the lower hybrid wave of figure 9.7. The electrostatic, large k limit, of this mode Ω_e at $\theta = 0$ gradually drops to the lower hybrid frequency at $\theta = 90°$.

At $\theta = 0$ the next two higher frequency curves, the plasma oscillation $\omega = \omega_{pe}$ and the L curve, cross. However, for θ slightly different from zero the crossing disappears, and the L curve at small frequency actually goes over to the plasma oscillation at higher frequencies. The plasma oscillation at low frequency goes over to the lower frequency light wave. When $\theta = \pi/2$ is reached, the upper of these two curves becomes the ordinary mode, while the lower curve connects to the curve starting at ω_L at $k = 0$ and going to ω_{UH} at $k = \infty$. All this is consistent with the motions of the electrostatic modes with θ at high k. Ω_i connects to zero, Ω_e connects to ω_{LH}, and ω_{pe} connects to ω_{UH}.

As stated above these are the relations when $\Omega_e < \omega_{pe}$. In general, which waves at $\theta = 0$ are connected to which at $\theta = \pi/2$ is established by their cutoff frequencies at $k = 0$, since these cutoff frequencies are independent of θ. The curves all start at these frequencies. The plasma wave at $\theta = 0$, which has ω_{pe} as its cutoff frequency, is connected to the ordinary wave at $\theta = 90°$. The R wave at $\theta = 0$ connects to the extraordinary wave at $\theta = 90°$, while the left wave connects to the upper hybrid mode.

This gives an adequate description of the five cold plasma waves. For more details we have to go to a numerical solution or to more detailed texts, such as Stix (1992) or Denise and Delcroix (1963).

9.7 The Cold Plasma Approximation

Our dispersion relations are based on the assumption that the plasma is cold. However, plasmas are never totally cold and it is necessary to ask for which of the waves the plasma approximation is sufficient. For this purpose it is useful to remember that in equilibrium the electrons and ions move along $\mathbf{B}_0$ with very little motion across $\mathbf{B}_0$.

The cold plasma assumption is valid if during the wave motion all the particles see approximately the same phase of the wave as a cold particle does. Then the current and charge response of all the particles is approximately the same. Consider the ions first. Now, during a significant fraction of a wave period, $1/\omega$, ions at any point z will spread out along z a distance $\Delta z \approx \Delta v/\omega$ and the range in phases of the wave that they will see will have a spread of $\Delta\Phi \approx k\Delta z \approx k\Delta v/\omega \approx kv_i/\omega$, where v_i is the thermal velocity of the ions. Further, if the frequency is less than the cyclotron frequency Ω_i, then the spread in position perpendicular to B will be $\Delta r_\perp \approx v_i/\Omega_i$. The gyroradius and the resulting spread in phase due to perpendicular motion will be $\Delta\Phi \approx k_\perp v_i/\Omega_i \approx k_\perp \rho_i$. If $\omega > \Omega_i$, then the spread in phase will be $k_\perp v_i/\omega$. Similar results hold for electrons.

Therefore, necessary conditions for the cold plasma assumption are of each particle,

$$k_\| v_\| < \omega \tag{53}$$

and

$$\begin{aligned} k_\perp \rho < 1, &\qquad \text{if } \omega < \Omega \\ k_\perp v < \omega, &\qquad \text{if } \omega > \Omega \end{aligned} \tag{54}$$

An additional condition arises when the frequency is close to either of the cyclotron frequencies. In the cold approximation all particles see the same wave frequency at the same time. But due to thermal motions they each see a different Doppler-shifted frequency, $\omega - k_\| v_\|$. Thus, when $\omega - \Omega$ is small some particles are closer to cyclotron resonances than others and their perturbation by the wave is sensitive to their unperturbed velocity. As a result, the dielectric tensor is particularly sensitive to thermal motions.

The cyclotron resonance denominator $\omega - \Omega$ in the cold plasma dielectric should be replaced by $\omega - k_\| v_\| - \Omega$ and the resonance is softened from the sharp peak of the cold plasma theory to a broadened one with width $\Delta\omega \approx k_\| v_\|$. For example, for parallel propagation the thermal contribution of the ions to the quantity L for left polarized waves becomes

$$L = \frac{\omega_{pi}^2}{\sqrt{2\pi}\, v_i} \int \frac{\exp -v^2/2v_i^2}{kv - (\omega - \Omega_i)} dv$$

where $v_i = \sqrt{T_i/M}$ is the thermal velocity. Here we use (Akhiezer et al. 1967, and problem 6 in chapter 10). It is easily seen that if $v_i = 0$, then this expression reduces to the ion part of the cold plasma result for L equation 19. On the other hand, if $\omega - \Omega$ is small, then we must have

$$kv_i \ll |\omega - \Omega_i| \tag{55}$$

as an additional condition for the cold plasma result to be accurate. If this is not the case the sharp cyclotron resonance of the cold plasma theory gets replaced by a softened resonance.

For $\theta = 0$, $k_\perp = 0$ and equation 54 is always satisfied. Away from the cyclotron resonances, the phase velocity of the waves is usually much larger than thermal velocity of the ions. In addition, below Ω_e the electrons have the local $\mathbf{E} \times \mathbf{B}$ velocity independent of their parallel velocity (if $kv < \Omega_e$ or $k\rho_e < 1$), so their response is indistinguishable from a that of a cold plasma. However, for waves with freqencies near the ion cyclotron frequency, k is large, equation 55 is violated, and the ion thermal motion causes the broadening of the cyclotron response. The large k limit of the ion cyclotron wave is sensitive to ion motion since the phase velocity goes to zero. The same is true for the electron cyclotron wave, which is the large k limit of the whistler.

For strictly perpendicular propagation, $k_\parallel = 0$, and we need to worry only about condition equation 54. However, for the lower hybrid wave at angles only a little different from 90° (by $\sqrt{m/M}$) the parallel phase velocity becomes comparable or less than the electron parallel velocity, and the cold plasma approximation equation 53 again fails.

For a general angle not near 90° we must consider the parallel phase velocity of the wave, ω/k_z, and compare it with the thermal velocity of the ions and electrons. There is a slope ω/k, which must exceed the ion and electron thermal velocities to satisfy equation 53. It is clearly satisfied for ions for all the waves except the ion cyclotron shear Alfven wave branch. If $v_e \gg v_A$, then the part of the magnetosonic branch fails for electrons. Further, all of the shear Alfven-ion cyclotron branch fails for ions. For modes with very large k the electrostatic modes all fail the cold plasma condition, and thermal effects must be included in their treatment. Such modes are the electron cyclotron end of the whistler, the ion cyclotron end of the shear Alfven wave, and the lower and upper hybrid ends of their branches, respectively.

What are the consequences of thermal effects? We anticipate the next chapter and quote the simplest results. There are actually two effects. The first effect is that, for electrons with $k_\parallel v_i \gg \omega$, $P = 1 - \omega_{pe}^2/\omega^2$ is replaced by $1/k^2\lambda_D^2$. (When ions are unmagnetized, then the ion terms in R and L are replaced by ω_{pi}^2/ω^2, as though $B_0 = 0$. Then we have to consider the second of conditions equation 54.) The second effect is a complex, out of phase, contribution to ϵ due to a resonance between the waves and the particles in resonance with them. One resonance occurs through the equality of the velocities and phase velocities, or, better, the parallel velocity of the particle with the parallel phase velocity, $\omega = k_\parallel v_\parallel$. A second resonance occurs because of a coincidence between the Doppler-shifted cyclotron frequency of a particle due to its parallel motion and the frequency of the wave $\omega - k_\parallel v_\parallel = \pm\Omega$. There is also a more general condition associated with harmonies of Ω_e and Ω_i.

These resonances allow an exchange of energy between the particles and waves that can go either way: some particles give energy to the waves and others take energy from the wave. If the particles are in local thermodynamic equilibrium, i.e., if they are Maxwellian, then the direction of the average energy transfer of the waves summed over all the resonant particle interactions is always such that the wave loses energy and the particles gain energy. This leads to a collisionless damping of the wave. (For non-Maxwellian distribution functions we sometimes have growth.) This resonance damping was first pointed out by Landau (1946) and goes under the name Landau damping. Another name for the first type of interaction, in which $\omega = k_\parallel v_\parallel$, is Cerenkov resonance. The second type, $\omega - k_\parallel v_\parallel = \pm\Omega$, is usually called cyclotron resonance.

9.8 Faraday Rotation and Magnetic Fields

Now let us return to the cold plasma approximation and consider the effects of a magnetic plasma on a light wave propagating through it. Let us consider the parallel propagation of the two upper modes, the R and L modes, at frequencies far above the cutoff. These are illustrated by radio waves emitted from pulsars. Their dispersion relations are given by equations 20 and 21. In the appropriate limit $\omega \gg \omega_{pe} \gg \Omega_e$ we have

$$n_\pm^2 = \frac{k_\pm^2 c^2}{\omega^2} = 1 - \frac{\omega_{pe}^2}{\omega(\omega \pm \Omega_e)} \approx 1 - \frac{\omega_{pe}^2}{\omega^2} \pm \frac{\Omega_e \omega_{pe}^2}{\omega^3}$$

or

$$k_\pm c \approx \omega - \frac{\omega_{pe}^2}{2\omega} \pm \frac{\Omega_e \omega_{pe}^2}{2\omega^2} \tag{56}$$

Assume that at $z = 0$ the wave is linearly polarized in the x direction so that

$$\mathbf{E} = \mathrm{Re}[(\hat{\mathbf{x}} + i\hat{\mathbf{y}})E_+ e^{-i\omega t} + (\hat{\mathbf{x}} - i\hat{\mathbf{y}})E_- e^{-i\omega t}] = E \cos\omega t \hat{\mathbf{x}} \tag{57}$$

where $E_+ = E_- = E/2$, and the amplitude associated with k_+ and k_- at $z = L$ is

$$\mathbf{E} = \mathrm{Re}[E_+(\hat{\mathbf{x}} + i\hat{\mathbf{y}})e^{i(k_+ z - \omega t)} + E_-(\hat{\mathbf{x}} + i\hat{\mathbf{y}})e^{i(k_- z - \omega t)}] \tag{58}$$

or

$$\mathbf{E} = \mathrm{Re}\left[E\left(\hat{\mathbf{x}}\cos\frac{\Delta k L}{2} - \hat{\mathbf{y}}\sin\frac{\Delta k L}{2}\right)e^{i(k_0 L - \omega t)}\right] \tag{59}$$

where $\Delta k = k_+ - k_- = \Omega_e \omega_{pe}^2/c\omega^2$ and $k_0 = (k_+ + k_-)/2 = \omega - \omega_{pe}^2/c\omega$, so that the **E** vector vibrates in a plane rotated by $-\Delta\theta$, where

$$\Delta\theta = \frac{\Delta k c}{2}\frac{L}{c} = \frac{\omega_{pe}^2 \Omega_e}{2\omega^2}\frac{L}{c}$$

$$= \frac{1}{2}\frac{4\pi e^2}{m}\frac{e}{mc} n_e B \frac{L}{c}\frac{\lambda^2}{(2\pi)^2 c^2}$$

$$= 0.9 n_e(cm^{-3}) B_0(\mu G) L(ps) \lambda^2(m) \tag{60}$$

(See Manchester and Taylor 1977 and Spitzer 1978.) Thus, we expect the electron density and interstellar magnetic field to rotate the plane of polarization of a given polarized source such as a pulsar or radio source. Note that the angle of rotation depends on the frequency or wavelength of the wave. At high frequencies $\Delta\theta = 0$, so the vibration is in the original plane of polarization at the source. Indeed, the slope of $\Delta\theta$ versus λ^2 is a constant, so plotting the observed angle versus λ^2 gives the product $n_e B L_0$.

If n_e and L are known, then we get a reliable estimate of B, provided these are constants. These quantities or, at any rate, their product, $n_e L$, is found from the dispersal in arrival time of the different frequencies of the pulsar signal. Indeed, the pulsar emits its signal in short bursts. The envelope of the burst propagates at the group velocity of the signal. For this purpose we can ignore the effect of the field and the angle of polarization. The reduced dispersion relation is, from equation 45,

$$\omega^2 = k^2 c^2 + \omega_{pe}^2 \tag{61}$$

or

$$\omega \approx kc + \frac{\omega_{pe}^2}{2kc} \tag{62}$$

so the group velocity v_g is

$$v_g = \frac{\partial\omega}{\partial k} = c\left(1 - \frac{\omega_{pe}^2}{2k^2c^2}\right) \approx c\left(1 - \frac{\omega_{pe}^2}{2\omega^2}\right) \tag{63}$$

The time of propagation for those parts of the pulse around ω is

$$t = \frac{L}{v_g} \approx \frac{L}{c}\left(1 + \frac{\omega_{pe}^2}{2\omega^2}\right) \tag{64}$$

so the delay relative to the arrival time of the highest frequencies is

$$\Delta t = \frac{L}{c}\frac{\omega_{pe}^2}{2\omega^2} = 0.05 L(\text{ps}) n_e(\text{cm}^{-3})\lambda^2(m)\ \text{sec} \tag{65}$$

and thus the derivative of Δt or t with respect to λ^2 gives the factor $L n_e$, which we need in the Faraday rotation formula, equation 60 (Manchester and Taylor 1977; Spitzer 1978).

9.9 Bremsstrahlung

One further result can be drawn from the treatment of light-wave-like modes. A general mechanism for thermal emission is bremsstrahlung, which occurs when an electron passes an ion. The inverse of this process is free-free absorption. If a wave happens to be passing when an electron encounters an ion, the wave can be absorbed.

The order of magnitude of the rate of absorption can be modeled by including a friction term, $-m\nu v$, on the right-hand side of the electron equations of motion, equations 2 and 4. We ignore the B field in this calculation since its effect is small. Bring the friction term to the left-hand side of equation 2 to get

$$-i\omega m\mathbf{v} + m\nu\mathbf{v}$$

We see that we can incorporate the drag term into the equations of motion in the form by replacing m by

$$m^* = m\left(1 + \frac{i\nu}{\omega}\right) \tag{66}$$

The electron mass m occurs in equation 45 through $\omega_{pe}^2 = 4\pi n_e e^2/m$, so the equation for the propagation of light rays in a "collisional" plasma now becomes

$$\omega^2 = k^2c^2 + \frac{\omega_{pe}^2}{1 + i\nu/\omega} = k^2c^2 + \omega_{pe}^2 - i\frac{\omega_{pe}^2\nu}{\omega}$$

or

$$\omega = kc + \frac{\omega_{pe}^2}{2kc} - \frac{i\omega_{pe}^2\nu}{2\omega kc} \tag{67}$$

ω is now complex. Its approximate solution is

$$\omega = \omega_1 + i\omega_2 = kc + \frac{\omega_{pe}^2}{2kc} + \frac{i\omega_{pe}^2\nu}{2\omega_1^2} \tag{68}$$

The wave evolves as

$$E = \hat{E}e^{ikc - i\omega_1 t - \omega_{pe}^2\nu t/2\omega_1^2}$$

so the amplitude of the wave damps at the rate $\omega_{pe}^2\nu/2\omega^2$ and its energy decays at the rate $\omega_{pe}^2\nu/\omega^2$. ν is the effective collision rate. (Because it is the collision rate of a rapidly oscillating particle in the oscillating electric field it is slightly different then the normal collision rate.) Now, by detailed balance, we must have that the absorption of the Planck spectrum $J(\omega)$ at this rate is equal to the emission $\epsilon(\omega)$, by bremsstrahlung of a thermal plasma

$$\epsilon(\omega) = J(\omega)\frac{\omega_{pe}^2\nu}{\omega^2} \tag{69}$$

This formula gives a correct order of magnitude estimate of the bremsstrahlung rate when a plausible choice for ν is made. The temperature dependence of the emission enters through J and ν. For $\hbar\omega \ll T$ we have $J(\nu) = 8\pi\nu^3 T/c^3$. Also, $\nu = n\sigma v_e = \sqrt{2}\pi n e^4/\sqrt{m}T^{3/2}$, where we take the effective cross section as $\sigma\pi(e^2/T)^2$ and $v_e = \sqrt{2T/m}$. Thus, equation 69 becomes

$$\epsilon(\omega) = \frac{\sqrt{2}}{\pi}\frac{e^6 n^2}{m^{3/2}c^3}\frac{n^2}{T^{1/2}} \tag{70}$$

and is independent of ω. Integrating this over ω from 0 to $\omega_{\max} = T/\hbar$ we see that

$$\int \epsilon(\omega)d\omega = \frac{4\sqrt{2}e^6}{\hbar m^{3/2}c^3}n^2 T^{1/2} = 0.87\ 10^{-25} n^2 T^{1/2} \frac{\text{ergs}}{\text{cm}^2\ \text{sec}} \tag{71}$$

where T is in kelvins. This result is actually about one-half of the true one, which is not surprising due to the roughness of the estimates that we have made (see Spitzer 1962; Dawson 1968).

9.10 Wave Energy

There is a simple formula for the energy of cold plasma waves (Stix 1992). First consider the normalization of the electric energy of a wave. Let the electric field of the wave be given by

$$E = \text{Re}\, \hat{E} e^{i\mathbf{k}\cdot\mathbf{r} - i\omega t} = \tfrac{1}{2}\left(\hat{E}^* e^{-ik\cdot r + i\omega t} + \hat{E} e^{ik\cdot r - i\omega t}\right) \tag{72}$$

where $\hat{\mathbf{E}}$ is a complex vector. In this complex notation we can write the square of the electric field energy as

$$\left\langle \frac{\mathbf{E}^2}{8\pi} \right\rangle = \frac{1}{8\pi}\left\langle \frac{1}{2}\left(\hat{\mathbf{E}}^* e^{-i\mathbf{k}\cdot\mathbf{r} - i\omega t} + \hat{\mathbf{E}} e^{+i\mathbf{k}\cdot\mathbf{r} + i\omega t}\right) \cdot \frac{1}{2}\left(\hat{\mathbf{E}}^* e^{-i\mathbf{k}\cdot\mathbf{r} - i\omega t} + \hat{\mathbf{E}} e^{+i\mathbf{k}\cdot\mathbf{r} + i\omega t}\right)\right\rangle \tag{73}$$

which on averaging over space and time reduces to

$$\left\langle \frac{\mathbf{E}^2}{8\pi} \right\rangle = \frac{1}{32\pi}\left\langle \hat{\mathbf{E}}^* \cdot \hat{\mathbf{E}} + \hat{\mathbf{E}} \cdot \hat{\mathbf{E}}^* \right\rangle = \frac{1}{16\pi}\langle \hat{\mathbf{E}}^* \cdot \hat{\mathbf{E}} \rangle \tag{74}$$

For two different complex vector amplitudes $\hat{\mathbf{E}}$ and $\hat{\mathbf{F}}$ we similarly would have

$$\langle \mathbf{E} \cdot \mathbf{F} \rangle = \tfrac{1}{4}\left\langle \hat{\mathbf{E}}^* \cdot \hat{\mathbf{F}} + \hat{\mathbf{E}} \cdot \hat{\mathbf{F}}^* \right\rangle \tag{75}$$

The other two terms are proportional to $e^{-2ik\cdot r + 2i\omega t}$ and $e^{2i\mathbf{k}\cdot\mathbf{r} - 2i\omega t}$ and these average to zero either over a period $2\pi/\omega$ or a wavelength $2\pi/k$ or both.

To derive the energy of a cold plasma wave we need to calculate the energy to set it up. We proceed as follows: Choose a wave with frequency ω_0 and wave number k_0 and a complex vector with mode amplitude E_0 at ω_0. The wave can be set up by imposing a proper external current $\mathbf{j}_s$. Choose the frequency ω of this external current with a wave number $\mathbf{k}_0$ to be close to, but not exactly equal to, ω_0. Then this current will produce an electric field given by

$$\mathbf{k} \times [\mathbf{k} \times \mathbf{E}(\mathbf{k}, \omega)] + \frac{\omega^2}{c^2}\boldsymbol{\epsilon}(\omega) \cdot \mathbf{E}(\mathbf{k}, \omega) = -i\frac{4\pi\omega}{c^2}\mathbf{j}_s(\omega) \tag{76}$$

(Notice that the value of $\mathbf{j}_s$ for $\omega = \omega_0$ is zero.)

We wish to apply an external current that increases from zero at $t = -\infty$ and grows as $e^{\gamma t}$. Thus, we take $\omega = \omega_0 + i\gamma$, with $\gamma \ll \omega_0$. Taylor expanding the left-hand side of equation 76 in ω with the zero-order term zero, we find

$$\mathbf{j}_s = -\frac{1}{4\pi}\frac{\gamma}{\omega}\frac{\partial}{\partial\omega}(\omega^2\boldsymbol{\epsilon} \cdot E_0) \tag{77}$$

The work done by the external current $\mathbf{j}_s$ is

$$W = -\int_{-\infty}^{0} < \mathbf{E} \cdot \mathbf{j}_s > e^{\gamma t} dt = \frac{\omega}{4} < \mathbf{E}_0^* \cdot \mathbf{j}_s + \mathbf{E}_0 \cdot \mathbf{j}_s^* > e^{2\gamma t} dt \tag{78}$$

or

$$W = \frac{\gamma}{2\gamma}\frac{1}{4}\frac{1}{4\pi}\left[\mathbf{E}_0^* \cdot \frac{\partial}{\omega\partial\omega}(\omega^2\boldsymbol{\epsilon}) \cdot \mathbf{E} + \mathbf{E}_0 \cdot \frac{\partial}{\omega\partial\omega}(\omega^2\boldsymbol{\epsilon}^*) \cdot \mathbf{E}_0^*\right] \tag{79}$$

where we have made use of equation 77. This expression reduces to

$$\mathcal{E} = W = \frac{1}{16\pi}\mathbf{E}_0^* \cdot \left[\frac{\partial}{\omega\partial\omega}(\omega^2\boldsymbol{\epsilon})\right]_{\omega=\omega_0} \cdot \mathbf{E}_0 \tag{80}$$

since, by the Hermetian character of $\boldsymbol{\epsilon}$, the two terms in equation 79 are equal. This is the desired expression for the energy of the wave.

For normalization purposes and to avoid discussion of the definition of amplitudes and rms values, the wave energies from equation 80 should always be compared with the electric field energy, equation 74. To illustrate the use of equation 80, let us consider three examples: A linearly polarized Alfven wave has $\epsilon = c^2/v_A^2$, so

$$\mathcal{E}_{\text{Alfven}} = \frac{1}{16\pi}\mathbf{E}_0^* \cdot \frac{\partial}{\omega\partial\omega}(\omega^2\frac{c^2}{v_A^2}) \cdot \mathbf{E}_0 = \frac{1}{8\pi}\frac{c^2}{v_A^2}|E_0|^2 \tag{81}$$

that is, $2c^2/v_A^2$ times the electric field energy, equation 74. Since $B = |kc/\omega|E = (c/v_A)E$ the wave energy is twice the perturbed magnetic energy. Since $v = cE/B$, the kinetic energy is $(nM/2)(cE/B)^2 = (c^2/v_A^2)(E^2/8\pi) = 2c^2/v_A^2$ times the electric energy so the wave energy equals the sum of the kinetic energy and the magnetic energy. For a circularly polarized Alfven wave $E_x = iE_y$ and $E^* \cdot \boldsymbol{\epsilon} \cdot E = (c^2/v_A^2)(\mathbf{E}_x|^2 + |E_y|^2)$, and the above results are identical.

For a linearly polarized electromagnetic wave $\boldsymbol{\epsilon} = 1$ and equation 80 gives

$$\mathcal{E} = \frac{1}{16\pi}\mathbf{E}_0^* \cdot \left(\frac{1}{\omega}2\omega\right) \cdot \mathbf{E}_0 = \frac{1}{8\pi}\mathbf{E}_0^* \cdot \mathbf{E}_0 \tag{82}$$

or twice the electric field energy, as it should. For a plasma oscillation, $\epsilon = 1 - \omega_{pe}^2/\omega^2$, and $\partial(\omega^2\epsilon)/\partial\omega = 2\omega$, we easily find that the wave energy is just twice the electrical energy.

The force exerted by the current on the plasma $\mathbf{F}$ must be such that $\int \mathbf{F} \cdot \mathbf{v}_\phi \, dt = \mathcal{E}_{\text{wave}}$ and is purely in the $\mathbf{k}$ direction, so the momentum of the wave is

$$\mathbf{P} = \frac{\mathbf{k}}{\omega}\mathcal{E}_{\text{wave}}. \tag{83}$$

9.11 Problems

1. (Moving plasma) Consider a cold plasma moving with velocity $\mathbf{U}$ along $\mathbf{B}_0$ in the z direction. Solve for the perturbed ion velocity from

$$M\left(\frac{\partial \mathbf{v}}{\partial t} + \mathbf{U} \cdot \nabla \mathbf{v}\right) = e\mathbf{E} + e\frac{\mathbf{U} \times \mathbf{B}_1}{c} + e\frac{\mathbf{v} \times \mathbf{B}_0}{c}$$

and the perturbed electron velocity from the its equation. Find the z component of the perturbed current from $j_z = n_0 e(v_z^i - v_z^e) + e(n_1^i - n_1^e)U$. Show that if $\mathbf{k}$ is parallel to $\mathbf{B}_0$, then the zz component of the dielectric tensor $\epsilon_{zz}(\omega)$ for the moving plasma can be obtained from the same component of the dielectric for the plasma at rest $\epsilon_{zz}^0(\omega)$ by replacing ω by $\omega' = \omega - \mathbf{k} \cdot \mathbf{U}$. Show that except for the 1 in the expressions for the transverse components, the dielectric tensor $\boldsymbol{\epsilon}(\omega)$ can be found from the same components of $\epsilon_0(\omega')$ multiplied by ω'^2/ω^2. Thus, if we neglect the 1, the dispersion relation is that for a cold plasma Doppler shifted by kU. (Including the 1 leads to the relativistic Doppler shift.)

Regard a thermal plasma as made up of a sum of plasmas with different $\mathbf{U}$'s. Then show that when the thermal spread is small compared to the phase velocity ω/k, the plasma may be considered as cold if ω is not close to either of the cyclotron frequencies. What is the condition, if it is?

2. (The magnetic energy of a wave) Show that the perturbed magnetic field in a wave is given from Faraday's law by $\mathbf{B}_1 = \mathbf{k}c \times \mathbf{E}/\omega$. By dotting equation 15 with $\mathbf{E}^*$ show that the perturbed magnetic energy can be written as

$$\frac{1}{16\pi}\mathbf{B}_1^* \cdot \mathbf{B}_1 = \frac{1}{16\pi}\mathbf{E}_1^* \cdot \epsilon \cdot \mathbf{E}_1$$

where the perturbed magnetic field is $\text{Re}(\mathbf{B}_1 e^{i\mathbf{k}\cdot\mathbf{r}-\omega t})$. Thus, show from equation 80 that the wave energy can be written

$$\mathcal{E} = \frac{1}{16\pi}\mathbf{B}_1^* \cdot \mathbf{B}_1 + \frac{1}{16\pi}\mathbf{E}_1^* \cdot \frac{\partial(\omega\boldsymbol{\epsilon})}{\partial\omega}\mathbf{E}_1$$

3. (The tenth-degree equation for ω) Show by clearing the denominator in equation 17 that the dispersion relation is a tenth-degree equation in ω. Calculate the coefficients of the tenth-degree term and the constant term. Proceed systematically in order to avoid undue complications in your calculations. Does either the highest coefficient or the constant coefficient ever vanish and if so what does it mean? (See section 9.4 on the number of waves.)

4. (The lower hybrid frequency) Assume that the lower hybrid frequency is large compared to the ion cyclotron frequency and small compared to the electron cyclotron frequency. Show by equating the denominator of equation 46 to zero that it can be written as

$$\omega_{\mathrm{LH}} = \omega_{pi} \frac{\Omega_e}{\sqrt{\omega_{pe}^2 + \Omega_e^2}}$$

Show that E_y and E_z are zero. Show that for perpendicular propagation near the lower hybrid frequency the ion current parallel to $\mathbf{k}$ is accurately given by neglecting any magnetic force. Show that the electron current is given by the polarization current discussed in chapter 2. Show that exactly at the lower hybrid frequency the components of the displacement current, the ion current, and the electron current all balance. If, further, $\omega_{pe} \ll \Omega_e$, show that the electron current is small and the other two currents balance, and $\omega_{\mathrm{LH}} \approx \omega_{pi}$. Show that in the opposite case, $\omega_{pe} \gg \Omega_e$, the displacement current is negligible, and the electron and ion currents balance and $\omega_{\mathrm{LH}} \approx \sqrt{\Omega_i \Omega_e}$ the hybrid cyclotron frequency.

5. (Polarization of an Alfven wave) For parallel propagation the Alfven modes ($\omega \ll \Omega_i$) are exactly circularly polarized. However, for $\theta \gg \sqrt{\omega/\Omega_i}$, they are linearly polarized the L mode going into the shear Alfven mode. Demonstrate this. One way to do this is to solve the dispersion relation for n^2 by the quadratic equation 17. For simplicity, you may keep only the P terms, since P is very large compared to S. Take the small θ limit of the solution and show that

$$n^2 \approx S(1 - \tfrac{1}{2}\theta^2) \pm \sqrt{S^2(\tfrac{1}{2}\theta^2)^2 + D^2}$$

Then from equation 16 show that

$$\frac{E_x}{iE_y} = \frac{S - n^2}{D} \approx x \mp \sqrt{x^2 + 1}$$

where $x = (S/D)(\theta^2/2)$. From $D/S = -(\omega/\Omega_i)$, it is easy to show that when $\theta \gg \sqrt{\omega/\Omega}$, either $E_x \ll E_y$ or vice versa according to the sign. Also, in the opposite limit, show that we get circular polarization. Track the sign through the calculation to show that $E_x \gg E_y$ for the sign corresponding to the L mode.

6. (Faraday rotation and time delay) Take a one-second pulsar 2 kpc away in the galactic plane. Take a uniform B of 2 μG, and a uniform

electron density of $n_e = 0.1\ \mathrm{cm}^{-3}$. What is the Faraday rotation angle and time delay at 100 MHz? At 1000 MHz? Could you measure the amount of rotation at 100 MHz without further electronics if your bandwidth was 20 MHz? Would you see any pulses at 100 MHz? If you were measuring the Faraday rotation of a Lyman alpha cloud at a large red shift, Z, you would have to include a factor of $1/(1+Z)^2$ because of this red shift. Why?

References

Akhiezer, A.I., A. I. Akhiezer, R. V. Polivin, A. G. Sitenko, and K. N. Stepanov. 1967. *Collective Oscillations in a Plasma*, p. 29, trans. H. S. H. Massey (trans. ed. by R. J. Tayler), MIT Press, Cambridge, MA.

Barkhausen, H. 1930. *Proceedings of the Institute of Radio Engineers* **18**, 1155.

Dawson, J. 1968. *Advances in Plasma Physics* **1**, 1.

Denise, J. F., and J. L. Delcroix. 1963. *Plasma Waves*, trans. M. Weinrich and D. BenDaniel, Interscience, Wiley, New York.

Eckersley, T. L. 1935. Nature, **135**, 104.

Krall, N. A., and A. W. Trivelpiece. 1973. *Principles of Plasma Physics*, p. 194, McGraw Hill, New York.

Landau, L. D. 1946. *J. Phys. USSR*, **10**, 25.

Manchester, R. N., and J. H. Taylor. 1977. *Pulsars*, pp. 129, 134, W. H. Freeman, San Fransisco.

Spitzer, Jr., L. 1962. *Physics of a Fully Ionized Gas*, p. 148, Interscience, New York.

Spitzer, Jr., L. 1978. *Physical Processes in the Interstellar Medium*, pp. 61, 65, Wiley (Interscience), New York.

Stix, T. H. 1962. *The Theory of Plasma Waves*, p. 19, McGraw Hill, New York.

Stix, T. H. 1992. *Waves in Plasmas*, American Institute of Physics, New York.

Chapter 10

COLLISIONLESS PLASMAS: THERMAL EFFECTS

10.1 Introduction

In the last chapter we studied waves under the approximation that the plasma was cold, so that the motions of all the electrons and ions in the wave field were identical. This would not be the case if the phase velocity of the waves were comparable to the thermal velocity of electrons, v_e. For example, the plasma oscillation has a frequency ω_{pe} and a phase velocity ω_{pe}/k. If k is large enough (short wavelengths), then ω_{pe}/k will be smaller than v_e, so different electrons will move differently. A more complex treatment of them is necessary and new physics emerges. The plasma oscillation itself will be qualitatively unchanged although its propagation and dispersion characteristics will be altered. Further, it will be damped due to resonant interaction between the wave and electrons traveling at the phase velocity of the wave.

To most easily examine the complications associated with thermal effects we will restrict ourselves to an unmagnetized plasma. For a cold, unmagnetized plasma we found two types of waves: electromagnetic waves and electrostatic waves. Electromagnetic waves propagate at near the speed of light and are essentially unaffected by thermal motions. It is the electrostatic waves that are most affected by thermal effects and it is these that are discussed in this chapter.

The plasma oscillation essentially retains its character. On the other hand, there is also a pair of electrostatic zero-frequency modes in a cold plasma. With thermal effects these modes develop a finite frequency and evolve into a pair of ion acoustic modes propagating in two different directions at the speed of sound.

We first consider the electrostatic plasma oscillation. It has such a high frequency that the ions are essentially unperturbed and we can restrict our attention to the electrons. For a proper description of the electrons in a wave we must deal with the electron distribution function $f(\mathbf{v}, \mathbf{x}, t)$ that keeps track of the number of electrons with different velocities. Its evolution is governed by the Vlasov equation

$$\frac{\partial f}{\partial t} + \mathbf{v} \cdot \nabla f - \frac{e}{m} \mathbf{E} \cdot \nabla_v f = 0 \qquad (1)$$

derived in chapter 8.

(For collisional plasmas, the distribution function is kept near a Maxwell distribution with three parameters, ρ, $\mathbf{v}$, T, so a fluid description involving these three quantities at each point is adequate. For a collisionless thermal plasma we need a three-dimensional function of **v** at each point. Thus, for a collisionless thermal plasma it is necessary to restrict ourselves to simpler problems, small-amplitude waves in a nearly homogeneous background.)

Since the waves are electrostatic, **E** is in the direction of propagation of the wave **k**, which we take as the x direction. Then we need only the Poisson equation to determine it. Let us linearize f, $f = f_0 + f_1$, and take **E** small. Any arbitrary homogeneous, time-independent function $f_0(\mathbf{v})$ will satisfy the zero-order Vlasov equation. To first order the Vlasov equation is

$$\frac{\partial f_1}{\partial t} + \mathbf{v} \cdot \nabla f_1 = \frac{e}{m}\mathbf{E} \cdot \nabla_v f_0 \tag{2}$$

For electrostatic modes this must be combined with Poisson's equation

$$\nabla \cdot \mathbf{E} = 4\pi(n_i - n_e)e = -4\pi n_{1e} e = -4\pi e \int f_1 d^3v \tag{3}$$

We seek a purely exponential solution of these equations

$$\mathbf{E}_1 = \mathrm{Re}\hat{\mathbf{E}}_{1x} e^{ikx - i\omega t} \tag{4}$$

$$f_1 = \mathrm{Re}\hat{f}_1 e^{ikx - i\omega t} \tag{5}$$

Since $\mathbf{k} \times \mathbf{E} = 0$, there is no perturbed **B**. Substitution of these exponential solutions into equations 2 and 3 yields

$$(-i\omega + ikv_x)f_1 = \frac{eE_x}{m}\frac{\partial f_0}{\partial v_x} \tag{6}$$

$$ikE_x = -4\pi e \int f_1 d^3v \tag{7}$$

The solution of equation 6 is

$$f_1 = i\frac{e}{m}E_x \frac{\partial f_0 / \partial v_x}{\omega - kv_x} \tag{8}$$

and substitution of this result into equation 7 yields

$$kE_x = -\frac{4\pi e^2}{m}E_x \int \frac{\partial f_0/\partial v_x}{\omega - kv_x} d^3v = \frac{4\pi e^2}{m}E_x \int \frac{\partial F_0/\partial v_x}{kv_x - \omega} dv_x \tag{9}$$

where F_0 is the reduced one-dimensional electron distribution function $F_0(v_x) = \int f_0 dv_y dv_z$. (Here F_0 is not the ion distribution, as is our usual convention.) The denominator of this integral vanishes at $v_x = \omega/k$, introducing a singularity into the integrand. Since ω/k is the phase velocity of the wave, the singularity is physically associated with particles that travel near this velocity. Since we expect a strong interaction between these particles and the wave, it is not surprising that something should go wrong at this velocity, and this is the origin of the singularity.

As we shall see, the interaction is a transfer of energy between the waves and these "resonant" particles. For a Maxwellian, F_0, this interaction results in a "collisionless" damping of the wave. This was first pointed out by Landau and is universally referred to as Landau damping (Landau 1946). Landau also showed how to best handle this damping by employing the Laplace transform method and contour integration. While this method is very efficient, it does tend to obscure the physical details. We will attempt to lay these out more simply when we discuss this damping.

Before proceeding to this more complex issue let us recover the cold plasma dispersion relation in the limit where the phase velocity of the waves, ω/k, is large compared to the thermal velocity, $v_e = \sqrt{T/m}$. In the cold plasma limit, $\omega = \omega_{pe}$, the phase velocity is ω_{pe}/k, and this becomes large as k becomes small. In this limit the numerator in equation 9, $\partial f_0/\partial v_x$, at the resonant velocity is exponentially small and can be ignored. Then, for the bulk of the particles, $v \sim v_e$ and the denominator $\omega - kv \approx \omega$, so we have

$$\frac{1}{\omega - kv} \approx \frac{1}{\omega}\left(1 + \frac{kv}{\omega} + \frac{k^2v^2}{\omega^2} + \frac{k^3v^3}{\omega^3}\right) \tag{10}$$

After substitution of this expansion into equation 9 only the second and fourth terms survive (for Maxwellian F_0), and after integrating by parts we get

$$kE = \frac{4\pi e^2 E}{m}\int\left(\frac{k}{\omega^2} + \frac{3k^3v^2}{\omega^4}\right)F_0 dv = \frac{4\pi ne^2}{m}\left(\frac{k}{\omega^2} + \frac{3k^3v_e^2}{\omega^4}\right)E \tag{11}$$

or

$$\omega^2 = \omega_{pe}^2 + \frac{3k^2v_e^2}{\omega^2}\omega_{pe}^2 \tag{12}$$

Since the second term must be small compared to the first, and the first by itself gives $\omega^2 = \omega_{pe}^2$, we can replace ω^2 in the second term by ω_{pe}^2. Thus, to the same accuracy we have

$$\omega^2 = \omega_{pe}^2 + 3k^2v_e^2 \tag{13}$$

If we let $k \to 0$, then $v_\phi = \omega_{pe}/k \to \infty$, and when $kv_e \ll \omega_{pe}$ we recover the cold plasma relation

$$\omega^2 = \omega_{pe}^2 \tag{14}$$

There is no dispersion of ω with k in equation 14, so to this order the group velocity is zero. The next correction in equation 12, the Bohm–Gross correction does give some dispersion (Bohm and Gross 1949). The group velocity is obtained by differentiating equation 12 with respect to k:

$$\omega\frac{\partial\omega}{\partial k} = 3kv_e^2 \tag{15}$$

Taking $\omega = \omega_{pe}$, on the left, we have

$$v_g = \frac{\partial \omega}{\partial k} = \frac{3kv_e^2}{\omega_{pe}} = 3k\lambda_D v_e \tag{16}$$

since $v_e/\omega_{pe} = \lambda_D$ is the Debye length. Dividing equation 15 by k, we can also write this conveniently as

$$v_g v_\phi = 3v_e^2 \tag{17}$$

Thus, for wavelengths of order the Debye length (actually $12\lambda_D$ because of the 3 and 2π) the group velocity is comparable to the electron thermal speed. It is proportional to k, so longer wavelength waves have higher phase velocities and lower group velocities. The original waves discovered by Langmuir (Tonks and Langmuir 1929) were found to have very small group velocities, so the wave packets did not seem to propagate and the plasma just shook like jelly. Since blood plasma has the same properties, this may be why they named the oscillations plasma oscillations.

10.2 Ion Acoustic Waves

If we inspect equation 9, we see that the numerator also vanishes as $v \to 0$, since $\partial F_0/\partial v = -(v/v_e^2)F_0$. Thus, another possible limit where we might discuss waves unaffected by the singularity is $\omega/k \ll v_e$. In this case we have

$$\frac{1}{\omega - kv} \approx -\frac{1}{kv} \tag{18}$$

Substituting this into equation 9 gives

$$kE = -\frac{4\pi e^2}{m} E \int \frac{\partial F_0/\partial v}{-kv} dv = -\frac{\omega_{pe}^2}{kv_e^2} E \tag{19}$$

which, of course, has no solution. However, our approximation is that $v_\phi \ll v_e$. Up till now we have neglected the ions because so far we have supposed that $v_\phi \gg v_e \gg v_i$. But, in this new limit, the ions are important.

Let us take this limit

$$kv_i \ll \omega \ll kv_e \tag{20}$$

Then we get the contribution from the ions, by analogy with equation 11, by replacing the electron mass m by the ion mass M. Adding this combination to right-hand side of equation 19 we have (on dropping the thermal contribution from the ions)

$$kE = \frac{k\omega_{pi}^2}{\omega^2} E - \frac{\omega_{pe}^2}{kv_e^2} E \tag{21}$$

or

$$1 = \frac{\omega_{pi}^2}{\omega^2} - \frac{\omega_{pe}^2}{k^2 v_e^2} \tag{22}$$

or

$$\omega^2 = \frac{\omega_{pi}^2}{1 + \omega_{pe}^2/k^2 v_e^2} = \frac{k^2 v_e^2 \omega_{pi}^2}{k^2 v_e^2 + \omega_{pe}^2} \tag{23}$$

or, alternatively,

$$\omega^2 = \frac{m}{M} \frac{k^2 v_e^2}{1 + k^2 \lambda_D^2} = \frac{k^2 c_s^2}{1 + k^2 \lambda_D^2} \tag{24}$$

where $c_s = \sqrt{m/M}\, v_e$ is the effective sound speed, and $\lambda_D = v_e/\omega_{pe}$ is the Debye length. This is the standard dispersion relation for ion acoustic waves. If $T_e \sim T_i$, then the ion thermal corrections are important. However, in this case, the assumption of equation 20 is not valid and, as we shall see, the wave is strongly damped by the ions. On the other hand, if $T_e \gg T_i$, then $c_s = \sqrt{T_e/T_i}\, v_i \gg v_i$, there are no resonant ions, and the wave can indeed propagate. To include the thermal effect of the ions, equation 24 should be multiplied by $(1 + 3T_e/T_i)$. This ion acoustic wave is the third MHD wave discussed in chapter 5.

10.3 The Dielectric Constant

Our discussion has been restricted in this chapter to collisionless plasmas. If collisions are present, we can have a sound wave even for $T_e = T_i$ if the collisional mean free path is shorter than the wavelength.

In our discussion of cold plasma waves we worked with the dielectric constant ϵ, while here we directly considered the electric field and the perturbed charge density. What is the relation between these two pictures? What is the dielectric constant for a thermal electron plasma? Here j is in the x direction and given by the continuity equation

$$ikj_1 + i\omega n_{e1} e = 0 \tag{25}$$

From equation 10 in chapter 9 we have

$$\epsilon = 1 + \frac{4\pi i j}{\omega E} = 1 - \frac{4\pi i n_1 e}{kE} \tag{26}$$

The Poisson equation 3 can be written

$$ikE\left(1 + \frac{4\pi n_1 e}{ikE}\right) = ik\epsilon E \tag{27}$$

Comparing this with equation 9, and adding the ion term, we have

$$\epsilon(\omega, k) = 1 + \frac{4\pi e^2}{km} \int \frac{\partial f_e/\partial v}{\omega - kv} d^3 v + \frac{4\pi e^2}{kM} \int \frac{\partial f_i/\partial v}{\omega - kv} d^3 v \tag{28}$$

The electrostatic dispersion relation for ω is thus,

$$\epsilon(\omega, k) = 0 \tag{29}$$

Can we understand the mechanism of the ion acoustic mode in simple terms? It turns out that in the limit on v_ϕ of equation 20, the electrons behave isothermally. If we substitute the approximation of equation 18 in equation 8 we have

$$f_1 = \frac{ieE}{kmv_e^2} f_0 = \frac{n_1}{n_0} f_0 \tag{30}$$

(because the ratio f_1/f_0 is independent of v), so the perturbed electron pressure is

$$p_1 = \frac{n_1}{n_0} p_0 = n_1 T_0 \tag{31}$$

The electron fluid equation (which is the velocity moment of the Vlasov equation) reads

$$n_0 m \frac{\partial \mathbf{v}_1}{\partial t} = -i\mathbf{k}p_1 - n_0 e\mathbf{E} \tag{32}$$

But the left-hand side is small, so

$$n_1 T_0 = p_1 = i\frac{n_0 eE}{k} = \frac{in_0 e(-ik\phi)}{k} = n_0 e\phi \tag{33}$$

where $E = -\nabla\phi$ and ϕ is the electrostatic potential. Thus,

$$\frac{n_1}{n_0} = \frac{e\phi}{T} \tag{34}$$

and the electron density distribution follows a Boltzmann law even though there are no collisions to enforce them (see problem 5). The Boltzmann distribution also follows from equation 30.

The linearized ion equation of motion, neglecting ion pressure, is

$$\rho\frac{\partial \mathbf{v}}{\partial t} = n_i e\mathbf{E} = n_e e\mathbf{E} = -\nabla p_{1e} = -(\nabla n_{1e})T_e \tag{35}$$

where, if $k\lambda_D \ll 1$ we can use $n_i \approx n_e$. We eliminate E by equation 32 and use the isothermality of the electrons, also included in equation 32. This is exactly the fluid equation that we combined with the continuity equation in chapter 5 to derive the sound wave.

If we combine this fluid equation with the continuity equation,

$$\frac{\partial n_i}{\partial t} + \nabla \cdot (n_0 \mathbf{v}_1) = 0 \tag{36}$$

we find after differentiating equation 35 with respect to time and substituting for $\partial n_i/\partial t$

$$\rho\frac{\partial^2 \mathbf{v}_1}{\partial t^2} = n_0 T_e \nabla(\nabla \cdot \mathbf{v}_1) \tag{37}$$

which is the wave equation with the propagation speed $c_s = \sqrt{n_0 T_e / n_e M} = \sqrt{T_e/M}$.

Notice that the driving force is the electron pressure. This force does not act directly on the ions but drives a tiny displacement between the ions and electrons that produces an **E** that balances ∇p_e. It is this **E** that then acts on the ions. Its force on the ions is equal to $-\nabla p_e$, so it appears that $-\nabla p_e$ does act directly on the ions. The slight motion of the electrons necessary to produce this **E** is by Poisson's equation $ikE = 4\pi \delta n e$, so with $n_0 e E = -ikp_{1e} = -ikn_{1e}T_e$ we have $\delta n = n_1 k^2 \lambda_D^2$, where n_1 is the perturbed electron density and δn is the difference between the perturbed ion density and the perturbed electron density. Thus, if $k^2\lambda_D^2 \ll 1$ we may safely assume that the two perturbed densities are equal. This is an example of charge neutrality and is usually the case. If we neglect $k^2\lambda_D^2$ in the denominator of equation 24, we get the same result.

10.4 Landau Damping

Let us now return to the purely electron perturbed Vlasov equation and the Poisson equation that gave us equation 9, to see how Landau handled the singularity in the denominator $\omega = kv$. (In this section we replace the one-dimensional distribution function F by f and return to our convention of letting a capital F represent ions.)

In line with our discussion after equation 9 we expect the wave to be damped due to the resonance particles at the singular velocity ω/k. On the other hand, the resonance electrons will be trapped by the potential of the wave and their number will not decrease with time (as we shall see), so it is not reasonable to expect that both E and f_1 will decrease exponentially. Thus, there should be no purely exponentially decreasing eigenmodes as would be the case if ω had a negative imaginary part in equation 9. (Such a complex ω would of course remove the singularity in the integral.)

Thus, we expect the wave to evolve with many harmonics ω, so E and f_1, would more appropriately have Fourier expansions rather that pure exponential behavior. But even this generalization is not sufficient since f_1, near resonance, does not decrease at large times, so it does not even have a Fourier expansion. The more appropriate expansion is by a Laplace transform.

Although this is a standard technique appropriate for initial value problems, we do not invoke it directly. Instead, let us follow a more naive approach to see we how we are led to it. Suppose we did not know about Laplace transforms. How should we go about solving the plasma oscillation problem?

We start our wave at $t = 0$ with amplitude $E(x, 0)$, $f_1(v, x, 0)$ and solve the Vlasov–Poisson system in time to find the evolution of E and f_1. We can certainly Fourier analyze in k and even consider a single mode k:

$$f_1(v, x, t) \sim e^{ikx}, \qquad E(x, t) \sim e^{ikx} \tag{38}$$

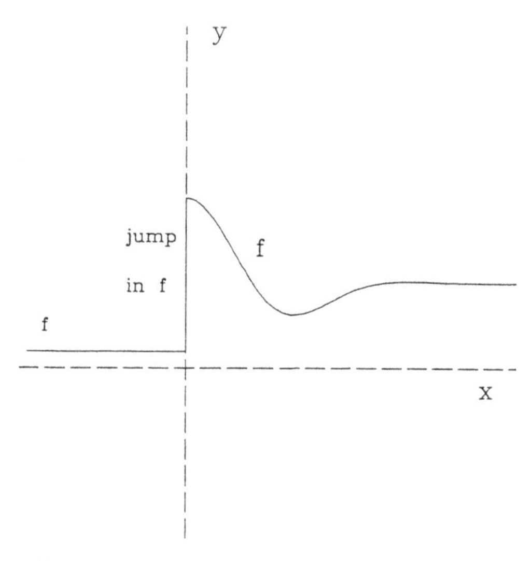

Figure 10.1. Extra term arising from the jump in f at $v = 0$.

with appropriate initial conditions at $t = 0$. Since we are interested only in $t > 0$, we take E and $f_1 = 0$ for $t < 0$ and let them jump to these initial values $E(0)$, $f_1(v, 0)$ abruptly at $t = 0$ as in figure 10.1.

Although, as discussed above, f_1 and E do not have Fourier integrals in time because f_1 does not decay in time, we can certainly multiply them both by functions $e^{-\lambda t}$. If λ is large enough, the resulting functions will have Fourier integrals. Therefore, let

$$f_\lambda(t, v) = f_1(t, v)e^{-\lambda t}$$

$$E_\lambda(t) = E(t)e^{-\lambda t} \tag{39}$$

Now we can Fourier analyze f_λ and E_λ in time

$$f_\lambda(v, t) = \int d\omega \bar{f}_\lambda(v, \omega)e^{-i\omega t}$$

$$E_\lambda(t) = \int d\omega \bar{E}_\lambda(\omega)e^{-i\omega t} \tag{40}$$

Once we find $\bar{f}_\lambda(v, \omega)$ and $\bar{E}_\lambda(\omega)$ we can, of course, recover $f(v, t)$ and E(t) by transforming and then multiplying by $e^{\lambda t}$, and

$$f_1(v, t) = e^{\lambda t} \int \bar{f}_\lambda(v, \omega)e^{-i\omega t} d\omega$$

$$E(t) = e^{\lambda t} \int \bar{E}_\lambda(\omega)d\omega e^{-i\omega t} d\omega \tag{41}$$

We find $\bar{f}_\lambda(\omega)$ and $\bar{E}_\lambda(\omega)$ from the Vlasov equation (2) and the Poisson equation (3). However, the f we are dealing with has a jump from zero to an initial value of $f(v,0)$ at the origin (see figure 10.1), so that this f satisfies the Vlasov equation only for $t \neq 0$. But, it does satisfy the Vlasov equation if we modify it by adding an inhomogeneous term $\delta(t)f(v,0)$ to the right-hand side.

Now substitute equations 41 into this modified Vlasov equation and the Poisson equation. Then the $\mathbf{v}\cdot\nabla$ operator becomes simply ikv_x, and $\mathbf{E}\cdot\nabla_v f_0$ becomes $e^{\lambda t}\int \bar{E}_\lambda e^{-i\omega t}(\partial f_0/\partial v)dv$. The $\partial/\partial t$ operator is slightly more complicated and gives $(\lambda - i\omega)\bar{f}_\lambda - (1/2\pi)f(v,0)$. The result is

$$\int \left[-i\omega \bar{f}_\lambda + \lambda \bar{f}_\lambda - \frac{f_\lambda(0,v)}{2\pi} + ikv\bar{f}_\lambda - \frac{e}{m}\bar{E}_\lambda \frac{\partial f_0}{\partial v} \right] e^{-i\omega t} d\omega = 0 \qquad (42)$$

This is satisfied only if each ω term in the integrand vanishes so $\bar{f}$ satisfies

$$-i\omega\bar{f}_\lambda + \lambda\bar{f}_\lambda + \frac{1}{2\pi}\bar{f}_\lambda(0,v) + ikv\bar{f}_\lambda = \frac{e}{m}\bar{E}_\lambda\frac{\partial f_0}{\partial v} \qquad (43)$$

The Poisson equation transforms into

$$ik\bar{E}_\lambda = -4\pi e \int \bar{f}_\lambda dv \qquad (44)$$

The solution for $\bar{f}_\lambda$ is

$$\bar{f}_\lambda = \frac{e}{m}\frac{i\bar{E}_\lambda \partial f_0/\partial v}{\omega + i\lambda - kv} + \frac{1}{2\pi}\frac{if_\lambda(0,v)}{\omega + i\lambda - kv} \qquad (45)$$

Substituting this into the Poisson equation we have

$$\bar{E}_\lambda + \frac{4\pi e^2}{km}\int \frac{\partial f_0/\partial v}{(\omega + i\lambda) - kv} dv \bar{E}_\lambda = \frac{4\pi e}{k}\int \frac{f_\lambda(0,v)}{(\omega+i\lambda) - kv}dv \equiv J(\omega + i\lambda) \qquad (46)$$

Comparing with the electron part of equation 28 we see that the factor multiplying E on the left-hand side of equation 46 is $\epsilon(\omega + i\lambda, k)$, so that we can write

$$\epsilon(\omega + i\lambda, k)\bar{E}_\lambda = J(\omega + i\lambda) \qquad (47)$$

where

$$\epsilon(\bar{\omega}) = 1 + \frac{4\pi e^2}{km}\int_{\mathrm{Im}\bar{\omega}=\lambda} \frac{\partial f_0/\partial v}{\bar{\omega} - kv} dv \qquad (48)$$

where $\bar{\omega} = \omega + i\lambda$.

Equation 47 gives us $\bar{E}_\lambda$ from which we can reconstruct $E_\lambda(t)$ and $E(t)$. Note that for $\lambda \neq 0$ the integrals are no longer singular. From equation 47 we have

$$E(t) = e^{\lambda t}\int d\omega e^{-i\omega t}\frac{J(\omega + i\lambda)}{\epsilon(\omega + i\lambda.k)} \qquad (49)$$

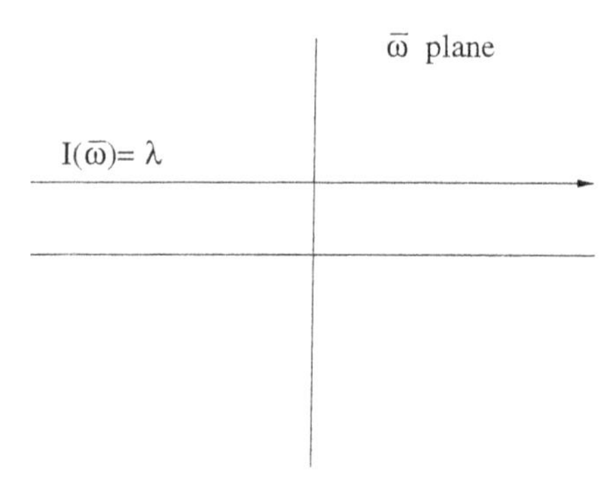

Figure 10.2. The contour of integration in the $\bar{\omega}$ plane

This is formally the same as the normal Fourier integral with the real transform variable replaced by complex variable $\bar{\omega} = \omega + i\lambda$, but the ω integration is still along the real axis. It is identical with the solution we get by the Laplace technique. If we transform the integration variable to $\bar{\omega} = \omega + i\lambda$, then

$$E(t) = e^{\lambda t} \int_{\text{Im }\omega=0} d\omega e^{-i\omega t} \frac{J(\bar{\omega})}{\epsilon(\bar{\omega}, k)} = \int_{\text{Im }(\bar{\omega})=\lambda} \frac{J(\bar{\omega})}{\epsilon(\bar{\omega}, k)} e^{-i\bar{\omega}t} \tag{50}$$

where the $\bar{\omega}$ integration is taken along the contour in figure 10.2, consisting of the straight line parallel to and above the real axis, Im $\omega = \lambda$.

$\epsilon(\bar{\omega}, k)$ is properly defined with no singularity in equation 48 when $\bar{\omega}$ is complex and in the upper half-plane. It is a well-defined analytic function, since the denominator of the integrand does not vanish. (If $f_0(v)$ is a Maxwellian, it, too, is an analytic function of v.)

Equation 50 is the solution for $E(t)$, but it is an awkward integral to perform since the exponential of the integrand grows at each point $\bar{\omega}$ at the rate λ. However, the entire integral is expected to decrease in time so there must be a great deal of cancellation. The situation is saved because the integrand is an analytic function of $\bar{\omega}$, so we can apply Cauchy's theorem to distort the contour of integration in the complex plane, as in figure 10.3. If we could distort the $\bar{\omega}$ contour into the lower $\bar{\omega}$ plane, the integrand would decrease exponentially at every point and this is what we want. However, to do this we must analytically continue $\epsilon(\bar{\omega}, k)$, as defined by the real v integral in equation 48, into the lower plane. If we try to continue the integrand in $\bar{\omega}$, we find that $\bar{\omega}/k$ crosses the real v axis and our singularity reappears. To avoid this we must make a second application of Cauchy's theorem to distort the velocity integral of equation 48, as in figure 10.4.

For any $\bar{\omega}$ in the lower half-plane, we can find the analytic continuation of ϵ by first distorting the velocity contour till it lies below this $\bar{\omega}/k$. Thus, $\epsilon(\bar{\omega}, k)$ is an analytic function of $\bar{\omega}$ everywhere, although its expression as a velocity integral depends on the particular value of $\bar{\omega}$ we choose. This is because the v contour cannot simultaneously lie below all values of $\bar{\omega}/k$. For

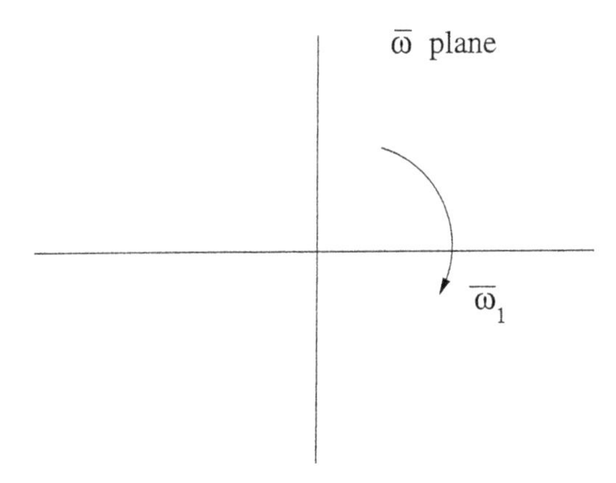

Figure 10.3. The analytic continuation of the integrand of $\epsilon(\bar{\omega}, k)$ into the lower half $\bar{\omega}$ plane

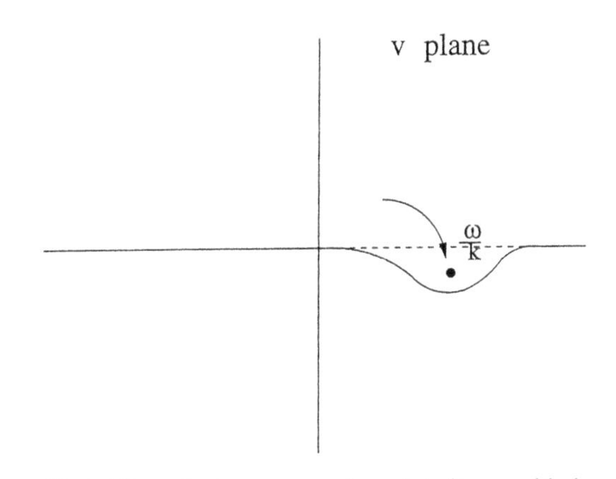

Figure 10.4. The velocity contour for $\epsilon(\omega)$ distorted below ω/k

any given v contour of integration $\epsilon(\bar{\omega}, k)$ is correctly given by equation 48 for all values of $\bar{\omega}$ such that $\bar{\omega}/k$ lies above this contour.

Now, given that $\epsilon(\bar{\omega})$ is analytic everywhere, we can apply Cauchy's theorem to the integral for $E(t)$, equation 50. We can distort the contour downward until the integrand encounters a singularity. This will happen when we reach an $\bar{\omega} = \bar{\omega}_1$ where $\epsilon(\bar{\omega})$ vanishes. Then we can loop the contour around $\bar{\omega}_1$ as in figure 10.5.

The vertical parts of the contour cancel, while the residue of the pole at $\bar{\omega}_1 = \omega_1 - i\mu$ is

$$2\pi i e^{-i\omega_1 t - \mu t} \left. \frac{J(\bar{\omega})}{\partial\epsilon/\partial\bar{\omega}} \right|_{\bar{\omega}=\bar{\omega}_1} \tag{51}$$

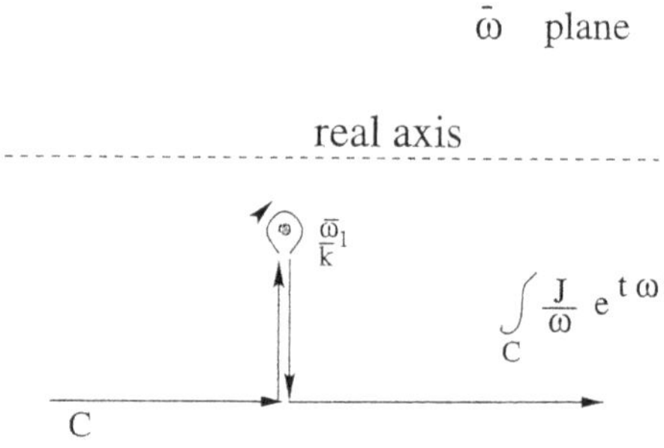

Figure 10.5. The contour looped over the singularity at $\bar{\omega}_1/k$

The time factor of the remaining part of the contour has a more negative imaginary part than this, and damps faster. Thus, $E(t)$ damps in time and for long times its behavior is dominated by the residue equation 51 at $\bar{\omega}_1$.

This discussion is a little complicated because we have to consider distorting the analytic v contour and the $\bar{\omega}$ contour. However, the result is simple. If ϵ has no zeros in the upper plane and a single zero in the lower plane at $\bar{\omega}_1$ or, at any rate, a zero at one point $\bar{\omega}_1$ with all the other zeros more negative (below the horizontal line in figure 10.5), then the long-time behavior of the E field is

$$E \sim e^{-i\bar{\omega}_1 t} = e^{-i\omega t - \mu t} \tag{52}$$

where $\bar{\omega}_1 = \omega - i\mu$. The solution for $\bar{\omega}_1$ such that $\epsilon(k, \bar{\omega}_1) = 0$ is just what we would naively expect, since it is normally what we would do to find the dispersion relation if there were no singularity to contend with. However, there are two differences from the nonsingular case. First, the mode is not a pure exponential with complex frequency $\bar{\omega}_1$, and, second, the definition of ϵ involves a distortion of the velocity contour that occurs in its definition. The velocity contour must lie below $\bar{\omega}_1/k$, where $\bar{\omega}_1$ is the real or complex root. Of course, to find $\bar{\omega}_1$ we need a rough idea of its location. In the case when the mode is weakly damped and $\bar{\omega}_1$ is close to the real axis this is comparatively easy to find.

Let us apply this result to our original problem of a plasma oscillation with resonant particles. However, let us choose k so that the wave is fast and the number of resonant particles is still small. Then, we can first neglect the effect of the resonant particles and use the zero-order solution for ϵ:

$$\epsilon \approx \epsilon_0(\omega_0) = 0 \tag{53}$$

From equation 13 this root is the Bohm–Gross root. Next, we examine ϵ near ω_0 more carefully. To continue ϵ to ω_0 and a little below we distort the velocity contour in equation 48 to a small semicircle below ω_0/k, as in figure 10.6. The integral along the real axis is real and very closely equal to the zero-order estimate. The integral around the semicircle is imaginary and

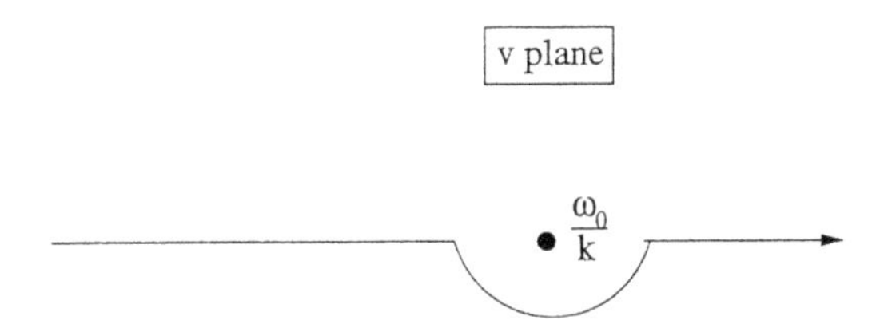

Figure 10.6. Velocity contour for a small imaginary part of ω

equal to

$$i\epsilon_2 = -\pi i \frac{4\pi e^2}{k^2 m} \left.\frac{\partial f_0}{\partial v}\right|_{\omega/k} \tag{54}$$

Thus, we expect a root ω of ϵ close enough to ω_0 that ω_0 is inside the semicircle and ϵ is properly defined by the integral. The equation for the root can be written

$$\epsilon_1(\omega_0 + \omega_1) + i\epsilon_2(\omega_0) = 0 \tag{55}$$

ϵ_2 is given by equation 54, and $\epsilon_1(\omega) = \epsilon_1(\omega_0) + \partial\epsilon/\partial\omega_0(\omega - \omega_0)$. Thus,

$$\omega - \omega_0 = \frac{-i\epsilon_2(\omega_0)}{\partial\epsilon_1/\partial\omega_0} \tag{56}$$

If we leave out the Bohm–Gross correction we can take

$$\epsilon_1 = 1 - \frac{\omega_{pe}^2}{\omega^2} \tag{57}$$

$\omega_0 = \omega_{pe}$, and

$$\frac{\partial\epsilon}{\partial\omega} = \frac{2\omega_{pe}^2}{\omega^3} = \frac{2}{\omega_{pe}} \tag{58}$$

Thus, denoting the imaginary part of ω by $i\gamma$ we have

$$i\gamma = \frac{\omega_{pe}}{2} \frac{4\pi^2 i e^2}{k^2} \left.\frac{\partial f_0}{\partial v}\right|_{\omega_0/k} \tag{59}$$

or

$$\gamma = \frac{2\pi^2 e^2}{k^2} \omega_{pe} \left.\frac{\partial f_0}{\partial v}\right|_{\omega_0/k} \tag{60}$$

If f is Maxwellian, then $\partial f/\partial v = -(v/v_e^2)f$, and

$$\frac{\gamma}{\omega_{pe}} = -2\sqrt{\pi}\frac{\omega_{pe}^3}{k^3 v_e^3} \exp -\left(\frac{\omega_{pe}^2}{2k^2 v_e^2}\right) = -\frac{2\sqrt{\pi}}{k^3 \lambda_D^3} e^{-1/2k^2\lambda_D^2} \tag{61}$$

This is very small if $k\lambda_D \ll 1$ or $\omega_{pe}/k \gg v_e$. Since $E \sim e^{-i\omega t + \gamma t}$ and γ is negative, the waves are damped, but the damping rate is small compared to ω, if $k\lambda_D \ll 1$.

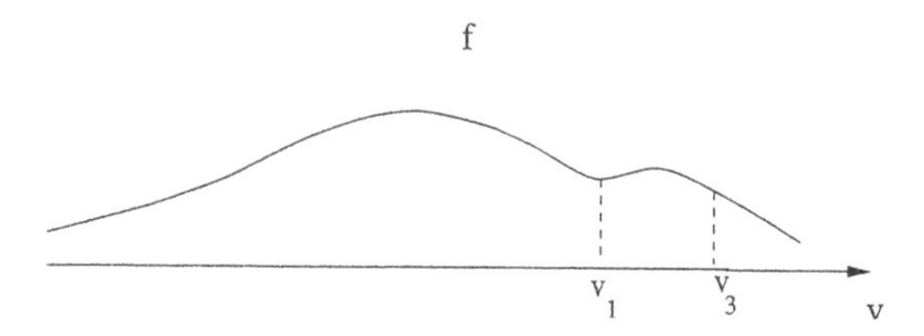

Figure 10.7. A bump in the Maxwellian distribution function

A little thought will show that if the distribution is approximately Maxwellian, but, in addition, has a bump in its distribution function between v_1 and v_2, then $\partial f_0/\partial v$ will be positive (see figure 10.7). Then waves propagating in the region of positive slope $v_1 < \omega/k < v_2$ will have Im $(\omega) > 0$, and the waves will grow at the rate given by equation 60. (For this case, Im $\omega > 0$ and there is no need to distort the velocity contour to properly define ϵ. In fact, the mode is purely exponential and there is no need at all for the Laplace transform or the contour distortion!)

Equation 56 has a further interpretation. The energy of an electrostatic wave is from chapter 9,

$$\mathcal{E}_{\text{wave}} = \frac{1}{16\pi}\omega\frac{\partial\epsilon}{\partial\omega}E^2 \tag{62}$$

Since the wave energy changes at the rate

$$\frac{d\mathcal{E}_{\text{wave}}}{dt} = 2\gamma\mathcal{E}_{wave} = \frac{-2\epsilon_2}{\partial\epsilon/\partial\omega}\frac{1}{16\pi}|E|^2\frac{\partial\epsilon}{\partial\omega} \tag{63}$$

we have

$$\frac{d\mathcal{E}_{\text{wave}}}{dt} = -\omega\epsilon_2\frac{|E|^2}{8\pi} \tag{64}$$

Since ϵ_2 is closely associated with the resonant particles we see that the wave growth comes directly from the energy of the resonant particles.

There is a heuristic way to understand why γ has the same sign as $\partial f_0/\partial v$ (Dawson 1961). The tendency of the wave is to slow down the resonant particles moving slightly faster than itself and to speed up those that are slower. If there are more of the latter ($\partial f_0/\partial v < 0$), then the wave loses more energy to slower particles than it gains from the faster particles and its energy decreases, i.e., it damps. Conversely, when the slope is positive it gains more energy from the faster particles than it loses to the slower particles and its energy grows. The result quoted this way can be confusing since the sign of the slope of f_0 does not always correlate directly with instability. We have restricted ourselves so far to positive-moving waves. For a negative-moving wave, $\partial f_0/\partial v > 0$ at its resonant velocity, v_r (see figure 10.8). But the more negative v particles are clearly more energetic and there are fewer of them, so the left-moving waves should also be damped. In fact, the ω_{pe} that appears in equation 60 should be the frequency of the wave, which is $-\omega_{pe}$ for a negative-moving wave. Thus, the heuristic picture is still valid,

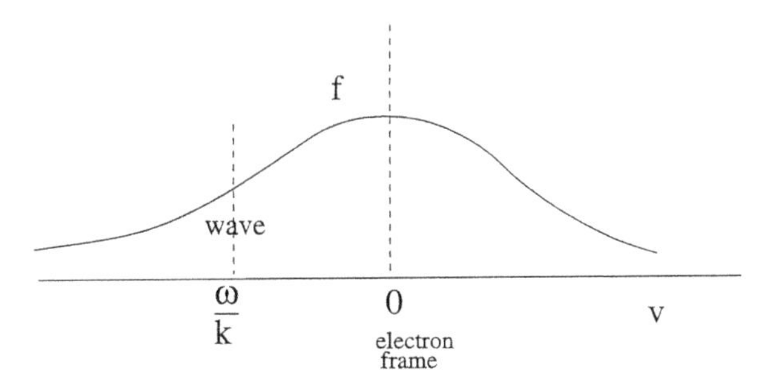

Figure 10.8. A left moving wave

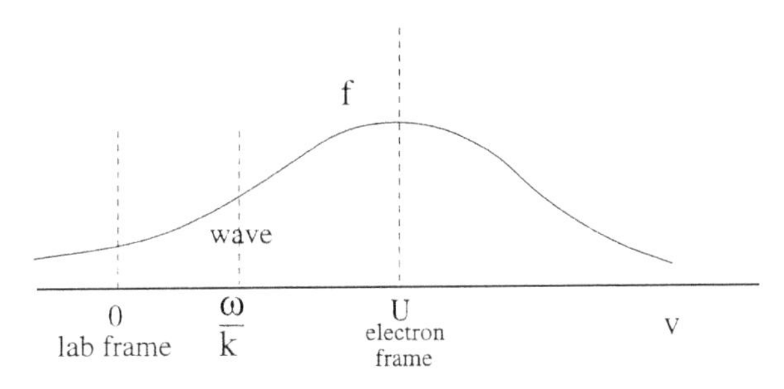

Figure 10.9. The left moving wave in a left moving frame

even though $\partial f_0/\partial v > 0$ at the resonant velocity, and the negative moving waves are damped.

However, if we go to a different velocity frame of reference the situation is still confusing, because we can find a frame where the left-moving wave in the laboratory frame is a right-moving wave in the frame with velocity U, as in figure 10.9. The contradiction is removed when we realize that the wave in this frame has negative energy. Thus, the energy that the more abundant particles give to the wave goes to increase its *negative* energy and consequently damps it. In fact, in the new laboratory frame (see problem 1 in chapter 9) $\omega = kU - \omega_{pe}$, and

$$\epsilon \approx 1 - \frac{\omega_{pe}^2}{(\omega - kU)^2} \tag{65}$$

so

$$\omega \frac{\partial \epsilon}{\partial \omega} = \omega \frac{2\omega_{pe}^2}{(\omega - kU)^3} = \frac{2\omega_{pe}^2 (kU - \omega_{pe})}{(-\omega_{pe})^3} < 0 \tag{66}$$

since the numerator is positive. Hence, in the moving frame the wave energy is actually negative.

Equation 54 can be easily generalized to include ions:

$$\epsilon_2 = \frac{4\pi^2 e^2}{k^2} \left[\frac{1}{M} \frac{\partial F_i(v)}{\partial v} + \frac{1}{m} \frac{\partial F_e(v)}{\partial v} \right]_{\omega/k} \tag{67}$$

For a plasma oscillation, the number of resonant ions at $v = \omega/k$ is exponentially small, since $\omega/k \gg v_i = \sqrt{T_i/M}$. However, for the ion acoustic wave there are many resonant ions. Substituting the ion and electron Maxwellians we get

$$\epsilon_2 = \frac{4\pi e^2}{k^2} \frac{n_0}{\sqrt{2\pi}} \left(\frac{v}{M v_i^3} e^{-v^2/2v_i^2} + \frac{v}{m v_e^3} e^{-v^2/2v_e^2} \right)_{v=\omega/k} \tag{68}$$

The electron exponential may be replaced by unity, so with $\omega = k\sqrt{T_e/M} = kc_s$

$$\epsilon_2 = -\frac{4\pi e^2}{k^2 T_e} \frac{n_0}{\sqrt{2\pi}} \left[\left(\frac{T_e}{T_i} \right)^{3/2} e^{-c_s^2/2v_i^2} + \sqrt{\frac{m}{M}} \right] c_s \tag{69}$$

The two factors in the brackets correspond to the amount of damping by the ions and electrons and we see that unless $c_s^2/2v_i^2 > \ln[(T_e/T_i)^{3/2}\sqrt{M/m}] = 3.8 + 1.5 \ln T_e/T_i$, the ion damping dominates the electron damping. That is, only if $T_i \ll T_e$ is the ion damping small. But in this case $c_s^2 = T_e/M$ and the criteria reduces to $T_e > 15T_i$. Thus, in most cases the ions dominate the damping.

This result is more interesting than we might at first think. In general, if there is a current in the plasma the electron distribution is shifted relative to that of the ions. Let us take the shifted electron distribution to be a shifted Maxwellian distribution with shift $u \ll v_e$. Then the bracketed expression is replaced by

$$\left(\frac{T_i}{T_e} \right)^{3/2} e^{-(T_e/2T_i)} - \sqrt{\frac{m}{M}} \frac{u - c_s}{c_s} \tag{70}$$

The electron interaction with the wave leads to growth if $u > c_s$ (Bernstein and Kulsrud 1961). However, the ion interaction damps the wave. If $u - c_s \approx c_s$, then, as before, the electrons can dominate only if $T_i < T_e/10$. If $T_i \approx T_e$ we need the second factor to be of order unity and thus $u \approx \sqrt{M/m} c_s \approx v_e$.

If the electrons do succeed in making the wave grow by their velocity shift, then the instability that arises is called the ion acoustic instability. However, it occurs only when either $T_i \approx T_e$ and the drift is as large as the electron thermal velocity or when $T_i < T_e/10$ and $u \approx 2c_s$. Thus, it is rather difficult to set up (at least in an unmagnetized plasma).

The reason that the ions usually dominate over the electrons is that the maximum velocity gradient of their distribution function is much larger for waves near c_s. This is because the resonant ions are near the maximum slope, while the resonant electrons are near the peak of their distribution function where the slope is near zero, $\partial f_e/\partial v \approx (c_s/v_e^2) f(0)$.

Let us estimate the growth or damping rate of ion acoustic waves in the various cases shown in figure 10.10. The energy of an ion acoustic wave is

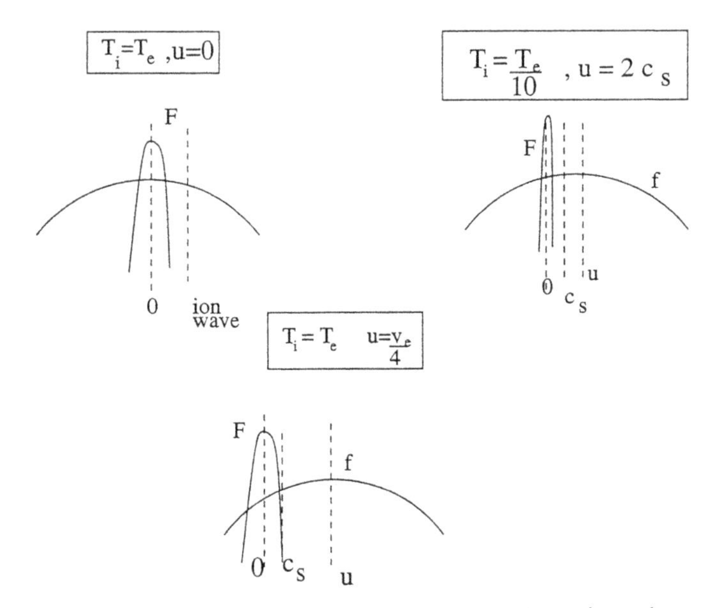

Figure 10.10. The ion acoustic wave for various values of u and T_e/T_i

twice the ion kinetic energy

$$\mathcal{E}_{\text{wave}} = 2\frac{M}{2}n\frac{|\tilde{v}|^2}{2} = \frac{ne^2}{2M\omega^2}|E|^2 \tag{71}$$

If $T_e \approx T_i$ and $u \ll v_e$, then we may take the bracketed expression in equation 69 to be of order unity, so

$$\epsilon_2 = \sqrt{\frac{1}{2\pi}\frac{4\pi n_0 e^2}{k^2 M c_s^2}} \tag{72}$$

Substituting this in equation 64 gives

$$2\gamma\mathcal{E}_{\text{wave}} = 2\gamma\frac{ne^2}{2M\omega^2}|E|^2 \approx -\omega\sqrt{\frac{1}{2\pi}\frac{4\pi n_0 e^2}{k^2 M c_s^2}}\frac{|E|^2}{8\pi} \tag{73}$$

or

$$\gamma = -\omega\sqrt{\frac{1}{8\pi}} \tag{74}$$

Thus, if $T_e \approx T_i$ and u is not too large, the wave is damped in about one wave period. In fact, ion acoustic waves (sound waves) will hardly propagate at all in a plasma with equal ion and electron temperatures. On the other hand, if $T_i \ll T_e$ we can neglect the ion Landau damping and keep only the electron damping. Then the bracketed expression is only about $\sqrt{m/M} \approx 1/40$ and the ion acoustic wave is much more weakly damped. It will propagate at least 40 periods with only a moderate decrease. If the electrons shift their velocity by u with $u - c_s \approx c_s$ it will grow weakly at the same rate.

However, the growth is weak only when compared to ω. In general, ω is very large compared to macroscopic rates and 40 periods is quite short. Thus, when we attempt to pass a current through a plasma with cold ions while increasing the current at a macroscopic rate, the ion acoustic mode will set in and grow rapidly. The consequence is actually not a blockage of the current, but rather a flattening of the electron distribution at c_s while the rest of the electrons successfully carry the current. If $u \gg c_s$, it turns out that, due to three-dimensional wave propagation, there is a serious blockage of the current. The consequences of instabilities due to resonant particle interaction are considered in the next chapter.

Landau damping and its inverse process, inverse Landau damping (i.e., growth), are probably the most characteristic properties of a plasma. They represent phenomena that are far from our natural experience when we consider fluid phenomena in astrophysics or in the laboratory. They usually lead to surprising and unexpected results. When Langmuir first passed a beam of electrons through a plasma in 1928 he found it did not pass through freely, but came to an abrupt stop in the plasma even though collisions were too weak to stop it. This result was referred to as the "Langmuir paradox" and its explanation remained a mystery for a long time. It was the first case of an instability associated with inverse Landau damping. The beam led to a very fast growth of plasma waves and a large reduction in its velocity.

10.5 Physical Picture of Landau Damping

It is unfortunate that the most direct approach to the evaluation of these damping rates leads us through a discussion of contour integration in the complex ω plane, since, thereby, any intuitive feeling for the physics of these processes is lost. Therefore, it seems useful to present an alternative picture that, while less efficient for calculation, does uncover underlying intuitive physics that is otherwise missed. These intuitive ideas are very useful when we proceed beyond the simplest models that yield to a simple analytic treatment, such as the damping of a single small-amplitude, monochromatic wave.

Before proceeding, let us make one remark: Although the Landau picture of seeking a zero of the dispersion relation $\epsilon(\omega) = 0$ as a complex function of ω by analytic continuation holds even if the damping is strong, the most familiar and easy case to treat is that in which the damping is weak and the root for ω is near the real axis. Then we can expand the equation $\epsilon = \epsilon_1 + i\epsilon_2 = 0$ in the limit that ϵ_2 is small compared to ϵ_1. The lowest-order approximation, ω_0, can be taken as a root of $\epsilon_1, \epsilon_1(\omega_0) = 0$, and its small correction $\omega_1 \approx i\epsilon_2/(\partial\epsilon_1/\partial\omega|_{\omega=\omega_0}$. Thus, implicitly we assume that the effect of the resonant particles is small.

This suggests starting the approximation earlier in the Vlasov–Poisson scheme. Solve for ω_0 and E as before, but in lowest order, neglecting the resonant particles that contribute to ϵ_2. Then treat their distribution function as a perturbation. We find that the wave acts on the resonant particles to bunch them, leading to a density perturbation in them that travels coherently with the unperturbed wave. Such a bunching leads to the emission of a secondary (growing) wave that is also coherent with the first unperturbed wave and either interferes destructively (damping) or constructively (growth) with the original wave. The secondary wave can be considered to be two waves generated separately by particles moving faster than the wave and particles moving slower. The part due to faster particles interferes constructively, while that due to the slower particles interferes destructively. If there are more slower particles than faster particles, then we have damping, and vice versa.

Let us carry out this picture explicitly. There are essentially two expansions. One is in the amplitude of the wave, which we consider infinitely small. The second is the fraction of resonant particles.

We divide the zero-order distribution into the nonresonant (a) and resonant (b) distribution functions

$$f^0 = f_a^0 + f_b^0 \tag{75}$$

and similarly

$$f^1 = f_a^1 + f_b^1 \tag{76}$$

with corresponding electric fields E_a and E_b. Let $f_a^0 + f_a^1$ and E_a be the solution to first order in the amplitudes, of the Vlasov–Poisson equations, corresponding to an undamped wave with no resonant particles. Let us use real notation and let

$$E_a = E^0 \cos(kx - \omega t) \tag{77}$$

Then the equation for f_b^1 is

$$\frac{\partial f_b^1}{\partial t} + v\frac{\partial f_b}{\partial x} = \frac{e}{m}E^0\frac{\partial f_b^0}{\partial v}\cos(kx - \omega t) \tag{78}$$

A particular solution is

$$f_b^1(v, x, t) = \frac{eE^0}{m}\frac{\partial f_b^0}{\partial v}\frac{\sin(kx - \omega t)}{kv - \omega} \tag{79}$$

A homogeneous solution of the left-hand side of equation 78 is

$$f_b^1 = \frac{eE_0}{m}\frac{\partial f_b^0}{\partial v}\sin k(x - vt) \tag{80}$$

We want the solution that vanishes at $t = 0$. This is

$$f_b^1 = \frac{eE^0}{m}\frac{\partial f_b^0}{\partial v}\frac{\sin(kx - \omega t) - \sin k(x - vt)}{kv - \omega} \tag{81}$$

Write the argument of the second sine as

$$(kx - \omega t) + (\omega t - kvt) = (kx - \omega t) - k(v - v_r)t \tag{82}$$

where $v_r = \omega/k$. Expanding this sine, we get

$$f_b^1 = \frac{eE^0}{m}\frac{\partial f_b^0}{\partial v}\left[\sin(kx-\omega t)\frac{1-\cos k(v-v_r)t}{k(v-v_r)}\right]$$

$$+\frac{eE^0}{m}\frac{\partial f_b^0}{\partial v}\left[\cos(kx-\omega t)\frac{\sin k(v-v_r)t}{k(v-v_r)}\right] \tag{83}$$

To find the density we integrate this over velocity. We are mainly interested in the part of the density in phase with E^0, the cosine part, so we temporarily drop the first part of f_b^1. For kv_et large compared with unity, the sine factor oscillates rapidly with v except for v near v_r, so we may treat $\partial f_b^0/\partial v$ as constant. The velocity integral of $\sin k(v-v_r)t/(v-v_r)$ is π, so the part of the perturbed resonant particle density coherent with $\cos(kx-\omega t)$ is

$$n_b = \frac{\pi e E^0}{mk}\left.\frac{\partial f_b^0}{\partial v}\right|_{v=v_r}\cos(kx-\omega t) \tag{84}$$

This is the bunched charge. (By equation 54 it is just $ke/4\pi$ times the imaginary part of the dielectric constant ϵ_2. This is not surprising considering the way in which it was derived.)

Now n_b is a bunched charge that should emit a plasma oscillation E_b^1 in addition to E^0. If we employ the dielectric constant for the nonresonant electrons ϵ_a to find this wave, we have

$$ik\epsilon_a(\omega,k)E_b = -4\pi n_b e \tag{85}$$

But, since $\epsilon_a(\omega,k) = 0$, this does not work. The inhomogeneous term is resonant with the homogeneous solution. The simplest way to remedy this is to take ω as the argument of n_b to be a value ω' slightly different than ω and let it approach ω.

Then for $\omega' \neq \omega$ an inhomogeneous solution is

$$E_b^1 = -\frac{4\pi e\hat{n}_b(\omega',k)}{k\epsilon_a(\omega',k)}\sin(kx-\omega' t) \tag{86}$$

where $\hat{n}_b$ is the amplitude of n_b in equation 84. But we want a solution that vanishes at $t=0$, so we subtract a constant times the homogeneous solution $\sin(kx-\omega t)$. Thus,

$$E_b = \frac{4\pi\hat{n}_b e\left[\sin(kx-\omega' t)-\sin(kx-\omega t)\right]}{k\epsilon_a(\omega',k)} \tag{87}$$

The limit $\omega' \to \omega$ is obtained by ℓ'Hôpital's rule. Take the ratio of the limit of the derivatives of the numerator ω and denominator as ω' goes to ω

$$E_b = \frac{4\pi e\hat{n}_b(\omega,k)t\cos(kx-\omega t)}{k\partial\epsilon_a/\partial\omega} \tag{88}$$

or

$$E_b = \gamma t E^0\cos(kx-\omega t) = \gamma t E_a \tag{89}$$

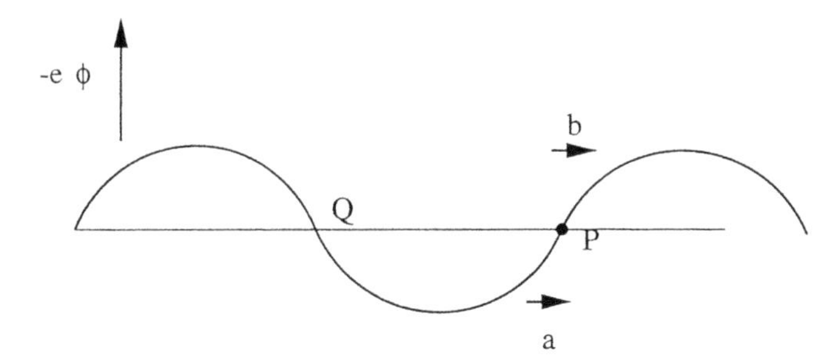

Figure 10.11. Faster electrons bunch at P and slower ones at Q

where γ is given by equation 60. We get this final result by employing equation 84 for n_b and equation 54 for ϵ_2.

The γt correction to E_a is just the first-order correction to the exponential. Thus,

$$E_a + E_b = E_a(1 + \gamma t) \approx E_a e^{\gamma t} \tag{90}$$

γ is, of course, negative for $\partial f_0/\partial v < 0$.

This verifies our picture. The other part of n_b, the $\sin(kx - \omega t)$ part, also emits a wave that can be shown to be $E^0 \Delta\omega t \sin(kx - \omega t)$, which is coherent with E_a. $E_a + E_b \sim \cos(kx - \omega t) + \Delta\omega t \sin(kx - \omega t) \sim \cos[kx - (\omega + \Delta\omega)t]$ and represents the correction in the solution of dispersion relation to the real part of ω.

It is interesting to examine the bunching of the resonant particles in the frame of the wave. At $t = 0$, the potential for the electrons $-e\phi = (E/k)$ $\sin(kx')$, where $x' = x - \omega t/k$.

Let us look at slow electrons in the wave frame near the point P in figure 10.11, where $kx' = 0$, i.e., where E is a maximum. For very small $v > 0$, an electron's position is decreased after a time t by $x_1 = e\langle E(x'')\rangle t^2/2$ from the position it would have occupied if E were absent. $\langle E(x)\rangle$ is the average of $E(x)$ along the unperturbed orbit. Let us compare this shift for an electron just arriving at P, electron a, with an electron b, which would have gone a little further in x. Inspection of the diagram shows that $\langle E(x')\rangle$ is larger for electron b, since the average of the slope of the potential is larger, so its shift backward is larger. This means that at t the electrons are closer together, so the density is greater.

A similar argument at Q, where $kx' = -\pi$, shows that the density is less. Thus, right-moving electrons tend to concentrate at P. The same argument (or a symmetry argument) shows that electrons with $v' < 0$ concentrate at Q. Thus, if there are more negative velocity electrons in the wave frame, then the part of the perturbed density of the resonant particles coherent with the electric field, i.e., proportional to $\cos(kx') = \cos(kx - \omega t)$, is a negative constant times $\cos(kx - \omega t)$. This is consistent with equation 84.

To summarize this intuitive picture, faster resonant particles concentrate at P and emit a wave that interferes constructively with the original wave E^a. Slower resonant particles concentrate at Q and emit a wave that interferes destructively with E^a. According to whether the fast or slow electrons predominate, the original wave grows or is damped. This is reminiscent of the concept of normal or inverted populations in an atom and the growth is similar to maser action in the latter case.

10.6 Types of Resonances

There are two types of resonance of particles with waves. The first is the Cerenkov resonance

$$\omega = kv \tag{91}$$

where the particle travels in near resonance with the wave. This is the case we have just discussed in detail in connection with Landau damping (or growth) of plasma oscillations. It can occur with electrons in a plasma oscillation, or with both electrons and ions in an ion acoustic wave.

The name Cerenkov comes from the phenomena of Cerenkov emission. This is the resonance in a refractive medium where light propagates slower than c and resonates with an energetic particle moving with a speed near c. This resonance then leads to the Cerenkov emission of light (Jackson 1999).

The wave generally propagates at an angle to the particle orbit and the condition is more correctly

$$kv\cos\theta = \omega \tag{92}$$

where θ is the angle between the wave vector k and the particle orbit. In a plasma the particle orbit is usually constrained to move along a background magnetic field.

The second type of resonance is the cyclotron resonance that occurs for magnetized particles when

$$\omega - k_z v_z \approx \pm\Omega \tag{93}$$

where z is along B. The resonance phenomenon is closely analogous to that of Landau damping. The resonance condition again depends on v_z. Faster particles tend to bunch their densities at a gyroangle such that there is a current rotating coherently with the electric vector rotating in the same direction as the particle. Their bunched charge or current then emits a secondary wave that interferes constructively with the original wave. Slower particles bunch at a gyroangle, which differs from E by 180 degrees, and they emit a wave that interferes destructively with the original wave. Accordingly, if the density of faster particles is greater (less) than the density of slower particles, then we have growth (damping).

The resonant interaction is twofold: it changes the amplitude of the wave and it rearranges the distribution function in the resonant region. The latter is usually the more important effect and will be treated in the next chapter on nonlinear processes. Note that resonant phenomena, when present, produce strong effects even for small amplitude waves. They dominate over slower phenomena such as collisions. For example, resonance can increase the resistivity of a plasma by a large amount and have a large effect on the phenomena of magnetic reconnection (see chapter 14).

10.7 The Drift Kinetic Equation

The MHD equations rely on collisions to keep the particles in the plasma localized, so we can evolve the plasma quantities by the conservation relations. That is, it is assumed that $\lambda \ll L$ and $\nu \gg 1/t$. There are phenomena where these conditions are not adequately satisfied, but if a magnetic field is present, the gyroradius may be so small that particles cannot move very far from each other in the direction perpendicular to the magnetic field. In this case, the kinetic behavior of the plasma is restricted to one-dimensional collisionless flows of the particles along each line. Qualitatively, we can think of the plasma as being divided into tubes along the field lines, such that all the particles in a tube stay the same, each particle acting like a bead on a string, which is actually the moving magnetic field line. However, the particles move freely and independently of each other along the tube. The perpendicular velocities are also restricted by the magnetic moment adiabatic invariant, and these velocities are axisymmetricly distributed about the line once the mean motion of the line is subtracted from them. The gyromotion also averages out so that the free variables for a particle are its parallel motion $q = \mathbf{v} \cdot \mathbf{b}$ and its position ℓ along the line L. The perpendicular motion of the line is still governed by MHD forces, pressure, magnetic tension, etc. How does this simplified picture of a magnetized collisionless plasma arise?

The derivation of such a picture is based on an expansion in the smallness of the gyroradius, the inverse of the gyrofrequency, the Debye length, and the inverse of the plasma frequency. This was first carried out by Chew, Goldberger, and Low in an unpublished report presented at a summer school at Los Alamos in 1955. If we examine these quantities, we see that limits arise formally if we simply consider the electronic charge e to be large and expands the distribution functions f the electric field $\mathbf{E}$ and the magnetic field $\mathbf{B}$ in $1/e$. Rather than provide a justification for this procedure we simply formally carry out this expansion and show that it leads to the limits we expect.

First, consider one of the Vlasov equations for either species, say, for the ions:

$$\frac{\partial f}{\partial t} + \mathbf{v} \cdot \nabla f + \frac{e}{M}\left(\mathbf{E} + \frac{\mathbf{v} \times \mathbf{B}}{c}\right) \cdot \nabla_{\mathbf{v}} f = 0 \qquad (94)$$

$\mathbf{E}$, $\mathbf{B}$, and f will be expanded in $1/e$, e.g., $f = f_0 + f_1/e + f_2/e^2 + \cdots$, etc. Then to lowest order in $1/e$, the Vlasov equation (94) is

$$\left(\mathbf{E}_0 + \frac{\mathbf{v} \times \mathbf{B}_0}{c}\right) \cdot \nabla_{\mathbf{v}} f_0 = 0 \tag{95}$$

Now, dotting this equation with $\mathbf{b}_0 = \mathbf{B}_0/B_0$ we get $(\mathbf{E}_0 \cdot \mathbf{b}_0)\partial f_0/\partial v_\parallel = 0$ and f_0 is constant in $v_\parallel$ for fixed $v_\perp$, which means that f_0 is finite at infinite $v_\parallel$. This is impossible and, therefore, we must have $E_{0\parallel} = \mathbf{E}_0 \cdot \mathbf{b}_0 = 0$. (Physically, any such unbalanced $E_{0\parallel}$ would lead to rapid acceleration, which would short out the parallel electric field to lowest order. $\mathbf{E}_\parallel$ can be nonzero in first order, however.)

Since $\mathbf{E}_{0\parallel} = 0$ we can write

$$\mathbf{E}_0 = -\frac{\boldsymbol{\alpha} \times \mathbf{B}}{c} \tag{96}$$

where $\boldsymbol{\alpha}(\mathbf{r}, t)$ is the usual $\mathbf{E} \times \mathbf{B}$ drift velocity, so equation 95 becomes

$$(\mathbf{v} - \boldsymbol{\alpha}_0) \times \mathbf{B}_0 \cdot \nabla f_0 = (\mathbf{s} \times \mathbf{B}_0) \cdot \nabla_{\mathbf{v}} f_0 = 0 \tag{97}$$

We have introduced $\mathbf{s}$, the perpendicular velocity in the moving frame, so that

$$\mathbf{v} = \boldsymbol{\alpha}_0 + \mathbf{s} + q\mathbf{b} \tag{98}$$

Now, equation 97 involves only the velocity variables $\mathbf{s}$, so introducing cylindrical coordinates in velocity space, $|s|$, ϕ, q, we easily see that equation 97 reduces to

$$\frac{\partial f_0(t, \mathbf{r}, |\mathbf{s}|, \phi, q)}{\partial \phi} = 0 \tag{99}$$

so that f_0 is independent of ϕ, but is still a function of t, $\mathbf{r}$, $|\mathbf{s}|$, and q, or

$$f_0 = f_0(t, \mathbf{r}, w, q) \tag{100}$$

where $w = s^2/2$. To lowest order all we can say is that f_0 is an arbitrary function of t, $\mathbf{r}$, w, q.

To find this function we proceed to next order in $1/e$:

$$\frac{\partial f_0}{\partial t} + \mathbf{v} \cdot \nabla f_0 - \frac{eB}{Mc}\frac{\partial f_1}{\partial \phi} = 0 \tag{101}$$

This equation is deceptively simple. Indeed, f_0 is a function of quantities that are functions of t, $\mathbf{r}$, and $\mathbf{v}$, while the left-hand side involves derivatives at fixed $\mathbf{v}$. Thus, to evaluate the left-hand side we need to employ the chain rule of differentiation. For example,

$$\left(\frac{\partial f_0}{\partial t}\right)_{\mathbf{r},\mathbf{v}} = \left(\frac{\partial f_0}{\partial t}\right)_{w,q,\mathbf{r}} + \left(\frac{\partial f_0}{\partial w}\right)_{\mathbf{r},q,t}\left(\frac{\partial w}{\partial t}\right)_{\mathbf{r},\mathbf{v}} + \left(\frac{\partial f_0}{\partial q}\right)_{\mathbf{r},w,t}\left(\frac{\partial q}{\partial t}\right)_{\mathbf{r},\mathbf{v}} \tag{102}$$

where we have

$$\left(\frac{\partial w}{\partial t}\right)_{\mathbf{rv}} = \mathbf{s} \cdot \left(\frac{\partial \mathbf{s}}{\partial t}\right)_{\mathbf{rv}} = \mathbf{s} \cdot \frac{\partial}{\partial t}\left(\mathbf{v} - \boldsymbol{\alpha} - \mathbf{v} \cdot \mathbf{bb}\right)_{\mathbf{rv}} = -\mathbf{s} \cdot \frac{\partial \boldsymbol{\alpha}}{\partial t} - q\frac{\partial \mathbf{b}}{\mathbf{s} \cdot \partial t} \tag{103}$$

and, similarly,

$$\frac{\partial q}{\partial t} = \mathbf{v} \cdot \frac{\partial \mathbf{b}}{\partial t} = (\boldsymbol{\alpha} + \mathbf{s}) \cdot \frac{\partial \mathbf{b}}{\partial t} \tag{104}$$

In this latter equation, $\mathbf{v} = \boldsymbol{\alpha} + \mathbf{s} + q\mathbf{b}$ has been inserted.

The resulting equation for f_1, after everything has been reduced to the independent variables t, $\mathbf{r}$, $\mathbf{s}$, and q, is very long. However, to be able to solve it for f_1 we must be certain that the left-hand side averaged over ϕ must vanish. Integrating this equation over ϕ and using $\oint d\phi \mathbf{s} = 0$, and $\oint \mathbf{s}\mathbf{s} d\phi / 2\pi = (\mathbf{I} - \mathbf{b}\mathbf{b})w$ we find an equation that involves only f_0,

$$\frac{\partial f_0}{\partial t} + (\boldsymbol{\alpha} + q\mathbf{b}) \cdot \nabla f_0 - -w \left(\nabla \cdot \boldsymbol{\alpha} - \mathbf{b} \cdot \nabla \boldsymbol{\alpha} \cdot \mathbf{b} + q \nabla \cdot \mathbf{b} \right) \frac{\partial f_0}{\partial w}$$
$$+ \left(-\mathbf{b} \cdot \left[\frac{\partial \boldsymbol{\alpha}}{\partial t} + (\boldsymbol{\alpha} + q\mathbf{b}) \cdot \nabla \boldsymbol{\alpha} \right] + w(\nabla \cdot \mathbf{b}) + \frac{e}{M} \mathbf{E}_{1\|} \right) \frac{\partial f_0}{\partial q}$$
$$= 0 \tag{105}$$

The independent variables are two velocity variables, q and w, three space variables, $\mathbf{r}$, and time t. The quantities $\boldsymbol{\alpha}$, $\mathbf{b}$, and $E_{1\|}$ depend on $\mathbf{r}$ and t, $\boldsymbol{\alpha}(\mathbf{r}, t)$, $\mathbf{b}(\mathbf{r}, t)$, and $E_{1\|}(\mathbf{r}, t)$, and the dependence of all but the last, comes from the space and time dependence of $\mathbf{E}_0$ and $\mathbf{B}_0$.

This equation becomes simpler and more transparent if we replace the independent variable w by the magnetic moment, $\mu = w/B$. This removes the $\partial f_0 / \partial w$ term,

$$\frac{\partial f_0}{\partial t} + (\boldsymbol{\alpha} + q\mathbf{b}) \cdot \nabla f_0 + \left[-\mathbf{b} \cdot \frac{d\boldsymbol{\alpha}}{dt} + w(\nabla \cdot \mathbf{b}) + \frac{e}{M} \mathbf{E}_{1\|} \right] \frac{\partial f_0}{\partial q} = 0 \tag{106}$$

where $d\boldsymbol{\alpha}/dt = \partial \boldsymbol{\alpha}/\partial t + \boldsymbol{\alpha} + q\mathbf{b} \cdot \nabla \boldsymbol{\alpha}$ is the convective derivative of $\boldsymbol{\alpha}$.

When we use $\nabla \cdot \mathbf{B} = \nabla \cdot (\mathbf{b}B) = B\nabla \cdot \mathbf{b} + \mathbf{b} \cdot \nabla B = 0$ we see that the $\mu B \nabla \cdot \mathbf{b}$ term equals $-\mu \mathbf{b} \cdot \nabla B$ and is just the magnetic mirror force on a particle. $\boldsymbol{\alpha}$ is the perpendicular motion of the plasma that can be thought of as the motion arising from the freezing of the plasma on the field lines moving with velocity $\boldsymbol{\alpha}$ by equation 96. If we were to further transform to the moving magnetic field line coordinates, we would find that equation 106 is actually a one-dimensional kinetic equation for the dependence of f_0 on q and ℓ, the distance measured along the line L for particles with a fixed μ.

From the coefficients of the Vlasov equation in the form of equation 106 we get the equation of motion of the particles in the small gyroradius limit:

$$\dot{q} = -\mathbf{b} \cdot \frac{d\boldsymbol{\alpha}}{dt} - \mu \mathbf{b} \cdot B + \frac{e}{M} E_{1\|} \mathbf{b} \tag{107}$$

$$\dot{\mathbf{r}} = \boldsymbol{\alpha} + q\mathbf{b} \tag{108}$$

The first term in $\dot{q}$ is the inertial term due to nonuniform motion of the field lines. It represents the acceleration from the line motion due to the fact that particles are constrained to stay on the line. The second term is the magnetic mirror force, and the third term is the force due to the first-order parallel electric field. The terms in $\dot{\mathbf{r}}$ are obvious.

Now, to complete the drift kinetic scheme, we must turn to Maxwell's equations and expand them in $1/e$ also. The Poisson equation is

$$\nabla \cdot \mathbf{E} = \Sigma 4\pi e_j \int f_{0j} d^3\mathbf{v} \tag{109}$$

where the subscript j refers to the species. To lowest order this gives

$$\Sigma 4\pi e_j \int f_{0j} d^3\mathbf{v} = 0 \tag{110}$$

which is just the charge neutrality condition.

Ampere's law (without displacement current) reads

$$\nabla \times \mathbf{B} = \Sigma 4\pi e_j \int f_{0j} \frac{\mathbf{v}}{c} d^3\mathbf{v} \tag{111}$$

which to lowest order is

$$\Sigma 4\pi e_j \int f_{0j} \frac{\mathbf{v}}{c} d^3\mathbf{v} = 0 \tag{112}$$

which is electric current neutrality. To complete Maxwell's equations we must go to first order in these equations:

$$\nabla \cdot \mathbf{E}_0 = \Sigma 4\pi e_j \int f_{1j} d^3\mathbf{v} \tag{113}$$

and

$$\nabla \times \mathbf{B}_0 = \Sigma 4\pi e_j \int f_{1j} \frac{\mathbf{v}}{c} d^3\mathbf{v} \tag{114}$$

and for this we must solve equation 10 for f_1. However, we need only the velocity moment of f_1 to get the current in equation 114 and we never need equation 113. We can therefore simply take the velocity moment of equation 101 or, what is really simpler, take the velocity moment of the exact equation 94 and take the first-order part. We do this for each species and add. The result is

$$\rho \left(\frac{\partial \mathbf{V}}{\partial t} + \mathbf{V} \cdot \nabla \mathbf{V} \right) + \nabla \cdot \mathcal{P} = \mathbf{j} \times \mathbf{B} \tag{115}$$

where we need only the pressure tensor $\mathcal{P}$ to lowest order. ρ is the ion density and $\mathbf{V}$ its mean velocity.

The pressure tensor in the $\boldsymbol{\alpha}$ frame is

$$\mathcal{P} = \Sigma m_j \int f_{0j} (\mathbf{s} + q\mathbf{b})(\mathbf{s} + q\mathbf{b}) d^3\mathbf{v} = p_\perp (\mathbf{I} - \mathbf{bb}) + p_\parallel \mathbf{bb} \tag{116}$$

where

$$p_\perp = \Sigma m_j \int f_{0j} w d^3\mathbf{v}, \quad p_\parallel = \Sigma m_j \int f_{0j} q^2 d^3\mathbf{v} \tag{117}$$

Only the perpendicular part of j is given by equation 115, but the parallel part can be obtained from $\nabla \cdot \mathbf{j} = 0$.

All this can be turned around to look more like MHD. Treat equation 115 as the fundamental equation rather than just the equation for j, and treat **V** as a fundamental variable. j is then given by Ampere's law,

$$\nabla \times \mathbf{B} = 4\pi \mathbf{j} \tag{118}$$

B is found from the induction equation together with $\mathbf{E} = -\boldsymbol{\alpha} \times \mathbf{B}/c = -\mathbf{V} \times \mathbf{B}/c$ so that

$$\frac{\partial \mathbf{B}}{\partial t} = \nabla \times (\mathbf{V} \times \mathbf{B}) \tag{119}$$

$\mathcal{P}$ is found from equations 116 and 117 in which $p_\perp$ and $p_\parallel$ arise from the two f_0's, which are given by the kinetic equations 106. Finally, in the kinetic equation $E_{1\parallel}$ is determined to satisfy the charge neutrality condition, equation 110. $\boldsymbol{\alpha}$ is, of course, given by the perpendicular part of **V**, which is now a fundamental dependent variable. We thus arrive at a complete picture that is an extension of the MHD picture. The main difference is that the pressure is anisotropic and is given by solving the one-dimensional kinetic equation, which properly takes into account the large heat flow along the lines in the absence of collisions.

This gyroradius expansion of the Maxwell–Vlasov system appears in many places in the literature. Two such places are Kulsrud (1964) and Kulsrud (1983). The basic physical ideas are most clearly presented in Kruskal (1960).

10.8 Problems

1. (The Z function) When the unperturbed distribution function is a Maxwellian the dielectric constant is usually rescaled and expressed in terms of the so-called Z function

$$Z(z) = \frac{1}{\sqrt{\pi}} \int_{-\infty}^{+\infty} \frac{e^{-u^2} du}{u - z}$$

where z is a complex variable. As in the text, the u integral is taken along the real axis when z is in the upper complex plane, and is distorted below z to carry out the analytic continuation. Show from equation 28 that the dielectric constant for a Maxwellian plasma can be written as

$$\epsilon = 1 - \frac{1}{2k^2\lambda_D^2}\left[Z'(z_e) + \frac{T_e}{T_i} Z'(z_i)\right]$$

where $\lambda_D^2 = T_e/4\pi ne^2$, Z' is the derivative of $Z(z)$ with respect to z, and

$$z_e = \frac{\omega}{\sqrt{2}kv_e}, \quad z_i = \frac{\omega}{\sqrt{2}kv_i}$$

with $v_e = \sqrt{T_e/m}$, $v_i = \sqrt{T_i/m}$.

Show that for large z, $Z \approx -1/z$, $Z' \approx 1/z^2$, and then show for electron plasma oscillations that as the electron temperature goes to zero ϵ reduces to the cold plasma result.

Show that

$$Z'(z) = -\frac{2}{\sqrt{\pi}} \int \frac{ue^{-u^2}du}{u-z}$$

and $Z'(0) = -2$. Derive the dispersion relation for ion acoustic waves in the limit $v_i \ll \omega/k \ll v_e$, from $\epsilon(\omega, k) = 0$ (Akhiezer et al. 1967; Stix 1962, 1992).

2. (The dielectric constant for a moving plasma) Take the velocity distribution function for a moving unmagnetized plasma as

$$f_e(\mathbf{v}) = f_{e0}(\mathbf{v} - \mathbf{U}), \quad f_i(\mathbf{v}) = f_{i0}(\mathbf{v} - \mathbf{U})$$

where the f_0's correspond to the same plasma at rest. Show from equation 28 that

$$\epsilon(\omega, \mathbf{k}) = \epsilon_0(\omega', \mathbf{k})$$

where $\omega' = \omega - \mathbf{k} \cdot \mathbf{U}$, and $\mathbf{k}$ is parallel to $\mathbf{U}$. (Compare with problem 1 of chapter 9.)

Take the ions at rest and ignore their contribution to ϵ. Consider two colliding beams of cold electrons with velocities $\pm U$ and each with density n. Show from $\epsilon = 0$ that the dispersion relation is

$$\frac{\omega_{pe}^2}{(\omega - kU)^2} + \frac{\omega_{pe}^2}{(\omega + kU)^2} = 2 \tag{120}$$

where $\omega_{pe}^2 = 4\pi ne^2/m$. Clear fractions to obtain a quadratic equation for ω^2. Solve it and show that one of the roots for ω is unstable if and only if $kU < \omega_{pe}$.

Arrive at the same conclusion by writing equation 120 as

$$f(\omega) = \frac{2}{\omega_{pe}^2}$$

and plotting $f(\omega)$ versus ω. Show that $f(\omega)$ has a local minimum at $\omega = 0$, and that if this minimum is less than the right-hand side, there are four real roots, but if it is greater, then there must be two complex conjugate roots.

3. (The electron beam instability in a type III solar radio burst) Consider a cold electron plasma of density n, and a low-density cold beam with velocity

U and density ϵn ($\epsilon \ll 1$). Show that the dispersion relation is

$$\frac{\omega_{pe}^2}{\omega^2} + \frac{\epsilon \omega_{pe}^2}{(\omega - kU)^2} = 1 + \epsilon$$

Show that this leads to a quartic equation for ω with real coefficients. Set the left-hand side equal to $f(\omega)$ and plot it versus ω. Again show that there is a single local minimum at approximately $\omega \approx kU$ and that if $kU < \omega_{pe}$ there must be complex conjugate roots, one of which is unstable.

Approximate this root to show that to lowest order $\omega = kU$ and to next order

$$\omega \approx kU \pm \frac{i\omega_{pe}\epsilon^{1/2}}{\sqrt{(\omega_{pe}/kU)^2 - 1}}$$

provided that kU is not too close to ω_{pe}. Show that when $kU - \omega_{pe} \approx \epsilon^{1/3}\omega_{pe}$ the imaginary part of ω is of order $\epsilon^{1/3}\omega_{pe}$. (A more precise estimate involves a cubic equation.)

Apply this to a beam of electrons propagating through the solar corona in a type III radio burst. Take $n = 10^9$ cm^{-3}, $\epsilon \approx 10^{-3}$, and $U = 10^{10}$ cm/sec and estimate the growth time for the fastest growing mode. What is the frequency of the mode? What do you expect the length of the electron beam to be (Boischot 1967)?

4. (An off-angle ion-acoustic mode) Consider an electrostatic ion wave in a strong magnetic field such that $\Omega_e \gg \omega_{pe}$. Consider that the electrons move only along the magnetic field **B** under the influence of the $\cos\theta$ component of the electron field of the wave, where **k** makes an angle θ with the **B** field. Show that problem 1 generalizes for unmagnetized ions to

$$\epsilon = 1 - \frac{2}{k^2\lambda_D^2}\left[Z'\left(\frac{z_e}{\cos\theta}\right) + \frac{T_e}{T_i}Z'(z_i)\right]$$

Take the limit $v_i \ll \omega/k \ll v_e \cos\theta$ and show that the dispersion relation for the waves is independent of θ

5. (The collisionless Boltzmann distribution) Make it plausible that in the adiabatic limit $\omega/k \ll v_e$ we expect a Boltzmann distribution for the perturbed electron density by considering a slowly growing sinusoidal potential. For any electron show that the change in its total energy $\epsilon = mv^2/2 - e\phi$ is small because it equals the integral of $(\partial\phi/\partial t)_x$ along its orbit and this averages out in the adiabatic limit. Then use the Liouville theorem

$$f(v, x, t) = f_0(v_0, t)$$

where $\epsilon = mv_0^2/2$ and where f_0 is the unperturbed Maxwellian distribution, to show that

$$f(v, t) = e^{e\phi/T} f_0(v)$$

Finally, carry out the velocity integral.

6. (The dispersion relation for an R wave in a thermal plasma) Calculate the dispersion relation for a parallel propagating right circular polarized wave ($E_y = iE_x$), in a Maxwellian plasma, by first showing that the perturbed Vlasov equation for ions, in cylindrical velocity coordinates, $v_\perp$, ϕ, v_z, is

$$-i\omega F_1 + ikv_z F_1 - \Omega_i \frac{\partial F_1}{\partial \phi} = \frac{e}{M}\left(E_x \frac{v_x}{v_\perp} + E_y \frac{v_y}{v_\perp}\right)\frac{\partial F_0}{\partial v_\perp}$$

$$= \frac{e}{M} e^{i\phi} E_x \frac{\partial F_0}{\partial v_\perp}$$

Solve for F_1 and show that

$$F_1 = i\frac{e}{M}\frac{e^{i\phi}\partial F_0/\partial v_\perp}{\omega - kv_z + \Omega_i} E_x$$

Integrate this to find $\mathbf{j}_1$ and show that

$$4\pi(j_x - ij_y) = \frac{4\pi e^2}{M}\int e^{-i\phi} v_\perp F_1 d^3v E_x$$

and also that $j_x + ij_y = 0$. Carry out the same calculation for the electrons and substitute into Maxwell's equations to show that the dispersion relation for the R mode can be written

$$\frac{k^2c^2}{\omega^2} = 1 + \frac{\omega_{pi}^2}{\sqrt{2}kv_i\omega} Z\left(\frac{\omega - \Omega_i}{\sqrt{2}kv_i}\right) + \frac{\omega_{pi}^2}{\sqrt{2}kv_e\omega} Z\left(\frac{\omega - \Omega_e}{\sqrt{2}kv_e}\right)$$

Take the cold plasma limit $Z(z) \approx -1/z$ to show that this reduces to $n^2 = R$.

References

Akhiezer, A.I., A. I. Akhiezer, R. V. Polivin, A. G. Sitenko, and K. N. Stepanov. 1967. *Collective Oscillations in a Plasma*, p. 29, trans. H. S. H. Massey (translation ed. R. J. Tayler), MIT Press, Cambridge, MA.
Bernstein, I. B., and R. M. Kulsrud. 1961. *Physics of Fluids* **4**, 1037.
Bohm, D., and E. P. Gross. 1949. *Physical Review* 75, 1851.
Boischot, A. 1967. In *Solar Radio Bursts* in Proceedings of the International School of Physics, Enrico Fermi, Course XXVIX, Plasma Astrophysics, p. 142, Academic Press, New York.
Chew, G. F., M. L. Goldberger, and F. E. Low. 1955. *Los Alamos Lecture Notes on Physics of Ionized Gases* LA2055.
Dawson, J. 1961. *Physics of Fluids* **4**, 869.
Jackson, J. D. 1999. *Classical Electrodynamics*, p. 637, 3rd ed., Wiley, New York.
Kruskal, M. D. 1960. In *Les Houches notes in Le th'e role des gas Neutre et ionis'es*, edited by C. DeWitt and J.-F. Detouef, pp. 251–274, Wiley, New York.
Kulsrud, R. M. 1964. Lecture Notes from Course XXV, Advanced Plasma Theory, International School of Physics, Enrico Fermi, Varenna, Italy, 1962, in *Proceedings of the International School of Physics*, pp. 54–96, Academic Press, New York.

Kulsrud, R. M. 1983. In *Handbook of Plasma Physics*, edited by R. N. Sagdeev and M. N. Rosenbluth, pp. 115–146, North-Holland, Amsterdam.
Landau, L. D. 1946. *Journal of Physics, USSR* 10, 25.
Stix, T. H. 1962. *The Theory of Plasma Waves*, p. 19, McGraw Hill, New York.
Stix, T. H. 1992. *Waves in Plasmas*, American Institute of Physics, New York.
Tonks, L., and I. Langmuir. 1929. *Physical Review* 33, 195.

Chapter 11

NONLINEAR PHENOMENA

11.1 Introduction

Up to this point, we have mainly treated small-amplitude waves and kept only linear parts of their physics. In this limit many waves can overlap without any interaction and they do not disturb the medium in which they propagate. (Exceptions are the discussion of shocks and steepening in chapter 6 and wave pressure in chapter 5.) If we proceed to second order in the wave amplitude, we find that the waves do affect the background, altering the distribution function of ions and electrons. This second-order interaction is called the wave–particle interaction, and is the subject of quasilinear theory. In the collisional case, waves affect the background by a wave pressure tensor or a ponderomotive force.

A second interaction that occurs for large wave amplitudes is the interaction of waves with each other. Waves of the same type interact to produce new waves of the same type and, more interestingly, waves of different types produce new waves either of one these types or of a third type. For example, an Alfven wave may interact with a sound wave to produce another sound wave or another Alfven wave or a magnetosonic wave. A single wave can split into two lower frequency waves. This is called wave decay. Even if a wave is subject to a small linear decay such processes may dominate if the amplitude is not extremely small. The set of such phenomena is called wave–wave interaction. A general wave kinetic equation that systematically describes all of these interactions is presented later in this chapter.

If the wave amplitudes are still larger, then the interaction between them can occur in a time of order or shorter than the wave periods. Then the interaction is designated as a strong or a turbulent interaction. In this case, we generally need a numerical simulation to properly understand it, though on occasion a qualitative picture works. A famous example of the latter is the Kolmogoroff treatment of homogeneous isotropic fluid turbulence (Chandrasekhar 1949). Another, more recent example is the Goldreich–Sridhar (Sridhar and Goldreich 1994; Goldreich and Sridhar 1995, 1997) theory of small-scale Alfven wave turbulence.

The formal systematic theory for strong turbulence is the direct interaction approximation (DIA) theory (Leslie 1973). Because of its numerical complexity this theory is difficult to carry out and so far its successes have been limited. At the moment we gain more insight into strong turbulence

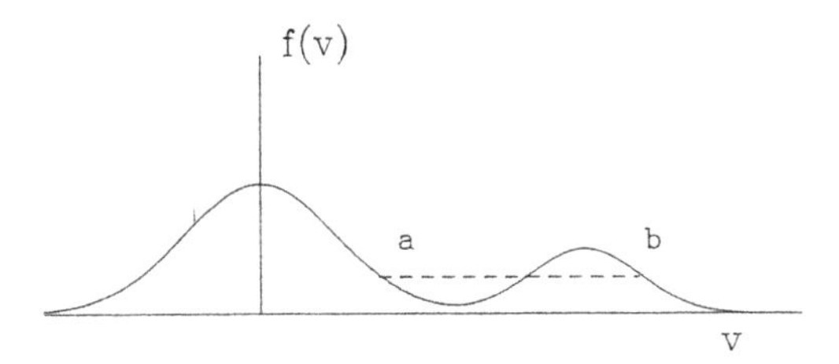

Figure 11.1. Maxwellian distribution with a bump in its tail

from the qualitative pictures, such as Kolmogoroff's and Goldreich and Sridhar's, than from these complex treatments. However, the increasing power of numerical computations of the fluid equations (or MHD equations) in conjunction with the new qualitative ideas will lead to more success in grappling with the tough field of strong turbulence.

Thus, we have three categories of nonlinear physics: wave–particle interactions, wave–wave interactions, and strong turbulence. We treat these in turn in this chapter, starting with wave–particle interactions.

11.2 Wave–Particle Interactions

When wave amplitudes are thermal, their effect on the particles is no more important than normal collisions. However, if the distribution function of the particles is such that the waves are linearly unstable they will grow to much larger than thermal amplitudes, and wave–particle effects will become very important. In fact, it is sometimes the case that the only way the exponential growth will be limited is by a modification of the particle distribution function by the wave–particle interaction until the waves becomes stabilized.

For example, consider the case of an electron distribution function in figure 11.1, with a bump in its tail.

Resonant electrons in the inverted slope region *ab* will lead to exponential growth of waves resonant with these electrons. As the wave grows, faster particles are slowed down and slower particles accelerated by these waves until the distribution function is flattened to the dotted line. The waves saturate and then damp either by wave–wave interactions or by propagation into a region where the slope of f is stable in the resonance region.

To properly treat wave–particle interactions for the case that the waves have a range of wave numbers and moderately large amplitudes, we must be able to statistically describe the waves. It is well known that if the mean square of the Fourier transform of an electric field is a smooth function of k, then the inverse transform is a series of wave packets. In our case, we expect these packets to fill a volume and each wave packet should have a random phase and position with respect to any other one (see figure 11.2).

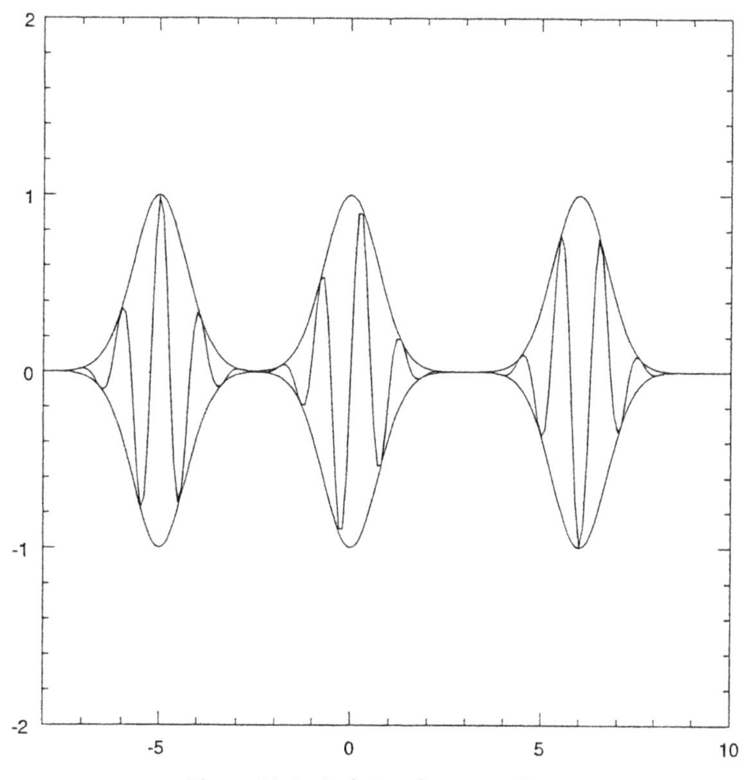

Figure 11.2. A chain of wave packets

At any given moment the way the wave fills the envelope of the wave packet is different for each packet. Further, the wave packets are of different amplitude, and their positions (characterized by the positions of their peak amplitude, say) are also randomly disposed. This is so because these wave packets arise by exponential growth from a thermal distribution that has these properties. (In turn, the thermal waves have this property because they are emitted by individual particles that endow them with these properties.)

The proper description of such a class of waves in Fourier space is called the "random phase approximations." This description of waves can arise as follows. For simplicity, we first treat these waves as one dimensional. We consider N identical electrostatic wave packets located between $-L/2 < x < L/2$. Let the position of the ith wave packet be x^i. Let the electric field of the ith packet at a given time t be

$$E^i = f(x - x^i, t) \tag{1}$$

where x^i is random. If the Fourier transform of f is

$$f(x, t) = \int f_k e^{ikx} dk \tag{2}$$

then

$$E^i = f(x - x^i, t) = \int f_k e^{ik(x-x^i)} dk \tag{3}$$

Thus, $E^i(k)$ the Fourier component of E^i is $f_k e^{-ikx^i}$ and the Fourier transform of the total $E = \Sigma_i E^i$ is

$$E(k) = \Sigma_i E^i(k) = \Sigma_i f_k e^{-ikx_i} \tag{4}$$

Now, since each x^i is random, the average of E_k is random and averages to zero. For this reason most physical effects depend on the square of $E(k)$, or rather the product $E^*(k')E(k)$, for two different k's averaged over the random positions x^i. Thus, we want the ensemble average of this product,

$$\begin{aligned}\langle E^*(k')E(k)\rangle &= \frac{1}{L^2}\Sigma_{i,j} \int_{-L/2}^{L/2} \int_{-L/2}^{L/2} dx^i dx^j E^{*i}(k')E^j(k) \\ &= \Sigma_i \frac{f^*(k')f(k)}{L} \frac{\sin \Delta k L/2}{\Delta k L/2} \\ &= \frac{N}{L} 2\pi f^*(k') f(k) \delta_L(k' - k) \end{aligned} \tag{5}$$

where $\delta_L(k) = \sin(kL/2)/2\pi k$ is a quasi-delta function of width $1/L$, and we write f_k as $f(k)$. Take L to be large compared to the length of the wave packets. Collecting the factors in front, we write the result as

$$\langle E^*(k')E(k)\rangle = I(k)\delta(k' - k) \tag{6}$$

where I is the "intensity" of the electric vector

$$I(k) = 2\pi N |f(k)|^2 \tag{7}$$

This is the random phase approximation for the Fourier transform of the one-dimensional statistical electric field. We have derived it for a set of identical wave packets, but it easily generalizes to a set of wave packets of different f's, so equation 6 is valid for any statistically homogeneous system of waves.

To form some idea of how the quasilinear representation is used let us first calculate the mean value of the square of the electric field at point x. Now,

$$E(x) = \int dk E(k) e^{ikx} \tag{8}$$

and the average of its square is

$$\left\langle E^2\right\rangle = \langle E^*(x)E(x)\rangle$$

$$= \int dk'dk \left\langle E^*(k')E(k)\right\rangle e^{i(k-k')x}$$

$$= \int dk'dk I(k)\delta(k'-k)e^{i(k-k')x}$$

$$= \int dk I(k) \tag{9}$$

so $I(k)/8\pi$ is the spectrum of the mean of the electrical field intensity.

Next, the correlation of E at two neighboring points x and $x' = x + \Delta x$ is

$$C(\Delta x) = \left\langle E^*(x+\Delta x)E(x)\right\rangle = \int dk'dk \left\langle E^*(k')E(k)\right\rangle e^{i(k-k')x - ik'\Delta x}$$

$$= \int dk I(k)e^{-ik\Delta x} \tag{10}$$

$C(\Delta x)$ is the Fourier transform of $I(k)$ and vice versa. Equation 9 is the special case of equation 10 for $\Delta x = 0$.

We have given the random phase expression of the electric field at a fixed time t. But it is also necessary to have it for a time-dependent electric field:

$$E(x,t) = \int dkd\omega e^{i(kx-\omega t)}E(k,\omega) \tag{11}$$

First, a remark is in order. Generally, the turbulence lasts for a long time, and the existence of its time Fourier transform is a problem. Just as we considered wave packets only inside a box $-L/2 < x < L/2$, let us truncate E in time outside of the interval $-T/2 < t < T/2$, where T is a time long enough for the wave packets to grow or decay by wave–particle instabilities or to be grown or damped by linear and nonlinear processes (see figure 11.3).

Thus, our one-dimensional space–time box is filled with wave packets, each of which is finite in space and time. We first take the wave packets to be identical functions of space and time, but to be randomly disposed both in space and time. Then the ith packet can be written as

$$E^i(x,t) = f(x-x^i, t-t^i) \tag{12}$$

where x^i and t^i are random.

It is easy to see, by analogy with the purely spatial case, that, in an ensemble average, $E(k,\omega)$ satisfy

$$\left\langle E^*(k',\omega')E(k,\omega)\right\rangle = I(k,\omega)\delta(\omega'-\omega)\delta(k'-k) \tag{13}$$

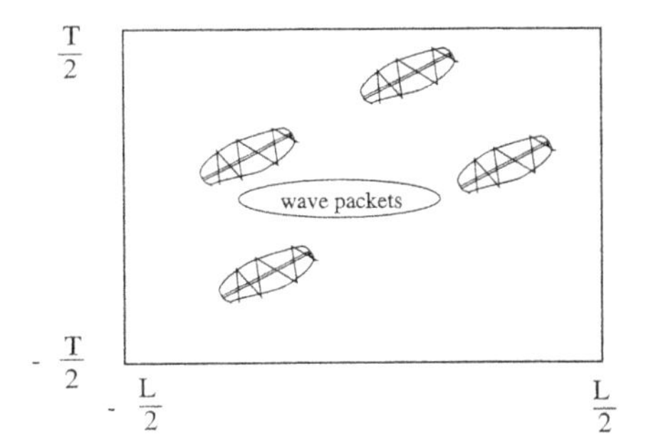

Figure 11.3. Wave packets randomly positioned in space-time

where $I(k, \omega)$ is a space–time spectra of the waves. Now, since the waves satisfy a dispersion relation $\omega = \omega(k)$, we have

$$I(k, \omega) = I(k)\delta[\omega - \omega(k)] \tag{14}$$

where $\delta(\omega)$ is a delta-like function with a breadth characteristic of the lifetime of the wave packet. Again, it is easy to see that these results apply even if the wave packets are nonidentical.

Now, let us use these results to see how this random turbulence evolves a distribution function $f(v, t)$ of electrons. The distribution function satisfies the Vlasov equation

$$\frac{\partial f}{\partial t} + v \cdot \nabla f = \frac{e}{m} E(x, t) \cdot \nabla_v f_0 \tag{15}$$

where $E(x, t)$ is the electric field of the turbulence, i.e, the sum of all the electric fields of the random wave packets. We assume that f_0 changes slowly in time, compared to the frequency of the waves. Thus, f_0 on the right-hand side of equation 15 can be taken as constant, and we can Fourier analyze this equation (still in only one dimension):

$$\frac{\partial f_1}{\partial t} + ikvf_1 = \frac{e}{m} \int dkd\omega E(k, \omega) e^{i(kx-\omega t)} \frac{\partial f_0}{\partial v} \tag{16}$$

or, choosing the solution that vanishes at $t = 0$,

$$f_1 = \frac{e}{m} \int E(k, \omega) \frac{e^{i(kx-\omega t)} - e^{i(kx-kvt)}}{-i(\omega - kv)} \frac{\partial f_0}{\partial v} dkd\omega \tag{17}$$

To next order

$$\frac{\partial f_2}{\partial t} = \frac{e}{m} E^* \frac{\partial f_1}{\partial v}$$

$$= \frac{e^2}{m^2} \frac{\partial}{\partial v} \int E^*(k', \omega') e^{-i(k'x-\omega' t)} E(k, \omega) e^{i(kx-\omega t)}$$

$$\times \frac{\partial f_0}{\partial v} \frac{1 - e^{i(\omega - kv)t}}{-i(\omega - kv)} dk' d\omega' dkd\omega \tag{18}$$

Ensemble averaging this equation and using equation 13 we get

$$\left\langle \frac{\partial f_2}{\partial t} \right\rangle = \frac{e^2}{m^2} \frac{\partial}{\partial v} \int I(k)\delta[\omega - \omega(k)]$$

$$\times \left[\frac{1 - \cos(\omega - kv)t}{-i(\omega - kv)} - \frac{i \sin(\omega - kv)t}{-i(\omega - kv)} \right] \frac{\partial f_0}{\partial v} dk \tag{19}$$

We need keep only the real part since the imaginary part must vanish. If t is large enough, $\sin(\omega - kv)t/(\omega - kv) \approx \pi\delta(\omega - kv)$, and we get

$$\left\langle \frac{\partial f_2}{\partial t} \right\rangle = \frac{\partial}{\partial v} \left[D(v) \frac{\partial f_0}{\partial v} \right] \tag{20}$$

where the diffusion coefficient $D(v)$ for particles with velocity v is given by

$$D(v) = \pi \frac{e^2}{m^2} \int I(k)\delta[\omega(k) - kv)]dk \tag{21}$$

The k integration can also be carried out by

$$\delta(\omega(k) - kv) = \frac{\delta(k - k_c)}{|\partial\omega(k)/\partial k - v|} = \frac{\delta(k - k_c)}{|v_g - v_\phi|} \tag{22}$$

where $v_g = \partial\omega(k)/\partial k$ is the group velocity at $k = k_c$, k_c is that value of k for which $\omega(k) = kv$, and $v_\phi = v$ is both the phase velocity and the particle velocity. Thus, we have

$$D(v) = \pi \frac{e^2}{m^2} \left. \frac{I(k)}{|v_g - v_\phi|} \right|_{\omega(k)=kv} \tag{23}$$

We can arrive at this result directly, up to a constant factor of order unity, by simple physical reasoning calculation. Since the packet velocity is different from the phase velocity with which the particles are resonant, the resonant particles pass through the wave packet with a relative velocity $v - v_g$ as they resonate with the waves in it. If the packet is of length L, the time of interaction Δt is approximately $L/|v_\phi - v_g|$. Let us assume that the spectrum $I(k)$ is peaked about k with width Δk. Then $L \approx 2\pi/\Delta k$. During the time interval in which the resonant particle passes through a given wave packet, it experiences a constant force eE, so it changes its momentum by $eE\Delta t$. Assume that the wave packets are space filling. Then during every time interval Δt it changes its velocity by

$$\Delta v \approx \pm \frac{eE}{m} \Delta t \tag{24}$$

Over a time interval t it passes $t/\Delta t$ wave packets. Summing the squares of Δv due to the $t/\Delta t$ wave packets we get, on average,

$$\left\langle \Sigma(\Delta v)^2 \right\rangle \approx \left(\frac{eE}{m} \Delta t \right)^2 \frac{t}{\Delta t} = \frac{e^2}{m^2} E^2 t \Delta t \tag{25}$$

By a random-walk argument an estimate of the diffusion coefficient is

$$D(v) = \frac{\langle \Sigma(\Delta v)^2 \rangle}{2t} = \frac{e^2}{2m^2} \frac{E^2 L}{|v - v_g|} = \frac{2\pi e^2}{2m^2} \frac{E^2}{\Delta k} \frac{1}{|v_g - v_\phi|} \tag{26}$$

But the total electric field intensity in the band Δk is $E^2 = I \Delta k$ from equation 9, so up to a factor of order unity this result of our simple physical calculation agrees with the more exact result of equation 23. Note that the result is independent of the bandwidth Δk.

If the packets are not tightly packed, but have a filling factor f, the intuitive result is reduced by f, since the wave packets are encountered by the particle, a factor f more slowly. However, the mean intensity $I(k)$ is also smaller by the same factor for fixed E in each wave packet, so the results still agree. If the spectrum is broader, i.e., $\Delta k \gg k$, then we must break it up into wave packets of bandwidth $\Delta k \sim k$, and only one of these sets of wave packets interacts with the resonant particle. Thus, equation 23 is still valid. Essentially, only the energy per unit k, $I(k)$, comes into the diffusion coefficients.

We can combine $\partial f_2/\partial t$ with $\partial f_0/\partial t$, even though f_2 is smaller than f_0, by saying f_0 is finite but has a slow evolution in time. The general quasilinear result, that which is usually quoted, is

$$\frac{\partial f}{\partial t} = \frac{\partial}{\partial v}\left[D(v)\frac{\partial f}{\partial v}\right] \tag{27}$$

So far we have restricted ourselves to one dimension. All the results generalize to three dimensions in a natural way (Kadomtsev 1965; Sagdeev and Galeev 1969). For the time-independent description, $\mathbf{I}(\mathbf{k})$ becomes a tensor function of a vector $\mathbf{k}$. We merely replace equation 6 by

$$\langle \mathbf{E}^*(\mathbf{k}')\mathbf{E}(\mathbf{k}) \rangle = I(\mathbf{k})\hat{\mathbf{k}}\hat{\mathbf{k}}\delta(\mathbf{k}' - \mathbf{k}) \tag{28}$$

Equation 21 is replaced by a three-dimensional integral over $\mathbf{k}$ and $I(\mathbf{k}, \omega)$ by the analog of equation 14. Then the three-dimensional diffusion tensor is

$$\mathbf{D}(\mathbf{v}) = \pi \frac{e^2}{m^2} \int \hat{\mathbf{k}} I(\mathbf{k}) \hat{\mathbf{k}} \delta\left[\omega(\mathbf{k}) - \mathbf{k}\cdot\mathbf{v}\right] d^3\mathbf{k} \tag{29}$$

$\mathbf{I}(\mathbf{k})$ is the actually three-dimensional spatial spectrum of E^2.

A three-dimensional wave can interact with a particle going in a different direction from $\mathbf{k}$, as in figure 11.4. The only requirement is that the argument of the δ function, $\omega(\mathbf{k}) - \mathbf{k}\cdot\mathbf{v} = 0$, vanish. That is to say, the resonant condition must be satisfied. However, equation 23 is no longer correct because of this. The three-dimensional equation for the evolution of f then reads

$$\frac{\partial f}{\partial t} = \nabla_{\mathbf{v}} \cdot [\mathbf{D}(\mathbf{v}) \cdot \nabla_{\mathbf{v}} f] \tag{30}$$

Now let us consider a magnetized plasma with a uniform magnetic field $\mathbf{B}_0$ in the z direction. The calculations are considerably more complicated

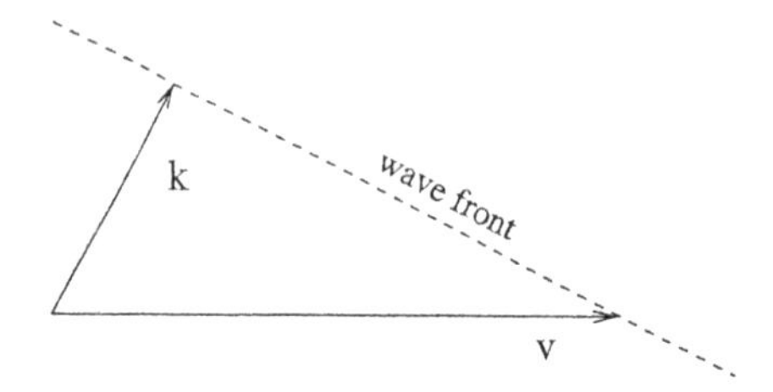

Figure 11.4. A particle interacting with a wave going in a different direction

in this case, so we only consider plane waves propagating in the z direction. The waves are the transverse waves of the R and L type discussed in chapter 9, and the longitudinal waves, which are either the ion acoustic waves or plasma oscillations. These latter two waves affect the particles in a manner independent of the magnetic field, and we have already discussed their wave–particle interaction.

Here we consider only the transverse waves and for definiteness consider the effect of the waves on ions. In an unmagnetized plasma, transverse waves propagate too fast to interact with the particles, but in the presence of a magnetic field they can propagate more slowly, and, further, they can interact by cyclotron resonance. These two conditions make wave–particle interactions of ions with transverse waves possible.

The perturbed Lorentz force of a wave on a particle is

$$\mathbf{F}_1 = e\left(\mathbf{E}_1 + \frac{\mathbf{v}\times\mathbf{B}_1}{c}\right) = e\mathbf{E}_1\left(1 - \frac{kv_z}{\omega}\right) + e\mathbf{k}(\mathbf{E}_1\cdot\mathbf{v}) \tag{31}$$

since $\mathbf{B}_1 = \mathbf{k}c \times \mathbf{B}_1/\omega$. Now, $\mathbf{E}_1$ varies as $e^{i(kz-\omega t)}$. Thus, an ion which moves along z at speed v_z sees the $\mathbf{E}_1$ field varying as $e^{i(kv-\omega)t}$, so that the frequency of the wave as seen by the ion is the Doppler-shifted frequency $\omega - kv_z$. In the frame of the particle

$$E_x = E_1\cos(\omega - kv_z)t$$
$$E_y = \pm E_1\sin(\omega - kv_z)t \tag{32}$$

where the upper sign refers to an R wave (right handed) and the lower sign to an L wave. On the other hand, due to its cyclotron motion the perpendicular components of the velocity are

$$v_x = v_\perp\cos(\Omega t + \phi)$$
$$v_y = -v_\perp\sin(\Omega t + \phi) \tag{33}$$

ϕ represents the phase shift between the wave and the ion's perpendicular motion.

Now let us take the z component of $\mathbf{F}_1$, the second term involving $\mathbf{E}\cdot\mathbf{v}$. Then

$$\mathbf{E}\cdot\mathbf{v} = E_1 v_\perp\cos[(\omega - kv_z \pm \Omega)t \pm \phi] \tag{34}$$

The resonant condition (the condition that the force not average to zero over a short period of time) is

$$\omega - kv_z \pm \Omega \approx 0 \tag{35}$$

The R wave cannot satisfy this condition unless $v_z > v_\phi = \omega/k$. The left wave will satisfy it for $v_z < v_\phi$.

If the waves are low-frequency Alfven waves with $\omega = kv_A$, then the condition becomes

$$k = \frac{\pm\Omega_i}{v - v_A} \tag{36}$$

For $v > v_A$ we need $k > 0$ or right-moving R waves for the upper sign, and $k < 0$ or left-moving L waves for the lower sign. This can be understood because the $v > v_A$ fast ion sees the right-moving R wave as a left-handed screw, though a stationary observer would see the **E** field in the R wave as rotating in a right circular direction.

Now, because the wave packets have a finite length, $L = 2\pi/\Delta k$, the resonant condition need not be satisfied exactly. If the time of passage of an ion through a wave packet is Δt, then we need only the resonant condition in the form

$$|(\omega - kv \pm \Omega)\Delta t| \leq 1 \tag{37}$$

Under this condition the force on an ion during the passage through a wave packet is roughly constant.

The change in v_z during the passage of a single resonant wave packet is roughly

$$\Delta v_z \approx \pm\frac{e}{M}\frac{kEv_\perp}{\omega}\Delta t \tag{38}$$

In analogy with the calculation for the nonmagnetized case, we have

$$\Sigma\frac{(\Delta v_z)^2}{2} = \left(\frac{e}{M}\frac{kEv_\perp}{\omega}\right)^2 t\Delta t \tag{39}$$

But

$$\Delta t = \frac{L}{|v - v_g|} = \frac{2\pi}{|v - v_g|\Delta k} \tag{40}$$

so

$$D_z = \Sigma\frac{(\Delta v_z)^2}{2t} \approx \pi\frac{e^2}{M^2\omega^2}\frac{k^2v_\perp^2}{|v - v_g|}\frac{E^2}{\Delta k} \approx \pi\frac{e^2}{M^2}I_\pm(k)\frac{k^2v_\perp^2}{\omega^2|v - v_g|} \tag{41}$$

where $I_\pm(k)$ is the one-dimensional intensity spectrum for the electric field in these plane transverse waves. It and the other factors of k and v_g are to be evaluated at the resonant k satisfying equation 35, where for the upper sign I is the intensity in the R waves and for the lower sign the intensity in the L waves. This expression may be written in terms of the magnetic field intensity using $B_1 = ckE_1/\omega$. Let

$$\frac{\langle B_1^2\rangle}{B_0^2} = \int J_\pm(k)dk \tag{42}$$

Then

$$D_z = \pi\Omega^2 J_\pm(k)\frac{v_\perp^2}{|v - v_g|} \tag{43}$$

So far, we have only discussed diffusion in v_z. There is also diffusion in $v_\perp$. However, since for a steady magnetized plasma f_0 is a function of $v_\perp$ and v_z only, we need not worry about diffusion in the third, gyrophase, coordinate. We can arrive at the full quasilinear diffusion equation for transverse wave–particle interactions by exploiting the fact that there is no electric field in the wave frame, and therefore the ion cannot change its energy in this frame. Thus, $\Delta v_\perp = v_z \Delta v_z / v_\perp$. Consequently, the diffusion equation in the wave frame must be of the form

$$\begin{aligned}\frac{\partial f_0(v_\perp, v_z)}{\partial t} &= \frac{\pi}{4}\Sigma_\pm\Omega^2 \int dk \left(-v_z\frac{\partial}{\partial v_\perp} + v_\perp\frac{\partial}{\partial v_z}\right) \\ &\quad\times \left(-v_z\frac{\partial f_0}{\partial v_\perp} + v_\perp\frac{\partial f_0}{\partial v_z}\right) J(k)\delta(-kv_z \pm \Omega)\end{aligned} \tag{44}$$

The constant has been adjusted to agree with a detailed kinetic calculation. It is seen that when f_0 is isotropic in $\mathbf{v}$, the right-hand side vanishes. The two factors in this equation are identical since they arise from the averages of $\Delta v_\perp \Delta v_\perp$, of $\Delta v_\perp \Delta v_z$, of $\Delta v_z \Delta v_\perp$, and of $\Delta v_z \Delta v_z$. so they must be symmetrical with each other. Thus, we have succeeded in deriving the diffusion equation in the wave frame.

To transform it back to the laboratory frame we simply replace v_z in it by $v_z - \omega/k$. Of course, this derivation is not totally rigorous the way it has been stated, since the wave frame depends on k and on whether the wave is the L wave or the R wave. However, by breaking the wave spectrum up into small ranges in k and into the two waves we can derive equation 44 for each of these pieces and transform to the laboratory frame and then sum. Thus, on replacing v_z by $v_z - \omega/k$ in this equation we arrive at the correct answer by a somewhat physical calculation without the bother of a full derivation.

The general form for the quasilinear diffusion equation for Alfven wave turbulence is

$$\frac{\partial f}{\partial t} = \frac{\partial}{\partial \mathbf{v}} \cdot \left(\mathbf{D} \cdot \frac{\partial f}{\partial \mathbf{v}}\right) \tag{45}$$

where the tensor $\mathbf{D}$ in $v_\perp$, $v_\parallel$ space is

$$\mathbf{D} = \frac{\pi}{4}\Omega^2 v_A^2 \Sigma_\pm \int d^3\mathbf{k} J(\mathbf{k})\delta(\omega - kv_z \pm \Omega)$$

$$\times \left[\left(1 - \frac{kv_z}{\omega}\hat{\mathbf{e}_\perp}\right) + \frac{kv_\perp}{\omega}\hat{\mathbf{e}_\mathbf{z}}\right]\left[\left(1 - \frac{kv_z}{\omega}\hat{\mathbf{e}_\perp}\right) + \frac{kv_\perp}{\omega}\hat{\mathbf{e}_\mathbf{z}}\right] \quad (46)$$

We will discuss applications of quasilinear theory for Alfven waves and cosmic rays in chapter 12, which will indicate its usefulness.

Let us now return to the electrostatic waves and calculate the rate of evolution of the total particle energy due to quasilinear diffusion by multiplying equation 30 by $mv^2/2$ and integrating over $\mathbf{v}$:

$$\frac{d}{dt}\int fm\frac{v^2}{2}d^3\mathbf{v} = \int \frac{mv^2}{2}\nabla_\mathbf{v}\cdot\left(\mathbf{D}\cdot\frac{\partial f}{\partial \mathbf{v}}\right)d^3\mathbf{v}$$

$$= -\int m\mathbf{v}\cdot\mathbf{D}\cdot\frac{\partial f}{\partial \mathbf{v}}d^3\mathbf{v}$$

$$= -\pi\frac{e^2}{m^2}\int \hat{\mathbf{k}}\cdot\mathbf{v}I(\mathbf{k})\hat{\mathbf{k}}\cdot\frac{\partial f}{\partial \mathbf{v}}\delta[\omega(\mathbf{k}) - \mathbf{k}\cdot\mathbf{v}]d^3\mathbf{v}d^3\mathbf{k}$$

$$= -2\int \gamma_k\frac{\mathbf{I_k}}{8\pi}\frac{\partial\epsilon}{\partial\omega}d^3\mathbf{k} \quad (47)$$

where for the last line we have used equations 60 in chapter 10. Thus, we see that if $\partial f/\partial v < 0$, then the rate of gain of the particle energy results from the rate of loss of the wave energy due to damping. If $\partial f/\partial v > 0$, then the waves are unstable and grow. They get their energy from the energy of the resonant particles. (Of course, it must be remembered that part of the wave energy is in the nonresonant particles.)

How efficient is wave particle diffusion? To estimate this let us return to one dimension and suppose that the waves are plasma oscillations. Let the ratio of their energy to the thermal energy of the plasma nT be ϵ. Let the waves occupy a bandwidth $\Delta k \sim k$; then we have

$$\frac{kI(k)}{8\pi} \sim \epsilon nT \quad (48)$$

and from equation 23 we get the effective collision rate at $v \approx v_e$

$$\nu_{\text{eff}}v_e^2 = \frac{\langle\Delta v^2\rangle}{2t} \approx \pi\frac{8\pi e^2}{m^2}\frac{\epsilon nT}{k|v_g - v_\phi|}$$

$$\approx 2\pi\omega_{pe}^2\frac{\epsilon T}{\omega_{pe}m} \approx 2\pi\epsilon\omega_{pe}^2 v_e^2 \quad (49)$$

where we approximate $k(v_g - v_\phi)$ as $kv_\phi = \omega \approx \omega_{pe}$. We see that the effective collision rate is of order $\epsilon\omega_{pe}$. If ϵ is not small this is a very large rate. If the wave energy is the thermal energy, kT/mode, we get $\nu_{\text{eff}} \approx \nu_{\text{coul}}$ the binary Coulomb collision rate.

We can ask why the effective collision rate is so large. This question is similar to the question of why collisional cross sections with high Z impurities are larger than those from protons by the factor of Z^2, rather than Z. In this last case, we see that a particle is simultaneously colliding with Z protons in the nucleus, so these collisions are not off random nucleons but are all simultaneous and coherent. The same occurs in a collision with a wave packet. Take the size of the wave packet to be comparable to λ_D. We have $\delta n/n = \sqrt{\epsilon}$ for the wave, so one is simultaneously colliding with $\approx \sqrt{\epsilon} n/k^3 \approx \sqrt{\epsilon} n \lambda_D^3$ protons (assuming that the wavelength is comparable to the Debye length). We have nonrandom collisions with these δn protons because of the resonance. Thus, the effective collision with the waves per electron is larger by $\epsilon n \lambda_D^3$ than that with a single electron.

So far we have considered only quasilinear diffusion of particles by electrostatic waves in an unmagnetized plasma, when the electric field is along k and is the only force on the particles, In this case, the Cerenkov resonance is the important one. A slight generalization to the electrostatic case occurs when a magnetic field is present that is strong enough to force the electrons to move along the magnetic field lines. In this case the quasilinear diffusion by waves propagating in a general direction is one-dimensional again. The resonance condition is that the particles travel with a constant phase of the wave. If the wave is proportional to $e^{i(\mathbf{k}\cdot\mathbf{r}-\omega t)}$ and the particle has a velocity v_z along z and no transverse motion, it satisfies

$$z = z_0 + v_z t$$

$$x = y = \text{const} \tag{50}$$

so it sees an electric field that varies as

$$\exp i(\mathbf{k}\cdot\mathbf{r} - \omega t) = \exp i(k_z v_z t - \omega t + k_z z_0)$$

and the resonance condition is $\omega - k_z v_z = 0$. The diffusion tensor of equation 29 is replaced by

$$\mathbf{D} = \int \frac{\pi e^2}{m} \hat{\mathbf{z}}\hat{\mathbf{z}} \int (\hat{\mathbf{k}}\cdot\hat{\mathbf{z}})^2 I(k)\delta[\omega(k) - k_z v_z] d^3\mathbf{k} d^3\mathbf{k} \tag{51}$$

This is suggested by figure 11.5. The case to which the figure applies is that in which electrons were restricted to move in the z direction and the ions have such a low cyclotron frequency compared to the wave frequency that they are essentially unmagnetized. In this case the electron diffusion coefficient $\mathbf{D}_e$ is equation 51, while the ion diffusion tensor $\mathbf{D}_i$ is equation 29.

11.3 Wave–Wave Interactions

When wave amplitudes are small we can superimpose them without any interaction. However, as we have frequently observed, in treating waves we

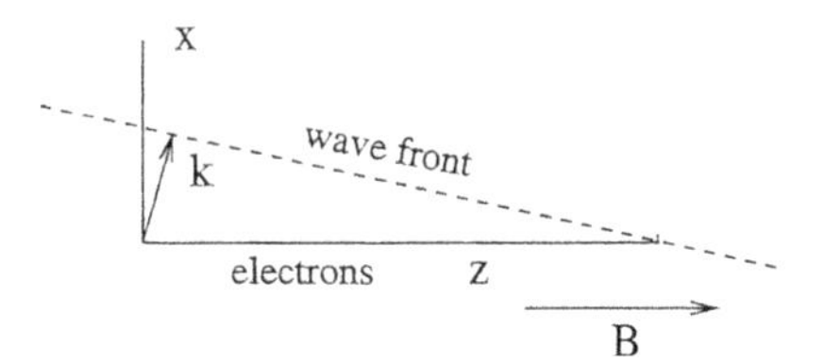

Figure 11.5. Electron moving along B and ions in a general direction

have been careful to neglect any nonlinear terms. For example, we have written the continuity equation as

$$\frac{\partial n_1}{\partial t} + \nabla \cdot (n_0 + n_1)\mathbf{v} \approx \frac{\partial n_1}{\partial t} + \nabla \cdot (n_0 \mathbf{v}_1) = 0 \tag{52}$$

and dropped the $n_1 \mathbf{v}_1$ term because it is small compared to the linear terms in our small-amplitude limit. For a single wave, the presence of such terms leads to a gradual evolution of the wave profile as we saw in chapter 6 on shocks and wave steepening.

But suppose we have two waves a and b, the a wave with

$$n_{1a} = \hat{n}_{1a} e^{ik_a x - i\omega_a t} \tag{53}$$

and the b wave with

$$n_{1b} = \hat{n}_{1b} e^{ik_b x - i\omega_b t} \tag{54}$$

Then the $\nabla \cdot (n_1 v_1)$ term is

$$i(k_a n_{1b} v_{1a} + k_b n_{1a} v_{1b}) e^{i(k_a + k_b)x - i(\omega_a + \omega_b)t} k_a \tag{55}$$

This part of the perturbed charge can be thought of as a source term, which will perturb the plasma. If the wave number $k_{ab} = k_a + k_b$ and frequency $\omega_{ab} = \omega_a + \omega_b$ satisfy a dispersion relation for a third wave, say $\epsilon(\omega_{ab}, k_{ab}) = 0$ for electrostatic waves, then the amplitude of this third wave will grow at the expense of the first two waves. Thus, the heretofore neglected quadratic terms offer the possibility for converting waves into each other. This is the wave–wave interaction. In any discussion of these interactions we get selection rules. For the conversion of wave a or b into wave c our a and b waves must satisfy

$$\omega_c = \omega_a + \omega_b$$

$$k_c = k_a + k_b \tag{56}$$

where a, b, and c waves must each satisfy the dispersion relation for waves. In general, two waves a and b will not secularly react because these relations are not satisfied. On the other hand, if equation 56 is not satisfied for some wave c, then there will still be a nonresonant interaction that will constitute a disturbance in the plasma. If there is a turbulent ensemble of waves a and b, some pair of these waves will satisfy equation 56 and interact. Equation 56 selects out a submanifold or subset of all possible pairs and it is these pairs that we are interested in.

Let us consider a particularly important and familiar example to illustrate these points, namely, the coherent scattering of a single light wave by an ensemble of waves that fluctuate the plasma density, for example, plasma oscillations. We take the single light wave to have the wave number $\mathbf{k}_0$, a high frequency so that $\omega_0 = k_0 c$, and a linearly polarized amplitude $\mathbf{E}_0$. Let $\mathbf{k}_0$ be in the z direction and $\mathbf{E}_0$ in the x direction

$$\mathbf{E} = \hat{x} E_0 e^{i\mathbf{k}_0\cdot\mathbf{r} - i\omega_0 t}$$

where as always we take the real part so the average of E^2 is $E_0^2/2$ (see chapter 19). Let the light wave pass through a box that is filled with waves characterized by their density perturbations. Thus, let

$$\delta n = \int n(\mathbf{k}, \omega) e^{i\mathbf{k}\cdot\mathbf{r} - i\omega t} d^3\mathbf{k} d\omega$$

Assume the random phase approximation,

$$\left\langle n^*(\mathbf{k}', \omega') n(\mathbf{k}, \omega) \right\rangle = I(\mathbf{k}, \omega)\delta(\mathbf{k}' - \mathbf{k})\delta(\omega' - \omega) \tag{57}$$

with

$$I(\mathbf{k}, \omega) = I(\mathbf{k})\delta[\omega - \omega(\mathbf{k})] \tag{58}$$

Just as we showed for E^2 in equation 9, we can write at a given position and time

$$\left\langle (\delta n)^2 \right\rangle = \int I(\mathbf{k}, \omega) d^3\mathbf{k} d\omega = \int I(\mathbf{k}) d^3\mathbf{k} \tag{59}$$

Now, as the light wave E_0, passes through the plasma it oscillates the electrons at $\mathbf{r}$ and t with a velocity

$$v_x = \frac{eE_0}{-i\omega_0 m} e^{i(\mathbf{k}_0\cdot\mathbf{r} - \omega_0 t)} \tag{60}$$

The current at $\mathbf{r}$ and t is then

$$\begin{aligned} j_x &= [n_0 + \delta n(\mathbf{r}, t)]\, v_x \\ &= e\hat{\mathbf{x}} \left[n_0 + \int n(\mathbf{k}, \omega) e^{i\mathbf{k}\cdot\mathbf{r} - i\omega t} d^3\mathbf{k} d\omega \right] \frac{eE_0}{-i\omega_0 m} e^{i\mathbf{k}_0\cdot\mathbf{r} - i\omega_0 t} \end{aligned} \tag{61}$$

We are not interested in the current due to the n_0 term. It merely alters the dispersion relation of the E_0 wave slightly. Let

$$\begin{aligned} \omega' &= \omega_0 + \omega \\ \mathbf{k}' &= \mathbf{k}_0 + \mathbf{k} \end{aligned} \tag{62}$$

Then

$$j_x = \int j_x(k', \omega') e^{i\mathbf{k}'\cdot\mathbf{r} - i\omega' t} d^3\mathbf{k} d\omega \tag{63}$$

where

$$j_x(k', \omega') = \frac{n(k, \omega)e^2 E_0}{-i\omega_0 m} \tag{64}$$

This current will produce a secondary electric field whose transverse part is

$$\mathbf{E} = \int \mathbf{E}(\mathbf{k}', \omega')e^{i\mathbf{k}'\cdot\mathbf{r}-\omega' t}d^3\mathbf{k}d\omega'$$

where $\mathbf{E}(\mathbf{k}', \omega')$ is perpendicular to $\mathbf{k}'$ and is in the $\hat{\mathbf{x}} - \hat{\mathbf{k}}'\hat{\mathbf{k}}' \cdot \hat{\mathbf{x}}$ direction. (It comes from the projection of $\hat{x}$ onto a plane perpendicular to $\mathbf{k}'$.)

We find the magnitude of $\mathbf{E}'(\mathbf{k}', \omega')$ from equation 15 in chapter 9,

$$\frac{\omega'^2}{c^2}E' - k'^2 E' = -\frac{4\pi i\omega' j_\perp(k', \omega')}{c^2} \tag{65}$$

or

$$(\omega'^2 - k'^2c^2)\mathbf{E}(\mathbf{k}', \omega') = \left(\frac{\omega'}{\omega_0}\right)\frac{4\pi n(k, \omega)e^2}{m}E_0 \sin\theta' \tag{66}$$

where θ' is the angle between $\hat{\mathbf{x}}$ and $\hat{\mathbf{k}}'$.

We are interested only in $E(k', \omega')$ for $\omega' \approx k'c$, since this is the resonantly scattered wave. We remove any singularity by starting $E(k', t) = 0$ at $t = 0$. Thus,

$$E(k', t) = \int \sin\theta' E_0 \frac{4\pi e^2\omega'}{m\omega_0}\left(\frac{e^{-i\omega' t} - e^{-ik'ct}}{\omega'^2 - k'^2c^2}\right)n(k, \omega)d\omega$$

$$= +\, i \sin\theta' E_0 \frac{4\pi e^2\omega'}{m\omega_0}\left(\frac{\omega'}{\omega_0}\right)\frac{\sin[(\omega' t - k'c)t/2]e^{-i(\omega'+k'c)t/2}}{(\omega' - k'c)/2 \times (\omega' + k'c)}n(k, \omega)d\omega \tag{67}$$

As t increases, the frequencies of $E(\mathbf{k}, t)$ are concentrated more sharply about $k'c$ by the sine factor. If this frequency is unequal to $\omega_0 + \omega$, then no emission occurs, demonstrating the selection rules, equation 62.

Now, let us evaluate $\langle ||E(\mathbf{r}, t)||^2\rangle = \int E^*(k''', t)E(k', t)e^{i(\mathbf{k}'-\mathbf{k}''')\cdot\mathbf{r}}d^3\mathbf{k}d^3\mathbf{k}''$ where $k''' = k'' + k_0$, $\omega''' = \omega'' + \omega_0$.

$$\left\langle |E(\mathbf{r}, t)|^2 \right\rangle = \left(\frac{4\pi e^2}{m}\right)^2 |E_0|^2 \sin\theta' \sin\theta''' \times$$

$$\int \left(\frac{\omega'\omega''}{\omega_0^2}\right)\left\langle n^*(k'', \omega'')n(k, \omega)\right\rangle e^{\frac{1}{2}[-i(\omega'-\omega''')t - i(\mathbf{k}'-\mathbf{k}''')\cdot\mathbf{r}]} \times$$

$$\frac{\sin(\omega' - k'c)t/2}{(\omega' - k'c)/2} \frac{\sin(\omega''' - k'''c)t/2}{(\omega''' - k'''c)/2} \times$$

$$\frac{1}{(\omega' + k'c)(\omega''' + k'''c)} d^3\mathbf{k}''d\omega''d^3\mathbf{k}d\omega \tag{68}$$

Making use of the delta functions arising from $\langle n^*(k'', \omega'')n(k, \omega)\rangle$ we reduce this to

$$\left\langle |E(\mathbf{r}, t)|^2 \right\rangle = \left(\frac{4\pi e^2}{m}\right)^2 |E_0|^2 \sin^2\theta' \times$$

$$\int I(k, \omega) \left(\frac{\omega'}{\omega_0}\right)^2 \frac{\sin^2(\omega' - k'c)t/2}{[(\omega' - k'c)/2]^2 \times (\omega' + k'c)^2} d^3\mathbf{k}d\omega \tag{69}$$

Now, replace the $\sin^2$ factor by $2\pi\delta(\omega' - k'c)$ times t. Then

$$\left\langle |E|^2 \right\rangle = t\frac{\pi}{2}\left(\frac{4\pi e^2}{m}\right)^2 |E_0|^2 \int \sin^2\theta' \frac{\omega'^2 I(k, \omega)}{(\omega_0 k'c)^2} \delta(\omega' - k'c)d\omega d^3\mathbf{k}$$

$$= t\frac{\pi}{2}\left(\frac{4\pi e^2}{m}\right)^2 |E_0|^2 \int \sin^2\theta' \frac{\omega'^2 I(k, \omega)}{(\omega_0 k'c)^2} \delta[\omega(k) + \omega_0 - k'c]d^3\mathbf{k} \tag{70}$$

(see Rosenbluth and Rostoker (1962) and problem 3). The delta function enforces the constraint

$$\omega' = \omega_0 + \omega(k) = |\mathbf{k} + \mathbf{k}_0|c \tag{71}$$

But if $k \approx k_0$ and $\omega \ll \omega_0$, then, since $\omega' \approx \omega_0$, we may conclude that $k'c \approx k_0 c$ and we expect k' to have nearly the same magnitude as k_0. Thus, we expect k' to lie very close to a sphere passing through $\mathbf{k}_0$ as in figure 11.6.

For every direction (scattering angle) $\mathbf{k}'$ we have a single $\mathbf{k}'$ and $\mathbf{k}$. The energy scattered in this direction is proportional to the energy of the scattering wave $I(\mathbf{k}) = I(\mathbf{k}' - \mathbf{k})$ as given by equation 70. Actually, ω is not exactly zero, so k' must be slightly larger than k and $\mathbf{k}'$ lies slightly outside of the sphere. $\omega' = \omega_0 + \omega_k$ so there is a slight shift in the scattered frequency equal to the frequency of the scattering density wave. As a consequence, the scattered wave should have slightly more energy, which comes from reducing the amplitude of the scattering wave (although this is not obvious from the derivation). This can be most easily seen by Doppler shifting to a frame where the ω wave has zero frequency and the ω_0 wave has the same frequency as the ω' wave. Then from the change in direction of these two

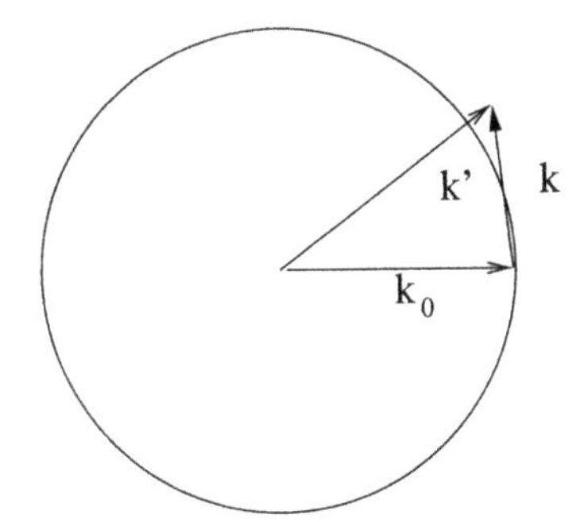

Figure 11.6. Scattering angle and the small shift in frequency

waves the momentum density of the ω_0 and ω' waves can be found. Then transforming back, the energy can be seen to increase, and thus, must come from the scattering wave **k**.

We have shown that the $\mathbf{E}'$ wave increases in energy as t, but this is because we have tacitly assumed a periodic box for our problem, so that energy accumulates as t. If the box actually allowed the scattered wave to escape, the coefficient of t would give the rate of scattering. How the scattering depends on angle is gotten by transforming the k integration to a k' integral by the usual Jacobian transformation.

Although we have approached our problem by giving E_0 and the density fluctuation spectrum $I(\mathbf{k})$ and asked for the amount of scattering, the problem is usually turned around. The density spectrum is unknown, E_0 is given, and the scattered power is measured. An important application of this is backscattering of radar pulse from the ionosphere. The turbulent spectrum of the ionosphere is thus determined (Perkins 1968).

For simplicity, we have taken the initial wave E_0 to be an electromagnetic wave and calculated the scattered electromagnetic power. But with a slight change of notation, we may take E_0 to be a plasma oscillation and the fluctuation also to be a plasma oscillation. Then if the selection rules can be chosen properly we can find the electromagnetic scattered signal. This seems to be happening in type II and type III radio bursts where beams of electrons create quite strong density fluctuations. The electromagnetic signal is observed to be at twice the plasma frequency. This is expected since the ω_0 and ω are near plasma frequency so that $\omega' \approx \omega_0 + \omega \approx 2\omega_{pe}$. However, $k' \approx 2\omega_{pe}/k$ must be much smaller than k_0 or k. Therefore, k and k_0 must be in nearly opposite directions to nearly cancel their sum. Of course, although we have considered the k_0 signal to be monochromatic any one of the density fluctuations can act as a k_0 signal, so that the integral in equation 70 should be actually be a double integral over k_0 and k (see Dawson 1968; Ginzburg and Zhelezniakov 1958).

11.4 Mode Decay

In the previous example we showed how nonlinear terms can lead to the addition of two modes to form a third mode. In this example, we started with an electromagnetic mode and a density mode and produced an additional mode, the scattered electromagnetic mode. It is possible for a single mode to produce two modes with lower frequency. The energy for this splitting comes from the original mode and reduces its amplitude. This is referred to as mode decay in analogy with the nuclear decay of a particle into other particles.

Rather than give a general treatment we will illustrate mode decay by a single example. We start with a moderately large-amplitude Alfven mode, A, propagating in the direction of an ambient field $\mathbf{B}_0$, which points along the z axis. We take its wave number and frequency to be k_A and ω_A. The mode is assumed to be polarized, with its velocity vector in the x direction.

In addition, we assume there is a very small-amplitude Alfven wave B also propagating along $\hat{z}$ with wave number k_B, frequency ω_B, and also polarized in the x direction, and a third wave, a sound wave C, also with a very small amplitude, with k_C and ω_C, and propagating in the $\hat{z}$ direction. Further, we assume that the sound speed is less than the Alfven speed. We will see that due to waves A and B, energy is nonlinearly transferred to mode C, and due to waves A and C, energy is transferred to mode B. As a result of these energy transfers wave A decreases its amplitude and waves B and C increase theirs. Thus, wave A decays into the two waves B and C. Wave A can be thought as the pump wave, pumping energy into B and C.

To keep matters simple we employ real notation. Further, we keep only a single nonlinear term for each of the waves B and C and neglect any nonlinear effect on wave A, so we assume that wave A does not change its amplitude significantly. This is the case until waves B and C have grown substantially.

We choose all the frequencies to be positive, and at first choose all the wave numbers to be positive. For wave A we have the linear equations

$$\frac{\partial}{\partial t}\delta v_A = \frac{B_0}{4\pi\rho}\frac{\partial}{\partial z}\delta b_A$$

$$\frac{\partial}{\partial t}\delta b_A = B_0\frac{\partial}{\partial z}\delta v_A \tag{72}$$

where δb_A and δv_A are the perturbed components in the x direction. A solution of these equations is

$$\delta b_A = b_A\cos(k_A z - \omega_A t) = b_A\cos\phi_A \tag{73}$$

where

$$\phi_A = k_A z - \omega_A t \tag{74}$$

and $\omega_A = k_A c_a$, where $c_a = B_0/\sqrt{4\pi\rho}$ is the Alfven speed.

The nonlinear equations for wave B are

$$\frac{\partial}{\partial t}\delta v_B = \frac{B_0}{4\pi\rho_0}\frac{\partial}{\partial z}\delta b_B - \frac{\partial}{\partial z}\left(\frac{B_0 \delta b_A}{4\pi\rho_0^2}\right)\delta\rho_C$$

$$\frac{\partial \delta b_B}{\partial t} = B_0 \frac{\partial}{\partial z}\delta v_B \tag{75}$$

There are other nonlinear terms but they can be shown to be smaller than the one we keep (see Sagdeev and Galeev 1969). This term couples the A wave and the C wave to the B wave through the variation of the density, which fluctuates the acceleration produced by the A wave.

If we drop the nonlinear term, the solution for the B wave can be written

$$\delta v_B = v_B \cos(k_B z - \omega_B t) = v_B \cos\phi_B \tag{76}$$

$$\delta b_B = -\frac{v_B}{c_a} B_0 \cos\phi_B \tag{77}$$

where the definition of $\phi_B = k_B z - \omega_B t$ is analogous to that of ϕ_A and $\omega_B = k_B c_a$. (Note that if k_B were negative and ω_B positive, the sign in equation 77 would be opposite.)

Finally, the nonlinear equations for the C wave are

$$\frac{\partial \delta v_C}{\partial t} = -\frac{c_s^2}{\rho_0}\frac{\partial \delta\rho_C}{\partial z} - \frac{\partial}{\partial z}\left(\frac{\delta b_A \delta b_B}{8\pi\rho_0}\right) \tag{78}$$

$$\frac{\partial \delta\rho_C}{\partial t} = -\rho_0 \frac{\partial \delta v_C}{\partial z} \tag{79}$$

where $c_s = \sqrt{\gamma p/\rho}$ is the speed of sound. A linear solution for the C wave is

$$\delta v_C = v_C \sin(k_C z - \omega_C t) = v_C \sin\phi_C \tag{80}$$

$$\delta\rho_C = \frac{v_C}{c_s}\rho_0 \sin\phi_C \tag{81}$$

(Similarly, if $k_C < 0$, then the sign in equation 81 is negative.) Now we take into account the nonlinear terms by allowing v_B and v_C to depend on time.

Substituting equations 73 and 79 into the nonlinear term of equation 75, we find

$$\frac{dv_B}{dt}\cos\phi_B = -\frac{\delta\rho_C}{4\pi\rho_0^2}\frac{\partial}{\partial z}(\delta b_A B_0)$$

$$= -b_A B_0 \frac{v_C\rho_0}{c_s}\frac{1}{4\pi\rho_0^2}\sin\phi_C\frac{\partial}{\partial z}\cos(\phi_A)$$

$$= \frac{b_A}{B_0}\frac{c_a^2}{c_s}k_A v_C \sin\phi_A \sin\phi_C$$

$$= \frac{b_A}{B_0}\frac{k_A c_a^2}{2c_s}v_C\left[-\cos(\phi_A+\phi_C)+\cos(\phi_A-\phi_C)\right] \qquad (82)$$

Substituting equations 73 and 77 into the nonlinear term of equation 78, we have

$$\frac{dv_C}{dt}\sin\phi_C = -\frac{\partial}{\partial z}\left(\frac{\delta b_A \delta b_B}{8\pi\rho_0}\right)$$

$$= \frac{b_A}{2B_0}c_a^2\frac{v_B}{c_a}\frac{\partial}{\partial z}(\cos\phi_A\cos\phi_B)$$

$$= \frac{b_A}{4B_0}c_a v_B\frac{\partial}{\partial z}\left[\cos(\phi_A+\phi_B)+\cos(\phi_A-\phi_B\right]$$

$$= -\frac{b_A}{4B_0}c_a v_B\left[(k_A+k_B)\sin(\phi_A+\phi_B)+(k_A-k_B)\sin(\phi_A-\phi_B)\right]$$

$$(83)$$

So far we have not specified the k's (or the ω's). If the arguments of the sines and cosines do not match, then the right-hand terms will create B and C modes at different k_B's and k_C's, but only in a nonresonant way so nothing will happen.

Let use assume that ω_A, ω_B, and ω_C are positive and for the moment assume that the corresponding k's are positive. We can thereby get matching in two different ways: First, choose $k_A + k_C = k_B$, in such a way that $\omega_A + \omega_B = \omega_B$ so that $\phi_A + \phi_C = \phi_B$ or $\phi_A - \phi_B = -\phi_C$. Then on the right-hand side of (82) the $\cos(\phi_A + \phi_C) = \cos\phi_B$ matches the left-hand side. Ignoring the other term we get

$$\frac{dv_B}{dt} = -\left(\frac{k_A b_A c_a^2}{2B_0 c_s}\right)v_C \qquad (84)$$

In equation 83, the $\sin(\phi_A - \phi_B) = \sin(-\phi_C) = -\sin\phi_C$ and we get

$$-\frac{dv_C}{dt} == \frac{b_A}{8B_0}k_C c_a v_B \qquad (85)$$

Eliminating v_C from these equations we get

$$\frac{d^2 v_B}{dt^2} = -\frac{1}{8}\left(\frac{b_A}{B_0}\right)^2 \frac{k_A k_C c_a^3}{c_s} v_B \tag{86}$$

In this case we see that v_B oscillates as does v_C and no net energy is transmitted to the B and C waves.

We cannot have $\phi_A + \phi_C = -\phi_B$, since all the k's are positive. However, another way of matching in these two equations comes when $\phi_A - \phi_C = \phi_B$ or $\phi_A - \phi_B = \phi_C$. In this case the equations for v_B and v_C become

$$\frac{dv_B}{dt} = \frac{b_A}{2B_0}\frac{k_A c_a^2}{c_s} v_A \tag{87}$$

$$-\frac{dv_C}{dt} = -\frac{k_A}{4B_0} k_C c_a v_B \tag{88}$$

and, again, the resulting equation for v_B is

$$\frac{d^2 v_B}{dt^2} = -\frac{1}{8}\left(\frac{b_A}{B_0}\right)^2 \frac{k_A k_C c_a^3}{c_s} v_B \tag{89}$$

In this case the B wave and the C wave also oscillate in amplitude, transferring energy back and forth between themselves but neither growing secularly.

The third possible way to match is $\phi_A - \phi_C = -\phi_B$ or $\phi_A + \phi_B = \phi_C$ and it is easy to see that this matching also leads to oscillation. Thus, as long as all the k's are positive we get oscillation of only the B and the C waves.

Let us now choose $k_A > 0$, $k_B < 0$, and $k_C > 0$, but all of the ω's still postive. Because of the change in sign of k_B the relation between δb_B and δv_B in equations 76 and 77 changes sign. As a consequence, equations 86 and 89 also change sign, and, indeed, all three possible matchings now lead to instability and growth of the energies of the B and C waves at the expense of the energy in the A wave. But is it really possible to satisfy the selection rules on the k's and the ω's?

$$k_C = k_A + k_B$$

$$\omega_C = \omega_A + \omega_B$$

For definiteness let us take the speed of sound c_s to be smaller than the Alfven speed c_a. Then these conditions for the second matching $\phi_A + \phi_C = -\phi_B$ lead to

$$k_A = k_C - |k_B|$$

$$\omega_A = \omega_C + \omega_B \tag{90}$$

These equations can be satisfied if $c_s < c_a$, and lead to the decay of the right-moving Alfven wave is into a lower-frequency, left-moving Alfven wave and a right-moving sound wave. (If we tried the other possibility, $k_C < 0$, $k_A > 0$, $k_B > 0$ we would find that the first two of the matching cases and the previously excluded case would be stable. The third case would be stable but it could be impossible because of our choice of the signs of the k's.)

Thus, the only decay situation is $k_A > 0$, $k_B < 0$, $k_C = k_A - k_B$, and k_B satisfies

$$k_B = -\frac{c_a - c_s}{c_a + c_s} k_A < 0$$

and the growth rate of the B and C modes, pumped by the A wave is

$$\gamma = \frac{1}{\sqrt{8}} \frac{b_A}{B_0} \sqrt{\omega_A \omega_C} \frac{c_a}{c_s} \tag{91}$$

The complete details of this problem with all the nonlinear terms is carried out in Sagdeev and Galeev (1969).

The total decay time of the A wave is of order that for the energy of the B wave to reach the energy for the A wave. (The C wave energy is different from the B wave by the factor $(c_s/c_a)^2$.) Thus, from equation 91, the decay time is of order

$$\sqrt{8} \frac{B_0 \ln(b_A/b_B)}{b_A \sqrt{\omega_A \omega_C}} \frac{c_s}{c_a}$$

and is inversely proportional to the square root of the energy in the A mode times the geometric mean of the period of the waves.

Actually, in the special case in which an Alfven wave decays into another (lower-frequency wave) and a sound mode, the sound mode is generally very strongly damped, at a rate γ_S. Thus, equations 87 and 88 should be written as

$$\frac{dv_B}{dt} = \alpha v_C \tag{92}$$

$$\frac{dv_C}{dt} = \beta v_B - \gamma_S v_C \tag{93}$$

We can usually drop the left-hand side of equation 93 and write

$$v_C = \frac{\beta v_B}{\gamma_S} \tag{94}$$

and thus (92) becomes

$$\frac{dv_B}{dt} = \frac{\alpha\beta}{\gamma} v_B = \frac{1}{2} \left(\frac{b_A}{2B_0}\right)^2 \frac{k_A k_C c_a^3}{\gamma_S c_s} v_B \tag{95}$$

so the growth rate of the B wave (and the C wave) is

$$\gamma = \frac{1}{2} \left(\frac{b_A}{2B_0}\right)^2 \frac{\omega_A \omega_C}{\gamma_S} \left(\frac{c_a}{c_s}\right)^2 \tag{96}$$

Then the rate of buildup of the B and C waves is slowed down by the damping of the acoustic wave and is quadratic in b_A rather than linear.

Using this result to modify equation 91, we see that the time to affect the A wave is of order

$$8\left(\frac{B_0}{b_A}\right)^2 \frac{\gamma_S}{\omega_A \omega_C}\left(\frac{c_s}{c_a}\right)^2 \ln\left(\frac{b_A}{b_B}\right) \tag{97}$$

Now, $\gamma_S \approx \omega_C$, so, except for the logarithmic factor, the effective nonlinear damping rate is of order

$$\gamma_{\text{NL}} = -\omega_A \left(\frac{b_A}{B_0}\right)^2 \beta \tag{98}$$

where $\beta = (c_s/c_a)^2 = \frac{1}{2}nT/(B^2/8\pi)$. Thus, large-amplitude waves damp faster.

One of the main effects of the nonlinear processes is to saturate an exponential instability. If mode A is exponentially unstable with growth rate γ_A, and mode B is also unstable with a different growth rate γ_B, then they cannot grow indefinitely but must be balanced by a nonlinear damping. In our case we have

$$\frac{\partial}{\partial t}\left(\frac{b_A}{B_0}\right)^2 = -\frac{b_A^2}{B_0^2}\frac{b_B^2}{B_0^2}\omega_a \beta \tag{99}$$

Adding the linear growth γ_A we have

$$\frac{\partial}{\partial t}\left(\frac{b_A}{B_0}\right)^2 = 2\gamma_A \left(\frac{b_A}{B_0}\right)^2 - \frac{b_A^2}{B_0^2}\frac{b_B^2}{B_0^2}\omega_a \beta \tag{100}$$

To get a balance we must have

$$\frac{b_B^2}{B_0^2} = \frac{2\gamma_A}{\beta \omega_A} \tag{101}$$

That is, saturation of the A mode puts a limit on b_B^2. The same thing happens to the saturation of the B mode. It puts a limit on the next D mode, etc. In general, when a range of modes are linearly unstable, equation 101 will lead to a balance between the sum of all the energies $\Sigma(b_B/B_0)^2$ shortward of A and the equation should be satisfied by all the wavelengths k_A. It should lead to a turbulent spectrum and this spectrum will depend on how γ_A varies with k_A. For very large k_A viscous effects come in and the spectrum truncates (e.g., γ_A becomes negative).

11.5 Nonlinear Landau Damping

So far we have discussed fluid nonlinear effects. The scattering of electromagnetic radiation by density fluctuations involves the macroscopic current. It does not involve individual particles. In this case two waves combined to form a third. Mode decay also involves only fluid properties and is macroscopic.

There is an important nonlinear process that involves coherent scattering of energy between just two waves. It is called nonlinear Landau damping since it involves resonance particles. In it, two waves A amd B interact to form a beat wave, and this beat wave resonantly interacts with particles traveling with its speed $v_r = (\omega_A - \omega_B)/(k_A - k_B)$ (Lee and Volk 1973; Kulsrud 1978). The resonant particles can extract energy from the sum of the energies of the two waves. The higher frequency wave with frequency ω_A gives up an energy $\hbar\omega_A$ with some of it, $\hbar\omega_B$, ending up in wave B and the difference going to the resonant particles. The process can reverse if $\partial f_0/\partial v$ is positive at v_r, and the B wave and the resonant particles give up energies $\hbar\omega_B$ and $\hbar(\omega_A - \omega_B)$, the sum of which, $\hbar\omega_A$, goes to the A wave. (In both these processes the number of "photons" seems to be preserved.)

Let us illustrate the process of nonlinear damping by considering the interaction of two circularly polarized Alfven waves, A and B, propagating in the same direction along B_0 (in the z direction) with frequencies ω_A and ω_B and wave numbers k_A and k_B. Then the perturbed fields are in the x, y plane

$$\begin{aligned} \delta B_x &= b_A \cos\phi_A + b_B \cos\phi_B \\ \delta B_y &= b_B \sin\phi_A + b_B \sin\phi_B \end{aligned} \tag{102}$$

where, as before, $\phi_A = k_A x - \omega_A t$, $\phi_B = k_B x - \omega_B t$. Then

$$(\delta b_\perp)^2 = b_A^2 + b_B^2 + 2 b_A b_B \cos(\phi_A - \phi_B) \tag{103}$$

and the field strength to second order is

$$|\mathbf{B}| = \sqrt{B_0^2 + (\delta b_\perp)^2} \approx B_0 + \frac{(\delta b_\perp)^2}{2B_0} = \frac{b_A b_B}{B_0}\cos(\phi_A - \phi_B) + \text{const} \tag{104}$$

The magnetic mirror force exerted on an ion with perpendicular velocity $v_\perp$ is

$$\begin{aligned} -\frac{M v_\perp^2}{2B}\frac{\partial |B|}{\partial x} &= \frac{M v_\perp^2}{2}(k_A - k_B)\left(\frac{b_A b_B}{B_0^2}\right)\sin(\phi_A - \phi_B) \\ &= F \sin(\phi_A - \phi_B) \end{aligned} \tag{105}$$

If we replace F by an effective electric force $F = eE_{\text{eff}}$, we see from equation 84 in chapter 10 that some ions resonate with the beat wave, i.e., those with

$$v_r = \frac{\omega_A - \omega_B}{k_A - k_B} = c_a \tag{106}$$

(since $\omega_A = k_A c_a$, $\omega_B = k_B c_a$, where c_a is the Alfven speed). According to equation 84 in chapter 10, ions with a given $v_\perp$ will be bunched into a density

$$n_1(v_\perp) = -\pi \frac{eE_{\text{eff}}}{Mk} \frac{\partial f}{\partial v_z}\bigg|_{v_z=v_r} \sin(\phi_A - \phi_B)$$
$$= \hat{n}_1(v_\perp)\sin(\phi_A - \phi_B) \tag{107}$$

where $k = k_A - k_B$ so that

$$\hat{n}_1(v_\perp) = \frac{-\pi F}{M(k_A - k_B)} \frac{\partial f}{\partial v_x}\bigg|_{c_a} \tag{108}$$

The force $F\sin(\phi_A - \phi_B)$ will do work on these particles with a given $v_\perp$ at a spatially average rate

$$\dot{\mathcal{E}}(v_\perp) = \left\langle \hat{n}_1 F c_a \sin^2(\phi_A - \phi_B)\right\rangle = \frac{\hat{n}_1 F c_a}{2} = -\frac{\pi}{2}\frac{F^2 c_a}{M(k_A - k_B)} \frac{\partial f}{\partial v_x}\bigg|_{c_a} \tag{109}$$

Now, as noted above and shown below

$$-\frac{\dot{\mathcal{E}}_A}{\omega_A} = \frac{\dot{\mathcal{E}}_B}{\omega_B} \tag{110}$$

where $-\mathcal{E}_A$ is the energy given up by wave A, and $\mathcal{E}_B$ is the energy received by wave B. The total energy given to all the resonant ions, $\mathcal{E}_{\text{tot}} = \int \dot{\mathcal{E}}(v_\perp) d^2 v_\perp$ is the difference of these two energies, so

$$\dot{\mathcal{E}}_{\text{tot}} = -\dot{\mathcal{E}}_A - \dot{\mathcal{E}}_B = -\dot{\mathcal{E}}_A\left(1 - \frac{\omega_B}{\omega_A}\right) = -\dot{\mathcal{E}}_A \frac{k_A - k_B}{k_A} \tag{111}$$

and the rate of damping of the A wave is thus

$$-\dot{\mathcal{E}}_A = \frac{k_A}{k_A - k_B}\dot{\mathcal{E}}_{\text{tot}} = -\int d^2 v_\perp \frac{\pi}{2}\frac{k_A c_a F^2}{M(k_A - k_B)^2} \frac{\partial f}{\partial v}\bigg|_{c_a} \tag{112}$$

But if $c_a \ll c_i = \sqrt{T/M}$, then

$$\frac{\partial f}{\partial v_z}\bigg|_{v_z=c_a} \approx - - n\frac{c_a}{v_i^2} f(0, v_\perp) = -n\frac{c_a}{\sqrt{2\pi}\, v_i^3} \frac{e^{-v_\perp^2/2v_i^2}}{2\pi v_i^2} \tag{113}$$

Substituting the expressions for F from equation 105 and for $\partial f/\partial v_z$ from equation 113 into equation 112 and carrying out the $v_\perp$ integration we get

$$\dot{\mathcal{E}}_A = -2\gamma_{\text{NL}}\mathcal{E}_A = -\frac{1}{4}\sqrt{\frac{\pi}{2}}\omega_A \frac{c_a}{v_i} n m v_i^2 \left(\frac{b_A b_B}{B_0^2}\right)^2 \tag{114}$$

But $\mathcal{E}_A = (b_A)^2/8\pi$, so

$$\gamma_{\text{NL}} = \frac{1}{4}\sqrt{\frac{\pi}{2}}\frac{v_i}{c_a}\left(\frac{b_B}{B_0}\right)^2 \omega_A \tag{115}$$

To verify equation 110, transform to the beat wave frame. Since the wave momentum is ω/k times its energy we can easily see that E_A/ω_A and E_B/ω_B

are invariants under this transformation. In the beat wave frame they must be equal, since in its own frame, the beat wave does no work on the ions.

Let us compare nonlinear Landau damping with mode decay as a method to nonlinearly damp an Alfven wave. First, note that nonlinear damping operates only if the plasma is collisionless. If the collision rate ν is larger than $\gamma_{\rm NL}$, then the ions cannot effectively nonlinearly Landau damp the beat wave. Second, in the case that the ion–sound wave is strongly damped (also a collisionless limit), the damping rates are both proportional to the amplitude squared $(b_B/B_0)^2$ of the wave into which energy is passing. Third, nonlinear Landau damping is most effective when the two Alfven waves are propagating in the same direction, while, as was shown for the mode decay, the Alfven waves must be propagating in opposite directions. For the important case of cosmic rays streaming through a background plasma, only waves propagating in the same direction in which the cosmic rays are streaming are unstable, while those propagating in the opposite direction are damped. This makes the mode decay process essentially inapplicable to this case.

We are interested in nonlinear damping mechanisms, since these are important saturation mechanisms for unstable waves. If the wave is unstable with linear growth γ, its energy satisfies the equation

$$\frac{d\mathcal{E}}{dt} = 2\gamma\mathcal{E} \tag{116}$$

as long as nonlinear terms are unimportant. Thus, the wave grows exponentially as $\mathcal{E} \sim e^{2\gamma t}$ until nonlinear terms actually do become important. Now, in general, if some waves are unstable, a whole class of waves are unstable so not only is a wave with a given wave number unstable but those with neighboring values of k are also unstable. Thus, as nonlinear effects saturate waves, the spectrum of wave energy settles down to a continuous function of k. So, for a nonlinear damping mechanism such as nonlinear Landau damping we can replace the nonlinear damping rate of wave A

$$\gamma_{\rm NL}(A) = \frac{1}{4}\sqrt{\frac{\pi}{2}}\frac{v_i}{v_A}\left(\frac{b_B}{B_0}\right)^2 \omega_A \tag{117}$$

by the rough approximation

$$\gamma_{\rm NL}(A) \approx \frac{1}{4}\sqrt{\frac{\pi}{2}}\frac{v_i}{v_A}\left(\frac{b_A}{B_0}\right)^2 \omega_A = \alpha\mathcal{E}_A \tag{118}$$

since wave B has an energy comparable with that of wave A. With this assumption we then have for the energy of wave A the approximate equation

$$\frac{d\mathcal{E}_A}{dt} = 2\gamma\mathcal{E}_A - \alpha\mathcal{E}_A^2$$

and as the wave develops a large amplitude we find

$$\mathcal{E}_A \rightarrow \frac{2\gamma}{\alpha} \tag{119}$$

We will apply these ideas in chapter 12 when we discuss the propagation of cosmic rays in the interstellar medium.

11.6 Particle Trapping

Another nonlinear mechanism for saturation of a growing collisionless mode such as a plasma oscillation is particle trapping (Kreuer and Dawson 1970). In discussing the physical explanation for Landau damping (and growth) in the last chapter we showed that in the small-amplitude limit the resonant particles tended to bunch into a density perturbation coherent with the electric field. But for larger amplitudes, we expect them to oscillate back and forth in the potential well and to effectively bunch into a perturbation coherent with the electric potential. The latter bunching is out of phase with the electric field and produces no growth or damping.

The time for this to happen varies inversely with the square root of the amplitude of the wave, while the linear bunching time is independent of the amplitude. The situation is such that the linear bunching happens almost instantaneously, while the nonlinear bunching takes a time τ_{bounce} of order of the time for a particle, that is nearly resonant, to oscillate in the potential of the wave.

If the original linear bunching is large enough that the Landau damping time $1/\gamma$ is short compared to the nonlinear bunching time τ_{bounce}, then the wave will change its amplitude before this nonlinear effect sets in. If the wave is linearly damped, then the nonlinear effect never catches up and we can ignore the nonlinear trapping effect. That is, if

$$\gamma \tau_{\text{bounce}} \gg 1 \tag{120}$$

then the ordinary linear effect applies.

On the other hand, a larger-amplitude wave has a shorter bounce time. Thus, if the wave is growing, its amplitude will become large enough that $\gamma t_{\text{bounce}} < 1$, the nonlinear bunching will dominate the linear bunching, and the wave will saturate. This happens at the amplitude when

$$\gamma \tau_{\text{bounce}} \approx 1 \tag{121}$$

Let us estimate τ_{bounce} by treating resonant particles as trapped in the bottom of the moving potential well and oscillating in the effective harmonic potential of the wave, i.e., let x be the distance measured from the bottom of the well in the wave frame (see figure 11.7). The force on the trapped particle is obtained by Taylor expanding the electric field and we have

$$m\ddot{x} = -eE(x) = -e\frac{\partial E}{\partial x}x = ek\hat{E}x \tag{122}$$

where $E = \hat{E}\sin x$, so the frequency of oscillation is

$$\omega^2 = \frac{ek\hat{E}}{m}$$

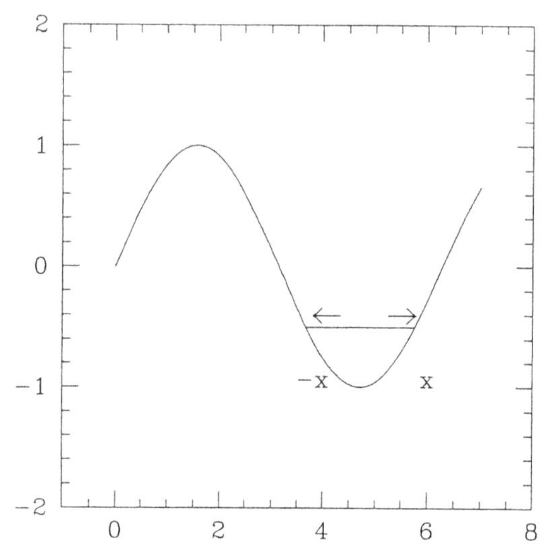

Figure 11.7. Particle trapped in the potential of an electrostatic wave

If τ_{bounce} corresponds to half an oscillation period, then

$$\tau_{\text{bounce}} = \frac{\pi}{\omega} = \frac{\pi}{\sqrt{ke\hat{E}/m}} \tag{123}$$

We see that, if E grows, τ_{bounce} gets shorter, but if E damps, then τ_{bounce} gets longer.

The above discussion applies to a monochromatic plane wave. When dealing with turbulent waves, we should consider that the particles pass through the wave packets with a relative velocity rate $|\mathbf{v} - \mathbf{v}_g|$. If turbulence is broadband, the wave packets have an effective length of order $1/k$. The question is, does the nonlinear bunching affect the particles in a time shorter than the time for them to pass through a wave packet, τ_{dc}, the decorrelation time,

$$\tau_{\text{dc}} = \frac{1}{k(|\mathbf{v} - \mathbf{v}_g|} \tag{124}$$

If this is shorter than τ_{bounce}, we may ignore the trapping effect. That is, if

$$\tau_{\text{dc}} \ll \tau_{\text{bounce}} \tag{125}$$

or

$$k(v_\phi - v_g) \gg \sqrt{keE/m} \tag{126}$$

then nonlinear bunching can be neglected and the linear theory is valid. If the wave turbulence is growing exponentially, then so is the right-hand side of this equation and eventually trapping does become important. (For trapping to be important not only must equation 125 be violated in the case of broadband turbulence, but equation 120 must be as well.) Of course, other nonlinear saturation mechanisms may come in first and limit the wave amplitude. If the turbulence has a narrow bandwidth Δk, the wave packets are larger and the k in the denominator of the right-hand side of equation 124 must be replaced by Δk. The trapping then may set in at a smaller amplitude. If $\Delta k = 0$, we are back to a monochromatic wave and equation 120 is the only criteria.

11.7 The Wave Kinetic Equation

We have illustrated the various nonlinear processes by exhibiting just a few examples of wave–wave interaction in simple contexts to familiarize the reader with the various possibilities, i.e., mode coupling, mode decay, and nonlinear Landau damping. There is a general equation that covers all these possibilities. Its derivation is given in the book of Kadomtsev (1965). We present it without derivation because the derivation is fairly complicated (see also Sagdeev and Galeev 1969). Most nonlinear equations for waves can be reduced to the following form (in Fourier representation):

$$(\omega - \omega_k - i\gamma_k)C(k,\omega) = \int V(k,\omega;k'\omega')C(k',\omega')C(k'',\omega'')dk'd\omega' \qquad (127)$$

where $C(k,\omega)$ is the amplitude of the wave, e.g., the electric field for plasma oscillations. The left-hand side represents the linear behavior of the wave with ω_k the frequency of a wave with wave number k, and γ_k its growth rate, The right-hand side includes all of the nonlinear terms with $V(k,\omega : k' : \omega')$ as the coupling coefficient. $k'' = k - k'$ and $\omega'' = \omega - \omega'$ are convenient notational abbreviations.

Equation 127 corresponds to a simpler case than that represented by equations 75 and 78 in which there are two types of waves, Alfven waves and sound waves. However, to cover this case, C can be treated as a vector with two components a and s and the coupling coefficient V as a tensor with components V_{aa}, V_{as}, V_{sa} and V_{ss}. In addition, we have to break up the cosines and sines into exponentials. V_{as} would be essentially the nonlinear term in equation 75, and V_{as} the nonlinear term in equation 78.

Kadomtsev assumes that the waves have a random origin so that the random phase approximation

$$\langle C^*(k',\omega')C(k,\omega)\rangle = I(k)\delta[\omega - \omega(k)]\delta(k'-k)\delta(\omega'-\omega) \qquad (128)$$

is valid to lowest order. $I(k)$ gives the spectra of the waves and is a slowly varying function of t whose time evolution is given by the "wave kinetic equation"

$$\frac{dI(k)}{dt} = 2\gamma_k I(k) - \pi \int [2v_{k,k'}v_{k',k}I(k)I(k'') - |v_{k,k'}|^2 I(k')I(k'')]$$

$$\times\ \delta(\omega(k) - \omega(k') - \omega(k''))dk' \tag{129}$$

where $v_{k',k}$ is an abbreviation for the symmetrized form of $V(k'\omega', k\omega)$

$$v_{k\omega,k'\omega'} = V(k\omega, k'\omega') + V(k\omega, k''\omega'') \tag{130}$$

The first term in the bracket of the wave kinetic equation corresponds to our mode decay example where the k wave gets energy from the k'' wave. The second term corresponds to our coherent scattering example where modes k' and k'' combine to drive the k mode, e.g., two strong plasma oscillations combine to produce an electromagnetic mode. In this wave plasma instabilities enable a plasma to produce intense radio emission (see Ginzburg and Zhelezniakov 1958). The delta function represents the selection rules of equation 56 in a compact form, since by definition $k'' = k - k'$.

The wave kinetic equation is extremely useful since it solves once and for all, all nonlinear problems that can be reduced to equation 127. It also handles the subtleties of the kinetics of waves once and for all. However, it does not handle kinetic problems such as nonlinear Landau damping. For this a more complex wave kinetic equation exists. We do not present it here but it is given in Kadomtsev's book. The same equations are available in the book of Sagdeev and Galeev, where they are presented from a somewhat different point of view. However, their resulting wave kinetic equations are completely equivalent to Kadomtsev's.

The formulation treated so far is called weak turbulence theory, since an expansion in the wave amplitude is still valid, although we have to go to fourth order to get a complete theory. For the theory to converge the amplitudes of the waves must be small enough that the wave–wave interactions do not force the waves to change their amplitude significantly during a single wave period. For this to be the case, the linear growth γ_k must be small. This is plausible for many problems since we generally start off with a stable system. This system then evolves due to external influences until γ passes through zero and becomes positive at a relatively slow rate. Then saturation is reached, which stops the evolution of the equilibrium while γ is still relatively small.

An example of strong turbulence occurs when the waves have low or zero frequencies such as happens for shear waves in hydrodynamics. In this case, there is no natural frequency with which to compare the mode interaction rates, and there is no easy analytic theory with which to treat the wave–wave interactions. The best procedures seem to be a qualitative treatment backed up by confirmation from numerical, or actual, experiments. Unfortunately, these experiments involve dimensionless parameters, such as the Reynold's number, that are moderate in size compared to actual astrophysical parameters, so that a rigorous comparison between the results and astrophysical problems is difficult.

11.8 Kolmogoroff Turbulence

Perhaps the most successful of the qualitative approaches to strong turbulence is the Kolmogoroff theory of isotropic, homogeneous, incompressible, unmagnetized, hydrodynamic turbulence. This theory is both very useful, easy to comprehend and use, and well supported both experimentally and by numerical simulations (Chandrasekhar 1949). Kolmogoroff argues as follows: Suppose some energy is put into a hydrodynamic fluid such as the interstellar medium by hot stars or supernova at a very large scale. This energy is generally put into the form of large-scale vortices. Due to the nonlinear term $\mathbf{v} \cdot \nabla \mathbf{v}$ in the equation of motion, each separate vortex breaks up into vortices of about one-half its size, at a rate of order $\mathbf{v} \cdot \nabla \approx k\tilde{v}$. That is, a vortex lasts for a time comparable to its "turnover time." There is a chain of such vortices extending from the outer scale, the large scale at which the energy is fed in, down to a very small scale where viscous effects destroy the vortices faster than they can transfer energy to still smaller scales. This small scale is called the inner scale.

The ratio of these two scales, the outer scale over the inner scale, is generally very large and the range in scales between them is called the inertial range. In the inertial range we expect the turbulence to look the same at all scales. As Kolmogoroff showed, this is borne out by the picture of the transfer of energy from any vortex to a smaller vortex at the eddy turnover rate.

Let the kinetic energy with wave numbers between k and $dk + dk$ be $\rho I(k)dk$. Thus,

$$\frac{\rho v^2}{2} = \frac{\rho}{2} \int I(k)dk \tag{131}$$

is the total kinetic energy density. The velocity v_k due to eddies in the range $dk \sim k$ is

$$v_k^2 = I(k)dk \tag{132}$$

In figure 11.8 consider that the energy is the range between $\sqrt{\epsilon}k$ and $k/\sqrt{\epsilon}$, where $\epsilon \leq 1$. Energy will be transferred into this range from larger eddies with k in the range between $\epsilon^{3/2}k$ and $\sqrt{\epsilon}k$ at the turnover rate of eddies in this larger range, $\approx \epsilon k v_{\epsilon k}$. In turn, energy in the first range will be transferred out to eddies of the next smaller range at the rate $\approx k v_k$. There will also be vicious dissipation of eddies at the rate $k^2 \nu$, where ν is the kinematic viscosity.

The balance equation for the energy kI_k is

$$\frac{\partial(kI)}{\partial t} = \epsilon k v_{\epsilon k}\left[\epsilon k I(\epsilon k)\right] - (k v_k)\left[kI(k)\right] - k^2 \nu k I(k) \tag{133}$$

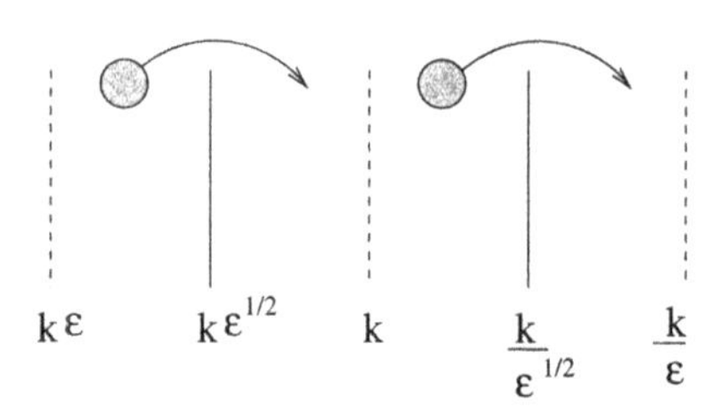

Figure 11.8. The turbulence transfer of energy to smaller scales

Treating the first two turns or the right as a logarithmic derivative $d/d\ln k$ but taking $\epsilon = 1$, we get

$$\frac{\partial(kI)}{\partial t} = -\frac{\partial}{\partial \ln k}\left(k\sqrt{kI}kI\right) - k^3\nu I \tag{134}$$

The boundary condition employed at the outer scale $k = k_0 = 2\pi/L$, where v_0 is the turbulent velocity there, is

$$k_0 I(k_0) = v_0^2 \tag{135}$$

We may solve for I in the steady state by letting

$$y = k^{5/2} I^{3/2} \tag{136}$$

Eliminating I from equation 134 and canceling a k we get

$$\frac{dy}{dk} = -\nu k^{1/3} y^{2/3} \tag{137}$$

or

$$\left(k^{5/2} I^{3/2}\right)^{1/3} = y^{1/3} = \alpha - \frac{\nu k^{4/3}}{4} \tag{138}$$

From the boundary condition equation 135, $\alpha = k_0^{1/3} v_0$ (since at k_0 we can neglect the viscous term). Therefore, replacing y in equation 138 by (136) and rearranging terms we get

$$I = \frac{k_0^{2/3} v_0^2}{k^{5/3}}\left[1 - \left(\frac{k}{k_0}\right)^{4/3} \frac{1}{4R}\right]^2 \tag{139}$$

where $R = v_0/k_0\nu$ is the hydrodynamic Reynold's number (up to a factor of 2π). The inner scale $k_{\max}$ occurs when the bracket vanishes or $k_{\max} = k_0(4R)^{3/4}$, so the inner scale is smaller than the outer scale by the factor $(4R)^{3/4}$, which is, essentially, the Reynolds number to the three-fourth's power. For $k \ll k_{\max}$, i.e., scales much larger than the inner scale, the bracket is unity and we have

$$I = \frac{k_0^{2/3} v_0^2}{k^{5/3}} \tag{140}$$

the well-known Kolmogoroff spectrum.

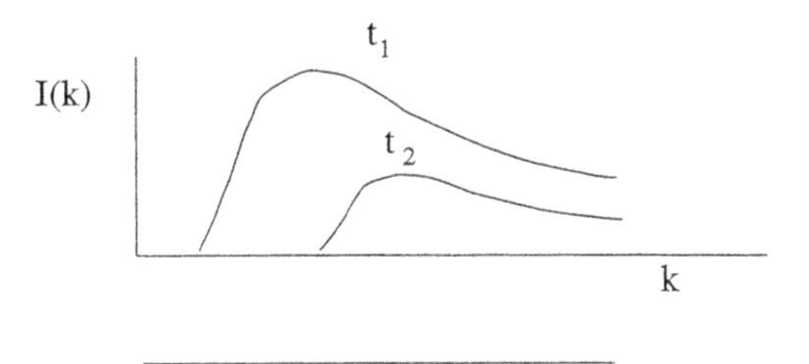

Figure 11.9. Kolmogoroff turbulence decaying in time after the source is turned off

Now, the power input Γ needed to sustain the turbulence in a steady state is the flux of energy carried down the turbulent scale by mode coupling and this is $\rho/2$ times the expression in the parentheses of equation 134:

$$\Gamma = \frac{\rho}{2}k^{5/2}I^{3/2} = \frac{\rho}{2}\left(k^{5/3}I\right)^{3/2} = \frac{\rho k_0 v_0^3}{2} \tag{141}$$

This is the energy that must be put in by the source of the turbulence to sustain the turbulence.

If the source is cut off, k_0 and v_0 will change and be functions of t. The turbulence will decay, but at a rate slower than the time it takes to adjust the spectrum to fit the slowly changing k_0 and v_0. k_0 will roughly double in the largest eddy turnover time $k_0 v_0(t)$, so that we have $d\log k_0/dt = k_0 v_0$ or

$$\frac{dk_0}{dt} = k_0^2 v_0.$$

The turbulent energy $\rho v_0^2/2$ will change at the rate $\Gamma(t)$ given by equation 141 so that

$$\frac{d}{dt}\frac{\rho v_0^2}{2} = -\frac{\rho k_0 v_0^3}{2} \tag{142}$$

From these two equations we have

$$k_0 v_0^2 = \text{const} \tag{143}$$

and

$$\frac{dv_0}{dt} = -\left(\frac{k_0 v_0^2}{2}\right) = \left(\frac{k_0 v_0^2}{2}\right)_{\text{initial}} \tag{144}$$

and v_0 decreases linearly in time (see figure 11.9). The time it takes to reach zero or equivalently the lifetime of the turbulence is thus $\approx 2/k_0 v_0$ or two initial turnover times of the largest eddy.

11.9 MHD Turbulence

In the presence of a strong, uniform, magnetic field, $\mathbf{B}_0$, incompressible turbulence is turned into MHD turbulence and changes its character entirely.

The ordinary hydrodynamic shear modes of unmagnetized turbulence are replaced by two of the three MHD modes, the slow and intermediate modes. The fast mode is compressible and does not enter. The transfer of energy to small scales is still by mode coupling, but because of the magnetic field, the transfer rate is not isotropic.

11.9.1 An Exact Solution

As has been emphasized by Goldreich and Sridhar, the striking difference between incompressible MHD turbulence and ordinary hydrodynamic turbulence is based on a remarkable, exact, nonlinear solution for Alfven waves propagating in a single direction along the magnetic field, say, rightward. We are already familiar with an exact linear solution for a right-moving Alfven wave (see equation 13 in chapter 5) in which the wave propagates linearly with an arbitrary profile in the transverse direction. We now show how this generalizes to the nonlinear case.

To show this Goldreich and Sridhar start with the ideal, incompressible equations, which are

$$\frac{\partial \mathbf{B}}{\partial t} = \nabla \times (\mathbf{v} \times \mathbf{B})$$

$$\frac{\partial \mathbf{v}}{\partial t} = -\mathbf{v} \cdot \nabla \mathbf{v} + \frac{\mathbf{B} \cdot \nabla \mathbf{B}}{4\pi\rho} - \frac{\nabla P}{\rho} \tag{145}$$

where $P = p + B^2/8\pi$ and $\nabla \cdot \mathbf{v} = \nabla \cdot \mathbf{B} = 0$. Take $\mathbf{B} = \mathbf{B}_0 + \mathbf{b}$, where B_0 is a constant field in the z direction. Then the equations can be written as

$$\frac{\partial \mathbf{b}}{\partial t} - \mathbf{B}_0 \cdot \nabla \mathbf{v} = \nabla \times (\mathbf{v} \times \mathbf{b})$$

$$\frac{\partial \mathbf{v}}{\partial t} - \frac{\mathbf{B}_0 \cdot \nabla \mathbf{b}}{4\pi\rho} = -\mathbf{v} \cdot \nabla \mathbf{v} + \frac{\mathbf{b} \cdot \nabla \mathbf{b}}{4\pi\rho} - \frac{\nabla P}{\rho} \tag{146}$$

These equations have an exact nonlinear solution, which is a right-moving wave. To find this solution take $\mathbf{v} = \mathbf{b}/\sqrt{4\pi\rho}$. This makes all of the terms on the right-hand side of these equations vanish except the ∇P term. Thus, we need only $\mathbf{v}$ and $\mathbf{b}$ to satisfy the linear equations

$$\frac{\partial \mathbf{b}}{\partial t} - v_A \frac{\partial \mathbf{b}}{\partial z} = 0$$

$$\frac{\partial \mathbf{v}}{\partial t} - v_A \frac{\partial \mathbf{v}}{\partial z} = -\frac{\nabla P}{\rho} \tag{147}$$

P is a free function in an incompressible plasma, and is usually taken to make $\nabla \cdot \mathbf{v}$ vanish. Indeed, we can see that this implies $\nabla^2 P = 0$, and for an isolated solution P must be a constant, and $\nabla P = 0$. A solution of the $\mathbf{b}$ equation is then

$$\mathbf{b} = \mathbf{b}(x, y, z - v_A t) \tag{148}$$

for any arbitrary function x, y, and $z - v_A t$, which satisfies $\nabla \cdot \mathbf{b} = 0$. Then $\mathbf{v} = \mathbf{b}/\sqrt{4\pi\rho}$ is also divergence free and satisfies its equation.

11.9.2 The Wave Interactions

Now let us see how this solution bears on the mode-coupling problem. First, treat the waves as of small amplitude, and Fourier analyze **b** and **v**:

$$\mathbf{b} = \int \mathbf{b}(k_z, x, y) e^{i(k_z z - \omega_{k_z} t)} \tag{149}$$

We only consider the Fourier harmonics in z, leaving the x and y dependence explicit. For definiteness we concentrate on the harmonics with positive ω. Now, from our exact solution we may conclude that if all the k_z's were positive there would be no coupling to smaller, parallel wavelengths. Thus, let us consider that some harmonics have negative values for k_z so that these waves will pass through positive k_z waves.

Consider the mode coupling of these pairs of waves with each other, through the nonlinear terms such as $\mathbf{v} \cdot \nabla \mathbf{v}$ and $\mathbf{b} \cdot \nabla \mathbf{b}$. For example, we have

$$\mathbf{b} \cdot \nabla \mathbf{b} = \int dk_z dk_z' \left(\mathbf{b}(k_z') \cdot \mathbf{b}(k_z)\right) e^{i\{(k_z' + k_z)z - [\omega(k_z') + \omega(k_z)]t\}} \tag{150}$$

where we have suppressed the x and y dependence in $\mathbf{b}(k_z')$ and $\mathbf{b}(k_z)$. As we have shown, the terms with the same sign of k_z do not contribute. The contribution from a term with positive k_z and negative k_z' will drive a further mode with parallel wave number $k_z - |k_z'|$ with a frequency of $(k_z + |k_z'|)v_A$ by our convention about considering only positive frequencies. Since the mode it is attempting to drive has a frequency $(k_z - |k_z'|)v_A$, these cannot match unless k_z' is very small. (It need not be exactly zero because there is some small nonlinear frequency that allows matching.) But for such nonlinear interaction the change in k_z is very small and so any cascade toward larger k_z is strongly suppressed. However, when we consider the transverse variations in these modes, we see that there will be a substantial cascade toward larger $k_\perp$.

This is the basis for the strong anisotropy in MHD turbulence. It is based on the existence of modes with very small k_z (Ng and Bhattacharjee 1996). Do such modes exist? The answer is yes, even if the turbulence consists of rather short wave packets. When two opposite such wave packets pass through each other they do not leave any velocity or magnetic disturbance behind, but they do produce a sideways displacement in the field lines. Because each wave packet tends to follow field lines this displacement acts as a zero-wavenumber mode. (In fact, if we look inside a wave packet we see that the spatial integral of the transverse **b** is actually nonzero, and this is the zero-frequency mode.)

It turns out to be clearer and easier to visualize the interactions of the modes through the line displacements in right-moving wave packets shearing left-moving wave packets that they pass through and vice versa. This shearing produces smaller transverse scales. This appears to be the easiest way to qualitatively treat the cascade and the turbulent spectrum. Let us now see how Goldreich and Sridhar (Goldreich and Sridhar 1995, 1997; Sridhar and Goldreich 1994) actually accomplish this.

11.9.3 The Goldreich–Sridhar Theory

Let us assume that the interstellar medium is filled with Alfven waves with a velocity spectrum

$$v^2 = \int E_3(\mathbf{k}) d^3\mathbf{k} = \int E_2(\mathbf{k}_\perp) d\mathbf{k}_\perp^2 \tag{151}$$

where $E_2(\mathbf{k}_\perp) = \int E_3(\mathbf{k}) dk_z$. We first examine small $k_\perp = 2\pi/\lambda$ modes, where λ is the transverse scale for which the turbulence is weak. This is actually the case for small $k_\perp$. Then, because of the suppression of the cascade in k_z, all the k_z's remain concentrated around the same reciprocal wavelength ℓ. Because of the freedom of the line displacements allowed in the interstellar medium, the net displacement of any of the lines in a given wave packet, with transverse scale $\lambda \sim 1/k_\perp$ and velocity v_λ.

$$\xi_1 = \int_{-\infty}^{\infty} \frac{v_\lambda}{v_A} dz \tag{152}$$

need not vanish and we expect it to be of order

$$\xi_1 \approx \ell \frac{v_\lambda}{v_A} \tag{153}$$

where $\ell \approx k_z^{-1}$ is the length of the wave packet. ξ_1 will vary transversely in the wave packet over the scale λ. ($v_\lambda^2 = k_\perp^2 E_2(k_\perp)$)

Consider a right-moving wave packet B encountering a left-moving wave packet C in figure 11.10. Wave packet B will attempt to follow the displacement of the lines in C, and it will end up sheared over the transverse scale λ by a relative amount ξ_1/λ. (In this case the three-wave interaction in which B and C produce a smaller scale structure in B does not vanish, because the sideways displacement of the lines in C has a $k_z = 0$ wave number, since it is a permanent displacement. Note that k_z does not change as a result of the interaction but $k_\perp$ does.)

Let the lines of force in C have random net displacements $\xi_C \approx \xi_1$ before the encounter and let wave packet B have a no net displacement. Then after the encounter some lines in B will have a displacement ξ_C in B, while other lines will have a different displacement. Thus, the wave packet B will be transversely distorted by wave packet C. If the left-moving wave packets are packed densely, then after moving a distance z, B will encounter of order $2z/\ell$ wave packets of type C, and will suffer of order $2z/\ell$ random transverse displacements $\pm\xi_1$ with random signs. The net square displacements of the

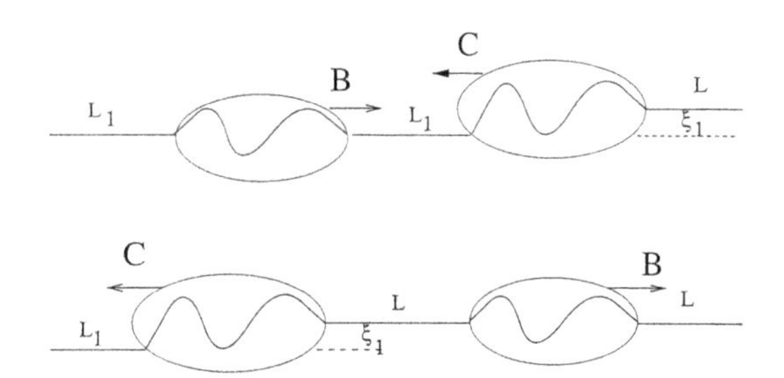

Figure 11.10. Wave packets B and C interacting

lines in B will then be random and of order

$$\boldsymbol{\xi}(z)^2 = \frac{2z}{\ell}\boldsymbol{\xi}_1^2 \approx \frac{z\ell v_\lambda^2}{v_A^2} \tag{154}$$

with transverse wavelength λ.

After moving a distance z_1 such that $\boldsymbol{\xi}(z_1) \approx \lambda$, the wave packet B will be distorted finitely, and its energy will be transferred to $\lambda/2$. The number of interactions for this to happen is

$$N = \frac{z_1}{\ell} = \frac{\lambda^2}{\ell^2(v_\lambda/v_A)^2} = \left(\frac{v_A\lambda}{v_\lambda\ell}\right)^2 \tag{155}$$

and the time for this transfer to occur is

$$t_\lambda = \frac{N\ell}{v_A} = \left(\frac{\lambda^2}{v_\lambda^2}\right)\frac{v_A}{\ell} \tag{156}$$

The rate of transfer of energy to larger $k_\perp$ is thus

$$\epsilon \approx \frac{v_\lambda^2}{t_\lambda} \approx \frac{v_\lambda^4}{\lambda^2}\frac{\ell}{v_A} \tag{157}$$

In a steady state this is a constant, and the spectrum for v_λ^2 is

$$v_\lambda^2 = \sqrt{\frac{\lambda^2 v_A \epsilon}{\ell}} \tag{158}$$

or v_λ is proportional to $\lambda^{1/2}$. That is,

$$v_\lambda \quad \propto \quad \lambda^{1/2} \tag{159}$$

or

$$E_2(k_\perp) \quad \propto \quad \frac{1}{k_\perp^2} \tag{160}$$

Notice that, since $v_\lambda/v_A \propto \lambda^{1/2}$, the waves become more linear as λ decreases. However, from equation 155 we see that

$$N \approx \lambda\sqrt{\frac{v_A^3}{\epsilon\ell^3}} \tag{161}$$

and the number of interactions becomes smaller as λ decreases. Eventually, we reach a λ, $\lambda_1 = \ell\sqrt{\epsilon\ell/v_A^3}$, below which N, the number of required interactions to transfer energy to the next smaller scale, would becomes less than one if our estimates presented above remained valid.

However, when N is of order or slightly less than one, the restrictions on the wave frequencies no longer hold, as we remarked earlier. We only have to satisfy the selection rule, $\omega_1+\omega_2 = \omega_3$, if the lifetimes of the waves are many wave periods. When $N = 1$ the waves last only one period, and we may ignore the frequency selection rule. This is the situation in strong turbulence. (It is also the situation in Kolmogoroff turbulence, which is always strong.) Thus, for λ smaller than λ_1, we expect k_z no longer to be held constant but to increase.

It is unreasonable to expect that N can continue decreasing with λ, since if $N < 1$, the wave packets would no longer live a full wave period. But the increase in k_z will increase the transfer rate if we assume a single wave packet interaction per transfer. This effectively reduces the magnitude of v_λ^2 to maintain the energy transfer at constant ϵ, since

$$\epsilon = k_z v_A v_\lambda^2 \tag{162}$$

The requirement that N is held at one is

$$N = 1 = \frac{k_z v_A \lambda}{v_\lambda} \tag{163}$$

(or alternatively $\xi_1 = \lambda$). Solving for k_z and v_λ from these two equations yields

$$v_\lambda = (\epsilon\lambda)^{1/3} \tag{164}$$

and

$$k_z v_A = \frac{\epsilon^{1/3}}{\lambda^{2/3}} \tag{165}$$

This balance between k_z and v_λ must be maintained. If k_z were to increase more rapidly as λ decreases, then v_λ would decrease (for fixed ϵ) and N would be larger than one, shutting off the k_z increase. If k_z were to increase more slowly than given by equation 165, v_λ would be larger and N would drop below one, which is implausible.

In summary, if $\lambda > \lambda_1$, then the turbulence is weak, k_z is constant, and v_λ decreases as $\lambda^{1/2}$. If $\lambda < \lambda_1$, then the turbulence is strong, k_z increases with a further decrease in λ as $\lambda^{-2/3}$, and v_λ decreases as $\lambda^{1/3}$.

This strong turbulence variation of v_λ is the same in Kolmogoroff hydrodynamic turbulence. If we could translate the Alfven wave spectrum into a density perturbation spectrum, either by entropy mixing or coupling to the slow mode, then the resulting spectrum would be consistent with that derived from observations of interstellar scintillation (Manchester and Taylor 1977).

11.10 Problems

1. (Elsasser variables) Introduce the Elsasser variables

$$\xi = \mathbf{v} + \frac{\mathbf{b}}{\sqrt{4\pi\rho}}, \qquad \eta = \mathbf{v} - \frac{\mathbf{b}}{\sqrt{4\pi\rho}}$$

into equation 146, and show that the nonlinear terms occur only as products of ξ and η. Thus, show that if $\eta = 0$, we get the exact equation for the right-moving wave

$$\frac{\partial \xi}{\partial t} - v_A \frac{\partial \xi}{\partial z} = -\nabla P$$

and if $\xi = 0$, we get the left-moving wave equation

$$\frac{\partial \eta}{\partial t} + v_A \frac{\partial \eta}{\partial z} = -\nabla P$$

By the incompressibility argument of the text show that $\nabla P = 0$. Thus, show that the general solution for right-moving waves is η, any divergence free vector function of x, y, and $z - v_A t$.

Identify the slow and intermediate right-moving waves in the incompressible limit by taking the linear solution $\eta \sim e^{i\mathbf{k}\cdot\mathbf{r}-\omega t}$, where $\mathbf{k}$ is in the x–z plane, and show that we get the intermediate wave when η is in the y direction, and the slow wave when η is in the x–z plane.

2. (Density statistics for randomly disposed particles) Treat the density of particles the same way that we treated the electric field. Assume that N particles are randomly disposed in a box of side L. For one particle at the origin (in analogy to one wave packet), set

$$n_1(\mathbf{r}) = \frac{1}{(2\pi)^3} \int e^{i\mathbf{k}\cdot\mathbf{r}} d^3k$$

so that

$$n_1(\mathbf{k}) = \frac{1}{(2\pi)^3}$$

Show that

$$\langle n_1(\mathbf{k}') n_1(\mathbf{k}) \rangle = \frac{N}{L^3} \frac{1}{(2\pi)^3} \delta(\mathbf{k}' - \mathbf{k})$$

Show that the density correlation function is

$$C(\Delta\mathbf{r}) = \langle n(\mathbf{r} + \Delta\mathbf{r}) n(\mathbf{r}) \rangle = \frac{N}{L^3} \delta(\Delta\mathbf{r})$$

Calculate N'^2 the average of the square of the number of particles in the box $-L'/2 < x < L'/2$, $-L'/2 < y < L'/2$, $-L'/2 < z < L'/2$, with $L' < L$, and show that

$$\langle N'^2 \rangle = \left\langle \int n(\mathbf{r}')d^3\mathbf{r}' \int n(\mathbf{r})d^3r \right\rangle = N\frac{L'^3}{L^3} = nL'^3$$

3. (Coherent scattering of light) Evaluate the radial part of the three-dimensional $\mathbf{k}$ integral equation 70, for the fractional coherently scattered rate of radiation $E_1^2/E_0^2 t$ making use of the delta function to show that $k = 2k_0 \cos\theta$, where θ is the angle between $\mathbf{k}$ and $-\mathbf{k}_0$. Assume that the spectrum of density fluctuations $I(k)$ is isotropic. Use vector algebra to evaluate $\sin^2\theta'$, where θ' is the angle between the x direction and the scattered direction $\mathbf{k}'$. Neglect the frequency of the scattering waves ω compared to the frequency of the incoming light wave ω_0. Show that the rate of scattering can be written as

$$R = \frac{E_1^2}{tE_0^2} = \frac{4\pi^2}{n^2}\frac{\omega_{pe}^4}{\omega_0^2}\frac{1}{c}k_0^2 \int_0^{\pi/2} \sin\theta d\theta I(2k_0\cos\theta)(\cos\theta - 2\sin^2\theta\cos^3\theta)$$

Now choose $I(k) = I_0$ for $k < 2k_0$ and zero elsewhere, where I_0 is a constant, and evaluate the θ part of the integral. By using equation 59

$$\overline{(\delta n)^2} = \int I_0 d^3\mathbf{k} = \frac{4\pi}{3}(2k_0)^3$$

show that R can be written as

$$R = \frac{3}{32\pi}\lambda_0^3\overline{(\delta n)^2}\sigma_T c$$

where $\lambda_0 = 2\pi/k_0$ is the wavelength of the incident light wave and $\sigma_T = (8\pi/3)(e^2/mc^2)^2$ is the Thompson cross section. (The density fluctuations are essentially on the scale of $\Delta x \approx \pi/k_o$.) Use the I_0 from problem 2 for uncorrelated electrons to show that this reduces to

$$R(\text{incoherent}) = n\sigma_T c$$

the standard result. (If you wish, you may carry out this result in order of magnitude ignoring the numerical constants.)

4. (Scattering of X-rays by interstellar grains) For high-energy photons passing through a grain, the atomic electrons respond as free electrons and we can ignore their binding energy. Regard the grain as a single wavelength density fluctuation of order unity. Take

$$I(k) = n^2\frac{\delta(k - k_g)}{4\pi k_g^2}$$

where $k_g = 2\pi/r_g$, where r_g is the grain radius. Show that $\overline{(\delta n)^2}$ is equal to n_0^2. With this $I(k)$ evaluate the expression for R in problem 3. Show that $\cos\theta \approx k_g/2k_0$ and that the scattering angle α between $\mathbf{k}'$ and $\mathbf{k}_0$ is of order k_g/k_0.

Substitute this $I(k)$ into the first expression for R and carry out the θ integral to show that if interstellar space were filled with grains, then

$$R(\text{grain}) = \frac{\pi}{4} \frac{\omega_{pe}^4}{\omega_0^3} \frac{k_0}{k_g}$$

But grains occupy only a fraction f of interstellar space, so R should be multiplied by the filling factor $f = n_{\text{grain}} V_{\text{grain}}$, where n_{grain} is the number density of grains in the interstellar medium and V_{grain} is the volume of a grain. Estimate f using the facts that the mass density of grains is about unity, grains possess approximately one percent of the interstellar matter, and the mean density of the interstellar medium is one hydrogen atom per cubic centimeter.

Next, estimate what fraction of the fluence of a gamma ray burst comes from grain scattering. Take a depth of 1 kpc (kiloparsec). Show that the ratio of the scattered fluence to the gamma ray fluence is of order

$$\epsilon \approx \frac{DfR}{c}$$

where $D \approx 1$ kpc is the path length of the gamma rays through the galaxy. Take X-rays of 10 keV and a grain radius of 1 μ and estimate ϵ. Finally, take into account that the path length of the scattered radiation is longer straight so that the scattered radiation is stretched over a time of order

$$\Delta t = \alpha^2 \frac{D}{2c}$$

Show this and estimate Δt.

5. (Short wavelength viscous cutoff for Kolmogoroff turbulence) Assume that the interstellar turbulence is unmagnetized, and has an input wavelength of 100 pc (parsecs) and that the turbulence velocity on this scale is 10 km/sec. Estimate the unmagnetized Coulomb viscosity, $\eta_{\|}$, from the results quoted in problem 1 of chapter 8 or directly from the Braginski equations. What is the cutoff scale for the turbulence? Take $n = 1$, $T_e = 1$ volt. Assume that there are 50 percent neutrals and that the neutral cross section is 10^{-16} cm^2 Reestimate the viscosity and the cutoff radius.

6. (Goldreich–Sridhar turbulence in the interstellar medium) Derive some of the critical Goldreich–Sridhar lengths scales for the interstellar medium. Take the input scale $\ell = 1$ pc, $v_A = 5$ km/sec, and the transverse velocity $v_\lambda = 1$ km/sec at $\lambda = L$.

What is ϵ? What is λ_1 the value at which N reaches one and the spectrum changes? What is v_λ when λ reaches 10^{12} cm?

Compute v_λ and ℓ_z when λ reaches the ion gyroradius. Take $B = 2 \times 10^{-6}$ G and $v_i = 10^6$ cm/sec.

Finally, it is known that an electron density fluctuation of 10^{-4} cm $^{-3}$ at $\lambda \approx 10^{10}$ cm might explain interstellar scintillation. Assume that at small scales the slow mode is in equipartition with the Alfven mode. Allow the slow mode to be compressible and take $\delta n_e / n_e = v_\lambda / v_A$ with $n_e \approx 10$ cm^{-3}. Will the resulting density fluctuations be large enough to explain the scintillations (Manchester and Taylor 1977)?

References

Chandrasekhar, S., 1949. *Astrophysical Journal* **110**, 327.
Dawson, J. 1968. *Advances in Plasma Physics*, **1**, 1.
Ginzburg, V. L., and V. V. Zhelezniakov. 1958. *Soviet Astronomy* **2**, 653.
Goldreich, P., and S. Sridhar. 1995. *Astrophysical Journal* **438**, 767.
Goldreich, P., and S. Sridhar. 1997. *Astrophysical Journal* **485**, 680.
Kadomtsev, B. B. 1965. *Plasma Turbulence*, Chapter II, p. 31, trans. L. C. Ronson (translation ed. W. G. Rusbridge), Academic Press, London.
Kreuer, W. L., and J. M. Dawson. 1970. *Physics of Fluids* **13**, 2747.
Kulsrud, R. M. 1978. In *Astronomical Papers Dedicated to Bengt Stromgren*, p. 323, edited by A. Reiz and T. Anderson, Copenhagen University Press, Copenhagen 1978.
Lee, M., and H. Volk. 1973. *Astrophysics and Space Sciences* **24**, 31.
Leslie, D.C. 1973. *Developments in the Theory of Turbulence*, Clarendon, Oxford, UK.
Manchester, R. N., and J. H. Taylor. 1977. *Pulsars*, p. 137, W. H. Freeman, San Francisco.
Ng, C.S., and A. Bhattacharjee. 1996. *Astrophysical Journal* **465**, 845.
Perkins, F. W. 1968. *Journal of Geophysical Research* 73, 6370.
Rosenbluth, M. N., and N. Rostoker. 1962. *Physics of Fluids* **5**, 776.
Sagdeev, R. Z., and A. A. Galeev. 1969. *Nonlinear Plasma Theory*, ed. T. M. O'Neil and D. L. Book, W. A. Benjamin, New York.
Sridhar, S., and P. Goldreich. 1994. *Astrophysical Journal* **432**, 612.

Chapter 12

COSMIC RAYS

12.1 Physical Properties of Cosmic Rays

It is a remarkable fact that our galaxy is filled with high-energy particles, cosmic rays. These cosmic rays have energies extending from 1 GeV up to above 10^{11} GeV. Their energy also extends downward from 1 GeV. This part of their spectrum is not well known because the flux reaching the top of the earth's atmosphere is seriously reduced by the fact that these cosmic rays must force their way through the outward flowing solar wind, which at these energies can push them outward and alter their flux.

Let the number of cosmic rays in the energy interval $d\epsilon$ be $N(\epsilon)d\epsilon$. Their energy spectrum, $N(\epsilon)$, is very nearly a perfect power law with exponent -2.7 from 1 to 10^6 GeV:

$$N(\epsilon) \sim \epsilon^{-2.7} \tag{1}$$

Above 10^6 GeV it remains a power law, but the exponent changes to 3.1 and then back to 2.7 above 10^9 GeV. It is believed these very high-energy cosmic rays are extragalactic and fill the universe uniformly. Their origin is believed to be in galactic sources.

Cosmic rays have some properties that can be produced only by plasma processes. We can show that binary Coulomb collisions between cosmic rays and the interstellar medium are completely negligible. In fact, their Coulomb cross section even at 1 GeV is $\sigma = 10^{-12}/\epsilon_{\mathrm{ev}}^2 = 10^{-12-18} = 10^{-30}\mathrm{cm}^2$. Their mean free path against such collisions is $1/n\sigma = 10^{30}$ cm (if we take $n = 1\ \mathrm{cm}^{-3}$). Their collision rate is $n\sigma v \approx 10^{-19.5}\ \mathrm{sec}^{-1}$, so any given cosmic ray would have a 1% chance of making such a collision during the entire Hubble time. However, it is known that cosmic rays spend only about three million years in the galaxy before they escape, so the chance of a Coulomb collision is even smaller (by 10^{-4}).

This lifetime for the cosmic rays in the galactic disk is derived from the chemical composition of the cosmic rays (Shapiro and Silverberg 1970). Their composition is actually close to the cosmic abundance, with two or three notable exceptions. First, the nuclei of hydrogen, H, are underabundant relative to the remaining cosmic ray nuclei. The fact that it is different can be partially understood from the fact that all cosmic rays are believed to be fully ionized, so their charge to mass ratio $Z/M \approx 1/2$ is the same for every cosmic ray species except H, for which $Z/M = 1$. Thus, we expect all nuclear species heavier than H to behave the same and, thus, to have a composition similar to the abundance distribution at their source.

Now, during the three million years a cosmic ray spends in the disk, it travels a distance equal to 10^{14} sec $\times 3 \times 10^{10}$ cm/sec $\approx 3 \times 10^{24}$ cm and passes through an amount of matter of 3×10^{24} cm $\times 1.6 \times 10^{-24}$ g/cm^3 = 5 g/cm^2. But ion nuclei are very fragile and would be broken up by spallation if the amount of mass passed through were much greater than 5 g/cm^2. The fact that Fe nuclei are only mildly depleted relative to the cosmic abundance indicates that 5 g/cm^2 is an order of magnitude upper limit for the cosmic ray lifetime.

On the other hand, cosmic rays have an appreciable fraction of Li, Be, and B in their composition, a much larger fraction than the cosmic abundance. It is inferred that these nuclei come from spallation of C and O cosmic rays, and the required amount of spallation indicates that the typical amount of mass passed through must not be much less than 3 g/cm^2. From these considerations the lifetime in the disk is determined to be about 3 million years.

The true lifetime of the cosmic rays observed at the earth, i.e., the interval of time between the initial acceleration of the cosmic ray and the time when it hits the earth's upper atmosphere, can be longer than this spallation estimate, because a cosmic ray after birth may leave the disk and reside in the much lower density halo for some time, where it would suffer very little spallation. It could then return to the disk and strike the earth. This true lifetime has been determined by the radioactive decay of the isotropic ^{10}Be (Shapiro and Silverberg 1970). Its true lifetime is found to be about 20 million years.

Thus, cosmic rays really have two lifetimes, 3 million years in the disk and 20 million in the galaxy. In any event, because the cosmic ray lifetime is so much shorter than the age of the galaxy it must be the case that the cosmic rays are constantly being produced in the galaxy perhaps by supernovae, rather than outside in intergalactic space. It is important to note that all the cosmic rays that we see have been accelerated (or created) in recent times, a time much shorter than the age of the universe.

The rate of production of cosmic rays per cubic centimeter per second is the density of cosmic rays in our neighborhood divided by the age "in the disk" = $10^{-10}/10^{14} = 10^{-24}$/cm^3/sec. The volume of the disk is roughly $\pi \times (10^4)^2 \times 10^2$ pc^{-3} = 10^{66} cm^3, so the production rate of our galaxy is $10^{66} \times 10^{-24}$ per cm^3/sec. Cosmic rays have an average energy of 5 GeV = 8×10^{-3} ergs, so the power of our galaxy in cosmic rays is about 10^{40} ergs/sec, a very significant power. It is about 10^{-4} times the total luminous power of our galaxy. All these facts are summarized clearly in the review article by Wefel (1987) (see also Ginzburg and Syrovatskii 1964).

The energy density of cosmic rays is 8×10^{-3} ergs $\times 10^{-10}$ cm^3 $\approx 10^{-12}$ ergs/cm^3. This is comparable with the magnetic energy, the turbulent energy, and the vertical gravitational energy. Thus, cosmic rays must play an important role in the local dynamics of the disk. In fact, they play a significant role in holding the interstellar medium up against gravity. How does this work if there are no Coulomb interactions between cosmic rays and the interstellar medium?

The support of the interstellar medium against gravity by the cosmic rays comes by means of the magnetic field. We first note that although the cosmic rays have very large energies and the magnetic field is very weak (3×10^{-6} G), it is still the case that their radius of gyration in the interstellar magnetic is very small compared to the half-thickness of the disk, H $\approx$ 100 pc = 3×10^{20} cm. The cyclotron frequency in the interstellar medium of a cosmic proton with energy γ GeV is

$$\Omega_{\text{cr}} = \frac{eB}{\gamma mc} = \frac{3\times10^{-2}}{\gamma}\text{sec}^{-1} \tag{2}$$

where γ is the relativistic factor $\gamma = \epsilon/Mc^2 = \epsilon$ (GeV). The velocity is $c = 3\times10^{10}$ cm, so the gyroradius is

$$\rho = \frac{c}{3\times10^{-2}}\epsilon\ (\text{GeV}) - 10^{12}\epsilon\ (\text{GeV})\ \text{cm} \tag{3}$$

The gyroradius of cosmic rays with energy ϵ GeV is small compared to the thickness of the galactic disk, $H \approx 200$ pc if $\epsilon < H/10^{12}$, or for $\epsilon < 10^9$. Thus, from equation 51 in chapter 2 we see that the cosmic-ray diamagnetic current is

$$j_{\text{cr}} = \frac{n_{\text{cr}}ev_D}{c} = \frac{\nabla p_{\text{cr}}}{B} = n_{\text{cr}}\langle\epsilon\ (\text{GeV})\rangle\frac{Mc^2}{3HB} \tag{4}$$

Suppose that the interstellar medium density is ρ. A current $j_g = -\rho g/B$ arises from the drift produced by the gravitational force. Finally, if the interstellar medium has a pressure gradient this gives rise to a diamagnetic current

$$j_p = \frac{\nabla p}{B} \tag{5}$$

Thus, the total current is

$$j_{\text{tot}} = j_{\text{cr}} + j_g + j_p = \frac{\nabla p_{\text{cr}} + \nabla p - \rho g}{B} \tag{6}$$

By Ampere's law this current is equal to $\nabla B/4\pi$, so we have

$$\frac{\nabla p_{\text{cr}} + \nabla p - \rho g}{B} = \frac{\nabla B}{4\pi} \tag{7}$$

or

$$\rho g = \nabla\left(p_{\text{cr}} + p + \frac{B^2}{8\pi}\right) \tag{8}$$

Thus, cosmic rays help to balance the gravitational force by adding their pressure to the plasma pressure. They appear to act directly on the interstellar medium but there is no direct interaction of individual cosmic rays with the interstellar medium.

Perhaps a better way to say this is that without cosmic rays, the gas is held up against g by its pressure and by magnetic pressure. The difference between the pressure support and gravitational force is supplied by the Lorentz force through a current through the gas. This current goes to decrease the magnetic field upward. But the actual decrease is due to the total current not just the gas current. The total current is the sum of the cosmic-ray current and the plasma current. But the cosmic-ray current is of such a sign, so as to reduce the rate at which B decreases (by itself alone it would lead to an increase in B). As a consequence, for a given decrease in B and in p, more mass can be supported against gravity.

It is often supposed that for this model to work, the magnetic pressure must be larger than the cosmic-ray pressure, since it is the agent that confines the cosmic rays against their pressure gradient. In fact, this is not the case. If the field is very small, support against gravity results from balancing the sum of ∇p_{cr} and ∇p against ρg, but if such balance is the case the above argument shows that the currents nearly cancel and, correspondingly, produce to a weak field, so that the net current is small. We can pictorially say that the cosmic rays exert a force on the field that is instantaneously (in a cyclotron period) passed on to the interstellar medium, and a weak field can thereby confine a high pressure of cosmic rays.

Another property of cosmic rays that arises from plasma processes is that of their high degree of isotropy. As pointed out in chapter 8, we find that cosmic rays are nearly isotropic at the earth, and from Liouville's theorem we can infer that they are isotropic in the interstellar space outside of the sun's heliosphere. The observed degree of isotropy is very high. The variation in intensity in the velocity frame of the local interstellar medium is of order 10^{-4}.

This result is at first very surprising. It is generally conceded that cosmic rays have their origin in violent events, and since we know that they leave the galaxy at some point it would be expected that they would stream from the source to the sink at a drift velocity v_D of order c. Even if there were many sources and sinks we would expect the streaming velocity to be at least $c/\sqrt{N}$, where N is the number of sources and not a very large number. This should make the degree of anisotropy of the cosmic rays δ no less than $v_D/c \approx 1/\sqrt{N}$.

The explanation for the lack of fast streaming is the result, to be demonstrated in this chapter, that if $v_D > v_A$, then Alfven waves in the interstellar medium are made unstable. These Alfven waves grow to saturation. The resulting perturbed magnetic fields in these Alfven waves interact with the cosmic rays by a quasilinear wave–particle interaction that changes their pitch angle and leads to a reduction of their drift velocity v_D. v_D will be reduced until it is less than v_A and the Alfven waves become stable. Since $v_A/c \approx 10^{-4}$ in the interstellar medium, this explains the small degree of anisotropy (Cesarsky 1980; Wentzel 1969; Kulsrud and Pearce 1969; Kulsrud and Cesarsky 1971).

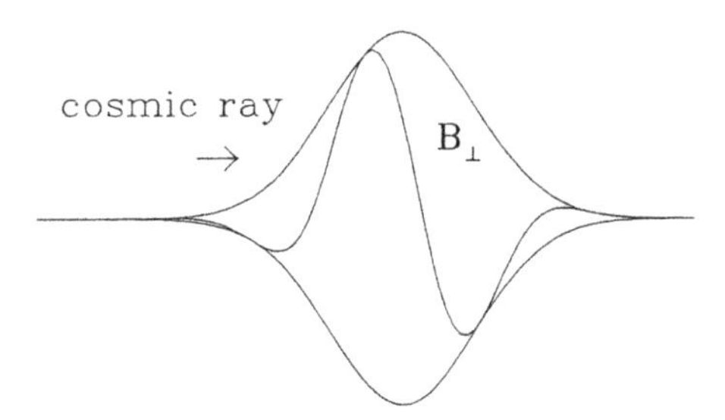

Figure 12.1. A cosmic ray and a wave packet

12.2 Pitch-Angle Scattering of Cosmic Rays by Alfven Waves

Let us first consider how a cosmic ray interacts with series of linearly polarized Alfven wave packets and leads to a random diffusion in its pitch angle (see figure 12.1). (This was worked out in chapter 11 for nonrelativistic particles.)

Suppose, first, that we have a single wave packet of length L and let us choose a velocity frame such that the wave packet is at rest. Let the wave packet length be about one period, so, for example, it could be a wave packet selected from a broad turbulent spectrum. Let the Alfven wave be polarized in the x direction.

Let the cosmic ray enter the wave packet from the left. The sinusoidal behavior of the perturbed magnetic field in the Alfven wave is (leaving out the wave packet envelope for the moment)

$$\delta B_{\perp} = \hat{x}\delta B \sin(kz - \omega t) \tag{9}$$

Now, the cosmic ray rotates about B_0 with frequency Ω, so its v_y component also has this frequency Ω:

$$v_y = v_{\perp} \sin(\Omega t + \phi) \tag{10}$$

where the ϕ takes into account the random phase between the cosmic ray and the wave. Since the z position of the cosmic ray is $z_0 + v_z t$, we get the z component of the Lorentz force at the cosmic ray from

$$\begin{aligned}(\mathbf{v} \times \mathbf{B})_z &= -ev_{\perp}\delta B_{\perp} \\ &= -e\hat{v}_{\perp}\delta B_x \sin(kz_0 + kv_z t)\sin(\Omega t + \phi)\end{aligned}$$

or

$$\begin{aligned}(\mathbf{v} \times \mathbf{B})_z = \frac{1}{2}e\hat{v}_{\perp}\delta B_x \{&\cos[(kv_z - \omega + \Omega)t + (kz_0 + \phi)] \\ - &\cos[(kv_z - \omega - \Omega)t + (kz_0 - \phi)]\}\end{aligned} \tag{11}$$

For $v_z > 0$ the first term has a high frequency in t and averages out. The second term has a lower frequency and may not average out if

$$k_z v_z - \omega - \Omega \approx 0 \tag{12}$$

If this is the case, the change in p_z due to this interaction is

$$\begin{aligned}
\Delta p_z &= e\int dt\left(\frac{\mathbf{v}\times\mathbf{B}}{c}\right)_z \\
&= \frac{1}{2}\frac{ev_\perp \delta B}{c}\frac{2\pi}{kv_z}\cos(kz_0 - \phi) \\
&\approx \pi\frac{ev_\perp \delta B}{c\Omega}\cos(\phi') \\
&= \pi\frac{e\gamma v_\perp}{c}\frac{\delta Bmc}{eB}\cos(\phi') \\
&= \pi p_\perp \sin\theta\left(\frac{\delta B}{B}\right)\cos(\phi')
\end{aligned} \tag{13}$$

where we have let the wave packet have a length $L = 2\pi/k$ and where $\phi' = kz_0 - \phi$ is the relative phase between the cosmic ray and the wave. We take the time τ for the cosmic ray to pass the wave packet equal to the wave period $\tau = 2\pi/(kv_z - \omega) = 2\pi/k(v_z - v_A) \approx 2\pi/kv_z$, since $v_A \ll v_z$.

Because in the frame of the wave the electric field is zero, the energy of the cosmic ray does not change. The only result of the interaction is a change in the pitch angle. Now $p_z = p\cos\theta$, so we have

$$\delta(pc\cos\theta) = -p\sin\theta\delta\theta = \pi p\sin\theta\left(\frac{\delta B}{B}\right)\cos\phi' \tag{14}$$

or

$$\delta\theta = -\pi\frac{\delta B}{B}\cos\phi' \tag{15}$$

$\delta\theta$ increases or decreases according to the relative phase of the interaction.

Suppose now that the wave packets are tightly packed and appear one after another, but with random phases. Then in a time t, t/τ such interactions occur. The square of the total change is the sum of the average of the squares of each interaction, i.e., $\langle\cos^2\phi\rangle = \frac{1}{2}$, so

$$\langle(\Delta\theta)^2\rangle = \Sigma\langle(\delta\theta)^2\rangle = \frac{t}{\tau}\frac{\pi^2}{2}\left\langle\left(\frac{\delta B}{B}\right)^2\right\rangle \tag{16}$$

or the pitch angle diffuses at a rate

$$\frac{\langle(\Delta\theta)^2\rangle}{2t} = \frac{\pi}{8}\Omega\left\langle\left(\frac{\delta B}{B}\right)^2\right\rangle \tag{17}$$

We have discussed the pitch-angle scattering of a right-moving cosmic ray ($v_z > 0$). If the cosmic ray is left moving and satisfies the resonant condition $kv_z - \omega + \Omega = 0$ instead of equation 12, then the second term in equation 11 averages out and the first term has the secular effect. Thus, for our linearly polarized Alfven wave packet, the pitch-angle scattering in the wave frame is identical for left- and right-moving resonant cosmic rays. This is a consequence of a linearly polarized wave being equivalent to a combination of a right circularly polarized wave and a left circularly polarized wave (which appear to rotate oppositely for $v_z > 0$ and the same for $v_z < 0$).

Ω is very large compared to the reciprocal of the time t for a cosmic ray to go a macroscopic distance: $\Omega = 3 \times 10^{-2}$ sec^{-1} and $t = L/c = 10^8$ sec for $L = 1$ pc. Therefore, if $\delta B/B$ is of order unity, the rate of diffusion is extremely fast. If we wish the pitch angle to diffuse by order unity in the time for the cosmic ray to go a parsec, we need only $\delta B/B \approx 10^{-3}$, an unobservably small amplitude.

Three remarks are worth making. The first is that, ignoring factors of order unity, we have

$$\delta\theta \approx \pm \frac{\delta B}{B} \tag{18}$$

for one interaction, where the + or − sign refers to the phase. On the other hand, $\delta B/B$ is the angle α that the perturbed magnetic field line makes with the mean field. We have chosen a wavelength λ for the Alfven wave such that

$$kv \approx \Omega \tag{19}$$

or

$$\frac{\lambda}{2\pi} \approx \frac{v}{\Omega} \approx r_L \tag{20}$$

the Larmor radius of the cosmic rays. If we had chosen λ much longer, the cosmic ray would adiabatically follow the field line and after the wave packet would adiabatically return to its previous pitch angle with no change. On the other hand, if $\lambda \ll r_L$, the cosmic ray would hardly be affected by the rapidly oscillating field that it would see. These two cases are illustrated in figure 12.2. The resonant condition $\lambda \approx r_L$ leads to an intermediate situation in between these two extremes. The cosmic ray would attempt to follow the field line as in case a, but would not do this completely as in case b, so it would end up with a changed pitch angle $\delta\theta$ of order α.

The second remark is as follows. We have assumed our wave packet to be of order one wavelength long. Let it be n wavelengths long. Then according to equation 13 the change in θ would be larger by n, so $\langle(\Delta\theta)^2\rangle$ in equation 17 would be larger by $n^2 t/\tau$. But τ is also longer by n, so the diffusion rate

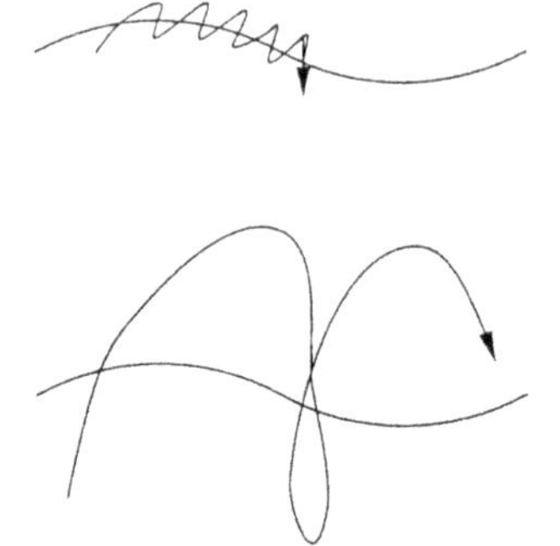

Figure 12.2. No change in pitch angle unless $\lambda \approx \rho$

equation 17 would be

$$\frac{\langle(\Delta\theta)^2\rangle}{t} = n\frac{\pi}{8}\Omega\left(\frac{\delta B}{B}\right)^2 \tag{21}$$

But for a smooth spectrum

$$\left(\frac{\delta B}{B}\right)^2 = \Delta k I(k) \tag{22}$$

and Δk is inversely related to the length of the wave packet (a precise calculation leads to $\Delta k = k/n = 1/nr_L$), so the result is that the diffusion coefficient is independent of n and depends only on $I(k)$ at the resonant wave number $k = 1/r_L$. In fact,

$$D_\theta = \frac{(\Delta\theta)^2}{2t} = \frac{\pi}{8}\Omega I \tag{23}$$

We have assumed that our Alfven wave packets are linearly polarized in the x direction. If they have both x and y components that are correlated in phase, i.e., they are circularly polarized, then only cosmic rays moving in one direction are affected. Right circularly polarized waves pitch-angle scatter with right-moving cosmic rays and left circularly polarized waves pitch-angle scatter left-moving cosmic rays. In compensation the pitch-angle scattering for the cosmic rays moving one way is twice as strong. Clearly, the cosmic ray will resonate with the wave if, in its frame, it sees a wave rotating in the left-handed sense (in space). Rather peculiarly, this sense corresponds to the wave rotating in the right-handed sense at a fixed point if the wave is propagating to the right. If the cosmic ray is moving to the left, all these senses of rotation change. For strictly parallel propagation Alfven waves are indeed circularly polarized. However, if the Alfven waves propagate at even a small angle to the mean field $> \sqrt{\omega/\Omega} \approx \sqrt{v_A/c}$, they are linearly polarized (see problem 5 in chapter 9). Thus, restricting ourselves to linear polarization is not a serious assumption and simplifies the number of different possibilities.

A third remark of considerable importance concerns the resonance condition

$$kv_z - k\cos\theta = \Omega \tag{24}$$

or

$$\frac{\lambda}{2\pi} = r_L \cos\theta \tag{25}$$

As θ approaches 90°, the wavelength of the resonant wave approaches zero and it turns out that, in general, the Alfven wave turbulence generated by instabilities is too weak to scatter the cosmic rays past pitch angles of 90°, although they can take their pitch angles close to 90°. This is known as the 90° pitch-angle scattering problem (Klimas and Sandri 1973; Goldstein 1976). It has been the subject of much discussion during the past decades. It turns out that this problem is resolved by the presence of a narrow boundary layer in pitch-angle space, through which the cosmic-ray pitch angle is carried by a secondary mechanism. We will come to this point later.

How do we connect the lifetime of the cosmic rays in the disk ($\approx 3 \times 10^6$ years or 3 g/cm^2) to the rate of pitch-angle scattering. Let us assume that the anisotropy of the cosmic rays has the observed value of $\delta \approx 3 \times 10^{-4}$. If the anisotropy corresponds to a drift velocity v_D, then

$$\frac{v_D}{c} = \delta \approx 10^{-4} \tag{26}$$

and $v_D \approx 10$ km/sec.

In three million years the bulk streaming velocity would move a freshly created group of cosmic rays a distance of 300 pc (parsecs), a distance barely comparable with the disk thickness. But we could argue that the mean value of δ in the galaxy could be a factor of ten larger than that observed at the sun. The relative height of the sun above the central phase of the galactic disk is about 10 percent. In this case the distance the cosmic ray can move could be larger than 3 kpc (kiloparsecs). If the mean free path for pitch-angle scattering through 90 degrees is λ, then the lifetime of the cosmic rays is

$$\frac{L^2}{c\lambda} = 3 \times 10^6 \text{ years} \tag{27}$$

which leads to $\lambda \approx 10$ pc. This leads to an estimate for $\delta B/B \approx 10^{-4}$.

Up to this point we have not discussed the source of the Alfven waves that are to do the scattering. There are a number of possible sources—interstellar turbulence, hot stars, moving magnetic stars, and so forth. But, surprisingly, the main source seems to be the cosmic rays themselves. A detailed kinetic calculation involving the Vlasov equation for the cosmic rays shows that if the cosmic rays are sufficiently anisotropic they will render Alfven waves carried by the interstellar medium unstable. If the anisotropy is due to a bulk drift velocity v_D, and if $v_D > v_A$, then the waves are unstable.

Before we enter into the detailed calculation of the expected growth rate for Alfven waves let us see if we can understand why cosmic rays should make Alfven waves unstable. It is important to be clear that at any point, the interstellar medium should be regarded as consisting of two components. The interstellar medium itself is a conducting collisional medium of low temperature and it is penetrated by a high-energy collisionless component, the cosmic rays. We regard these two components superimposed on each other, the first described by ideal MHD equation and the second, the cosmic ray component, described by the collisionless Vlasov equation (taken in relativistic form). Now, suppose that there are some small-amplitude, right-moving Alfven waves present. As seen above, the waves lead to a pitch-angle diffusion of the cosmic rays (in the wave frame) at the rate

$$D_\theta = \frac{(\Delta\theta)^2}{2t} \approx \Omega\left(\frac{\delta B}{B}\right)^2 \tag{28}$$

where we ignore constants of order unity. Let the cosmic rays have a drift velocity v_D. In the frame of the wave the drift velocity is $v_D - v_A$. As a result of the pitch-angle diffusion (ignoring the 90° pitch-angle problem) the cosmic rays would become isotropic in the frame of the wave in a time

$$\tau = \frac{1}{D_\theta} = \frac{1}{\Omega(\delta B/B)^2} \tag{29}$$

Before the scattering, the cosmic rays would have a linear momentum

$$n_{\rm cr} m v_D \tag{30}$$

per unit volume, where m is the mean relativistic energy of a cosmic ray divided by c^2, or the mean relativistic mass of the cosmic ray. After the scattering, the momentum is

$$n_{\rm cr} m v_A \tag{31}$$

so the rate of the loss of momentum is

$$\frac{dP_{\rm cr}}{dt} = \frac{nm(v_D - v_A)}{\tau} \tag{32}$$

But this corresponds to a rate of gain of wave momentum,

$$2\gamma\frac{(\delta B)^2}{v_A 8\pi} \tag{33}$$

Equating these rates and taking $m = \Gamma_{\rm cr} M$, where $\Gamma_{\rm cr}$ is the mean value for the cosmic ray γ's, and M is the rest mass of the cosmic ray, gives

$$\begin{aligned} 2\gamma &= \frac{n_{\rm cr}\Gamma_{\rm cr}(v_D - v_A)\Omega(\delta B/B)^2}{(\delta B)^2/8\pi v_A} \\ &= n_{\rm cr}\frac{8\pi M}{B^2}(v_D - v_A)v_A\Omega_0 \end{aligned} \tag{34}$$

or

$$\gamma = \frac{n_{cr}}{n}\left(\frac{v_D - v_A}{v_A}\right)\Omega_0 \tag{35}$$

the factor $(\delta B/B)^2$ canceling. Ω_0 is the nonrelativistic cyclotron frequency. In this chapter the symbol γ appears in three different contexts: the relativistic $\gamma = 1/\sqrt{1 - v^2/c^2}$, the growth rate of waves, and the ratio of specific heats. We denote the first by γ_R, the second without any distinguishing subscript, and the third by γ_A.

Our qualitative argument for the growth of the Alfven waves and the pitch-angle scattering of the cosmic rays should convince the reader that these effects are real and strong. The growth rate has the factor Ω_0, which is large, but it is reduced by $n_{\rm cr}/n$. The scattering rate also has the large Ω factor, but it is reduced by $(\delta B/B)^2$, which can be small. An estimate of γ is found from $n_{\rm cr} = 10^{-10}$, $n = 1$, $v_A = 3 \times 10^6$. Let us assume $v_D - v_A \approx v_A$. Then

$$\gamma = \Omega_0 \frac{n_{\rm cr}}{n} = 3 \times 10^{-2} \times 10^{-10} = 3 \times 10^{-12}\ \text{sec}^{-1} \approx 10^{-4}\ \text{years}^{-1} \quad (36)$$

so the waves grow in a time of order 10^4 years, which is short compared to the life of the cosmic rays for $v_D = 2v_A$. The wave amplitude $\delta B/B$ then increases until the pitch-angle scattering reduces the drift velocity to $v_D \approx v_A$. At this point the streaming produced by the sources and sinks comes into balance with pitch-angle scattering, and a steady state between these two effects is reached.

The lesson to be learned from these comparatively fast rates associated with the Alfven wave instability is that, in general, it is very difficult for any significant density of energetic particles to flow rapidly through a plasma. Their bulk flow velocity will always be reduced to about the Alfven speed.

Although equation 35 gives the growth rate of Alfven waves due to cosmic rays, there are also damping processes for the Alfven waves. One damping process occurs if the interstellar medium is only very partially ionized. In this case neutrals do not follow the field lines motion as closely as the ions. Generally, for the Alfven waves resonant with cosmic rays, $\omega > \nu$, so the neutrals hardly follow the lines at all and the damping rate is

$$\Gamma = \tfrac{1}{2}\nu_{\rm in} \quad (37)$$

(see problem 4 in chapter 5). This is plausible since a collision of an ion with a stationary neutral destroys its directed kinetic energy and half of the energy of an Alfven wave is in this kinetic energy. This damping is called ambipolar damping. Thus, if $\Gamma \gg \gamma_2$, where γ_2 is the linear growth rate for $v_D = 2v_A$, then we expect a balance when $v_D - v_A = (\Gamma/\gamma_2)v_A$ and the cosmic rays can actually drift considerably faster than the Alfven speed.

As just mentioned, the amplitude of the waves must be such that the pitch-angle diffusion balances the tendency of flow from source to sink to produce further anisotropy. If there is ambipolar damping and no nonlinear damping, the wave amplitude would settle down to just the value needed to get a balance:

$$\gamma(v_D) = \Gamma \quad (38)$$

However, there is also nonlinear Landau damping of the waves (equation 115 in chapter 11) (see Kulsrud 1978; Lee and Volk 1973):

$$\gamma_{\mathrm{NL}} \approx \omega \left(\frac{\delta B}{B}\right)^2 \frac{v_i}{v_A} = -\alpha_{\mathrm{NL}} \left(\frac{\delta B}{B}\right)^2 \tag{39}$$

To balance this additional damping $\gamma(v_D)$ must be somewhat larger than Γ,

$$\gamma(v_D) = \Gamma + \alpha_{\mathrm{NL}} \left(\frac{\delta B}{B}\right)^2 \tag{40}$$

This equation, combined with the pitch-angle scattering requirement on $\delta B/B$ to sustain the drift velocity v_D, determines both v_D and $\delta B/B$.

In our qualitative discussions of the propagation and pitch-angle scattering of the cosmic rays, and the growth and damping of the waves, we have treated the cosmic rays as a unit with a single relativistic energy band and the waves as having a single band of wave numbers, chosen resonant with the cosmic rays. The cosmic rays have a large range of energies and a cosmic ray with a given energy resonates with Alfven waves with a specific wavelength resonant with the energy. As a matter of fact, cosmic rays with the same energy and different pitch angles themselves resonate with different waves, but in our qualitative treatment we may treat cosmic rays within the same logarithmic band of energies as having pitch angels not to close to 90° and resonating with the same logarithmic band of wavelengths. This allows us to make qualitative statements that are simple to formulate and not too far from the truth.

If the plasma is fully ionized, as is the case in the coronal regions of the interstellar medium, then Γ is zero and the growth of the Alfven waves must be balanced entirely by nonlinear Landau damping. We then have

$$\gamma(v_D) = \gamma_{\mathrm{NL}} \tag{41}$$

The result of balancing these pitch-angle scattering and growth processes at a given energy band and corresponding resonant wavelength leads to a value for the drift velocity that bears on the lifetime of the cosmic rays at that energy. We cannot determine the lifetime till we know the escape process of cosmic rays from the galaxy as well as the distribution and strengths of the sources of cosmic rays. However, if we assume that the sources and sinks of the cosmic rays are at the same place for different energies we can determine how the lifetime of the cosmic rays depends on energy. (Waves of larger wavelength are resonant with higher energy cosmic rays.) The actual dependence of the lifetime on energy is known by composition analysis up to an energy of 1000 GeV. It has been found that the lifetime decreases roughly as $\epsilon^{-0.7}$. If the entire interstellar medium were partially ionized and homogeneous, the predicted lifetime would be proportional to $\epsilon^{-1.5}$. This follows from $\gamma \sim n_{\mathrm{cr}} \sim \epsilon^{-1.5}$, equations 35 and 38, neglecting any nonlinear damping of the waves. If the plasma is totally ionized, then ambipolar damping is

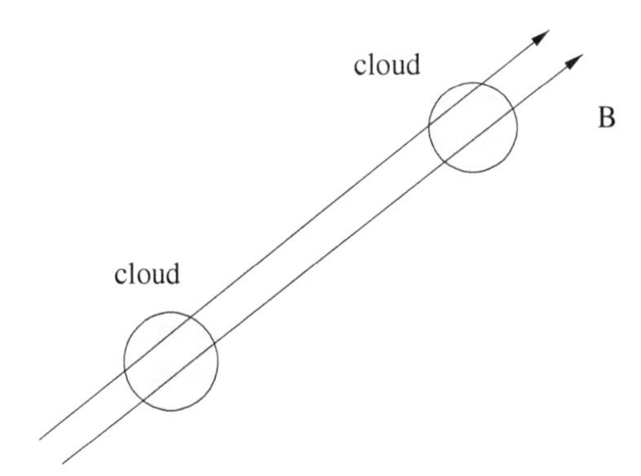

Figure 12.3. Cosmic rays passing through clouds

zero, and from equations 35 and 39, we get a lifetime proportional to $\epsilon^{-0.75}$ (Cesarsky and Kulsrud 1981).

We know that the bulk of the mass of the interstellar medium is in the clouds, while the majority of the volume is in the coronal region and is ionized (see figure 12.3). We have to combine these regions together in a sensible and physically correct way to determine the propagation and spallation of cosmic rays. It turns out that the cosmic rays drift rapidly through clouds and so the slow drift through the coronal and warm regions predominate in determining their lifetimes. The amount of cloud material passed through is proportional to mass density times c times the lifetime in the coronal region.

The amount of spallation in clouds has nothing to do with the drift velocity of cosmic rays as they pass through the clouds. This is analogous to absorption of starlight in clouds which is independent of the cloud volume. Therefore, for the theoretical determination of the lifetime, we can essentially ignore partially ionized regions and use the fully ionized result for the observed lifetimes. The mass in the clouds is only of importance for evaluating the amount of spallation and relating it to the cosmic ray abundances (see problem 3).

We conclude the semiquantitative discussion of the plasma physics of cosmic rays by a hint as to the solution of the problem of scattering through the pitch angle of 90°. It turns out that the region of pitch angles about 90° where pitch angle diffusion fails is extremely narrow, of order $\Delta\theta \approx 10^{-4}$. Cosmic rays with such a low velocity as $v_{Dz} = (\Delta\theta)c$ are easily mirrored by nonresonant inhomogeneities in the field strength B. These inhomogeneities are actually provided by the Alfven waves that are resonant with cosmic rays at a finite pitch angle. The boundary region does delay the pitch-angle scattering from positive to negative v_z by a factor equal to the logarithm of $\Delta\theta$, so the drift velocity is larger than originally predicted by this logarithmic factor. However, the net result is not a serious modification of the above qualitative discussion.

Although the above semiquantitative picture of the cosmic-ray–interstellar medium interaction does lead to the correct growth and quasilinear pitch-angle scattering rate, there is enough indeterminacy in our estimate that, to arrive at precise numerical coefficients for the processes, we should carry out a more systematic calculation. We have not presented any calculation for an instability involving the cyclotron resonance in this book. Therefore, since the cosmic ray instability is an important example of such an instability, we now present the calculation of its growth rate in some detail.

12.3 The Cosmic-Ray Alfven-Wave Instability

We treat the cosmic rays by the relativistic version of the Vlasov equation. Instead of a phase space involving the position x and velocity v as coordinates, it is more appropriate to use a phase space based on the position x and momentum p as coordinates of the cosmic rays. For simplicity, we will restrict ourselves to cosmic-ray protons.

The equation for the momentum $\mathbf{p}$ of a cosmic ray, in an electromagnetic field is

$$\frac{d\mathbf{p}}{dt} = e\left(\mathbf{E} + \frac{\mathbf{v} \times \mathbf{B}}{\mathbf{c}}\right) \tag{42}$$

$\mathbf{p}$ is related to $\mathbf{v}$ by $\mathbf{p} = M\mathbf{v}/\sqrt{1 - v^2/c^2} = m\gamma_R\mathbf{v}$. In this section m will denote the rest mass of a proton. Let the number of cosmic rays in a box $x < x' < x + dx, \ldots p_x < p_x', p_x + dp_x$ be

$$dN = f(\mathbf{p}, \mathbf{x})d^3\mathbf{p}d^3\mathbf{x} \tag{43}$$

Then the flow velocity in momentum space is $\mathbf{v}$ in the $\mathbf{x}$ directions, $d\mathbf{p}/dt$ in the $\mathbf{p}$ directions. The continuity equation in momentum phase space is

$$\frac{\partial f}{\partial t} + \nabla_{\mathbf{x}} \cdot (\mathbf{v}f) + \nabla_{\mathbf{p}} \cdot \left[e\left(\mathbf{E} + \frac{\mathbf{v} \times \mathbf{B}}{\mathbf{c}}\right)f\right] = 0 \tag{44}$$

But $\nabla_{\mathbf{x}} \cdot \mathbf{v} = 0$, $\nabla_{\mathbf{p}} \cdot \mathbf{E} = 0$ and

$$\nabla_{\mathbf{p}} \cdot (\mathbf{v} \times \mathbf{B}) = (\nabla_{\mathbf{p}} \times \mathbf{v}) \times \mathbf{B} = 0 \tag{45}$$

It is easy to see that $\nabla_{\mathbf{p}} \times \mathbf{v} = (\partial v/\partial p)\nabla_{\mathbf{p}} \times \mathbf{p} = 0$. Thus, Vlasov equation reduces to the Liouville equation

$$\frac{\partial f}{\partial t} + \mathbf{v} \cdot \nabla f + e\left(\mathbf{E} + \frac{\mathbf{v} \times \mathbf{B}}{c}\right) \cdot \nabla_{\mathbf{p}} f = 0 \tag{46}$$

which says that f is constant along a cosmic-ray trajectory no matter how complicated $\mathbf{E}$, $\mathbf{B}$, and the cosmic-ray orbit are.

Take the background field $\mathbf{B}_0$ in the z direction. Assume that there is a monochromatic Alfven wave propagating in the positive $\mathbf{z}$ direction and linearly polarized in the $\mathbf{x}$ direction. Let its amplitude E be small.

We wish to find the linear response of the cosmic rays to this field and express it in terms of the contribution of the cosmic rays to the imaginary part of dielectric tensor ϵ. This cosmic ray response is complex, with the imaginary part of ϵ coming from those cosmic rays in cyclotron resonance with the Alfven wave. We assume that the unperturbed cosmic-ray distribution function, $f_0(\mathbf{p})$, is shifted to the right from an isotropic distribution and that it has a drift velocity v_D. Remember that ϵ is defined by

$$\frac{-i\omega\epsilon}{c}\cdot\mathbf{E}=\frac{-i\omega}{c}\mathbf{E}+\frac{4\pi}{c}\mathbf{j} \tag{47}$$

or

$$\epsilon\cdot\mathbf{E}=\mathbf{E}+\frac{4\pi i}{\omega}\mathbf{j} \tag{48}$$

For an Alfven wave, $\mathbf{E}$ has the form

$$\mathbf{E}=\mathrm{Re}\left(\hat{\mathbf{E}}e^{ikz-i\omega t}\right) \tag{49}$$

From the induction equation $(-i\omega/c)\mathbf{B}_1=-i\mathbf{k}\times\mathbf{E}$ or

$$\mathbf{B}_1=\frac{\mathbf{k}c\times\mathbf{E}}{\omega} \tag{50}$$

Therefore, the linearization of equation 46 is

$$\begin{aligned}-i\omega f_1+ikv_zf_1+\frac{e}{c}(\mathbf{v}\times\mathbf{B}_0)\cdot\nabla_{\mathbf{p}}f_1&=-e\left[\mathbf{E}+\frac{\mathbf{v}\times(\mathbf{k}\times\mathbf{E})}{\omega}\right]\cdot\nabla_{\mathbf{p}}f_0\\&=-e\left[(1-\frac{kv_z}{\omega})\mathbf{E}+\frac{\mathbf{v}\cdot\mathbf{E}}{\omega}\mathbf{k}\right]\cdot\nabla_{\mathbf{p}}f_0\end{aligned} \tag{51}$$

It is convenient to introduce cylindrical coordinates $p_\perp$, ϕ, p_z, in $\mathbf{p}$ space. f_0 depends only on $p_\perp$ and p_z and is independent of ϕ, but f_1 depends on ϕ as well as $\mathbf{x}$ and $\mathbf{p}$. In these coordinates we have

$$\frac{e}{c}\mathbf{v}\times\mathbf{B}_0\cdot\nabla_{\mathbf{p}}f_1=-\frac{ev_\perp B_0}{cp_\perp}\frac{\partial f_1}{\partial\phi}=-\Omega\frac{\partial f_1}{\partial\phi} \tag{52}$$

where $\Omega=v_\perp B_0/cp_\perp=eB_0/\gamma_R mc$ is the relativistic cyclotron frequency.

On the right-hand side of equation 51, $\mathbf{E}\cdot\nabla_p f_0=E\cos\phi\,\partial f_0/\partial p_\perp$ and $\mathbf{E}\cdot\mathbf{v}=Ev_\perp\cos\phi$. With these simplifications, equation 51 reduces to

$$\begin{aligned}(-i\omega+ikv_z)f_1-\Omega\frac{\partial f_1}{\partial\phi}&=-e\left[\left(1-\frac{kv_z}{\omega}\right)\frac{\partial f_0}{\partial p_\perp}+\frac{kv_\perp}{\omega}\frac{\partial f_0}{\partial p_z}\right]E\cos\phi\\&=-eAE\cos\phi\end{aligned} \tag{53}$$

where

$$A=\left(1-\frac{kv_z}{\omega}\right)\frac{\partial f_0}{\partial p_\perp}+\frac{kv_\perp}{\omega}\frac{\partial f_0}{\partial p_z}$$

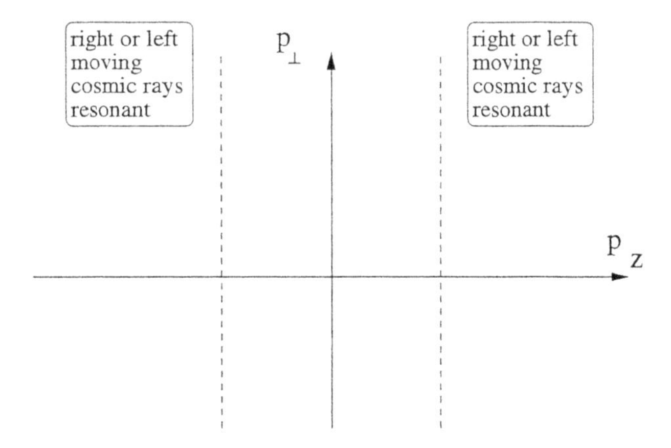

Figure 12.4. Resonant line in momentum space

A is an abbreviation for the bracketed expression in equation 53, the coefficient of $\cos\phi$ on the right-hand side. Equation 53 is a simple differential equation in ϕ for f_1 whose solution is

$$f_1 = -eE\frac{1}{2}\frac{iAe^{i\phi}}{\omega - kv_z + \Omega} - eE\frac{1}{2}\frac{iAe^{-i\phi}}{\omega - kv_z - \Omega}$$
$$= f_+ + f_- \tag{54}$$

The right-moving resonant cosmic rays are included in the first term and occur for p's such that

$$-\omega + kv_z = \Omega = \frac{\Omega_0}{\gamma_R} \tag{55}$$

where $\Omega_0 = eB/mc$ is the nonrelativistic cyclotron frequency. The left-moving resonant cosmic rays are included in the second term and satisfy $-\omega + kv_z = -\Omega$. If we drop ω as small, these conditions reduce to

$$mk\gamma_R v_z = kp_z = \pm m\Omega_0 \tag{56}$$

or a condition involving only p_z, since Ω_0 is independent of p (see figure 12.4).

The perturbed resonant current from the right-moving cosmic rays $\mathbf{j}_{1r}$ comes from f_+ alone and is

$$\mathbf{j}_{1r} = e\int (f_+ + f_-)v_\perp(\cos\phi\hat{\mathbf{x}} + \sin\phi\hat{\mathbf{y}})d^3\mathbf{p}$$

$$= -\frac{ie^2}{2}\int\left(\frac{v_\perp Ae^{i\phi}}{\omega - kv_z + \Omega} + \frac{v_\perp Ae^{-i\phi}}{\omega - kv_z - \Omega}\right)(\cos\phi\hat{\mathbf{x}} + \sin\phi\hat{\mathbf{y}})d^3\mathbf{p}E$$

$$= \frac{-ie^2}{4}\int\left(\frac{v_\perp A}{\omega - kv_z + \Omega}\right)d^3\mathbf{p}(\hat{\mathbf{x}} + i\hat{\mathbf{y}})E$$

$$+ \frac{-ie^2}{4}\int\left(\frac{v_\perp A}{\omega - kv_z - \Omega}\right)d^3\mathbf{p}(\hat{\mathbf{x}} - i\hat{\mathbf{y}})E \tag{57}$$

We have restricted ourselves to a purely shifted cosmic-ray distribution so that A is the same for both the resonant right-moving and the left-moving cosmic rays. As a consequence, the sum of the y currents vanishes, while the two x currents (that are in the same direction as $\hat{\mathbf{E}}$) are equal. In this case our assumption of a linearly polarized Alfven wave is justified.

(On the other hand, if the distribution is symmetric in p_z, then A is odd and the x currents cancel while the y currents add. Since E_x produces y currents and no x currents the wave cannot be linearly polarized but must be circularly polarized. For the rest of our detailed calculation, we consider only the case of the shifted cosmic-ray distribution. Thus, we keep only the x currents.)

The electromagnetic equations reduce to

$$\mathbf{k}\times(\mathbf{k}\times\mathbf{E}) = -\frac{\omega^2}{c^2}\mathbf{E} + 4\pi\mathbf{j}_1(-i\omega) = -\epsilon\frac{\omega^2}{c^2}\mathbf{E} \tag{58}$$

or

$$\frac{k^2c^2}{\omega^2} = \epsilon_0 + \epsilon_{xx}^{\mathrm{cr}} \tag{59}$$

and

$$\epsilon_{xx}^{\mathrm{cr}} = i\frac{4\pi j_{1x}}{\omega E} = 2\frac{e^2}{4}\frac{4\pi}{\omega}\int\frac{Av_\perp}{\omega - kv_z + \Omega}d^3\mathbf{p} \tag{60}$$

where the 2 comes from the sum of the + and − currents.

In the p_z integral there is the usual resonant denominator, which we can replace by $-\pi i\delta(p_z - m\Omega_0/k - m\gamma_R\omega/k)$. From the fluid equation for the background plasma, $\epsilon_0 = c^2/v_A^2$. To lowest order in equation 59 we drop the cosmic-ray term as small and get the usual Alfven wave dispersion relation:

$$\omega_0 = kv_A \tag{61}$$

To next order the cosmic-ray contribution to ϵ then produces a change in ω away from ω_0. That is, $\omega = \omega_0 + \omega_1$. Substituting this into equation 59 and

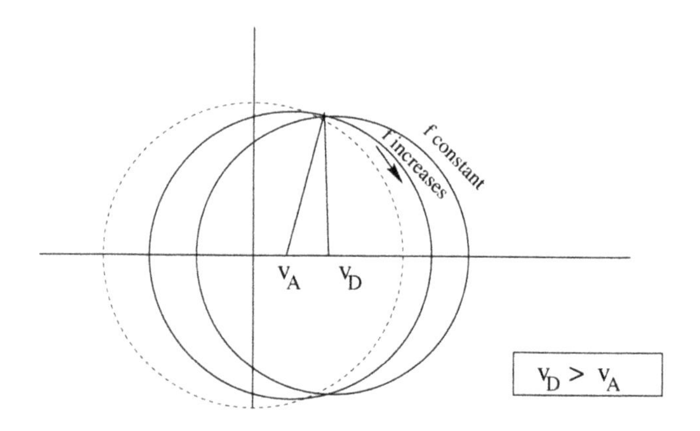

Figure 12.5. Position of the constant f surfaces relative to the pitch and scattering curves and instability

keeping ω_1, only on the left-hand side, we get

$$-\frac{2\omega_1 k^2 c^2}{\omega_0^3} = -\frac{2\omega_1}{\omega_0}\frac{c^2}{v_A^2} = \epsilon_{xx}^{\mathrm{cr}} \tag{62}$$

or from equation 60, γ the imaginary part of ω_1 is given by

$$\gamma = \mathrm{Im}(\omega_1) = -\frac{\omega_0}{2}\mathrm{Im}(\epsilon_{xx}^{\mathrm{cr}})$$

$$= \mathrm{Im}\left[-\frac{e^2}{4}4\pi\frac{v_A^2}{c^2}\int v_\perp A(-\pi i)\delta\left(kv_z - \Omega - \omega\right) d^3\mathbf{p}\right] \tag{63}$$

Substituting the expression for A, given after equation 53, in this we have

$$\gamma = \pi^2 e^2 \frac{v_A^2}{c^2}\int v_\perp \left[\left(1 - \frac{kv_z}{\omega}\right)\frac{\partial f_0}{\partial p_\perp} + \frac{kv_\perp}{\omega}\frac{\partial f_0}{\partial p_z}\right]\delta(kv_z - \Omega)d^3\mathbf{p} \tag{64}$$

The sign of γ determines whether the wave damps or grows. It depends on the bracketed expression, which is gradient of f_0 along the curve

$$\frac{dp_\perp}{dp_z} = \frac{\omega - kv_z}{kv_\perp} \tag{65}$$

This is a circle in momentum space centered about the point $p_z = mv_A$, $p_\perp = 0$.

In other words, in the wave frame it is the circle along which pure pitch-angle scattering occurs. If f_0 increases along this circle in the direction in which p_z increases, then we see from figure 12.5 that the contribution from this circle tends toward instability. The point at which the gradient is relevant is, of course, the point where cosmic rays are resonant.

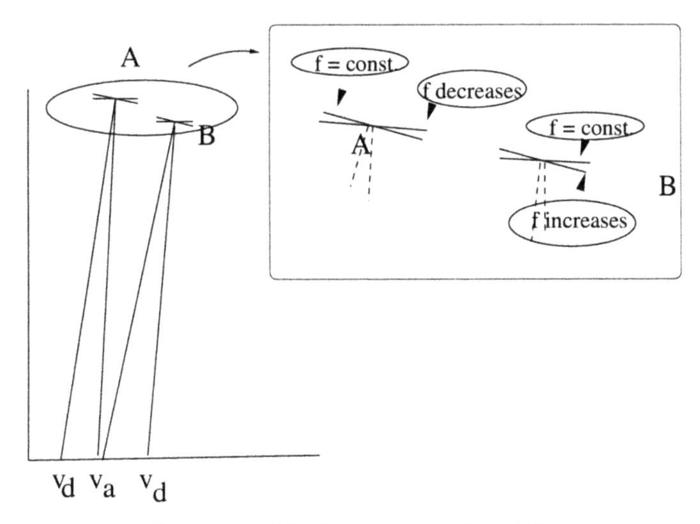

Figure 12.6. The three cases for instability

We can see that if the distribution is isotropic in the laboratory frame, then f_0 is constant on circles about the origin. Such circles are cut by the relevant circle centered on v_A more flatly, so as we move toward larger p_z along the v_A circle, we move toward larger p. Since f_0 decreases with increasing p, the situation is stable.

We can also see that because the circles about the origin cut the v_A circles at a small angle, even a small v_D shift in the constant f circles can lead to instability, since the cosmic rays are isotropic in this drift frame. We easily see from the figure 12.5 and in slightly more detail in figure 12.6 that we get instability for $v_D > v_A$ and stability for $v_D < v_A$. If $v_D < 0$, all right-moving waves are stable. However, by symmetry, the left-moving waves are unstable if $v_D < -v_A$.

Let us evaluate ω_1 given by equation 64 for a cosmic-ray distribution that has a bulk velocity v_D and is isotropic in a frame moving with this velocity. Denoting momentum in the velocity frame by primes we have

$$p'_\perp = p_\perp$$

$$p'_z = p_z - \frac{v_D \epsilon}{c^2} \tag{66}$$

where we drop v_D^2/c^2 in the $\sqrt{1 - v_D^2/c^2}$ of this transformation. ϵ is the energy in the laboratory frame. Then, if $F(p')$ is the distribution function in the moving frame, the distribution in the laboratory frame is

$$f(p_\perp, p_z) = F(p') = F\left(\sqrt{p_\perp'^2 + p_z'^2}\right) \tag{67}$$

Here we use the fact that the momentum distribution is a relativistic invariant. Evaluating the term in brackets on the right-hand side of equation 64, we find that it reduces to

$$\left[\left(1-\frac{v_z}{v_A}\right)\frac{\partial f_0}{\partial p_\perp}+\frac{v_\perp}{v_A}\frac{\partial f_0}{\partial p_z}\right]=\frac{dF}{dp'}\frac{p_\perp}{p'}\left(1-\frac{v_D}{v_A}\right) \tag{68}$$

where we drop higher-order terms in v_D and v_A. Thus, to lowest order in v_D and v_A,

$$\Gamma_1=-i\omega_1=\pi^2e^2\frac{v_A^2}{c^2}\left(1-\frac{v_D}{v_A}\right)\int\frac{dF}{dp'}\frac{p_\perp^2}{p}\delta(kp_z-m\Omega_0)d^3p \tag{69}$$

This integral is the same if we replace Ω by $-\Omega$, which justifies our remarks that for a shifted distribution the x components of $\mathbf{j}_1$ add while the y components cancel.

Since $dF/dp'<0$, the sign of Γ is determined by the factor $1-v_D/v_A$, and we find, as expected, that if $v_D>v_A$, then $\Gamma>0$. To evaluate the growth rate to lowest order, we we set $p'=p$ in the argument of F,

$$F=\frac{a}{p^r} \tag{70}$$

($r=4.7$ for a -2.7 power law energy spectrum.) The integral I in equation 69 is

$$I=\int\frac{dF}{dp}\frac{p_\perp^2}{p}\delta(kp_z-m\Omega_0)d\phi dp_z p_\perp dp_\perp=-2\pi ra\int_{mc}^{\infty}\frac{p_\perp^2}{p^{r+2}}\frac{1}{k}p_\perp dp_\perp \tag{71}$$

where in the last integral $p=\sqrt{p_1^2+p^2}$, where $p_z=p_1=m\Omega_0/k$.

We assume that the resonant value of p_z, p_1 is greater than mc and take $p_1=q_1mc$, $p=qmc$ to get

$$I=\frac{2\pi}{k}\frac{ra}{(mc)^{r-2}}\int_{q_1}^{\infty}\frac{q^2-q_1^2}{q^{r+1}}dq \tag{72}$$

Let us compare this with the number of cosmic rays with $p>p_1$:

$$N_{\text{cr}}(p>p_1)=\frac{4\pi a}{(mc)^{r-3}}\int_{q_1}^{\infty}\frac{1}{q^r}q^2dq \tag{73}$$

The ratio of these two expressions

$$\frac{I}{N(p>p_1)}=\frac{C_r}{kp_1} \tag{74}$$

where

$$C_r=\frac{r-3}{r-2} \tag{75}$$

so

$$\Gamma = \frac{\pi^2 e^2}{kp_1}\frac{v_A^2}{c^2} N_{cr}(p > p_1) C_r \frac{v_D - v_A}{v_A}$$

$$= \frac{\pi}{4}\Omega_0 \frac{N_{cr}(p > p_1)}{n} C_r \frac{v_D - v_A}{v_A} \tag{76}$$

(see Kulsrud and Cesarsky 1971). For a spectral index of 2.7, $r = 4.7$ and $C_{4.7} = 1.7/2.7 = 0.6$, so $(\pi/4)C_{4.7} = 0.5$. If we drop the numerical factors in equation 76, we will still get the correct answer up to the factor of one-half.

In the absence of cosmic rays, parallel propagating Alfven waves are nearly degenerate. Whether the eigenmodes are circularly polarized or linearly polarized depends on the cosmic-ray distribution, which removes the degeneracy. If the cosmic-ray distribution is shifted to the right then the unstable eigenmode is a linearly polarized Alfven mode.

Consider this case, and let the wave move to the right. Then both right-moving and left-moving cosmic rays resonate with the wave, diffused by it along a circle in momentum space in the wave frame. Along this circle, cosmic rays with positive and slightly greater p_z than the resonant p_z, $M(v_z = \Omega/k)$, will give momentum to the wave, while cosmic rays with slightly smaller p_z will take momentum away from the wave. Thus, instability will occur if there are more of the larger p_z cosmic rays than of the smaller p_z about this resonance.

For a distribution shifted to the right in the wave frame, this is the case. But there are also left-moving resonant cosmic rays at $v_z = -\Omega/k$. These also contribute to the instability in same manner as the positive p_z ones. Namely, the shifted distribution also has cosmic rays along the circle with more cosmic rays with (algebraically) larger p_z than cosmic rays with more negative p_z. Thus, they also contribute to the instability. For left-moving waves, both classes of resonant cosmic rays give stability.

The possibility for resonance of both positive and negative cosmic rays occurs because a linear polarized wave is made up of both right and left circularly polarized components. What about a symmetrically distorted distribution, say, a prolate distribution with more p_z than $p_\perp$? In this case, if the wave were linearly polarized, then E_x would produce a j_y but no j_x, so that we must have a circularly polarized wave. A right-moving, right circularly polarized wave has **E** rotate spatially along a left-hand screw, so that right-moving cosmic rays see a left-rotating E field, and left-moving cosmic rays see a right-rotating E field. Thus, this wave resonates only with right-moving cosmic rays. Likewise, only left-moving cosmic rays resonate with a right-moving, left-circularly-polarized wave. Thus, for symmetrical cosmic-ray distribution and a right-moving, right-circularly-polarized Alfven wave there is only one class of resonant cosmic rays to check, to see if there are more faster p_z cosmic rays than slower ones (along the circle in the wave

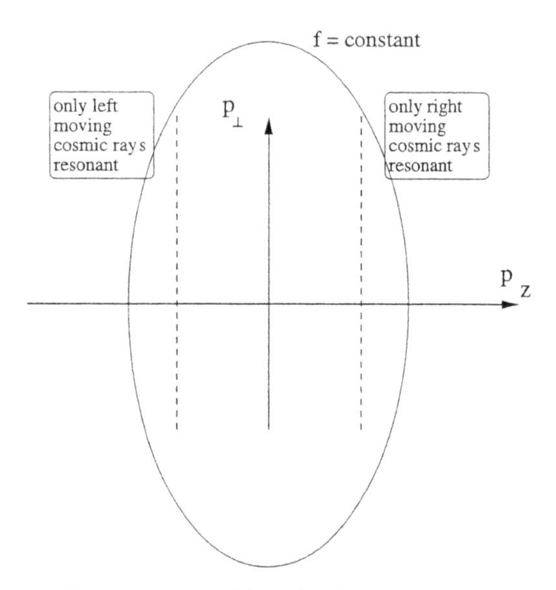

Figure 12.7. An oblate distribution function

frame), namely, the right-moving cosmic rays. For a prolate distribution (larger p_z than $p_\perp$) it is easy to check that right-moving, right-circularly-polarized waves and left-moving, left-circularly-polarized waves are unstable, while the other two possibilities are stable. For an oblate distribution (p_z smaller than $p_\perp$) the opposite is the case. The two unstable types of waves for the prolate case are now stable and the two stable types are unstable. It is easier to apply these simple physical arguments to determine instability rather than to try to remember all the many different cases, or to get them out of the formulas.

The symmetric case of a cosmic-ray distribution arises when we consider the evolution of an energetic cosmic-ray electron distribution as it loses energy by synchrotron radiation (see figure 12.7). Since the cosmic rays with larger $p_\perp$ lose energy fastest, we expect a prolate distribution. But once the prolateness anisotropy δ is larger than v_A/c, then the circularly polarized Alfven waves become unstable. These waves then lead to a pitch-angle diffusion, which restores isotropy at least to order v_A/c. Thus, the cosmic rays with smaller $p_\perp$ scatter nearly isotropically into larger $p_\perp$ and radiate their energy. The distribution radiates nearly all of its energy through this mechanism, not just its initial perpendicular energy.

12.4 Quasilinear Diffusion of Cosmic Rays

We have seen that as cosmic rays resonate with Alfven waves they change their pitch angles θ in a diffusive way. In the frame of the waves, there is no electric field and they do not change their energy. That is, in the wave frame they diffuse in pitch angle on a constant energy surface. This is the case if the waves are all moving in the same direction. (For waves propagating in both directions of course there is no frame in which the electric field vanishes, and the cosmic rays will change their energy in all frames. However, if the waves are excited by a cosmic-ray distribution drifting to the right, only right-moving waves are excited and there is no acceleration or deceleration in the frame moving to the right with the Alfven speed.)

Let us assume that we have a given population of right-moving waves. For definiteness, take them linearly-polarized in either the x or the y direction. Let them have fluctuating magnetic fields whose Fourier transforms satisfy the random phase approximation

$$\frac{\langle \delta \mathbf{B}^*_{k'} \delta \mathbf{B}_k \rangle}{B_0^2} = \mathbf{I}(k)\delta(k' - k) \tag{77}$$

where $\mathbf{I}(k)$ is a tensor in x and y. If the x linear polarized waves and y linearly polarized waves have the same spectra, then $\mathbf{I}$ is diagonal in x and y. (If we wish to deal with circularly polarized waves, $\mathbf{I}$ would be Hermetian.)

Now, if we substituted a particular set of waves into the Vlasov equation 51, we would find to first order that f_1 is given by equation 54. Substituting this f_1 back into equation 51 we are able to find the long time evolution of f_2 due to the collection of waves. Ensemble averaging over all possible sets of waves we find the quasilinear evolution of the mean distribution of cosmic rays f is given (in the Alfven frame) by

$$\frac{\partial f}{\partial t} + v_z \frac{\partial f}{\partial z} = \frac{\partial}{\partial \mu}\left[(1 - \mu^2) D_\mu \frac{\partial f}{\partial \mu}\right] \tag{78}$$

where

$$D_\mu = \frac{\pi}{4}\Omega k I(k_r) \tag{79}$$

where

$$k_r = \frac{\Omega}{\mu v} = \frac{1}{\mu r_L} \tag{80}$$

and where at Larmor radius $r_L = v/\Omega$, k_r is the wave number of the Alfven waves directly resonant with cosmic rays with momentum p_z, and μ is the cosine of the pitch angle. These formulas are easily derived from equation 46 in chapter 11.

Since

$$\left(\frac{\delta B}{B}\right)^2 = \int I(k)dk, \tag{81}$$

D comes from the contribution to $(\delta B/B)^2$ of waves in a band of width k about the resonant region. We see that equation 28, arrived at by a qualitative argument, disagrees slightly from this result. The disagreement comes

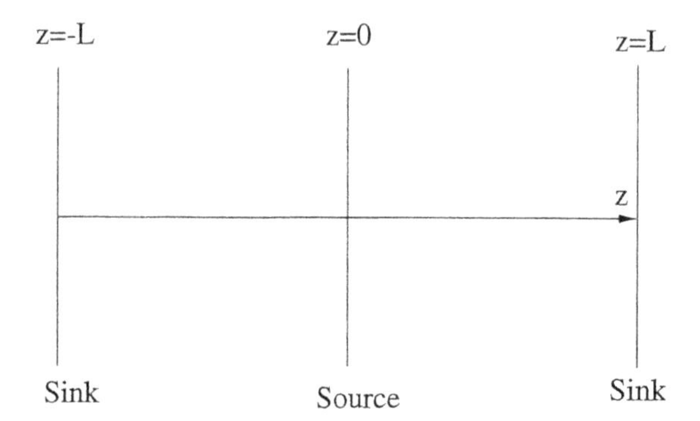

Figure 12.8. The source and sinks for the model

from the fact that the bandwidth for the definition of δB in this argument is a little smaller than k. It can be shown that the rate of gain of momentum by the waves is just that being lost by the cosmic rays, due to this diffusion. We use momentum balance between the cosmic rays and the waves rather than energy, since momentum is independent of frame, while energy is frame dependent. (In fact, waves have no energy in their rest frame and the cosmic rays do not change energy in this frame.)

It should be noted that, as μ approaches zero, i.e., θ approaches 90°, k_r goes to infinity and so waves of infinitely small wavelength are necessary to resonantly pitch-angle scatter the cosmic rays past $\mu = 0$. We show how we can get around this singularity in the following model propagation problem.

12.5 A Model for Cosmic-Ray Propagation with Sources and Sinks

Let us apply all these results to the problem of the propagation of cosmic rays through a fully ionized medium from a given source to a given sink, presumably a place in the interstellar medium where the galactic field is so distorted that cosmic rays can escape from the disk. We assume a one-dimensional model in z with the single (planar) source at $z = 0$ and with two sinks at $z = \pm L$ (see figure 12.8). For more details of this problem and its solution see Felice and Kulsrud (2001).

By the symmetry of the problem we may restrict ourselves to the region $0 < z < L$. Again we assume that the magnetic field is in the z direction. As the cosmic rays stream from the source to the sink they produce Alfven waves by the instability. These waves will grow to a certain amplitude at which they become stabilized by nonlinear Landau damping. In a steady state, they then will scatter the cosmic rays at such a rate that the streaming velocity determined by the flow rate from source to sink is just balanced by this scattering.

For a given source strength S, the cosmic-ray density $N_{\rm cr}$ builds up until the above balance is reached. We wish to find this density $N_{\rm cr}$, since $N_{\rm cr}/S$ will provide a measure for the lifetime of the cosmic rays. We characterize the waves by the local value in k space of

$$\left(\frac{\delta B}{B}\right)^2 = kI = \mathcal{E}(k) \tag{82}$$

Since the waves are driven unstable by a cosmic-ray distribution drifting to the right, only waves with $k > 0$ are excited and $\mathcal{E}$ vanishes for negative k. (However, recall that rightward-propagating, linearly polarized waves resonate with cosmic rays going both to the left and to the right.)

We want to find $\mathcal{E}(k)$ as a function of k. We also want to determine the distribution of the $f(\mu, z)$ at a given position z as a function of pitch angle μ. For simplicity, we measure the pitch angle in the wave frame. We approximate the $\mathbf{v} \cdot \nabla f$ term in equation 78 by

$$v_z \frac{\partial f}{\partial z} = -\frac{\mu v f(\mu)}{L} \tag{83}$$

where $f(\mu)$ is $f(\mu, z)$ at $z = L/2$. Then we can write equations 78 and 79 as

$$\frac{\partial f}{\partial t} - \frac{\mu v f}{L} = \Omega \frac{\pi}{4} \frac{\partial}{\partial \mu} \left[\mathcal{E}(\mu)(1-\mu^2)\frac{\partial f}{\partial \mu}\right] \tag{84}$$

$\mathcal{E}(\mu) = \delta B/B$ is the relative magnetic fluctuation energy for waves at $k(\mu) = \Omega/v\mu$ in resonance with the cosmic rays with a given μ, so that

$$\mathcal{E}(\mu) = \mathcal{E}[|k(\mu)|] = \mathcal{E}\left(\frac{\Omega}{v|\mu|}\right) \tag{85}$$

(For convenience, we parameterize the waves by the value of μ, the cosine of the pitch angle of 1-GeV cosmic rays, with which they are interacting. Similarly, we parametrize the growth rate $\gamma(\mu)$ by the growth rate of the same waves. Note that $\mathcal{E}(-\mu) = \mathcal{E}(\mu)$. From now on we replace the cosmic-ray velocity v by c.)

From equation 64, the growth rate $\gamma(\mu)$ for these waves depends on the μ derivative of f, and is

$$\gamma(\mu) = \gamma(k) = \pi^2 e^2 \frac{v_A^2}{c^2} \int \frac{1-\mu^2}{p v_A} \frac{\partial f}{\partial \mu} v^2 \frac{\delta(\mu - \mu_c)}{kv} d^3 p \tag{86}$$

where $\mu_c = \Omega/kv$. We now approximate the cosmic-ray distribution function by factoring out the angular dependence in f and writing

$$f(\mu, p) \approx N_{\rm cr}(p > p_1) F(\mu) \tag{87}$$

to simplify the calculation. With this approximation the expression for γ reduces to

$$\gamma(\mu) = \frac{\pi}{4}\mu(1-\mu^2)\Omega_0 \frac{c}{v_A} \frac{N_{\rm cr}(p > p_1)}{n^*} \frac{\partial F}{\partial \mu} \tag{88}$$

where $p_1 = mc$, and n^* is the background density of ions. But $\gamma(\mu)$ must balance the nonlinear Landau damping rate,

$$\gamma_{\rm NL} = 0.3\omega\frac{v_i}{v_A}\mathcal{E} = 0.3\frac{kv_A}{v_A}v_i\mathcal{E} = 0.3\frac{\Omega}{\mu}\frac{v_i}{c}\mathcal{E} \tag{89}$$

where we use the resonance condition $k = (\Omega/c\mu)$. Thus, we have

$$\gamma(\mu) = 0.3\frac{\Omega}{\mu}\frac{v_i}{c}\mathcal{E}(\mu) \tag{90}$$

The problem has been reduced to finding $F(\mu)$, the cosmic-ray angular distribution function, and $\mathcal{E}(\mu)$, the relative energy of the Alfven waves resonant with the cosmic rays at the pitch angle μ, versus μ. The pitch-angle trouble at small μ is reflected in the fact that at small μ, $\gamma \sim \mu$ and $\gamma_{\rm NL} \sim 1/\mu$, so $\mathcal{E}$ decreases with μ, while the pitch-angle diffusion in equation 84 goes to zero.

Before addressing the problem of finding the exact solution for $F(\mu)$, we first form a rough idea as to the size of the drift velocity of the cosmic rays and the wave amplitude. Assume that $\mathcal{E}$ is constant in μ and that F is of the form

$$F = 1 + 3\frac{v_D}{c}\mu \tag{91}$$

Then, since $\langle\mu^2\rangle = \frac{1}{3}$, $\langle\mu c F(\mu)\rangle = v_D$. Now, multiply equation 84 by μc and integrate by parts. Neglecting terms of order $(v_D/c)^2$, we get

$$\frac{\partial v_D}{\partial t} = \frac{c^2}{3L} - \pi\mathcal{E}\Omega v_D \tag{92}$$

Let us seek a steady state. From equation 92,

$$v_D = \frac{2c^2}{3\pi\Omega L\mathcal{E}} = \frac{2r_L}{3\pi L}\frac{c}{\mathcal{E}} \tag{93}$$

with $r_L = c/\Omega$.

$\mathcal{E}$ is determined by equation 89. Substituting F from equation 91 into equation 88 to find γ, and substituting this and v_D from equation 93 into equation 89. and taking $\mu = \frac{1}{2}$ in γ and $\gamma_{\rm NL}$, we get

$$\frac{9\pi}{32}\Omega\frac{v_D}{v_A}\frac{N_{\rm cr}}{n^*} = 0.6\Omega\frac{v_i}{c}\mathcal{E} \tag{94}$$

Eliminating $\mathcal{E}$ from equations 93 and 94 we find

$$v_D = 0.38\sqrt{v_i v_A\frac{n^*}{N_{\rm cr}}\frac{r_L}{L}} = 0.38 v_{D0} \tag{95}$$

and eliminating v_D we get

$$\mathcal{E} = 0.57\sqrt{\frac{r_L}{L}\frac{N_{\rm cr}}{n*}\frac{c^2}{v_i v_A}} = 0.57\mathcal{E}_0 \tag{96}$$

If we take for the coronal region of the interstellar medium, $B = 3 \times 10^{-6}$ G, $L = 1$ kpc, $N_{\text{cr}} = 10^{-10}$ cm^{-3}, $n^* = 10^{-3}$ cm^{-3}, and $v_i = 10^7$ cm/sec, we get $v_A = 2.1 \times 10^6$ cm/sec, and

$$v_D = 10^5 \text{cm/sec} \tag{97}$$

for the drift velocity of the cosmic rays in the wave frame. But this is small compared to the Alfven speed of 2.1×10^7 cm/sec. Thus, cosmic rays propagate with the waves at the Alfven speed, and their lifetime is thus L/v_A, or three million years. The energy dependence of v_D, is proportional to N_{cr}. $N_{\text{cr}} \sim \epsilon^{-1.5}$, so $v_D \sim \epsilon^{0.75}$, and when the energy is large enough for v_D to exceed v_A, the lifetime is proportional to $\epsilon^{-0.75}$, not very far from the observed value.

This is a crude estimate of the behavior of our model problem. We have assumed that $F = \text{const}(1 + 3\mu v_D/c)$, which is not a very good approximation for small μ, and we used this assumption to determine the evolution of F. However, it does give a rough magnitude for v_D. The corresponding value for $\mathcal{E}$ is 10^{-6}, which is in line with our previous estimate.

Now let us derive the actual dependence of F and $\mathcal{E}$ on μ. Return to equation 84 and integrate it with respect to μ:

$$F\frac{1-\mu^2}{2}\frac{v}{L} = \frac{\pi}{4}\Omega\mathcal{E}(\mu)(1-\mu^2)\frac{\partial F}{\partial \mu} \tag{98}$$

The constant of integration on the left-hand side is chosen to be consistent with the vanishing of the right-hand side at $\mu \pm 1$. Thus,

$$\frac{\pi}{2}\Omega\mathcal{E}(\mu)\frac{\partial F}{\partial \mu} = F\frac{c}{L} \tag{99}$$

The growth rate γ is given by equation 88. Thus, setting $\gamma = \gamma_{\text{NL}}$, we get

$$\frac{\pi}{4}\mu(1-\mu^2)\Omega_0\frac{N_{\text{cr}}}{n^*}\frac{c}{v_A}\frac{\partial F}{\partial \mu} = 0.3\Omega\frac{\mathcal{E}}{\mu}\frac{v_i}{c} \tag{100}$$

Solving equations 99 and 100 for $\partial F/\partial \mu$ and $\mathcal{E}(\mu)$ by eliminating $\partial F/\partial \mu$, we have

$$\mathcal{E} = 1.29\sqrt{\frac{c^2}{v_i v_A}\frac{r_L}{L}\frac{N_{\text{cr}}}{n^*}}\mu\sqrt{(1-\mu^2)} = 1.29\mathcal{E}_0\mu\sqrt{1-\mu^2} \tag{101}$$

where we have replaced F by 1. Then we get for $\partial F/\partial \mu$

$$\frac{\partial F}{\partial \mu} = 0.49\sqrt{\frac{v_i v_A}{c^2}\frac{n^*}{N_{\text{cr}}}\frac{r_L}{L}}\frac{1}{\mu\sqrt{1-\mu^2}} = 0.49\frac{v_{D0}}{c}\frac{1}{\mu\sqrt{1-\mu^2}} \tag{102}$$

where $\mathcal{E}_0$ and v_{D0} are the crude estimates for the drift velocity and the magnetic intensity given in equations 95 and 96. $\partial F/\partial \mu$ is very small compared to unity, so that F is nearly constant over most of μ, and therefore it is legitimate to replace it by unity above. However, it is divergent at small μ.

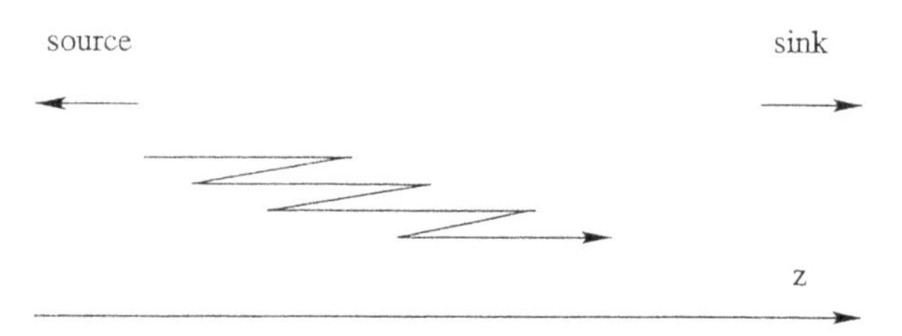

Figure 12.9. Why the mean speed of a cosmic ray is less than c

If pitch-angle scattering were the entire story of how μ changes, then this divergence indicates that the cosmic rays would never get past $\mu = 0$. Physically they would be stuck at positive μ and flow directly from the source to the sink at a speed of $c/3$. They would scatter down to a pitch angle μ_1 and then return to a finite value of pitch angle. F would be proportional to $[\ln(\mu/\mu_1)]/[\ln(1/\mu_1)]$, where μ_1 would be determined by how small a pitch angle could be reached in the time $3L/c$ the cosmic rays would live as they pass from source to sink. The bulk velocity can be reduced to a small value only if there is back and forth scattering across $\mu = 0$, as in figure 12.9.

Actually, we expect that they would get across $\mu = 0$, because when the cosmic rays have very small μ, they still have a finite $v_\perp^2$, and, thus, any small inhomogeneity in $|B|$ should exert a large enough force to change μ through zero. But the Alfven wave packets that pitch-angle scatter the cosmic rays at finite μ themselves have a second-order variation in field strength, $\delta|B|$, even though the first-order $\mathbf{B}_1$ is perpendicular to $\mathbf{B}_0$. In fact,

$$\delta|B| = \frac{B_1^2}{B_0} \tag{103}$$

The wavelength is $\lambda = 1/k = r_L$, so the time τ it would take a magnetic well with this depth to reverse τ is gotten from

$$\frac{dp_z}{dt} = k\delta|B|\frac{vp}{2B}\cos kz \tag{104}$$

or from

$$\gamma_R\frac{d^2z}{dt^2} = \gamma_R\frac{k\delta|B|v^2}{2B}\cos kz \tag{105}$$

so

$$\tau \approx \frac{B}{\Omega\delta|B|} = \frac{1}{\sqrt{\mathcal{E}_1}\Omega} \tag{106}$$

since $kv \approx \Omega$. (The most important waves for mirroring are those that resonate with cosmic rays of finite pitch angle for which $\mathcal{E}(\mu) \approx \mathcal{E}_1$.) The reversal cannot happen unless the well is deep enough to actually reflect the cosmic rays, that is, unless

$$\mu^2v^2 < \frac{\delta|B|}{B}v^2 = \frac{(\delta B)^2}{B_0^2}v^2 = \mathcal{E}_1v^2 \tag{107}$$

or a cosmic ray is reflected if $\mu < \mu_c$ where

$$\mu_c < \frac{\delta B}{B} = \sqrt{\mathcal{E}_1} \tag{108}$$

A further condition is that τ should not be greater than the pitch-angle quasilinear scattering time $\tau_{\mathrm{QL}} = \mu^2/D(\mu)$ to change μ by an amount comparable with itself μ_c. ($D(\mu)$ is the quasilinear diffusion coefficient.) At μ_c this time is

$$\tau_{\mathrm{QL}} \approx \frac{\mu^2}{\Omega\mathcal{E}(\mu)} \approx \frac{\mu}{\Omega\mathcal{E}_1} \tag{109}$$

since $\mathcal{E}$ is essentially linear in μ, except for the factor $\sqrt{1-\mu^2}$. But with equation 106 this condition becomes

$$\frac{1}{\tau} = \Omega\sqrt{\mathcal{E}_1} \geq \frac{\Omega\mathcal{E}_1}{\mu} \tag{110}$$

or

$$\mu \geq \mu_c \tag{111}$$

In other words, for $\mu > \mu_c$ the the mirroring is faster than pitch-angle scattering but the mirror is not deep enough to reflect the particles or to change their pitch angle.

When μ_c is reached, the two rates become comparable. Numerical computations of the behavior of $\mathcal{E}$ and $\partial F/\partial\mu$ in the narrow boundary layer below μ_c show that the two effects are approximately equal below μ_c until at a smaller μ, $F(\mu)$ flattens out and only the mirroring persists. The resulting $F(\mu)$ is plotted in figure 12.10 as a function of μ/μ_c and it is seen that at small μ the Alfven waves are no longer unstable. However, the distribution function continues into the negative values of μ and the particles are successfully reflected. The details of the calculation of $F(\mu)$ below μ_c are found in Felice and Kulsrud (2001), where a detailed boundary layer analysis is given.

We can estimate the drift velocity from

$$\langle v\rangle = \frac{v\int F\mu d\mu}{\int F d\mu} = \frac{1}{2F}\int_{-1}^{1}\frac{1-\mu^2}{2}c\frac{\partial F}{\partial\mu}d\mu = \frac{0.382}{2}\left(\ln\frac{1}{\mu_c}+\frac{1}{2}\right)v_{D0} \tag{112}$$

where

$$\mu_c = \sqrt{\mathcal{E}_1} \approx \sqrt{\mathcal{E}_0} \approx 10^{-3} \tag{113}$$

so $v_D \approx 3.3v_{D0}$. ($\partial F/\partial\mu$ is taken as finite below μ_c so this region contributes very little to the integral.) Thus, the correction to cosmic-ray propagation by the 90° pitch-angle problem is significant but not overwhelming.

Although it may appear from equation 112 that the main contribution to the drift velocity comes from $\mu \approx \mu_c$, this is not the case. μ_c is where we match the positive μ and negative μ parts of F through the boundary layer,

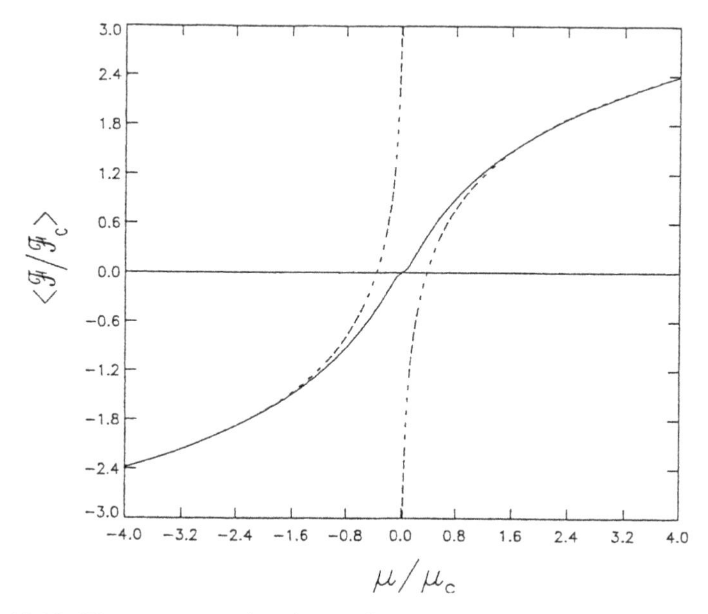

Figure 12.10. The cosmic ray distribution function near $\theta = 90$ $\mathcal{F}$ stands for the $F(\mu) - F(0)$ normalized to its variation in the boundary layer $\mu < \mu_C$. the dotted line represents the analytic solution without mirroring, Equation (102), and the solid line is the true distribution including mirroring.

where F has only a small variation. If μ_c were smaller F would tend to the dotted curves, which would make $F(1)$ larger. The lifetime is proportional to the difference between $F(1)$ and $F(-1)$, which is proportional to the difference in the times that the cosmic ray spends going left and going right or to the net rate of advance of the cosmic ray to the right. The variation from a constant F is greatly exaggerated in this figure since it is actually of order v_D/c.

12.6 Cosmic-Ray Pressure and Energy

Early in this chapter we showed how the cosmic rays could exert a macroscopic force perpendicular to a magnetic field, for example, when the cosmic rays participated in exerting a macroscopic force through their pressure on the interstellar medium, helping to support it against the gravitational force. Cosmic rays can also exert a force along the magnetic field on the background plasma if their pressure varies along the magnetic field B_0, which we again take in the z direction (Wentzel 1971). Let us see how this comes about by a semiqualitative argument. This exercise provides an excellent example of the concepts with which we have been dealing.

If the cosmic-ray pressure $P_{\rm cr}$ varies in z, this will give rise to a bulk drift velocity v_D relative to the velocity of the waves v_A. This drift velocity produces a growth in the waves and therefore an increase in their momentum. This momentum is essentially the momentum of the background plasma. At first, it is directly associated with the waves, but when the waves damp the momentum is transferred to the mean motion of the plasma. This increase in the plasma momentum represents the force that the cosmic rays transfer to the plasma. (This increase of the momentum of the plasma comes from the intrinsic momentum of the cosmic rays, i.e., their drift velocity. As the cosmic rays accelerate the plasma, their pressure gradient drops and so does their bulk velocity where the momentum originally resides.)

If the pitch-angle scattering rate in the wave frame is ν and the pressure gradient scale of the cosmic rays is L, then the bulk drift of the cosmic rays v_D (relative to the rest frame) is given by

$$v_D - v_A = \frac{c^2}{3\nu L} \tag{114}$$

To see this write equation 84 as

$$-\frac{v\mu F}{L} = \nu\frac{\partial}{\partial\mu}\left(\frac{1-\mu^2}{2}\right)\frac{\partial F}{\partial\mu} \tag{115}$$

Integrating this over μ, taking F as zero order on the left-hand side, gives

$$-\frac{v}{L}\left(\frac{\mu^2-1}{2}\right)F = \nu\left(\frac{1-\mu^2}{2}\right)\frac{\partial F}{\partial\mu} \tag{116}$$

or

$$\frac{\partial F_1}{\partial\mu} = \frac{v}{\nu L}F_0\mu \tag{117}$$

We treat ν as a constant as a convenient approximation. F_1 is small compared to F_0 if $\lambda = c/\nu \ll L$. The velocity moment gives

$$v_D - v_A = \frac{\int F_1\mu v d^3p}{\int F d^3p} = \frac{v^2}{3\nu L} = \frac{c^2}{3\nu L} \tag{118}$$

where we integrate by parts to make use of equation 117 and find the average of μ^2 to be $\frac{1}{3}$.

Now the finite value of $v_D - v_A$ gives a growth rate of order

$$\gamma = \frac{\pi}{4}\Omega_0\frac{v_D - v_A}{v_A}\frac{N_{\rm cr}}{n} \tag{119}$$

to the waves (see equation 76), so that their momentum increases at the rate

$$\frac{dP_{\text{wave}}}{dt} = 2\gamma \frac{(\delta B)^2}{4\pi} \frac{1}{v_A}$$

$$= 2\frac{\pi}{4}\Omega_0 \frac{v_D - v_A}{v_A} \frac{(\delta B)^2}{4\pi} \frac{N_{\text{cr}}}{n} \frac{1}{v_A}$$

$$= \frac{1}{8}\Omega_0 \frac{c^2}{3\nu L} \frac{1}{v_A^2} (\delta B)^2 \frac{N_{\text{cr}}}{n} \tag{120}$$

But from equation 28

$$\nu = \frac{\pi}{2}\Omega \left(\frac{\delta B}{B}\right)^2 \tag{121}$$

Substituting this and $v_A^2 = B^2/4\pi n M$ into equation 120 we get

$$\frac{dP_{\text{wave}}}{dt} = \frac{\Gamma_{\text{cr}}}{3} \frac{N_{\text{cr}} M c^2}{L} \tag{122}$$

But $\Gamma_{\text{cr}} N_{\text{cr}} = \mathcal{E}/Mc^2$, where $\mathcal{E}$ is the cosmic-ray energy, which is three times the cosmic-ray pressure P_{cr} so

$$\frac{dP_{\text{wave}}}{dt} = \frac{P_{\text{cr}}}{L} \approx -\nabla P_{\text{cr}} \tag{123}$$

We have given a qualitative argument that the cosmic rays exert a force equal to their pressure gradient along the field as well as across the field. A more precise argument involves taking a p_z moment of equation 78 and integrating over all momentum space. If we make use of equation 64 for the growth rate, we arrive at a precise derivation for the relation of the cosmic-ray pressure gradient and the growth rate of the wave momentum:

$$-\frac{\partial}{\partial t}\int p_z F d^3p - \frac{\partial}{\partial z}\int v_z p_z F d^3p = \int 2\gamma(k)\frac{\mathcal{E}_k}{v_A} d^3\mathbf{k} \tag{124}$$

The first term is the loss of cosmic-ray momentum (related to v_D). The second term is the cosmic-ray pressure. The right-hand side represents the growth of the wave momentum due to cosmic-ray streaming.

(γ is only that the part of the wave growth γ_{cr} due to the cosmic rays. The total growth or damping must include the damping term Γ as well. γ_{cr} may be balanced by Γ so the wave amplitudes are statistically constant. In this case, the Γ term generally transforms the wave momentum to the mean velocity of the background plasma. The plasma already has the momentum when it is the wave form; it is just coherent with the waves at this time. The important thing for the cosmic-ray force is the γ_{cr} term.)

We must accept the concept that the cosmic rays do interact macroscopically with the background plasma both across and along B_0. This is true in spite of the fact that there is no collisional interaction between them in the usual sense. Along the field the interaction is through coherent wave–particle collisions. The cosmic rays collide only with the waves.

If there were no other forces to balance the cosmic-ray force, then the cosmic-ray pressure gradient would accelerate the background plasma. If this were the case, cosmic rays would do work on the plasma:

$$\frac{d\mathcal{E}}{dt} = -\mathbf{v}_D \cdot P_{\text{cr}} \tag{125}$$

But this work must come from the cosmic-ray energy.

We can derive an equation for the evolution of the cosmic-ray pressure or energy from analogy with ordinary pressure p with adiabatic index γ_A. Recall from equation 36 in chapter 4 that for an ideal fluid

$$\frac{\partial p}{\partial t} = -\mathbf{V} \cdot \nabla p - \gamma_A p \nabla \cdot \mathbf{V} \tag{126}$$

This equation is consistent with energy conservation of pressure and kinetic energy. The rate of work the pressure does by acceleration is

$$\frac{d\mathcal{E}_K}{dt} = -\mathbf{V} \cdot \nabla p \tag{127}$$

Thus, the total change of the pressure energy is

$$\begin{aligned} \frac{d}{dt} \int \frac{p}{\gamma_A - 1} d^3\mathbf{x} &= \int \left(\frac{-\mathbf{V} \cdot \nabla p}{\gamma_A - 1} - \frac{\gamma_A p}{\gamma_A - 1} \nabla \cdot \mathbf{V} \right) d^3\mathbf{x} \\ &= -\int \frac{\gamma_A}{\gamma_A - 1} (p\mathbf{V}) \cdot d\mathbf{S} + \int \mathbf{V} \cdot \nabla p d^3\mathbf{x} \end{aligned} \tag{128}$$

The first term on the right-hand side is enthalpy flow and is the pdV work done on the outside $p\mathbf{v} \cdot d\mathbf{S}$ plus the escape of thermal energy $p\mathbf{V} \cdot d\mathbf{S}/(\gamma_A - 1)$. The last term is the work done by the acceleration. This is consistent with energy conservation only if equation 126 is satisfied.

Now, since $\gamma_A = \frac{4}{3}$ for relativistic cosmic rays, the work done by acceleration is $\mathbf{V} \cdot \nabla p$ and the escape of energy and pdV work done is $\mathcal{E}V \cdot Ds + pV \cdot dS = 4pV \cdot dS$, consistent with $\gamma_A/(\gamma_A - 1) = 4$. Thus, the cosmic-ray pressure must satisfy equation 126 with $\gamma_A = \frac{4}{3}$, i.e.,

$$\frac{\partial P_{cr}}{\partial t} + V \cdot \nabla P_{\text{cr}} = -\frac{4}{3} (\nabla \cdot V) P_{\text{cr}} \tag{129}$$

This equation is valid under the assumption that the mean free path of cosmic rays c/ν is small compared to any macroscopic length L. The term on the right-hand side represents adiabatic compression and expansion, which changes the cosmic-ray energy. Equation 129 implies that following the bulk motion of the cosmic rays, their pressure is proportional to $\rho^{4/3}$. This may be different than the bulk motion of the plasma, since the cosmic rays can slip through the plasma at a velocity v_D that may be comparable to or larger than the plasma velocity $\mathbf{V}$. However, if $\mathbf{v}_D - \mathbf{V}$ is small compared to V, we assume that the cosmic rays are locked to the plasma and behave as an ideal gas.

Let us assume that $|\mathbf{v}_D - \mathbf{V}| < V$, so that we can think of the cosmic rays as locked to the plasma, and so that the cosmic ray pressure behaves as that of an ideal gas with a γ_A of $\frac{4}{3}$. The cosmic-ray energy per unit volume ϵ then varies as $n^{1/3}$. If we trace back through our arguments, we see that this is also true in the average for each particle. That is, $\epsilon \sim n^{1/3}$ or in terms of the fluid motion

$$\frac{d\epsilon}{dt} = -\frac{\epsilon}{3}\nabla \cdot V \tag{130}$$

Our kinetic equation for quasilinear diffusion was derived assuming that the background velocity was constant in space, so that in the frame of the waves, ϵ remains constant during pitch-angle scattering. Equation 130 says that if $\mathbf{V}$ varies in space, then the cosmic rays change their energy and this must be taken into account in the cosmic-ray equation. Thus, the more general kinetic equation for cosmic rays is

$$\frac{\partial f}{\partial t} + \mathbf{v} \cdot \nabla f = -\nabla \cdot (D\mathbf{nn} \cdot \nabla f) + \frac{1}{3}(\nabla \cdot \mathbf{V}) p \frac{\partial f}{\partial p} \tag{131}$$

This equation refers to the evolution of f at a fixed point in phase space, so it is indeed valid even for nonzero $\mathbf{v}_D - \mathbf{V}$. It is derived analytically in Skilling (1975) and it will be applied in the next section to shock acceleration.

12.7 Fermi Acceleration and Shock Acceleration of Cosmic Rays

In 1949 Fermi (Fermi 1949) developed an origin theory for cosmic rays that involved a new theory of acceleration of high-energy particles. He envisioned that cosmic rays in interstellar space would collide with moving clouds and in the collision the energy of the clouds would be gradually transferred to the cosmic rays (see figure 12.11). Imagine that a cosmic ray with energy ϵ and momentum p is traveling along a magnetic field and encounters an incoming cloud with velocity u, which we treat as a moving mirror. For simplicity, take $p_\perp$ zero.

In the frame of the cloud the cosmic-ray energy and momentum are

$$\epsilon' = \epsilon + up$$

$$p' = p + \frac{u\epsilon}{c^2} \tag{132}$$

(if $u \ll c$). After colliding with the cloud p' is reversed to $-p'$ and ϵ' is unchanged. Transferring back to the rest frame, we find that the new energy

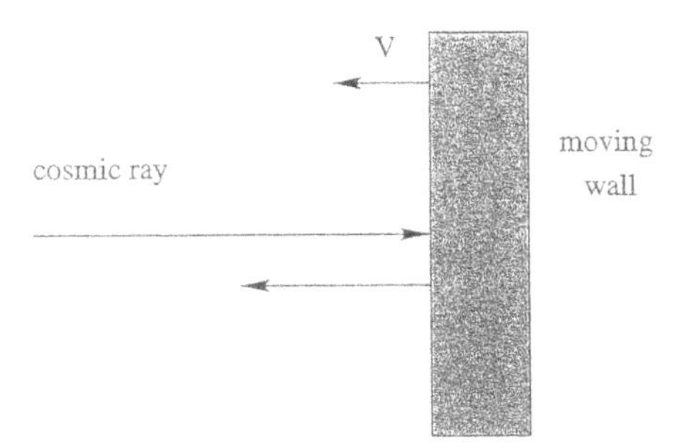

Figure 12.11. A cosmic ray reflected from a moving wall with $V = u$

ϵ'' is

$$\epsilon'' = \epsilon' - (-p')u = \epsilon' + up'$$
$$= \epsilon + 2up = \epsilon\left(1 + \frac{2vu}{c^2}\right) \tag{133}$$

where we drop u^2 terms. If the cosmic-ray velocity $v \approx c$, this gives

$$\delta\epsilon = \frac{2vu\epsilon}{c^2} \approx \frac{2u}{c}\epsilon \tag{134}$$

The cosmic ray after the encounter is then moving to the left. It will encounter a second cloud with a similar velocity u.

Let the velocity of the ith cloud be u_i. Then, after each encounter $\ln\epsilon$ will change by u_i/c, so after N encounters

$$\Delta\ln\epsilon = \Sigma\frac{u_i}{c} \tag{135}$$

If the cloud velocities are random, then the sum over the u's averages to zero. Fermi argued that even for random velocities the encounters of the cosmic ray with approaching clouds would dominate over receding clouds by another factor of u/c because the rate of encounters is proportional to $c + u$ and $c - u$, respectively.

Thus, after N random collisions the mean change in the logarithm is

$$\Delta\ln\epsilon = \Sigma\frac{u_i^2}{c^2} = NB \tag{136}$$

where B is the average change per collision. If the mean time between encounters is τ, then the energy after a time t would be

$$\epsilon = \epsilon_1 e^{Bt/\tau} \tag{137}$$

where ϵ_1 is the initial energy.

Now, Fermi was aware that cosmic rays probably have a finite lifetime in the interstellar medium. In his theory he started with n_0 cosmic rays and assumed that the number would decay exponentially with time, so between t and $t + dt$ he would have

$$dn = -n_0 e^{-t/T} \frac{dt}{T} \tag{138}$$

cosmic rays and they would have the energy of equation 137. If cosmic rays are injected continually at energies ϵ_1 and at a rate n_0, then equation 138 gives the distribution of their lifetimes, for at $t = 0$ a cosmic ray would have a lifetime between t and $t + dt$ if it was injected during the interval $-t - dt$ and $-t$ and survived the length of time t. This also gives the distribution of their energies if we relate t and ϵ by equation 137. The resulting distribution in energies is thus

$$n_0 e^{-t/T} \frac{dt}{T} = n_0 \left(\frac{\epsilon_1}{\epsilon}\right)^{\tau/BT} \frac{dt}{T d\epsilon} d\epsilon \tag{139}$$

which yields

$$dn \sim \frac{\epsilon_1^{\tau/BT}}{\epsilon^{1+\tau/Bt}} d\epsilon \tag{140}$$

a power law distribution ϵ^{-r} with exponent r:

$$r = 1 + \frac{\tau}{BT} \tag{141}$$

For interstellar clouds, $B \approx 10^{-8}$. Now, to agree with the observed cosmic-ray spectrum r must be of order unity. If $T = 10^6$ years, then an encounter would have to occur every 10^{-2} years if r is to be of order one. This is very much faster than can be inferred from observations. Thus, although Fermi's ideas were sound, the origin theory failed because of the numbers.

A second type of acceleration occurs if the clouds are always approaching the cosmic ray. This would be the case when a cosmic ray gets trapped between two approaching clouds on the same line of force. Then $B = 2u/c$ and the acceleration is much more effective. A time of $\tau = 10^2$ years would make $r = 2$. However, if a cosmic ray is trapped between approaching cloud for a finite time interval, it would later be expected to be trapped between two receding clouds and would lose energy at the same rate. Thus, on the average, the cosmic rays would not gain energy. So this second form of acceleration, which is called first-order Fermi acceleration, also does not work in the interstellar medium. However, it does lead to a new idea for cosmic-ray acceleration, acceleration in shocks, and this does work.

Before discussing shock acceleration we must take note of a further point. The exponent r involves three parameters: τ, the time between encounters; T, the lifetime of the cosmic rays; and B, the relative acceleration per encounter $\Delta\epsilon/\epsilon = 2(u^2/c^2)$ for second-order Fermi acceleration or $2(u/c)\epsilon$ for first-order acceleration. There seems to be no reason why they should be related. Yet, for $r \approx 2$ they must satisfy $\tau/TB \approx 1$. For the interstellar medium this could happen only by a marvelous coincidence. However, it turns out that in shock acceleration this relation between τ, T, and B holds automatically and leads to $r \approx 1$ in a natural way.

After Fermi's theory for the origin of cosmic rays was abandoned, people almost universally came to the conclusion that the origin was associated with supernovae. Supernovae seemed to be the only places where a sufficient amount of energy is released to account for the enormous power, 10^{40} ergs/sec, that is required to resupply the galactic cosmic rays every few million years. Also, supernovae are violent enough to be responsible for such high particle energies. However, it is very difficult for cosmic rays to come directly from the remnant or from a pulsar arising at the center of the explosion. This is because such cosmic rays are trapped in the expanding supernova bubble, and their energy is lost adiabatically by decompression according to equation 130 during the enormous expansion of the bubble (see problem 1).

For example, if the cosmic rays came from a radius of 10^{12} cm at the beginning of the explosion and the remnant expanded to 1 pc, then the cosmic-ray energies would be decreased by 3×10^6, so that the initial energies would have to be incredibly large (10^{15} eV) to end up with the GeV energies that cosmic rays are observed to have. Thus, if supernovae are the source of cosmic rays, their acceleration of the cosmic rays has to occur after the remnant has already expanded, to avoid this adiabatic deceleration.

However, after a supernova remnant has expanded, its energy is still around in the form of thermal energy of the background interstellar medium. The hot thermal material drives shocks into the undisturbed interstellar medium. In this way the shock can deliver energy to the cosmic rays, and any energy given to them is resupplied to the shock by the hot medium. At the same time, adiabatic expansion no longer decelerates the cosmic rays, since the further expansion is small.

How does a shock actually accelerate cosmic rays? The answer is by the first-order Fermi acceleration produced by the upstream and downstream flows around the shock. In the frame of a strong hydrodynamic shock, the upstream fluid velocity is $u_- \approx V_S$, the shock speed, and the downstream velocity is $u_+ = V_S/4$. The cosmic ray crosses the shock front unaffected by the shock mechanism, as in figure 12.12. It then bounces off the downstream fluid, losing an energy $2u_+\epsilon/c = (2V_S/4)\epsilon/c$, reflects back across the shock front, and then bounces off the upstream flow, gaining an energy $2u_-\epsilon/c = 2V_S\epsilon/c$. After these two reflections the cosmic ray gains an energy $2 \times \frac{3}{4}(V_S/c)\epsilon/c$. As the cosmic ray bounces back and forth it sees converging fluid velocities as Fermi originally supposed acceleration to happen. (Of course, the shock situation differs from that of approaching clouds. In the shock case the upstream fluid actually turns into the downstream fluid.) Then for the shock the value of Fermi's B is $\frac{3}{2}V_S/c$.

What could make the cosmic ray bounce back and forth across the shock? This can happen if there is a magnetic field crossing the shock perpendicular to the shock front and such that $v_A \ll V_S$. In this situation, Alfven waves

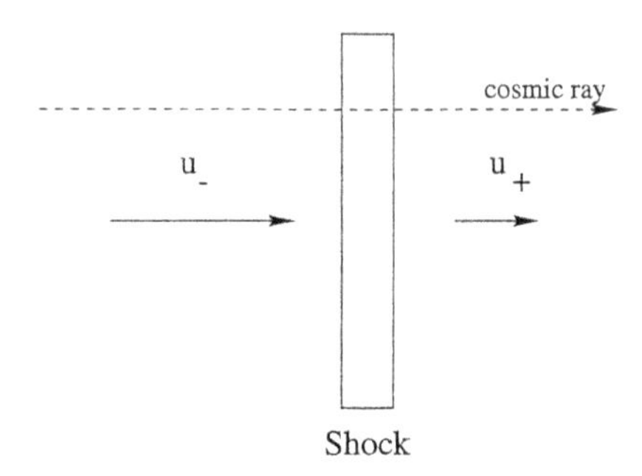

Figure 12.12. A cosmic ray crossing a strong shock

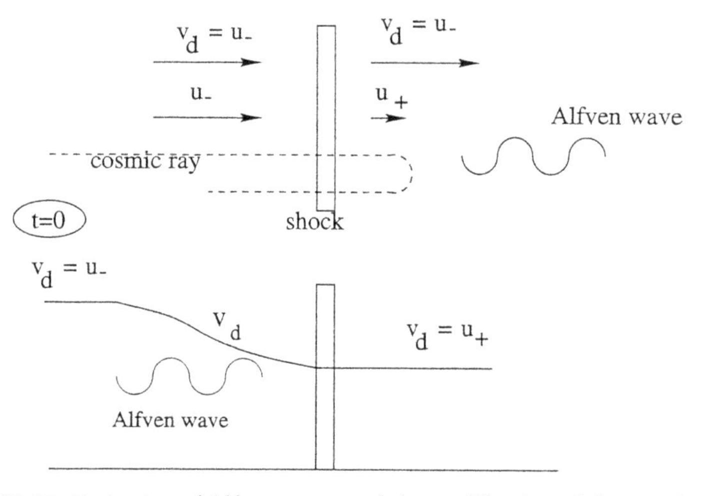

Figure 12.13. Excitation of Alfven waves and the modification of the cosmic ray distribution function

are driven unstable and these waves are the mechanism that scatters the cosmic rays. Suppose, as in figure 12.13, that the cosmic rays came in from upstream with a mean bulk velocity equal to u_-. When the fluid and cosmic rays cross the shock front the fluid is decelerated to u_+, but the cosmic-ray streaming velocity remains the same, u_-, so in the downstream region the relative velocity $u_- - u_+ = \frac{3}{4}V_S \gg v_A$ is usually enough to drive Alfven waves unstable if the shock speed is larger than four-thirds times the Alfven speed. Once the waves are strong encough, they reflect the cosmic rays and slow down their bulk velocity to u_+. But then the cosmic rays are reflected back into the upstream region where their relative velocity is again large enough to create upstream Alfven waves.

Now, as the cosmic rays bounce back and forth across the shock they are gradually drawn downstream with a mean velocity $\approx u_+$ or u_-. The actual (not the bulk) velocity of the cosmic rays with which they bounce back and forth is c, so they are going $\approx c/u$ times faster. This means that

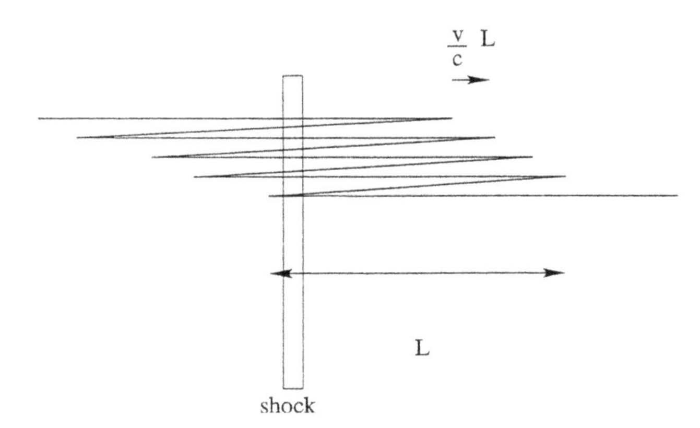

Figure 12.14. The cosmic ray cross the shock $\approx c/v_A$ times

they cross the shock c/u times, as in figure 12.14. Applying Fermi's theory we take $\tau = 2\ell/c$, while from the diagram $T = 2\ell/u$ and $B = u/c$. Thus, $BT/\tau \approx (u/c)(c/u) \sim 1$. Note that if u is larger, then the cosmic ray has fewer bounces but gains more energy per bounce. Thus, the parameters τ, T, and B automatically adjust to make the exponent $r \sim 1$ in Fermi's theory. Conditions are just such as to make shock acceleration a viable origin for cosmic rays. (For a comprehensive discussion of shock acceleration of cosmic rays see Blandford and Eichler (1987).)

Of course, those young supernovae that have small bubbles do not contribute much, since not many cosmic rays pass through them. Many cosmic rays pass through those older supernovae that have large bubbles, but for them the shock is weaker and V_s is smaller than v_A. It is only intermediate-age supernovae with moderate-size bubbles that contribute to cosmic-ray acceleration. Fortunately, there is a sufficient number of them that in the cosmic-ray lifetime of 3 million years any given cosmic ray will pass through several of them.

Our qualitative description may not be sufficiently convincing because of the uncertainties in the approximations. Fortunately, Blandford and Ostriker (1978) and others give a much more precise quantitative treatment based on Skilling's equation 131. In this equation applied, in the shock frame, $\nabla \cdot \mathbf{V} = (u_+ - u_-)\delta(x)$, if we take the shock at $x = 0$. In steady state with this result equation 131 reads

$$v\frac{\partial f}{\partial x} - \frac{\partial}{\partial x}\left(D\frac{\partial f}{\partial x}\right) = \frac{1}{3}(u_+ - u_-)\delta(x)p\frac{\partial f}{\partial p} \tag{142}$$

where D is gotten from equation 121. We solve this equation in the upstream ($x < 0$) and downstream ($x > 0$) regions separately (where the right-hand side is zero) and then match by taking $f(p)$ as continuous at the origin and getting a second condition by integrating equation 142 over a very thin region about $x = 0$:

$$-D\left(\frac{\partial f_+}{\partial x} - \frac{\partial f_-}{\partial x}\right) = \frac{1}{3}(u_+ - u_-)p\frac{\partial f}{\partial p} \tag{143}$$

Equation 143 represents the conservation of cosmic-ray flux. This discontinuity in flux is balanced by a sudden flux in the momentum direction.

Upstream we have

$$u_-\frac{\partial f}{\partial x} - \frac{\partial}{\partial x}\left(D\frac{\partial f}{\partial x}\right) = 0 \tag{144}$$

We assume $f(x, P) \to f_-(p)$ and $\partial f/\partial x \to 0$ as $x \to -\infty$ and $f(x, p) \to f_+(p)$ and $\partial f/\partial x \to 0$ as $x \to +\infty$. The upstream solution of equation 144 is

$$f = f_- + (f_0 - f_-)e^{xu_-/D} : \quad x < 0 \tag{145}$$

where $f_0 = f(p, 0)$.

Downstream we have

$$u_+\frac{\partial f}{\partial x} - \frac{\partial}{\partial x}\left(D\frac{\partial f}{\partial x}\right) = 0 \tag{146}$$

The solution of this equation is a constant since the $e^{xu_+/D}$ homogeneous part of the solution blows up as $x \to \infty$,

$$f(p, x) = f_0(p) : \quad x > 0 \tag{147}$$

Substituting equations 145 and 147 into equation 143 we have

$$(f_0 - f_-)u_- = \frac{u_+ - u_-}{3}p\frac{\partial f_0}{\partial p} \tag{148}$$

f_- is the incoming cosmic-ray distribution function and is regarded as known, so this equation is a differential equation for the downstream distribution function f_0. Notice that D has dropped out of the equation. (We have treated D as constant in x, but if it varies with x and depended on p we would still get the same differential equation for f_0.) Defining $s = u_-/u_+$, where s is the compression ratio of the shock (i.e., $n_+ = sn_-$), we can write equation 148 as

$$p\frac{\partial f_0}{\partial p} + qf_0 = qf_- \tag{149}$$

where $q = 3s/(s - 1)$. The solution of this equation is

$$f_+ = f_0 = \frac{q}{p^q}\int_0^p f_- p'^{(q-1)} dp' \tag{150}$$

If f_- is very steep in p, corresponding to mostly low-energy particles, then f_0 has a power law distribution in p with index $r = q$. For a strong shock $s = 4$ and $q = 4$. Transforming to a 1D scalar distribution the number of cosmic rays with $|p|$ between $|p|$ and $|p|+d|p|$ is proportional to $|p|^{-2}$. If we set $\epsilon = cp$, the energy distribution is ϵ^{-2}, whose index $n = 2$ is a little flatter than the observed index of 2.7. However, some of the approximations are too ideal, and there are additional losses that we have not considered, which would steepen the spectrum. For example, we assume that $V_S \gg v_A$ and further that the shock is strong. Altogether, it is extremely gratifying that the shock acceleration, which fits the astrophysical situation so well, leads automatically to the nearly correct power law distribution, which, before the introduction of the idea of shock acceleration, seemed such a mystery.

12.8 Problems

1. (Cosmic-ray trapping by a supernova) Suppose that a supernova dumps 10^{50} GeV cosmic-ray protons into its remnant when the radius of the remnant is L. Suppose the lines through the remnant are open, so without a plasma effect the cosmic rays would easily escape along the lines. Assume that the expansion velocity of the remnant is a constant $U = 10^3$ km/sec. Discuss whether, with the plasma interaction, the cosmic rays can escape before the remnant doubles in size. For this take the remnant to be a cubic box of side L aligned along the field lines and treat the escape problem one dimensionally. Take the simplest equations for the wave growth and diffusion, ignoring all numerical factors in the range of one. Namely, take the equations as

$$\frac{d\mathcal{E}}{dt} = (\gamma - \Gamma_{\mathrm{NL}}\mathcal{E})\mathcal{E}, \quad \gamma = \frac{v_D}{v_A}\frac{n_{\mathrm{cr}}}{n^*}\Omega$$

$$v_D = \frac{c^2}{\nu L}, \quad \nu = \Omega\mathcal{E}, \quad \Gamma_{\mathrm{NL}} = \Omega\mathcal{E}\frac{v_i}{c}$$

where $\mathcal{E} = (\delta B)^2/B^2$ and v_D is the drift velocity of the cosmic rays out of the box.

Show that $\mathcal{E}$ satisfies

$$\frac{d\mathcal{E}}{dt} = \left(\frac{c^2}{Lv_A}\frac{n_{\mathrm{cr}}}{n^*}\right) - \left(\frac{v_i}{c}\Omega\right)\mathcal{E}^2$$

Take parameters of 10^{50} cosmic rays in the cubic volume so that $N_{\mathrm{cr}} = 10^{50}/L^3$, $B = 10^{-5}$ G, $v_i = U = 10^8$ cm/sec, $n^* = 1/\mathrm{cm}^3$, so $v_A \approx 2 \times 10^6$ cm/sec. (For your crude estimate you may take L, n_{cr}/n^*, v_A, and Ω as constants.) Let L_0, the initial value of L, be a parameter and let $dL/dt = U$.

Start with $\mathcal{E}$ small enough that the nonlinear term is small compared to the linear growth time and $v_D = c$. How long before saturation is reached? For what values of L_0 is $v_D < U$ when saturation is reached? For what range of values of L_0 is the distance drifted up till this saturation time, $\int v_D dt$ less than L_0?

2. (Pitch-angle scattering near 90 degrees) Consider a 1-GeV cosmic ray in a mean field of 3×10^{-6} G that is pitch-angle scattered by waves of intensity $\mathcal{E}$ that would lead to a mean free path of 1 pc. For what value μ_c of μ is it mirror trapped? What is the oscillation period in seconds in the trap? What is the pitch-angle scattering rate at μ_c for changing μ by a factor of 2, if $\mathcal{E}$ is linear in μ.

3. (Spallation in clouds) Suppose the cosmic-ray spatial diffusion coefficient in clouds, D_c, is much higher than it is in the coronal region, D_0. To determine the amount of spallation in each region, solve a one-dimensional diffusion equation in a region $-L < z < L$, where the region $-a < z < a$ is occupied by a cloud and the rest of the region by the coronal gas. Let there be absorbers of the cosmic rays at $\pm L$. For an initial condition, assume an eigenvalue solution for $n_{\rm cr} \sim e^{-\lambda t}$:

$$-\lambda n_{\rm cr} = D(z)\frac{\partial^2 n_{\rm cr}}{\partial z^2}$$

Solve for λ to lowest order in the small quantities D_0/D_c and a/L. Give an argument to justify that the total rate of spallation is

$$\alpha \int \rho n_{\rm cr} dz$$

where α is a constant depending on the spallation cross sections, and ρ is the interstellar medium density.

Assume an eigenfuction of the diffusion equation and calculate the total spallation of the cosmic rays and show that it is

$$\frac{2\alpha L}{\lambda}(M_c + \frac{2}{\pi}M_0)n_{\rm cr}(0)$$

where $n_{\rm cr}(0)$ is the initial value of the cosmic-ray density at the origin, M_c is the total cloud mass, and M_0 is the total mass in the coronal region. Show that the spallation in the cloud is essentially independent of the diffusion coefficient in the clouds, provided that it is large.

4. (Acceleration by magnetic pumping) If the interstellar magnetic field pulsates in strength it cannot change the cosmic-ray energy without scattering. Is it plausible that there is sufficient interstellar scattering of the cosmic rays to produces significant acceleration if we choose a realistic estimate of the parameters?

Take a uniform sinusoidally varying magnetic field $\mathbf{B} = \hat{\mathbf{z}}B_0(1+\epsilon \sin\omega t)$, where $\epsilon \ll 1$. Take the diffusion equations for the perpendicular cosmic-ray momentum $p_\perp$ and the parallel momentum $p_\parallel$ as

$$\frac{dp_\perp}{dt} = \frac{\dot{B}}{2B_0}p_\perp + \nu(p_\parallel - p_\perp)$$

$$\frac{dp_\parallel}{dt} = -\nu(p_\parallel - p_\perp)$$

Find the solution to first order in ϵ. Then calculate the work done on the cosmic rays by the varying B field, $\langle(\dot{B}/B_0)p_\perp\rangle$. Let ω be of order kv_A with

$v_A = 10^6$ km/sec. Take $k = 10^{-17}$ cm^{-1} and take $\epsilon = 0.1$ Take $\nu = \Omega\mathcal{E}$ from the results from cosmic-ray propagation in the interstellar medium, with $B_0 = 2\ \mu$G and $\mathcal{E} = 10^{-6}$. Estimate how long it will take to accelerate the cosmic rays this way. Do you think that this is a viable acceleration mechanism?

5. (Fermi acceleration in the interstellar medium) Show that Fermi's second-order acceleration by scattering off of the moving interstellar clouds cannot represent the origin of cosmic rays by calculating the amount of this acceleration. Take the velocity of the clouds as 10 km/sec and let them be spaced at 100 pc along a given field line. Show that we expect an acceleration of $4v^2/c^2\epsilon$ per cloud encounter, where ϵ is the relativistic energy of the cosmic ray. Take into account that the acceleration is sometimes positive and sometimes negative, depending on whether the cloud is approaching or receding. The probability of approaching is larger than that of receding by the factor $(c + v)(c - v)$. How much acceleration can we get this way in a million years? (When Fermi proposed his theory the lifetime of cosmic rays was thought to be 60 million years. Would this have been long enough? Fermi originally considered the cloud motions to be much larger and the clouds closer together.)

References

Blandford, R., and D. Eichler. 1987. *Physics Reports* **154**, 1.

Blandford, R. D., and J. P. Ostriker. 1978. *Astrophysical Journal Letters* **221**, 29.

Cesarsky, C. J. 1980. *Annual Reviews of Astronomy and Astrophysics* **18**, 289.

Cesarsky, C. J., and R. M. Kulsrud. 1981. In IAU Symposium 94, *Origin of Cosmic Rays*, p. 251, ed. G. Setti, G. Spadi, and A. W. Wolfendale, D. Reidel, Dordrecht.

Felice, G. M., and R. M. Kulsrud. 2001. *Astrophysical Journal* **553**, 198.

Fermi, E. 1949. *Physical Review* **75**, 1169.

Ginzburg, V. L., and S. I. Syrovatskii. 1964. *The Origin of Cosmic Rays*, trans. H. S. H. Massey and D. Ter Haar, Pergamon Press, New York.

Goldstein, M. L. 1976. *Astrophysics Journal* **204**, 900.

Klimas, A. J., and G. Sandri. 1973. *Astrophysics Journal* **180** 937.

Kulsrud, R. M. 1978. In *Astronomical Papers Dedicated to Bengt Stromgren*, p. 323, ed. A. Reiz and T. Anderson, Copenhagen University Press, Copenhagen.

Kulsrud, R. M., and C. J. Cesarsky. 1971. *Astrophysical Letters* **8**, 189.

Kulsrud, R. M., and W. P. Pearce. 1969. *Astrophysical Journal* **156**, 445.

Lee, M., and H. Volk. 1973. *Astrophysics and Space Science* **24**, 31.

Shapiro, M. M., and R. Silverberg. 1970. *Annual Review of Nuclear Science* **20**, 323.

Skilling, J. 1975. *Monthly Notices of the Royal Astronomical Society* (MNRAS), **172**, 557.

Wefel, J. P. 1987. In *Genesis and Propagation of Cosmic Rays*, p. 1, ed. M. M. Shapiro and J. P. Wefel, D. Reidel, Dordrecht.

Wentzel, D. 1969. *Astrophysical Journal* **156**, 303.

Wentzel, D. 1971. *Astrophysical Journal* **163**, 503.

Chapter 13

ASTROPHYSICAL DYNAMOS

13.1 Introduction

Magnetic fields are observed in almost all astronomical objects—planets, stars, the interstellar medium, galaxies, and clusters of galaxies. Exceptions are Venus and possibly the space in between galaxies, i.e., the entire universe. The lifetime for such fields against resistive decay is very long, for, according to Equation 32 in chapter 3,

$$T_{\text{decay}} = \frac{L^2}{\eta c / 4\pi} \tag{1}$$

and the very large value of L leads to very long lifetimes. We find that these lifetimes are 10^{10} years for stars, 10^{26} years for the the galactic disk, and even longer for entire galaxies. However, T_{decay} is only about 10^5 years for the earth. People at first considered there to be no problem about the sustainment of fields in all objects except the earth and other planets. If the field existed originally, it would be sustained for the larger objects, by induction, and would last for times longer than the age of the universe.

On the other hand, this very long decay time presents a serious problem if the universe or the galaxies emerging in it were born without a magnetic field. Presumably, it takes the same long time, of order T_{decay}, to build the field up against induction. Further, because the sun's magnetic field reverses in polarity every eleven years, the validity of equation 1 comes into doubt. When the effect of velocities on the magnetic field are included it is possible that the field is pushed into a state where, because L is smaller, its resistive lifetime might be much shorter than that given by equation 1. The effect of velocities in changing magnetic fields is called dynamo action. Thus, the important question is, can a magnetic field be created and sustained by such dynamo action? In this connection an important and sophisticated theory was developed to handle the behavior of MHD dynamos.

Even before these serious problems arose, the question of the sustainment of the earth's field for times very long compared to its 10^5-year decay time was already in the forefront of MHD research. It has long been appreciated that the earth has a conducting spherical iron core with a radius of 3500 km, or slightly greater than half of the earth's radius. Its outer part is an MHD fluid. In the resistive MHD equation for B

$$\frac{\partial \mathbf{B}}{\partial t} = \nabla \times (\mathbf{v} \times \mathbf{B}) + \frac{\eta c \nabla^2 \mathbf{B}}{4\pi} \tag{2}$$

the relevant timescale of the last term on the right-hand side is of order 10^5 years. Thus, in the first term on the right, which is called the dynamo term, velocities of order $V \approx R/T_{decay} \approx 3 \times 10^8/3 \times 10^{12} \approx 10^{-4}$ cm/sec are needed to balance it in a steady state. Such velocities are believed to exist in the convection that carries heat from the solid core through the liquid zone to the mantle and outward.

There is a concentration of radioactive U and Th in the central core, that produces heat to drive such convection. (It is now believed that a more important source of heat is the release of energy by a phase transition when the liquid iron freezes onto the solid core.) The necessary amount of convective energy needed for the dynamo is consistent with observations. The theoretical question is: Is there some plausible axisymmetric velocity pattern that makes the two terms on the right-hand side of equation 2 balance? For many years people could not find such velocities. The reason why was shown by Cowling in 1934 in his antidynamo theorem (Cowling 1934). This theorem, called Cowling's theorem, forced people to look for nonaxisymmetric velocity patterns.

In 1955 Parker found a qualitative nonaxisymmetric set of such velocities, which, although he did not precisely specify them, had strong plausibility for solving the dynamo problem (Parker 1955). Because nonaxisymmetric solutions are difficult to handle, successful attempts to numerically verify Parker's ideas or to discover other solutions had to wait until more recent times.

These efforts finally led to the famous α–Ω mean field dynamo theory of Steenbeck, Krause, and Radler (1966). This theory involved equations for the mean field called the mean field dynamo equations. Solutions of these equations led to strong support for Parker's qualitative theory for the earth's field. Other solutions applied to the sun's dynamo problem in the convection zone that appeared to have considerable success in explaining the solar cycle. Finally, Parker (1970, 1971a, b, 1979), Vainshtein and Ruzmaikin (1972), and others were able to develop a theory for the origin of the magnetic fields seen in the galactic disk.

Before proceeding to the description of these theories we first distinguish between two essentially different types of dynamos. The first type is called the slow dynamo. By means of it the field is sustained against resistive decay, which is short compared to the life of the object. The earth's dynamo is the prototypical example of this type of dynamo. The dynamo is able to create the earth's magnetic field on the decay timescale, to sustain it, and and to reverse it.

The second type of dynamo is the fast dynamo, which can create a magnetic field on a timescale short compared to T_{decay}, and also reverse it in the same short time. Here, the dynamo is faced with the problem of flux conservation. The fast dynamo cannot create net flux, but it can, for example, start with a given flux and by natural fluid motions create twice as much flux in one direction and a flux equal to the original flux in the opposite direction. This is possible by ideal motions since the total flux is not changed. Then

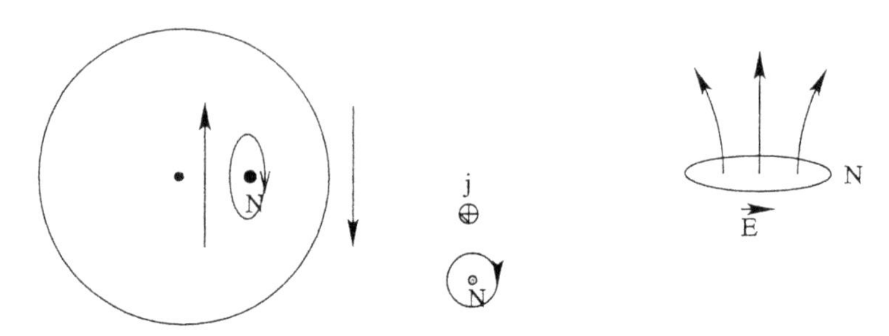

Figure 13.1. The neutral point N

by expelling the backward flux from the object, the remaining forward flux is doubled.

Thus, the fast dynamo can operate in the absence of any resistivity at all. The fast dynamo is believed by many workers to be responsible for the creation and amplification of the galactic field. The question of the solar field and its reversal in the solar cycle is intermediate between these two cases. The effective decay time may possibly be reduced to 11 years in the outer regions of the sun by magnetic reconnection processes so that the solar dynamo may be considered slow. Alternatively, the very long decay time given by equation 1 could be valid. In this case, the solar dynamo must be a fast dynamo. In treatments of the sun's magnetic field the mean field theory does not distinguish between the two types of dynamos, except through boundary conditions. However, whether the solar dynamo is fast or slow depends mainly on astrophysical questions such as magnetic reconnection inside the sun, which is problematical, and solar surface phenomena, such as mass ejections and flares, which could be related to to the escape of flux if the dynamo is truly fast.

13.2 Cowling's Theorem

Suppose we have an axisymmetric time-independent velocity and magnetic field that solve equation 2. This is equivalent to finding a time-independent solution of the two equations, Ohm's law

$$\mathbf{E} + \frac{\mathbf{v} \times \mathbf{B}}{\mathbf{c}} = \eta \mathbf{j} \tag{3}$$

and the induction equation

$$\nabla \times \mathbf{E} = -\frac{1}{c}\frac{\partial \mathbf{B}}{\partial t} \tag{4}$$

(from which equation 2 is derived). Now, assuming that the terrestrial magnetic field is axisymmetric, there must a point N where the poloidal field vanishes, as we can see from figure 13.1.

For example, if $\mathbf{B}$ is up–down symmetric about the equator, then the radial component of $\mathbf{B}$, B_r, is zero in the equatorial plane. If the magnetic field points upward across the equatorial plane in the vacuum region outside the sphere, then it must point downward on the equatorial plane near the axis of the earth. This follows because there can be no net flux crossing this plane. That is, at some point N, $B_r = 0$, and B_z must change sign. The poloidal field lines must surround N so there must be a toroidal current j_ϕ there. Taking the toroidal component of equation 3 at N, where $\mathbf{B} = 0$, we then have,

$$E_\phi + \left(\frac{\mathbf{v} \times \mathbf{B}}{c}\right)_\phi = E_\phi = \eta j_\phi \neq 0 \tag{5}$$

so E_ϕ must be nonzero at N. But by equation 4, $2\pi r_N E_\phi$ is the time rate of change of the poloidal flux threading the axisymmetric circle through N. But because B is time independent, the flux is also, and this is a contradiction.

For a more mathematical proof of Cowling's theorem, introduce the poloidal flux function, equation 88 in chapter 4, ψ with $\psi = 0$ on the axis:

$$\mathbf{B} = -\frac{\hat{\phi}}{r} \times \nabla\psi + B_T\hat{\phi}$$

$$E_\phi = -\frac{1}{c}\frac{\partial \psi}{\partial t} \tag{6}$$

and the toroidal component of equation 2 becomes

$$-\frac{1}{r}\frac{\partial \psi}{\partial t} - \mathbf{v} \cdot \nabla\psi = -\frac{\eta c}{4\pi r}\left(r\frac{\partial}{\partial r}\frac{1}{r}\frac{\partial \psi}{\partial r} + \frac{\partial^2 \psi}{\partial z^2}\right) = -\frac{\eta c \Delta^* \psi}{4\pi r} \tag{7}$$

We choose the arbitrary constant in ψ to be zero at the origin so that ψ is unique. Then $\partial\psi/\partial t = 0$, since $\mathbf{B}$ is time independent. Since ψ vanishes at the origin and at infinity, there is at least one point N where ψ has a minimum or maximum. Near this point

$$\mathbf{v} \cdot \nabla\psi = -\frac{\eta c}{4\pi r}\Delta^*\psi \tag{8}$$

If $\nabla^2\psi \neq 0$ at N, then since $\nabla\psi = 0$ at N we have a contradiction. If $\nabla^2\psi$ vanishes at N, and if it vanishes to order n, then $\nabla\psi$ vanishes to order $n+1$ and the terms in equation 8 still cannot match near N, so again we get a contradiction.

13.3 Parker's Model for the Earth's Dynamo

For a long time Cowling's argument stymied people in their search for a dynamo solution, since no one felt able to handle three-dimensional nonaxisymmetric flows, although Elsasser made some notable attempts (Elsasser

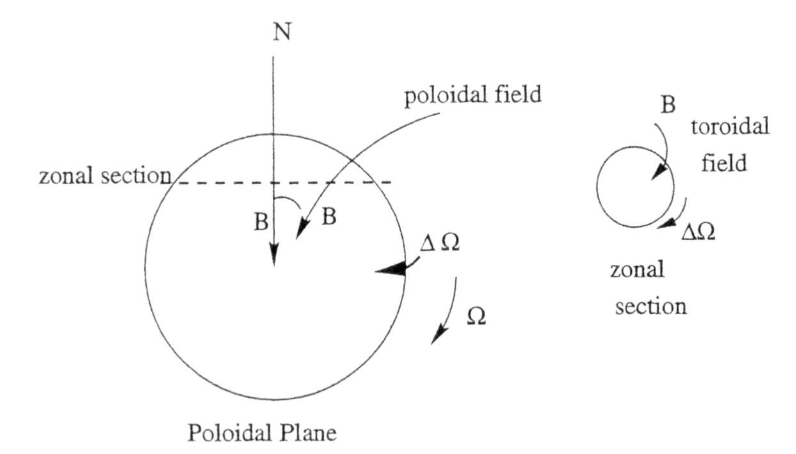

Figure 13.2. Toroidal shearing of the poloidal field projected onto a sectional cut perpendicular to the earth's axis

1946). It was certainly appreciated that there are three-dimensional Coriolis forces acting on the convection flows inside the earths core, but Parker was the first person bold enough to demonstrate the role they must play in a dynamo. It was further appreciated that the earth's liquid core does not rotate rigidly. Conservation of angular momentum forces the rising convective cells to develop a smaller angular velocity as they get farther from the earth's axis (see figure 13.2). Thus, because of this variable rotational velocity the earth must have a toroidal magnetic field B_T. In fact, from equation 97 in chapter 4, with $B_T = 0$,

$$\frac{\partial B_T}{\partial t} = \rho \mathbf{B}_p \cdot \nabla \Omega \tag{9}$$

(where ρ is a cylindrical coordinate), we see that one must develop.

In the northern hemisphere $B_\rho < 0$, where ρ is the cylindrical coordinate. Since Ω varies inversely with ρ, there must be an eastward toroidal field. It should develop over a decay time $T_{\mathrm{decay}} \sim L/\eta$ so $B_T \approx \rho \mathbf{B_p} \cdot \nabla \Omega T_{\mathrm{decay}}$. In the southern hemisphere B_ρ reverses signs and B_T is in the westward direction. Thus, the toroidal field is easily produced from B_p. The real problem is developing B_p from B_T.

Parker's outstanding achievement was to show that the rotating convection cells accomplish this. He starts with a pure toroidal field in the eastward direction (as would happen in the northern hemisphere) and looks at the bottom of a convection cell (see figure 13.3). As the convection flow converges toward the axis at the bottom of a convection cell, the fluid involved decreases its effective moment of inertia and it rotates faster. That is, in the rotating frame it rotates in the same direction as the earth, counterclockwise and horizontal inward flow of the convection cell is directed to the right of the cell axis. It also rises upward.

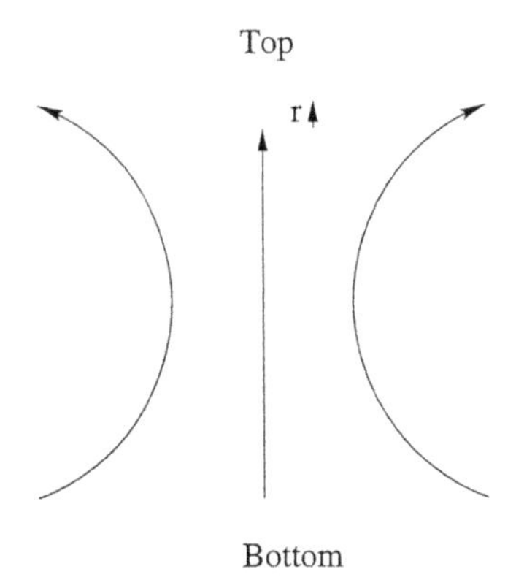

Figure 13.3. A convection cell

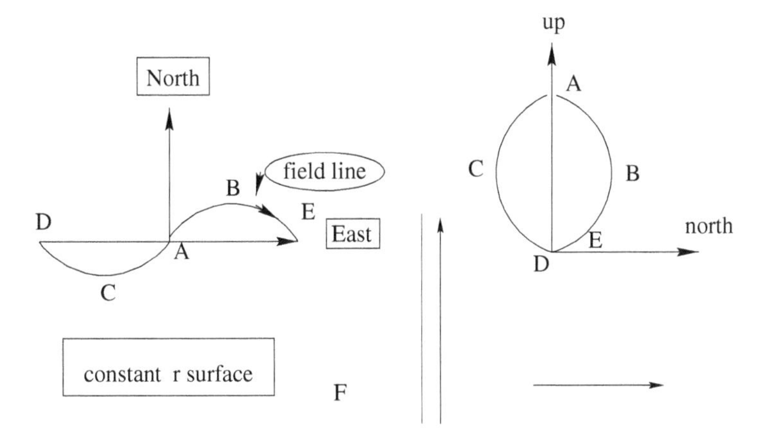

Figure 13.4. Parker's model for the creation of poloidal flux

Take a piece of the toroidal line in the eastward direction. It is bent by the rotational component of the convection, as in figure 13.4. Now the inner part of the line nearest the axis rises up. Consider the projection of the resultant line on the north-up plane where the points $ABCDE$ on the line in figure 13.4a move to the same points in the north-up plane of figure 13.4b. We then get a loop of poloidal flux that is counterclockwise about the westward direction. The line still has its toroidal part.

If we carried out the same operation on the upper parts of the convective pattern of figure 13.3, the outward motions would be deflected in a clockwise direction. However, a little thought will show that the outer parts of the line are lifted higher than the inner parts. Thus, after projection of the distorted line onto the north-up plane we would also get a loop with the same sense as in figure 13.4b. We can also see that a downward convection cell also gives a loop with the same sense and both at its top and its bottom. Similarly, any convection in the southern hemisphere acts on a toroidal field, of the opposite sign, to yield poloidal magnetic loops with the same sense as those in the northern hemisphere.

All these loops in the up-north plane (or the up-south plane in the southern hemisphere) are loops of poloidal flux. Each one has a toroidal current of the same sign, so that we can find the averaged poloidal field by summing the currents of all loops. But these currents have the same sign as the currents associated with the poloidal field we started with. Finally, all the irregularities associated with the statistical collection of loops are diffused out by resistivity.

In summary, we start with a poloidal field like the earth's dipole field, that is, upward outside, downward inside the core. The differential rotational velocity of the earth then stretches this field to make toroidal field in the eastward direction in the northern hemisphere and in the westward direction in the southern hemisphere. Then the combination of convection and Coriolis forces twists this toroidal field into poloidal loops whose contribution to the toroidal current are all of the same sign and of the sign that reinforces the original poloidal field. If the convective flow velocities are of the proper magnitude, then the poloidal field can be reinforced at the same rate at which it decays and we arrive at a steady state.

If the convective flows were faster than the critical value needed to grow the field at the same rate at which it resistively decays, then B would grow and start affecting the convective flows and their patterns until they would slow to such a point that the steady state would be reached. Thus, the magnetic field strength is self-controlling and leads to a unique value.

If the convective cells with their Coriolis forces were such that the cells rotated more than 180°, then we would get a destruction of the original field rather than a reinforcement. Such a situation would lead to an oscillatory variation of the field, probably related to the observed reversals of the earth's field. However, because of the feedback of the field on the convection this would be quite complicated. Recent calculations by Glatzmaier and Roberts (2000) indicate that these reversals do happen, but in a complicated and chaotic way that is not predictable. Actual observations of magnetized fossils show that the earth's field has reversed many times in the past, but in an irregular way. The average period of reversal is several hundred thousand years. The irregular reversals appear to be consistent with the reversals seen in the numerical simulations.

One last comment. We have not worried about flux conservation because resistive decay is important. However, we could imagine what would happen if the resistivity were absent. In this case, Parker's model would lead to growth of the field and we could ask how flux conservation is satisfied. The answer is that boundary conditions at the earth's mantle are such that negative flux would escapes into the earth's mantle, which is an isolator, and in which it would be destroyed, leaving behind increased positive flux.

13.4 The Mean Field Dynamo Theory

Parker's theory represented a big step forward toward understanding the puzzle of how to sustain the earth's magnetic field. However, because of its qualitative approach and the intrinsic numerical complication of treating the interaction of statistical convection cells in a rotating body, the solution was not verified numerically, and its significance was not generally appreciated. Parker not only had attacked a three-dimensional problem but had considered a turbulent velocity flow rather than a smooth one. Moreover, he had chosen flows that had a high probability of being realistic. The original challenge had been only to see if any velocity at all, realistic or not, could lead to a sustained magnetic field.

The dynamo community was better satisfied when a more systematic formalism for treating these random flows was introduced in 1966 by Steenbeck et al. (1966). This is the well-known mean field dynamo theory, which expresses the time evolution of the mean magnetic field in terms of the statistics of the velocity field.

The main assumptions in the theory are that, first, the turbulent scales are small compared to the large scale of the magnetic field; second, the turbulent velocities have a short correlation time; third, for simplicity, the theory is specialized to statistically isotropic and incompressible velocities. An essential ingredient of the theory is that the statistics are allowed to be noninvariant under reflections, so that cyclonic flows are allowed. This is necessary because the magnetic field is a pseudovector and cannot be changed by a velocity field whose statistics are invariant under reflection.

Consider an incompressible velocity field at two neighboring points $\mathbf{r}$ and $\mathbf{r}'$ and at times t and t'. Then ensemble average the tensor product of $\mathbf{v} = \mathbf{v}(\mathbf{r})$ and $\mathbf{v}' = \mathbf{v}(\mathbf{r}')$ over all positions differing by $\boldsymbol{\rho} = \mathbf{r}' - \mathbf{r}$ and times differing by $\tau = t' - t$. This average, the velocity correlation function, depends only on these differences. Further, this product is invariant under all rotations, but not reflections. The most general form of such a correlation is

$$\langle \mathbf{v}'\mathbf{v} \rangle = A(\rho, \tau)\mathbf{I} + B(\rho, \tau)\boldsymbol{\rho}\boldsymbol{\rho} + C(\rho, \tau)\boldsymbol{\rho} \times \mathbf{I} \tag{10}$$

To see this, choose a coordinate system with $\boldsymbol{\rho}$ along $\mathbf{x}$. It is clear that $\langle \mathbf{v}'\mathbf{v} \rangle$ does not have a nonzero xy or an xz component. By rotations about the $\hat{\mathbf{x}}$ axis we can see that the yy component must be equal to the zz component $= A$, and the yz component must equal minus the zy component, $= C$. Then, setting the xx component equal to $A + B$ yields the above result. Further, because $\mathbf{v}$ is divergent free, we must have

$$\frac{\partial}{\partial \mathbf{r}'} \cdot \langle \mathbf{v}'\mathbf{v} \rangle = \nabla_{\boldsymbol{\rho}} \cdot \langle \mathbf{v}'\mathbf{v} \rangle = \nabla A + 4B\boldsymbol{\rho} + \boldsymbol{\rho} \cdot \nabla B \boldsymbol{\rho} = (A' + \rho^2 B' + 4\rho B)\hat{\boldsymbol{\rho}} = 0 \tag{11}$$

where primes denote ρ derivatives, and this relates $A(\rho)$ and $B(\rho)$. Finally, since $\langle \mathbf{v}'\mathbf{v}\rangle$ is even in ρ, A and B are even in ρ while C is odd. (That is, in an expansion about $\boldsymbol{\rho} = 0$, A and B have only even terms in ρ, and C has only odd terms.)

We have assumed that A, B, and C depend only on ρ and t. But this is only true locally. In general, they vary with position on a larger scale. A and B represent Parker's convection cells, while C represents the rotation of these cells due to the Coriolis force. That is, C represents the cyclonic feature of the convection. If different cells rotated randomly, clockwise and counterclockwise in Parker's theory, then there would be no net poloidal flux generated by the convection. The extent to which a poloidal field is generated is the extent to which cyclonic rotations exceeds anticyclonic rotations. But, in addition, in Parker's theory C varies slowly with position since motions at the bottom of a convection cell had the opposite sense to those at the top.

Let us examine this a little closer. Let $\mathbf{v}$ be in the upward x direction at the center of a convection cell. Then Parker showed that at the bottom of the convection cell the average of $\rho_y v_z' - \rho_z v_y'$ is > 0 and this represents the counterclockwise cyclonic rotation. Also on the axis of the cell $v_x > 0$. We will see that this implies that $C > 0$. At the top the rotation $\rho_y v_z - \rho_z v_y < 0$, but we still have $v_x > 0$, so C has changed sign. For Parker's case C has to change sign to produce the net poloidal flux. For the downward cell we still have $C < 0$ at the top and $C > 0$ at the bottom. In the southern hemisphere C will have the opposite sign, being positive at the top of the convective cells and negative at the bottom. This turns out to be just the correct parity to produce the net toroidal field when it is appreciated that the toroidal field reverses between hemispheres.

We make one final remark before proceeding to the actual derivation of the mean field equations. Equation 10 is reminiscent of the random phase approximation in equation 6 in chapter 11. We can imagine that the fluid turbulence consists of many structures, such as eddies or convection. Let us consider a single structure $\mathbf{v}(\mathbf{r}, \mathbf{t})$ that is randomly positioned in space and time, i.e., $\mathbf{v}_i = \mathbf{v}(\mathbf{r} - \mathbf{r}_i, t - t_i)$, and rotates randomly but with the same cyclonic sense in any local region. Then the net average correlation function, $\langle \mathbf{v}', \mathbf{v}\rangle$, for this model leads to equation 10. Now, superimpose many different copies of this structure, each also randomly positioned and rotated. We still get equation 10. The supposition is that the velocities from such a model will lead to the correct evolution for an averaged or mean field. This is a natural assumption often made in physics, and often difficult to justify.

13.4.1 Derivation of the Mean Field Equations

To derive the mean field equations we start with random velocities whose statistics are given by equation 10. (For the moment ignore resistivity.) The basic equation for the evolution of $\mathbf{B}$ is

$$\frac{\partial \mathbf{B}}{\partial t} = \nabla \times (\mathbf{v} \times \mathbf{B}) \tag{12}$$

Let us average this over our ensemble of representations of our model:

$$\frac{\partial \langle \mathbf{B} \rangle}{\partial t} = \nabla \times (\langle \mathbf{v} \times \mathbf{B} \rangle) \tag{13}$$

Consider the expression under the curl:

$$\langle \mathbf{v}' \times \mathbf{B} \rangle$$

Let us suppose that the change in the mean $\mathbf{B}$, $\mathbf{B}_0$, over a correlation time is small, so that we may take $\mathbf{B}_0$ independent of t for times of order a correlation time. We then have the macroscopic field $B_0 = \langle \mathbf{B} \rangle$ plus a random small $\mathbf{B}_1$

$$\mathbf{B} = \mathbf{B}_0 + \mathbf{B}_1 \tag{14}$$

$\mathbf{B}_1$ results from the effect of the velocities $\mathbf{v}$ acting on $\mathbf{B}_0$ during a correlation time. From equation 12 we have

$$\mathbf{B}_1(t) = \int_{-\infty}^{0} \nabla' \times (\mathbf{v}' \times \mathbf{B}'_0) dt' \tag{15}$$

where $\mathbf{v}' = \mathbf{v}(\mathbf{r}', \mathbf{t}')$. After the ∇' gradient is taken we set $\mathbf{r}' = \mathbf{r}$. Then

$$\mathbf{B}_1(\mathbf{r}, 0) \times \mathbf{v}(\mathbf{r}, 0) = \int_{-\infty}^{0} \nabla' \times (\mathbf{v}' \times \mathbf{B}'_0) \times \mathbf{v}(\mathbf{r}, \mathbf{t}) dt'$$

$$= \int_{-\infty}^{0} [\mathbf{B}'_0 \cdot \nabla'(\mathbf{v}' \times \mathbf{v}) - (\mathbf{v}' \cdot \nabla')(\mathbf{B}'_0 \times \mathbf{v})] dt' \tag{16}$$

Now take the ensemble average and again set $\mathbf{r}' = \mathbf{r}$. Note that

$$\langle \mathbf{v}' \times \mathbf{v} \rangle = C(\boldsymbol{\rho} \times \mathbf{I})^* = -2C\boldsymbol{\rho} \tag{17}$$

where the $*$ over I means take the cross-product of the left and right sides of the tensor in the parentheses. Then,

$$\mathbf{B}_0 \cdot \nabla' \langle \mathbf{v}' \times \mathbf{v} \rangle = -2B_0 \cdot \nabla'(C\boldsymbol{\rho}) \rightarrow -2C_0 B_0 \tag{18}$$

where after differentiation we set $\rho = 0$ and we define $C_0 = C(0, \tau)$. Also

$$\langle \mathbf{v}' \cdot \nabla(\mathbf{B}'_0 \times \mathbf{v}) \rangle = -A\nabla \times B_0 + C \text{ terms} \tag{19}$$

After we set $\rho = 0$, the B and C terms vanish and $A = A_0 = A(0, t)$. Thus,

$$\langle \mathbf{B}_1 \times \mathbf{v} \rangle = -2 \int_0^{\infty} C(0, \tau) d\tau \mathbf{B}_0 - \int_0^{\infty} d\tau A(0, \tau) \nabla \times \mathbf{B}_0 \tag{20}$$

Setting

$$\alpha = +2 \int_0^{\infty} C(0, \tau) d\tau \tag{21}$$

and

$$\beta = \int_0^\infty A(0, \tau)d\tau \tag{22}$$

we have

$$\langle \mathbf{v} \times \mathbf{B} \rangle = \alpha \mathbf{B}_0 - \beta \nabla \times \mathbf{B}_0 \tag{23}$$

and from equation 13

$$\frac{\partial \langle \mathbf{B} \rangle}{\partial t} = \nabla \times (\mathbf{V} \times \mathbf{B}_0) - \nabla \times \alpha \mathbf{B}_0$$

$$- \nabla \times [\beta(\nabla \times B_0)] + \frac{\eta c}{4\pi} \nabla^2 \mathbf{B}_0 \tag{24}$$

where we have replaced $\langle \mathbf{B} \rangle$ by $\mathbf{B}_0$ and restored the resistive term by hand. $\mathbf{V}$ is the mean velocity, usually a rotational velocity. If we assume that β is constant in space and use $\nabla \cdot \langle \mathbf{B} \rangle$ is zero, we can write the mean field dynamo equation as

$$\frac{\partial \mathbf{B}}{\partial t} = \nabla \times (\mathbf{V} \times \mathbf{B}) + \nabla \times (\alpha \mathbf{B}) + \left(\beta + \frac{\eta c}{4\pi}\right) \nabla^2 \mathbf{B} \tag{25}$$

(An alternative formal derivation of these equations is given in the book of Ruzmaikin et al. (1988), which is devoted to the problem of the galactic dynamo. A less formal derivation is given in the book of Parker (1979) and in Parker (1970, 1971a, b), where some of the physical problems related to the galactic magnetic field are discussed.)

The β term can be interpreted as the turbulent mixing term and is often referred to as turbulent resistivity. The convection cells mix up the positive and negative lines of force and thus reduce the mean field, which is equal to the field averaged over some scale. This mixing by itself cannot actually destroy the magnetic energy. If there is sufficient resistivity, these fluctuations will be destroyed (as is the case for the slow dynamo). However, if η is small, then a substantial random field deviating from the mean field is developed by this β term.

For the fast dynamo case, $\beta \gg \eta c/4\pi$. If the $\eta c/4\pi$ term is neglected, then the dynamo equations really refer to an ideal fluid, and flux must be conserved by the mean field theory. It is possible that if the field is finely enough mixed, magnetic reconnection can further merge the plus and minus field and destroy the magnetic energy. However, on the very large scale of the galactic disk, reconnection is problematical (see chapter 14).

It is important to have a more physical picture of the quantities α and β than simply their expression in terms of A_0 and C_0. First, α is related to quantity called kinetic helicity

$$\langle \mathbf{v} \cdot (\nabla \times \mathbf{v}) \rangle \tag{26}$$

Let us set

$$\int_{-\infty}^{0} A(0, \tau)d\tau = A(0, 0)\tau_c \tag{27}$$

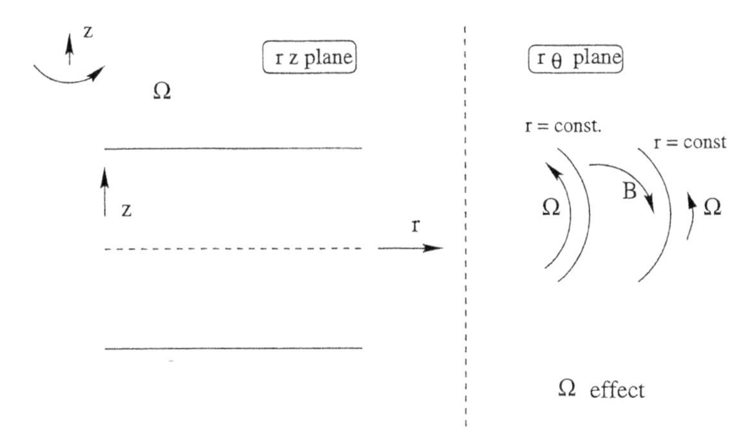

Figure 13.5. The galactic disk dynamo

where τ_c is an effective correlation time for A and similarly for C in equation 10. Then we have from equation 21

$$\langle \mathbf{v}' \cdot \nabla \times \mathbf{v} \rangle_{\boldsymbol{\rho}=0,\tau=0} = -\nabla' \cdot \langle \mathbf{v}' \times \mathbf{v} \rangle = 6C(0,0)$$

so

$$\alpha = -\tfrac{1}{3}\tau_c \langle \mathbf{v} \cdot (\nabla \times \mathbf{v}) \rangle \tag{28}$$

Similarly, $\langle \mathbf{v}' \cdot \mathbf{v} \rangle = 3A(0,0)$ so

$$\beta = \tfrac{1}{3}\tau_c \langle v^2 \rangle \tag{29}$$

The quantity β is clearly related to a random walk for fluid elements, since $x^2 = (v_x \tau_c)^2 t/\tau_c = \frac{1}{3}\mathbf{v}^2 \tau_c t = \beta t$. α is related to the amount of rotation times the height reached by one of Parker's cells since $\hat{\mathbf{z}} \cdot \nabla \times v = 2d\theta/dt$, so $\langle z\Delta\theta \rangle = (v_z \tau_c d\theta/dt \tau_c)t/\tau_c = \frac{1}{2} v_z |\nabla v| \tau_c t = \alpha t$.

13.4.2 The Growth Rate of Dynamo Modes in the Galactic Disk

To show how a magnetic field would evolve by the mean field theory we will consider an important example, the evolution of the galactic magnetic field in the galactic disk (see figure 13.5) (Ruzmaikin et al. 1988; Kulsrud 1999). Since the disk thickness h is only several hundred parsecs, while its radius R is of order 10–20 kpc, we approximate the disk as very thin. Then in cylindrical coordinates the z variation is much larger than the radial or axial variation and, to lowest order, we keep only z derivatives and drop B_z. Then the dynamo equation 25 reduces to

$$\frac{\partial B_r}{\partial t} = -\frac{\partial}{\partial z}(\alpha B_\theta) + \beta \frac{\partial^2 B_r}{\partial z^2} \tag{30}$$

$$\frac{\partial B_\theta}{\partial t} = B_r r \frac{\partial \Omega}{\partial r} + \beta \frac{\partial^2 B_\theta}{\partial z^2} \tag{31}$$

where in the B_θ equation $\mathbf{V} = \Omega \hat{\theta}$ is the galactic rotation. Since $V_\theta = r\Omega$ is constant throughout most of the disk, $r d\Omega/dr = -\Omega$ and we treat Ω as a constant parameter. We assume that the mode is localized in radius. Since in the B_θ equation the Ω term generally dominates the α term, we drop the α term. We also drop the true resistivity term.

The functional dependence of α and β is not very well known, so for simplicity we take β as constant. However, α is odd in z. This is clear since, when an eddy rises in a decreasing density gradient, the fluid in it expands radially from its axis, and its moment of inertia increases. In a nonrotating inertial frame, the galactic rotation that it had before the expansion slows down, and it thus rotates clockwise in the galactic rotation frame. Thus, $(\mathbf{v} \cdot \nabla \times \mathbf{v}) < 0$ for this eddy and α is positive. A downward-moving eddy below the $z = 0$ midplane also rotates in the clockwise sense, but v_z is negative so $(\mathbf{v} \cdot \nabla \times \mathbf{v}) < 0$ is positive and α is negative. We expect the eddy rotation to increase away from the central plane of the disk so that α should increase with z. We thus take α to be linear in z,

$$\alpha = \alpha_0 \frac{z}{h} \tag{32}$$

where h is the half-thickness of the disk.

Let us introduce the vector potential $A(z)$ for B_r

$$B_r = -\frac{\partial A}{\partial z}$$

and choose $A(0) = 0$ as the gauge. Then equation 30 can be integrated with respect to z. We assume exponentially growing solutions $B_\theta \sim e^{pt}$, $A \sim e^{pt}$ and obtain

$$pA = \frac{\alpha_0}{h} z B_\theta + \beta \frac{\partial^2 A}{\partial z^2} \tag{33}$$

$$pB_\theta = \Omega \frac{\partial A}{\partial z} + \beta \frac{\partial^2 B_\theta}{\partial z^2} \tag{34}$$

Equations 33 and 34 represent a coupled fourth-order system of linear equations. To complete the system we need boundary conditions at $z = \pm h$. The generally accepted boundary conditions are the so-called vacuum boundary conditions that set $B_r = B_\theta = 0$ at $z = \pm h$, which correspond to no field outside the disk. Thus,

$$B_r = -\frac{\partial A}{\partial z} = 0 \quad \text{at } z = \pm h \tag{35}$$

The equations are symmetric in z so we can have either of two solutions. First, there is the even solution, B_θ even, A odd:

$$B_\theta(z) = B_\theta(-z), \quad A(z) = -A(-z) \tag{36}$$

This is called the quadrupole solution. The second solution is the odd solution, B_θ odd and A even in z. This is called the dipole solution. But the galactic B_θ field is even in z, so we discuss only the quadrupole solution. (For the dipole solution see problem 1.) Several numerical solutions of these equations are given in Ruzmaikin et al. (1988).

Functions that satisfy these boundary conditions and the quadrupole symmetry are

$$\begin{aligned} B_\theta &= \hat{B} \cos k_0 z \\ A &= \hat{A} \sin k_0 z \end{aligned} \tag{37}$$

with $k_0 = \pi/2h$. Equation 34 is satisfied by this choice if

$$(p + k_0^2 \beta)\hat{B} = k_0 \Omega \hat{A} \tag{38}$$

but equation 33 is not. However, if we replace zB_θ term in it by $a \sin k_0 z$, which has a similar shape, we get a reasonable approximation to the solution.

Since we are more interested in the net effect of the various terms on the magnetic field and since an exact analytical solution is complicated, we make this approximation. We choose a so that the averages of αB_θ and $a \sin k_0 z$ agree when multiplied by $\sin k_0 z$ (this is really the first term in a Fourier series):

$$a \int_{-h}^{h} \sin^2 k_0 z dz = \frac{\alpha_0}{h} \hat{B} \int_{-h}^{h} z \sin k_0 z \cos k_0 z dz$$

The result is

$$a = \frac{\alpha_0 \hat{B}}{\pi} \tag{39}$$

With $a \sin k_0 z$ replacing the first term on the right-hand side of equation 34, we have

$$(p + k_0^2 \beta)\hat{A} = \frac{\alpha_0}{\pi} \hat{B} \tag{40}$$

Combining this equation with equation 38 we have a solution if

$$(p + k_0^2 \beta)^2 = \frac{k_0 \alpha_0 \Omega}{\pi} \tag{41}$$

or

$$p = -k_0^2 \beta \pm \sqrt{\frac{k_0 \alpha_0 \Omega}{\pi}} \tag{42}$$

We see that if

$$\frac{k_0 \alpha_0 \Omega}{\pi} > \beta^2 k_0^4 \tag{43}$$

p is positive for the upper sign and we have growth.

The critical condition for growth is usually expressed in terms of the "dynamo number,"

$$D = \frac{\alpha_0 \Omega h^3}{\beta^2} \tag{44}$$

and the approximate condition for instability (from our solution) is $D > \pi^4/8 = 12.2$. This is quite close to the exact numerical value of 13 (see Ruzmaikin et al. 1988). We expect the dynamo instability to be favored by large α/h and Ω and to be unfavored by large effective turbulent resistivity β/h^2. The dynamo number D represents this feeling very well because it can be written so that the numerator is the product of α/h and Ω and the denominator is the square of β/h^2.

Let us estimate α and β and the growth rate for the interstellar medium. The turbulent velocities are found be of order 10 km/sec, from the observed random motions of the clouds. Clouds are expected to collide every 10^7 years, which gives the velocity correlation time τ_c. Thus, since 10 km/sec = 10 pc/million years,

$$\beta = \frac{\tau_c}{3} v^2 = \frac{10}{3} \times 10^2 = 333 \text{ pc}^2/\text{million years} \tag{45}$$

After an eddy has increased its size substantially as it rises it will rotate at the rate $-\Omega$ in the galactic frame so that its angle of rotation is $\Delta\theta = \Omega\tau_c = 2\pi \times 10/250 \approx 0.2$ radians, and thus

$$\alpha = \frac{v}{2}(\Delta\theta) \approx \frac{10}{2} 0.2 \approx 1.25 \quad \text{km/sec}$$

Thus, the dynamo number for the interstellar medium is

$$D = \frac{\alpha \Omega h^3}{\beta^2} = \frac{1.25(2\pi/250)(3 \times 100)^3}{(333)^2} = 7.6 \tag{46}$$

The calculation has been carried out with one million years as the unit of time and one parsec as the unit of length. We take $h = 300$ pc. Our estimate of the interstellar medium dynamo number is actually less than the critical value, but supernovae and superbubbles are believed to be sufficiently powerful to intensify the cyclonic motions and increase α enough to produce instability. Assuming that D has twice the critical value we can estimate the growth rate. In equation 43 the ratio of the second term to the first is D/D_{crit} and we get

$$p = \frac{\beta}{h^2} = \frac{333}{(300)^2} = \frac{1}{270 \quad \text{million years}} \tag{47}$$

Based on these estimates it is not unreasonable to expect the growth time to be in the range of several hundred million years. If it were 200 million years, then in 10^{10} years the magnetic field would grow from its initial value by a factor $e^{50} = 10^{20}$. If the growth time were 500 million years, the factor

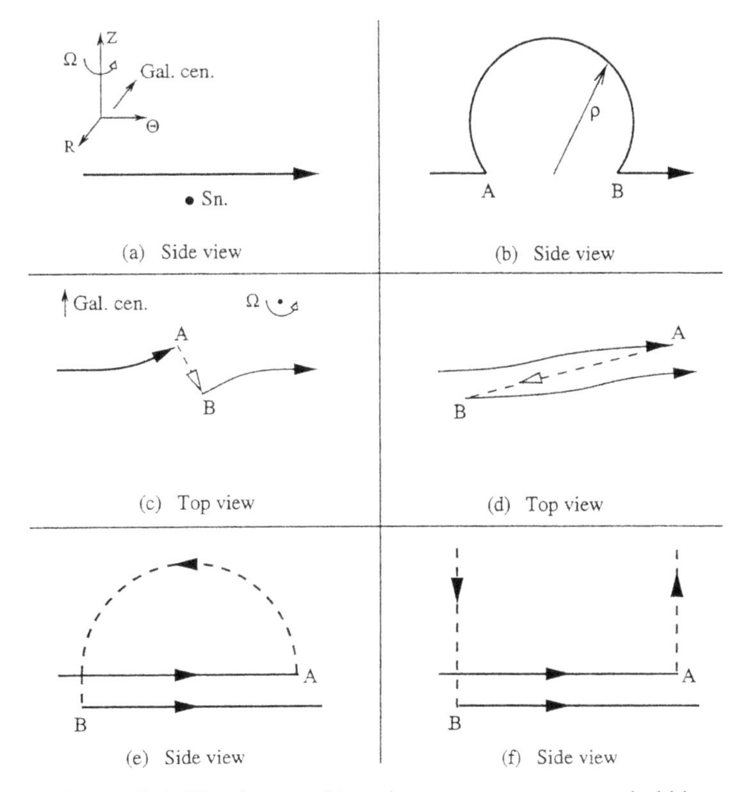

Figure 13.6. The dynamo driven by a supernova or superbubble

would be $e^{20} = 10^{8.6}$. Even for the latter case we would expect that an initial field of 10^{-14} G could be amplified to its present value of 3×10^{-6} G.

The general consensus is that the observed turbulence is too weak to drive the dynamo and it is important to include more violent events, such as superbubbles. Such powerful large-scale events produce flow patterns with scales comparable to the thickness of the galactic disk. Such large-scale turbulent motions violate the small-scale assumption invoked in the derivation of the mean field dynamo theory. However, it should still give a rough feeling for whether the dynamo is unstable and how fast it is. As a matter of fact, it is possible to dispense entirely with the mean field theory and go directly to single events represented by these superbubbles. For these events we can follow the lines directly and see how the dynamo works in detail.

The picture of the dynamo driven by supernova or superbubble is represented in figure 13.6 (see Kulsrud 2000). Start with a single toroidal line of force passing through the center of the superbubble as in figure 13.6a. The superbubble will take a small sphere on the line and expand it enormously, increasing its moment of inertia (figure 13.6b). As a consequence of conservation of angular momentum, the bubble will become nonrotating in an inertial frame and therefore rotate backward in the galactic frame at the angular rate Ω. If it rotates by $\pi/2$, the foot points of the magnetic line will also rotate into the position of figure 13.6c, while the line connecting them will stretch over the superbubble. Now, since the foot points are displaced radially with respect to each other they will see different values for the galactic rotation, with the inner point rotating faster. The parts of the field still in the galactic plane will have be sheared relative to each other and will overlap, creating more positive flux in the disk, as in figure 13.6d. Meanwhile, the piece of the line of force connecting them has been lifted high out of the plane of the galaxy by the superbubble as in figure 13.6e. According to the assumptions of the vacuum boundary conditions this piece of the line should be lifted entirely out of the galactic disk (figure 13.6f). Since total flux must be conserved, the extra positive flux developed in the disk must be compensated by the negative flux that is now way out of the galactic disk and no longer counted among the galactic field lines. We would conclude that the superbubble has generated positive flux in the disk, in accordance with what we expect from a dynamo, although the total flux, including that removed to infinity, is really the same.

We see from this picture that for the dynamo to generate net flux in the disk, it is vital for a part of the line of force to escape from the disk. If it falls back into the galactic disk, there will be no increase in disk flux at all.

Unfortunately, the piece of the line that is usually expected to freely escape is embedded in the expanding plasma of the superbubble explosion. It is generally agreed that superbubble does not have enough power to fully lift the matter out of the galaxy, so the matter actually does fall back (Rafikov and Kulsrud 2000). The only possible hope for an escape of the line of force is that the matter on it slip down along the line, unloading it. This would allow the line to become light enough to escape. However, such a slippage is hardly likely when the field is weak, since it would exert a negligible force on the plasma. Thus, the matter in the explosion should behave in exactly the same way as if the field were not there at all and there should be no unloading by mass slipping down the field lines.

This criticism of the superbubble theory for the amplification of the field applies only to the quadupole mode in which a net total flux is produced in the disk. The dipole mode does not change the total flux and could potentially be unstable, even with closed boundary conditions and no flux escape. However, the critical dynamo number for the dipole mode is considerably larger, and it is unlikely to be unstable. Thus, the fact that the vacuum boundary conditions are inapplicable when flux freezing is considered is a very serious objection to the galactic dynamo theory for the origin of the galactic field.

When the field is stronger, it is possible that it would be able to control the plasma and force it to slip downward along the lines, allowing the negative

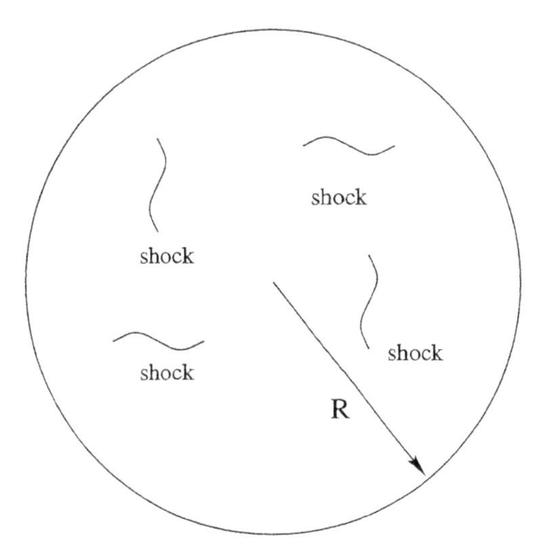

Figure 13.7. Shocks creating turbulence

flux to escape. But if the field were this strong, we would not need dynamo amplification in the first place!

13.5 Protogalactic Origin of the Magnetic Field

The plasma that forms our disk very likely has already been through an era of strong turbulence. The presence of this turbulence in this early plasma, when it formed a large protogalaxy, has been found in the very extensive numerical simulations of Cen and Ostriker (1993). In fact, the energy in the turbulence is found to be comparable to the entire gravitational binding energy of the baryonic plasma in the gravitational well of the dark matter protogalaxy. As the cosmic matter collapses into the protogalaxy due to a large-scale gravitational instability, it is relatively cold and many small-scale gravitational unstable wave packets form, as in figure 13.7. These wave packets grow to a nonlinear limit to form shocks of small spatial extent, which produce intense hydrodynamic vortices (Ryu et al. 1994).

It turns out, as we shall show, that these vortices generate magnetic fields from scratch by a mechanism known as the Biermann battery (Biermann 1950). These fields are quite weak, of order 10^{-18} G, and exist on all scales. The turbulence then acts on these magnetic fields to amplify them by dynamo action to much stronger fields. At the same time the effective scale of the magnetic fields grows from a small scale up to a scale comparable to the scale of the entire protogalaxy.

In the dynamo process there is no change in the total magnetic flux, but the field direction is not necessarily regular. In some regions one sign of flux may prevail, and in other regions the opposite sign. Thus, for example, the region that compresses into the disk near the sun could have net positive toroidal flux, and the region that compresses into the next spiral arm at a smaller radius could have the opposite sign of the toroidal flux. In this way a field somewhat similar to the present field could develop.

The protogalactic dynamo is much less elegant than the α–Ω dynamo since the fields involved in the protogalaxy are more incoherent. But, as pointed out in the last section, the α–Ω dynamo cannot work, at least for quadrupole modes, unless negative flux is expelled from the galactic disk. There is great difficulty in expelling such flux in the z direction because it is loaded with plasma that is bound to the disk by the strong gravitational field. However, topologically, it could work if the flux could be expelled in the radial direction to a region of different signed flux. This is a very slow process in the disk because the turbulent eddies are on a scale small compared to the radius of the galaxy. On the other hand, if the magnetic field were a dipole field reversing across the midplane of the galaxy there would be no net toroidal flux, and there would be no need for the expulsion at all. Unfortunately, the dipole configuration, which, in any event, is more difficult to produce, does not agree with the observations.

By contrast to the galactic disk dynamo, the eddies in the protogalaxy are on a much larger scale and could possibly accomplish this transfer of flux in radius. There is no large-scale systematic α and Ω, but on the scale of a single large eddy there is a local systematic rotation and a local mean helicity.

Thus, the ideas developed for the disk could apply to subregions of the protogalaxy, and accomplish the creation and amplification of magnetic fields. These rough ideas are not as widely accepted as the more popular and elegant α–Ω dynamo theory. But, as the difficulties pile up against the galactic disk α–Ω theory, we may be driven to take the protogalactic ideas seriously.

We will now discuss the behavior of magnetic fields in the protogalaxy—first, the Biermann battery; second, nonhelical turbulent amplification; third, the inverse cascade of magnetic fields to large scale; and, finally, the compression of the field as the protogalaxy collapses into the galactic disk. Of course, these ideas are of wider application, but discussing them in the definite example provided by the protogalaxy will make their explanation clearer.

13.5.1 The Biermann Battery

The problem of starting up a magnetic field from zero is an important one, since there need be no field at all at the beginning of the universe. The start-up problem would seem to be a formidable one in light of the strong constraint on flux conservation. If flux were strictly conserved and B was initially zero, then the flux through any region must remain zero, and B could never change from a zero value. Even resistivity could not help here, for consider the magnetic differential equation

$$\frac{\partial \mathbf{B}}{\partial t} = \nabla \times (\mathbf{v} \times \mathbf{B}) + \frac{\eta c}{4\pi} \nabla^2 \mathbf{B} \tag{48}$$

with the initial condition $\mathbf{B} = 0$ everywhere. Then no matter what $\mathbf{v}$ is, $\mathbf{B} = 0$ will continue to trivially satisfy this equation, and since the solution is unique, no magnetic field can be created.

However, the equation has been derived from Ohm's law in the form

$$\mathbf{E} + \frac{\mathbf{v} \times \mathbf{B}}{c} = \eta \mathbf{j} \tag{49}$$

which is not strictly accurate, since several terms have been omitted. In fact, the more general form of Ohm's law is actually the same as the equation of motion for electrons, equation 95 in chapter 8:

$$nm \left(\frac{\partial \mathbf{v}_e}{\partial t} + \mathbf{v}_e \cdot \nabla \mathbf{v}_e \right) = -n_e e \left(\mathbf{E} + \frac{\mathbf{v}_e \times \mathbf{B}}{c} \right) - \nabla p_e - \nu m (\mathbf{v}_e - \mathbf{v}_i) \tag{50}$$

The simple form of Ohm's law was obtained by simplifying this equation, first by dropping the left-hand side because the electron mass density is small. The last term on the right-hand side gives rise to the ηj term, which is always totally negligible on large scales. The ∇p_e term is also very small if the gyroradius is small. But if $\mathbf{B} = 0$, then the gyroradius is infinite and this term must be kept.

Thus, drop the inertial and resistive terms and divide equation 50 by ne to get

$$\mathbf{E} + \frac{\mathbf{v} \times \mathbf{B}}{\mathbf{c}} = -\frac{\nabla p_e}{n_e e} \tag{51}$$

Next take the curl of this equation and combine it with $\partial \mathbf{B}/\partial t = -c\nabla \times \mathbf{E}$ to obtain

$$\begin{aligned} \frac{\partial \mathbf{B}}{\partial t} &= \nabla \times (\mathbf{v} \times \mathbf{B}) + \nabla \times \left(\frac{c \nabla p_e}{n_e e} \right) \\ &= \nabla \times (\mathbf{v} \times \mathbf{B}) - c \frac{\nabla n_e \times \nabla p_e}{n_e^2 e} \end{aligned} \tag{52}$$

The last term in this equation is the Biermann battery term. It is an inhomogeneous term, so it does not vanish when B does. Thus, it can generate a magnetic field from zero. However, it is generally zero, since p_e is generally a function of n_e. That is, $p = p(n_e)$, and therefore ∇p_e and ∇n_e are parallel and their cross product is zero.

To break this barytropic constraint, some rotational motion is necessary. Such a rotational motion arises in the protogalaxy from the shocks of finite extent that are generated in it during its rather chaotic collapse (Ryu et al. 1994). Consider the shock front in figure 13.8 that extends only a finite distance and that is propagating into a zero pressure medium. Then, behind

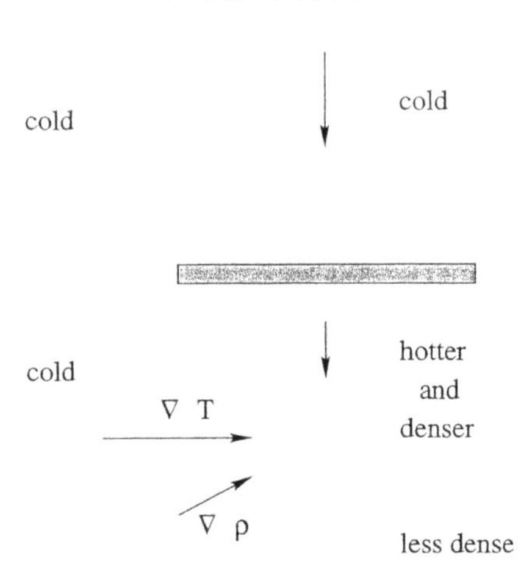

Figure 13.8. Physics of Biermann battery

the shock p_e is high (and, of course, so is n_e). However, after the plasma density expands back to its original value downstream, p_e is still nonzero, because it as been heated by the shock. On the other hand, in the region on the left, unaffected by the shock, p_e is still zero. In this way shocks of limited spatial extent break the barytropic relation, $p_e = p_e(n_e)$, which says that p_e is a unique function of n_e. Hence, such shocks enable the Biermann battery mechanism to operate.

Even if the plasma is only partially ionized, we can still express equation 52 in terms of p, the total plasma pressure, and ρ, the total fluid density. The temperature of the ions and electrons is the same as the temperature of the neutrals. Therefore, if we set $n_e = n_i = \chi(n_i + n_n)$, where n_n is the neutral density, and χ is the degree of ionization, we have

$$\frac{p_e}{p} = \frac{n_e}{n_e + n_i + n_n} = \frac{n_e}{(1+\chi)n} = \frac{n_e M}{\rho(1+\chi)} \tag{53}$$

where $n = n_n + n_i$, so the ratio of p_e to p is a constant times n_e/ρ (if χ is constant). Substituting this into equation 52, multiplying by e/Mc, and assuming χ is constant, we get

$$\frac{\partial \mathbf{\Omega}}{\partial t} = \nabla \times (\mathbf{v} \times \mathbf{\Omega}) - \frac{\nabla \rho \times \nabla p}{(1+\chi)\rho^2} \tag{54}$$

where $\mathbf{\Omega} = e\mathbf{B}/Mc$ is a vector in the $\mathbf{B}$ direction with a magnitude equal to the ion cyclotron frequency.

It is useful to compare this with the equation for the fluid vorticity $\boldsymbol{\omega} = \nabla \times \mathbf{v}$. Let us divide the fluid equation of motion by ρ and take the curl

$$\nabla \times \frac{\partial \mathbf{v}}{\partial t} + \nabla \times (\mathbf{v} \cdot \nabla \mathbf{v}) = -\nabla \times \frac{\nabla p}{\rho} \tag{55}$$

Now, $\mathbf{v} \cdot \nabla \mathbf{v} = -\mathbf{v} \times (\nabla \times \mathbf{v}) + \nabla \mathbf{v}^2/2$, so equation 55 can be written

$$\frac{\partial \boldsymbol{\omega}}{\partial t} = \nabla \times (\mathbf{v} \times \boldsymbol{\omega}) + \frac{\nabla \rho \times \nabla p}{\rho^2} + \nu \nabla^2 \mathbf{v} \tag{56}$$

where we have added the viscous term $\nabla \times (\nu \nabla^2 \mathbf{v})$.

Now, without the viscous term equations 54 and 56 are identical inhomogeneous differential except for a factor of $-1/(1+\chi)$. It is believed that initially $\boldsymbol{\Omega}$ was zero before the protogalaxy formed. We also assume that $\mathbf{B}$, and thus $\boldsymbol{\Omega}$, are zero. With these initial conditions, we see that $-\boldsymbol{\Omega}(1+\chi)$ and $\boldsymbol{\omega}$ satisfy the identical equations and initial conditions, so that everywhere and for all time we have

$$\boldsymbol{\Omega} = -\frac{\boldsymbol{\omega}}{1+\chi} \tag{57}$$

Since from the numerical simulations we know that vorticity is nonzero in the protogalaxy, the magnetic field B must also be nonzero. Note that as the ionization changes from zero to 1, the factor $1/(1+\chi)$ changes from 1 to $\frac{1}{2}$, so that the result is not sensitive to the degree of ionization, provided, at least, that there are enough electrons to carry the current, $\mathbf{j} = \nabla \times \mathbf{B}/4\pi$ (an extremely weak condition). Equation 57 has been confirmed by the numerical simulations of structure formation (Kulsrud et al. 1997).

Now we expect that the turbulence that develops when a protogalaxy forms will have a Kolmogoroff spectrum (chapter 11, section 11.7). Based on Kolmogoroff theory and equation 57, how big do we expect the magnetic field strength developed by the Biermann battery to become?

For an estimate take a protogalaxy with 10^{11} solar masses of baryonic material and 10^{12} solar masses of dark matter. Assume that the protogalaxy collapses to a radius of 100 kpc, at which it is balanced against the gravitational field of the dark matter by thermal and turbulent pressure. Now, before collapse, the plasma temperature was negligibly small. Both the thermal energy and the turbulent energy come from the shocks and adiabatic compression. The shocks produce kinetic energy (turbulence) and thermal energy in approximately equal amounts, and by the virial theorem a static balance is achieved when their sum is one-half of the gravitational binding energy. Thus, we have

$$kT = \frac{1}{4}\frac{GM}{R} m_{\mathrm{H}}, \qquad \frac{\rho v^2}{2} = \frac{GM}{4R}\rho \tag{58}$$

which gives $T = 10^2$ eV, $v = 2 \times 10^7$ cm/sec, and $v_{\mathrm{th}} = 1.5 \times 10^7$ cm/sec. The baryonic density is given by

$$n m_{\mathrm{H}} \frac{4\pi}{3} R^3 = 10^{11} M_{\odot} \tag{59}$$

or $n = 10^{-3}$ cm^{-3}. The Coulomb cross section is $10^{-12}/10^4 = 10^{-16}$ cm^{-2}, so the mean free path is $\lambda = 1/n\sigma = 10^{19}$ cm. The turbulence develops with an outer scale of order $R' \approx R/3 \approx 10^{23}$ cm, so the hydrodynamic Reynold's number is

$$\mathrm{Re} = \frac{R'v}{\nu} = \frac{R'v}{\lambda v_{\mathrm{th}}} = 1.3\frac{R'}{\lambda} = 1.3 \times 10^4 \tag{60}$$

The vorticity on the largest scale is

$$\omega_{\mathrm{large}} = \frac{v}{R'} = 2 \times 10^{-16}\ \mathrm{sec}^{-1} \tag{61}$$

Therefore, the magnetic field on the largest scale is

$$B = \frac{\Omega}{10^4} = 10^{-4}\frac{\omega}{1+\chi} = 2 \times 10^{-20}\ \mathrm{G} \tag{62}$$

taking $\chi = 1$. But for Kolmogoroff turbulence $\omega \sim k\tilde{v} \sim kk^{-1/3} = k^{2/3}$. Also $k_{\mathrm{max}} = (Re)^{3/4}k_{\mathrm{outer}}$, so the vorticity on the smallest scale is

$$\omega_{\mathrm{small}} = \omega_{\mathrm{large}}(\mathrm{Re})^{1/2} \tag{63}$$

so the magnetic field on the smallest scale, $r = R/\mathrm{Re}^{3/4} = R'/1.2 \times 10^3 \approx 8.2 \times 10^{19}$ cm, is larger than B_{large} by ~ 125:

$$B_{\mathrm{small}} \sim 2.5 \times 10^{-18}\ \mathrm{G}$$

Magnetic fields of such a small strength are usually considered to be seed fields that provide initial conditions for stronger amplification mechanisms than the Biermann. This field strength is reached after one turnover time of the largest eddy, since this is the time to establish the Kolmogoroff spectrum.

13.5.2 The Protogalactic Dynamo

The details of the operation of the Biermann battery are insensitive to the model we take for the protogalaxy. However, the magnetic fields it develops are very weak and can only serve as seeds for the dynamo action we expect from the first term on the right-hand side of equation 54. To discuss dynamo action in the protogalaxy with any precision we need a great more detail about the structure and evolution of the protogalactic turbulence than is presently available even from numerical simulations.

Therefore, for lack of this information we adopt a simple scenario for the behavior of the turbulence and the dynamo action. We assume that the protogalaxy collapses to a sphere with virialization radius 100 kpc, and at this time is supported by thermal and turbulent energy in equipartition with the gravitational binding energy. As the plasma cools and the turbulence decays the protogalaxy then collapses to a smaller sphere of radius 20 kpc and finally to the present galactic disk of radius 20 kpc and half-thickness 200 pc.

During the life of the turbulence, which we take to have a Kolmogoroff spectrum, the magnetic field is exponentially amplified on all scales from a seed field produced by the Biermann battery. This amplification is well understood and is described below in the section on small-scale fields (Kulsrud et al. 1997; Kulsrud and Anderson 1992). Each eddy amplifies magnetic energy on its own scale. The smaller scale eddies amplify faster so that all eddies below a certain scale amplify the fields to a saturated value, which is that value at which the magnetic energy at that scale is equal to the turbulent energy at the same scale.

As the field grows it is initially very tangled, but when it becomes strong enough, the tension unwinds it and decreases the field strength at each scale. This happens while the turbulence is strong, but continues as the turbulence decays and for the period of time after the turbulence has disappeared, until the disk forms and there is no longer room for the unwinding. The resulting field is further amplified by compression as the protogalaxy collapses into the disk.

Let us examine this picture semiquantitatively. The vorticity spectrum $J(k)$ is given by

$$\omega^2 = \int J(k)dk \tag{64}$$

The vorticity at scale k, $\omega(k)$, is given by k times the velocity in the Kolmogoroff spectrum, so from equation 139 in chapter 11,

$$\omega^2(k) = kJ(K) = k_0^{2/3}k^{4/3}v_0^2 \tag{65}$$

so that

$$\omega = \omega_0 \left(\frac{k}{k_0}\right)^{2/3} \tag{66}$$

where $\omega_0 = k_0 v_0$ is the vorticity at the largest scale, k_0. Also, let

$$B^2 = \int M(k)dk \tag{67}$$

and

$$B^2(k) = kM(k) \tag{68}$$

where $B(k)$ is the magnetic field strength at k. Now, as we will see in equation 88, $B(k)$ will increase due to dynamo action as

$$\frac{dB(k)}{dt} = \omega(k)B(k) : \quad \text{dynamo} \tag{69}$$

This equation involves the tacit assumption that each eddy generates magnetic energy that is concentrated in the same wave number range. This assumption involves considerable difficulties that are presently under active discussion and it is by no means certain that it is valid. Nevertheless, we make it to explore the potentialities of the protogalactic dynamo for the origin of magnetic fields.

However, as the magnetic field strengthens it also becomes tangled and it will unwind under its tension. So, as it increases due to the dynamo action, it also decreases due to unwinding at roughly the Alfven rate $kv_A(k)$, so

$$\frac{dB(k)}{dt} = -\frac{kB^2(k)}{\sqrt{4\pi\rho}} : \quad \text{unwinding} \tag{70}$$

Thus, assuming that there is no coupling between scales, we have an approximate equation that includes both dynamo growth and unwinding:

$$\frac{dB(k)}{dt} = \omega(k)B(k) - \frac{kB^2(k)}{\sqrt{4\pi\rho}} \tag{71}$$

We assume that $B(k)$ is given initially by the field produced by the Biermann battery,

$$B_i(k) = \frac{Mc}{e}\omega(k) \tag{72}$$

The solution of equation 71 with initial conditions equation 72 is

$$B(k,t) = \frac{B_i(k)e^{\omega(k)t}}{1 + [kB_i(k)/\omega(k)\sqrt{4\pi\rho}]e^{\omega(k)t}} \tag{73}$$

For small t, the second term in the denominator is small and B grows exponentially. For larger k and for larger t the second term becomes more important than the first and $B(k,t)$ approaches a constant saturated field,

$$B_S(k) = \frac{\sqrt{4\pi\rho\omega(k)}}{k} \tag{74}$$

However, if the turbulence turns off at $\omega_0 t_1 = \tau_1$, then $B(k,t)$ does not saturate for smaller k. Let

$$B_1(k) = B\left(k, \frac{\tau_1}{\omega_0}\right) \tag{75}$$

Then the unwinding will continue for τ_2 more turnover times until the protogalaxy collapses into the disk and during this time $B(k,t)$ will satisfy equation 71 with only the second term on the right-hand side. Taking the initial condition at $t = \tau_1/\omega_0$ as $B(k,t) = B_1(k)$, the solution at $t = (\tau_1 + \tau_2)/\omega_0$ is

$$B_2(k) = B\left(k, \frac{\tau_1 + \tau_2}{\omega_0}\right) = \frac{B_1(k)}{1 + \tau_2 k B_1(k)/k_0 B_0 \tau_2} \tag{76}$$

where

$$B_0 = \frac{\sqrt{4\pi\rho\omega_0}}{k_0} \tag{77}$$

is the saturated field at k_0. For small k the B_1 field is too weak to unwind and $B_2(k) = B_1(k)$, but for larger k, $B_2(k) \approx kB_0/k\tau_2$ and is independent of $B_1(k)$.

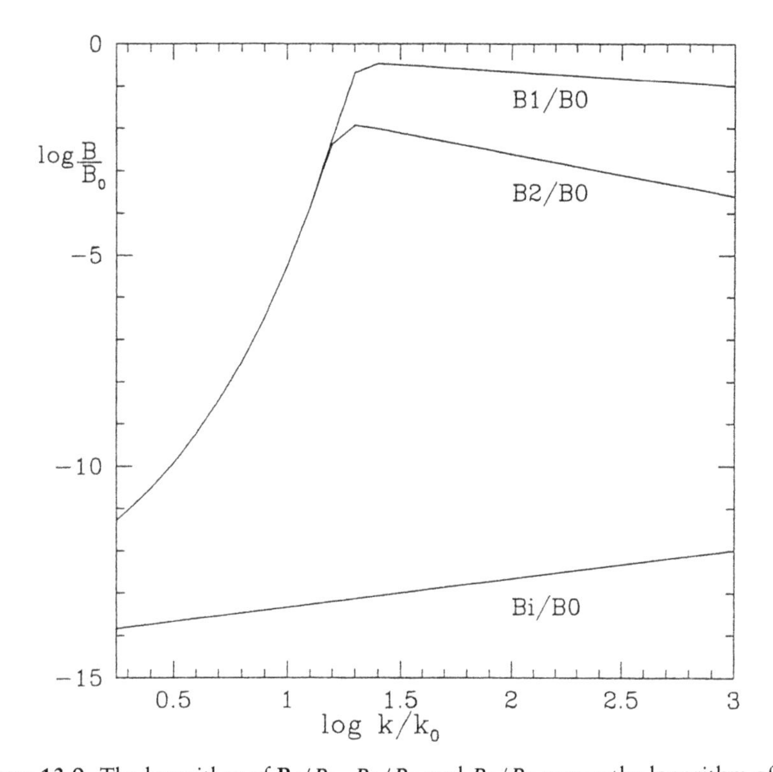

Figure 13.9. The logarithm of $\mathbf{B}_i/B_0$, B_1/B_0 and B_2/B_0 versus the logarithm of k/k_0

The logarithm of three magnetic field strengths—$B_i(k)$, the initial field from the Biermann battery; $B_1(k)$, the field after the turbulence turns off; and $B_2(k)$, the field after unwinding stops—are plotted in figure 13.9 as a function of the logarithm of k/k_0. The τ parameters in the figure are $\tau_1 = 4$, $\tau_2 = 4$. The value of B_0 for our parameters is 2×10^{-6} G.

The general expected features are seen in this plot. Above a certain k the B_1 field saturates and below this value the field is still growing when the turbulence stops. After the turbulence stops the field decreases for the larger values of k and approaches a function independent of B_1, while for smaller k the field does not unwind. The result is that the field strength peaks at a value of k corresponding to about a thirtieth of the size of the protogalaxy.

When the protogalaxy collapses into a disk, whose radius we take as 20 kpc, the field is amplified by compression. First, as in figure 13.10, imagine collapse to a smaller sphere of radius 20 kpc. At this point, the field $B_2(k)$ amplifies by a factor 25. Then it collapses to a thin disk, say of semiheight 200 pc. During this collapse, the field amplifies by an additional factor of 100, so the disk field is stronger than $B_2(k)$ by a total factor of 2500. At the same time the horizontal wave numbers decrease by 5.

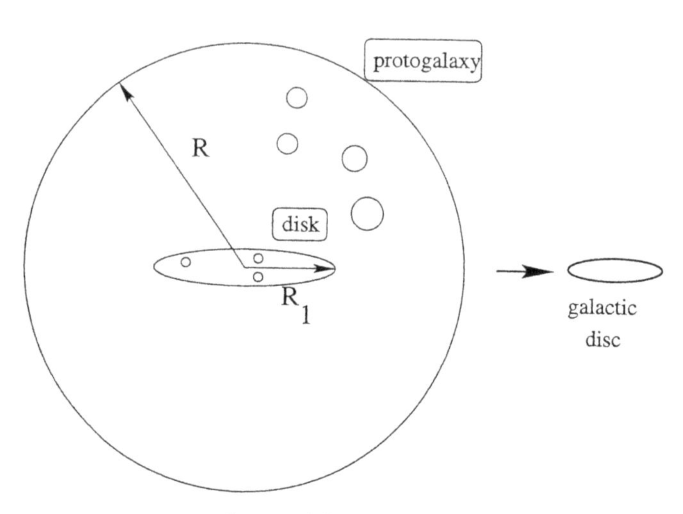

Figure 13.10. Collapse of the protogalaxy to galactic disk

The spectrum of the resulting disk field is shown in figure 13.11, where Λ is the horizontal wave number in units of inverse kiloparsecs. The ordinate is actually $[\Lambda M(\Lambda)]^{1/2}$, so it is the square root of the magnetic energy in a logarithmic interval.

We see that the resulting field is very similar in scale and strength to the presently observed value. These results depend somewhat sensitively on the choice of τ_1, the time during which the turbulence amplifies the field, but only weakly on τ_2. However, they do show that for a not unreasonable choice τ_1 the protogalatic turbulence can produce a field consistent with observations. This happens without invoking any disk dynamo.

13.6 Small-Scale Fields

In the derivation of the mean field equation only the evolution of the mean field is considered, since it is the mean field that most influences the properties of the interstellar medium. It is tacitly assumed that deviations from the mean field are small. This is probably true of the large-scale field, but when we consider small-scale fields it is the mean square field and its spectrum in space that are of importance. Further, in the protogalaxy there is no mean field at all. In this case the theory of small-scale fields applies to all scales. Equation 69, which says that the rate of growth of magnetic fields is equal to the vorticity at each scale, is based on the results of this section.

We start with the Fourier analysis of the magnetic field in space:

$$\mathbf{B}(\mathbf{r}) = \int \mathbf{B}(\mathbf{k}) e^{i\mathbf{k}\cdot\mathbf{r}} d^3\mathbf{k} \tag{78}$$

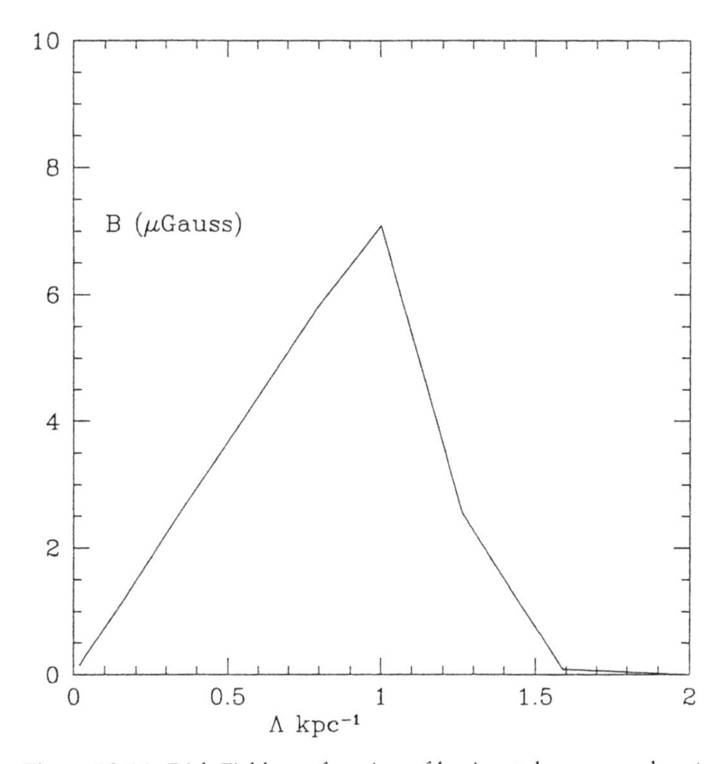

Figure 13.11. Disk Field as a function of horizontal wave number Λ

We think of each $\mathbf{k}$ harmonic as a mode. These modes evolve in time by the Fourier transform of equation 2 (without the resistivity term):

$$\frac{\partial \mathbf{B}(\mathbf{k})}{\partial t} = \int \nabla \times \left[\mathbf{v}(\mathbf{k}') \times \mathbf{B}(\mathbf{k}'')\right] d^3\mathbf{k}' \tag{79}$$

where $\mathbf{k}'' = \mathbf{k} - \mathbf{k}'$. Assume that the velocities are describable by a random phase approximation; then

$$\begin{aligned}\left\langle \mathbf{v}^*(\mathbf{k}', \mathbf{t}')\mathbf{v}(\mathbf{k}, \mathbf{t})\right\rangle = &\, J(k)\left(\mathbf{I} - \hat{\mathbf{k}}\hat{\mathbf{k}}\right)\delta(\mathbf{k}' - \mathbf{k})\delta(t' - t) \\ &+ \bar{J}(k)(i\mathbf{k} \times \mathbf{I})\delta(\mathbf{k}' - \mathbf{k})\delta(t' - t)\end{aligned} \tag{80}$$

where the second term is the helicity term of the turbulence. This equation is the Fourier transform of equation 10. In this section we assume that the field is always so weak that the velocity spectrum is unaffected. This is called the kinematic assumption. The philosophy behind this assumption is that the main goal is to show that we can generate magnetic fields of appreciable strength. Therefore, it is reasonable to start with weak fields that do not affect the velocities and see if the unaffected velocities can make them strong.

Now, apply quasilinear theory to equation 79 to determine the evolution of the magnetic energy spectrum. From equation 67 we have

$$\mathcal{E}_M = \frac{B^2}{8\pi} = \frac{1}{8\pi}\int d^3k|\mathbf{B}(\mathbf{k})|^2 = \frac{1}{8\pi}\int M(k)dk \tag{81}$$

and $M(k)$ evolves according to a mode-coupling equation.

We are primarily interested in small scales of the magnetic field and it can be shown (Kulsrud and Anderson 1992) that the helicity spectrum $\bar{J}(k)$ is small compared to $J(k)/k$, so we give the mode-coupling equation without the effect of helicity. (There is actually no net helicity in the protogalaxy.) The derivation of this equation is given in Kulsrud and Anderson (1992) and in Kraichnan and Nagarajan (1967) and we do not repeat it here. The derivation is close to that of the quasilinear diffusion equation of chapter 11.

$$\frac{\partial M(k)}{\partial t} = \int K(k,k')M(k')dk - 2k^2\beta M(k) \tag{82}$$

where $K(k',k)$ is the kernel coupling the k' modes and the k modes,

$$K(k,k') = 4\pi^2k^4\int d\theta \sin^2\theta \frac{k^2 + k'^2 - kk'\cos\theta}{k''^2}J(k'') \tag{83}$$

where $\mathbf{k}'' = \mathbf{k} - \mathbf{k}'$ and θ is the angle between $\mathbf{k}$ and $\mathbf{k}'$. The scalar magnitude of $\mathbf{k}''$ is $k''^2 = k^2 + k'^2 - 2kk'\cos\theta$, and β is the α–Ω dynamo coefficient,

$$\beta = \frac{2\pi}{3}\int J(k'')d^3k'' \tag{84}$$

This mode-coupling equation has many interesting properties, which are discussed in Kulsrud and Anderson. Two of the most interesting properties are the energy evolution and the small-scale spectrum.

We can calculate $\partial\mathcal{E}_M/\partial t$ by integrating equation 82 over k and manipulating the double integral. The result is

$$\frac{d\mathcal{E}_M}{dt} = 2\gamma\mathcal{E}_M \tag{85}$$

where

$$\gamma = \frac{2\pi}{3}\int k^2 J(k)d^3\mathbf{k} \tag{86}$$

This result should be compared with the expression for the square of the vorticity $\langle\omega^2\rangle$. Let $\boldsymbol{\omega}(\mathbf{r},t) = \int \boldsymbol{\omega}(\mathbf{k},t)e^{i\mathbf{k}\cdot\mathbf{r}}d^3\mathbf{k}$. Then from equation 80 we see that

$$\left\langle\boldsymbol{\omega}^*(\mathbf{k}',t')\cdot\boldsymbol{\omega}(\mathbf{k},t)\right\rangle = 2k^2J(k)\delta(t'-t)\delta(\mathbf{k}'-\mathbf{k}) \tag{87}$$

Set $t' = t$ in this equation. Then

$$[\omega(\mathbf{r},0)]^2 = 2\int\frac{k^2J(k)}{\tau_k}d^3\mathbf{k} = \int\tilde{\omega}_k^2\frac{dk}{k}$$

where τ_k is the width of the delta function, $\delta(0) = 1/\tau_k$, and $\tilde{\omega}_k^2$ is the contribution to the square of the vorticity from a band of width $\Delta k = k$ about k. τ_k is the decorrelation time, which is equal to the eddy turnover time $1/\tilde{\omega}_k$. Therefore,

$$\gamma = \frac{\pi}{3} \int \tilde{\omega}_k^2 \tau_k \frac{dk}{k} = \frac{\pi}{3} \int \tilde{\omega}_k \frac{dk}{k} \tag{88}$$

Suppose that we can neglect the coupling between velocities and magnetic fields at different k's. Then, consider the case that the turbulence is confined to a bandwidth $\Delta k = k$ about k. In this case we see from equation 88 that the magnetic energy in the same bandwidth grows at the rate

$$2\gamma_k = 2\frac{\pi}{3}\omega_k$$

which agrees with equation 69 up to the factor of $\pi/3 = 1.05$.

It is plausible to neglect the coupling between different k's in the inertial range of the Kolmogoroff spectrum. However, in the kinematic limit the bulk of the growing magnetic energy is at wave numbers shortward of the inner scale, i.e., for k's larger than $k_{\max}$. Thus, for this range of magnetic energy we must go back to the full mode-coupling equation, equation 82. However, we can treat $\mathbf{k}''$ as small and expand the mode-coupling kernel in equation 83 in it. The result is a differential equation for $M(k, t)$:

$$\frac{\partial M(k,t)}{\partial t} = \frac{\gamma}{5}\left(k^2 \frac{\partial^2 M}{\partial k^2} - 2k\frac{\partial M}{\partial k} + 6M\right) - k^2 \frac{\eta c M}{4\pi} \tag{89}$$

where the last term is the resistivity term. Without the resistive term the general solution of this equation can be expressed in terms of a Green's function

$$M(k,t) = \int_{-\infty}^{t} M(k_{\text{ref}}, t-\tau) G(\frac{k}{k_{\text{ref}}}, \tau) d\tau \tag{90}$$

where k_{ref} is a k at which we assume that we know M as a function of t, and

$$G(k,\tau) = \sqrt{\frac{5}{4\pi}} \frac{k^{3/2} \ln k}{\gamma^{1/2}\tau^{3/2}} \exp\left(-\frac{5\ln^2 k}{4\gamma\tau} + \frac{3}{4}\gamma\tau\right) \tag{91}$$

A plot of $G(k)$ is given in figure 13.12. The essential features of this solution are included in G. The solution increases as $k^{3/2}$ till a peak is reached in $kG(k)$ at k_p where $k_p \sim e^{\gamma t}$.

The spectrum increases in time as $e^{(3/4)\gamma t}$ at any fixed k. If we assume that $M(k,t) \sim e^{(3/4)\gamma t}$ and include the resistive term, we find that the exact solution of equation 89 is

$$M(k,t) = \text{const } \times k^{3/2} K_0\left(\sqrt{\frac{5\eta}{2\pi\gamma}}k\right) e^{(3/4)\gamma t} \tag{92}$$

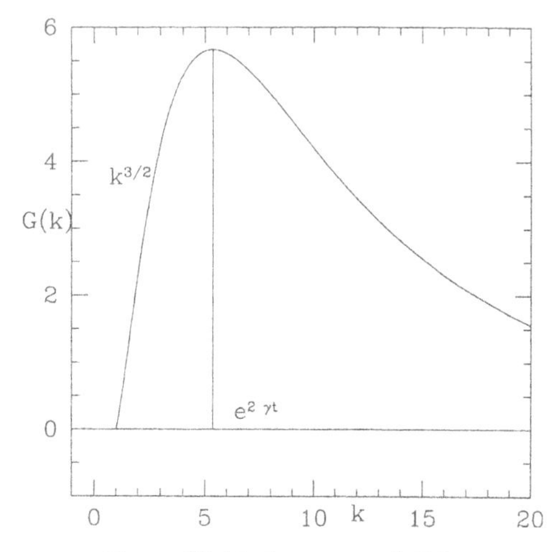

Figure 13.12. Spectrum of $G(k)$

where K_0 is the zero-order Bessel function of the second type. Thus, the spectrum first spreads rapidly toward the resistive scale $\sqrt{\eta/\gamma}$ with a growth, at any fixed k, of $\sim e^{(3/4)\gamma t}$ but with the range in k increasing fast enough that the integral (i.e., the total magnetic energy) increases as $e^{2\gamma t}$. When it reaches the resistive wave number $k_r \approx \sqrt{\gamma/\eta}$ it stops spreading in k. At this time $M(k)$ becomes a fixed function of k that grows everywhere in k space as $e^{(3/4)\gamma t}$, and whose k integral also grows more slowly as $e^{(3/4)\gamma t}$. (The original growth as $e^{2\gamma t}$ came from the exponential rate of spread in k of the excited modes. When the resistive scale is reached, we get the slower growth for the integral.)

For galactic and protogalactic scales the resistive scale is so small that it is usually the case that the spectrum saturates the turbulence before the resistive scale is reached. Unwinding then occurs as fast as the winding and the resistive scale is never reached. Finally, it should be noted that in equation 88 for a Kolmogoroff spectrum the largest contribution comes from the largest k, so the smallest eddy amplifies the magnetic energy fastest and is the first eddy that saturates.

13.7 Problems

1. (The dipole solution) Consider the solution for the dipole eigenfunction of equations 33 and 34. To find the eigenvalue try the approximate solution

$$A = \hat{A}(1 - \cos 2k_0 z)$$

$$B_\theta = \hat{B} \sin 2k_0 z$$

Show that these satisfy equation 34 if

$$(p + 4k_0^2\beta)\hat{B} = 2k_0\Omega\hat{A}$$

but do not solve equation 33 exactly. Show that by approximating the zB_θ term by $2a\sin^2 k_0 z = a(1 - \cos 2k_0 z)$ (which has the same shape in z as A), where the constant a is determined by $\int_{-h}^{h} zB_\theta dz = \int_{-h}^{h} a(1 - \cos 2k_0 z)dz$, equation 33 is satisfied in the mean if

$$(p + 4k_0^2\beta)\hat{A} = \alpha_0\hat{B}$$

(but only if an additional constant is added to the vector potential A). Thus, obtain the approximate equation for p

$$(p + 4k_0^2\beta)^2 = \frac{2k_0\alpha_0\Omega}{\pi}$$

What is the critical value D_c for the dynamo number D, defined in equation 44, for instability? How does this compare with D_c for the quadrupole solution? Is it larger? (Note that the dipole solution, if it were unstable, would have the advantage of not being subject to the escape problem.)

2. (Power necessary to drive the galactic α–Ω dynamo) Assume that the galactic dynamo works according to theory but take the flux escape problem seriously. Convince yourself that for every doubling of the flux, an amount of flux (the negative flux) equal to about one-third of the total flux without regard to sign must be removed to infinity if total flux is conserved. Namely, appreciate that one positive unit of flux can be converted by the dynamo to two positive units and one negative unit of flux and this negative unit must be discarded. Then argue by flux freezing that at the same time one-third of the interstellar matter must be removed to infinity, thereby requiring an energy equal to the escape energy.

Take the total mass of the interstellar medium at any time to be 10^9 solar masses. Take the escape velocity to be 400 km/sec. Require that the flux doubles in 500 million years. With these numbers show that the power to drive the dynamo is thus about 3×10^{40} ergs/sec for the galaxy. Estimate the supernova kinetic power as 10^{49} ergs per supernova and that the frequency of supernova in the galaxy is one per 30 years. Is there enough power in the supernovae to drive the dynamo? Do superbubbles (multiple supernovas) help in this energy balance?

3. (The growth rate of small-scale fields) The growth rate of small-scale fields γ is given by equation 88, where for a Kolmogoroff turbulence the spectrum is approximately equal to $k\tilde{v}$, where $\tilde{v} \sim k^{-1/3}$. Take the largest scale of the turbulence to be the radius of a protogalaxy and take the turbulent energy $\tilde{v}$ at the largest scale to be in equipartition with the gravitational binding energy of the protogalaxy (that is, $m_H v_0^2/2 = GMm_H/R$, where M is the mass of the protogalaxy and m_H is the mass of a hydrogen atom. Also take the temperature to be in equipartition (that is the Jeans temperature $T_J \approx m_H v_0^2$). Roughly estimate the amount of growth in a dynamic free-fall

time $\tau_D \approx R/v_0$; that is, estimate $\gamma\tau_d$, keeping only dimensional factors and show that

$$\gamma\tau_d \approx \frac{e^2}{Gm_{\rm H}^2}\sqrt{\frac{m_{\rm H}}{M}}$$

To do this you need the hydrodynamic Reynold's number, which involves the kinematic viscosity $\nu \approx \ell v_{\rm th}$, where ℓ is the mean free path $\ell \approx 1/n\sigma$, where σ is the Coulomb cross section. Note that this quantity depends only on the mass of the protogalaxy. What is the rough value for this amount of growth if the gravitational object has the galactic mass, say 10^{11} solar masses, or one solar mass, or the mass of a galactic cluster, say, 1000 galactic masses?

4. (The small-scale spectrum) Show that if $\eta = 0$ and if all the energy of the magnetic field turbulence is on the smallest scale so that $M(k)$ satisfies the differential equation 89, then the total magnetic energy $\int M(k)dk$ increases at the rate 2γ. Show that the Green's function solution

$$M(k) = G(k, t)$$

is a solution of the differential equation. Show that, if we ignore the $\ln k$ factor in G, the peak of $kM(k,t)$ varies as $e^{\gamma t}$. Take the parameters of Kolmogoroff turbulence in the interstellar medium to be a largest scale of 100 pc and v_0 at the scale to be 10 km/sec. Take the interstellar density as one per cubic centimeter and a temperature of one electron volt. Estimate the viscous cutoff scale. Estimate γ the growth rate of the small-scale magnetic field. The small-scale field will saturate when its total energy equals the turbulent energy at the viscous scale. If the initial small-scale magnetic energy is concentrated at the viscous scale and corresponds to a field of 10^{-14} G, then will the field reach the resistive scale given by the Bessel function cutoff in equation 92 before it reaches saturation? Finally, show that equation 92 is actually a solution of the differential equation 89 when resistivity is included.

References

Biermann, L. 1950. *Zeitschrift fur Naturforschung* **5a**, 65.
Cen, R., and J. P. Ostriker. 1993. *Astrophysical Journal* **417**, 404.
Cowling, T.G. 1934. *Monthly Notices of the Royal Astronomical Society* (MNRAS) **94**, 39.
Elsasser, W. M. 1946. *Physical Review* **69**, 106.
Glatzmaier, G. A., and P. H. Roberts. 2000. *Review of Modern Physics* **72**, 1081.
Kraichnan, R. H., and S. Nagarajan. 1967. *Physics of Fluids* **10**, 859.
Kulsrud, R. M. 1999. *Annual Review of Astronomy and Astrophysics* **37**, 37.
Kulsrud, R. M. 2000. *The Origin of Magnetic Fields*, Proceedings of the International School of Physics, Enrico Fermi, Course CXLII, Plasmas in the Universe, p. 167, ed. B. Coppi and A. Ferrari, IOS Press, Amsterdam.
Kulsrud, R. M., and S. W. Anderson. 1992. *Astrophysical Journal* **396**, 606.
Kulsrud, R. M., R. Cen, J. P. Ostriker, and D. Ryu. 1997. *Astrophysical Journal* **480**, 481.

Parker, E. N. 1955. *Astrophysical Journal* **122**, 293.

Parker, E. N. 1970. *Astrophysical Journal* **160**, 383.

Parker, E. N. 1971a. *Astrophysical Journal* **163**, 255.

Parker, E. N. 1971b. *Astrophysical Journal* **163**, 279.

Parker, E. N. 1979. *Cosmical Magnetic Fields*, p. 543, Clarendon Press, Oxford, UK.

Rafikov, R. R., and R. M. Kulsrud. 2000. *Monthly Notices of the Royal Astronomical Society (MNRAS)* **314**, 834.

Ruzmaikin, A. A., A. M. Shukurov, and D. Sokoloff. 1988. *Magnetic Fields of Galaxies*, Chapter VII, p. 171, Kluwer, Dordrecht.

Ryu, D., J. P. Ostriker, H. Kang, and R. Cen. 1994. *Astrophysical Journal* **428**, 1.

Steenbeck, M., P. Krause, and R. Radler. 1966. *Zeitschrift fur Naturforschung* **21a**, 369.

Vainshtein, S. I., and A. A. Ruzmaikin. 1972. *Soviet Astronomy* **16**, 365.

Chapter 14

MAGNETIC RECONNECTION

14.1 Introduction

Flux freezing is a very strong constraint on the behavior of magnetic fields in astrophysics. As we show in chapter 3, this implies that lines do not break and their topology is preserved. The condition for flux freezing can be formulated as follows: In a time t, a line of force can slip through the plasma a distance

$$\ell = \sqrt{\frac{\eta c t}{4\pi}} \tag{1}$$

If this distance ℓ is small compared to δ, the scale of interest, then flux freezing holds to a good degree of approximation.

However, there are important situations where δ is so small that it is comparable with ℓ, and then the assumption of flux freezing no longer holds. In this case, a pair of lines can break and reattach themselves to each other, as in figure 14.1. This is called magnetic reconnection. Several reviews of the magnetic reconnection process are Biskamp (1994), Kulsrud (1998), Priest and Forbes (2000), and Vasyliunas (1975).

There are two specific examples in which it is inferred from observations of the changing topology of the magnetic fields that magnetic reconnection must be occurring. The first example is the interaction of the earth's magnetosphere with the solar wind. The second example is the sudden release of energy in the solar atmosphere in the form of solar flares. In both of these cases a clear change in the topology of the observed magnetic field lines is seen.

In the first example, the magnetosphere, the fast-moving solar wind impacts on the surface of the magnetosphere, causing the two distinct plasmas to be brought into close contact and leading to a very narrow boundary between their magnetic fields (see figure 14.2). One plasma is the solar wind

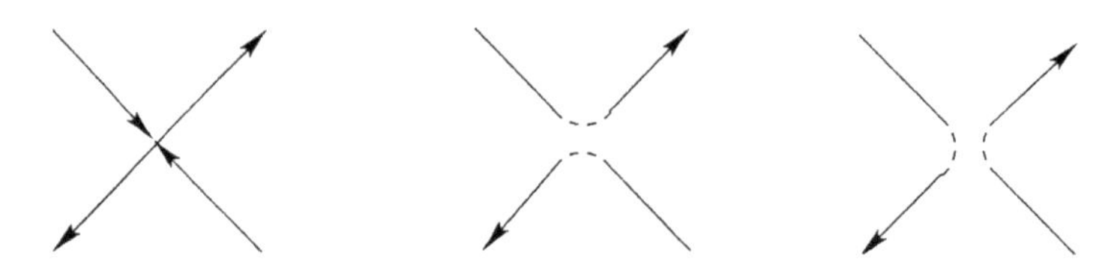

Figure 14.1. Lines breaking and reconnecting

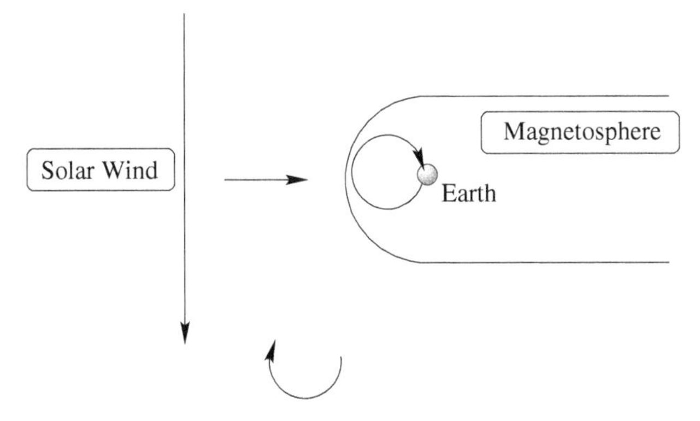

Figure 14.2. Contact between the solar wind and the magnetosphere

impacting on the earth's magnetosphere, and the other plasma is the magnetospheric plasma itself (see figure 14.2). When they are brought into contact at the magnetopause, their field lines are pressed together so closely that a substantial fraction of them break and reconnect. That is, the solar wind lines break and the magnetospheric lines break and they reconnect to each other across the magnetopause surface.

This produces newly reconnected lines of force that cross the magnetospheric surface on the north side of the magnetosphere, as in figure 14.3, where the letters A, B, C, and D indicate the time sequence for a single line of force. These lines then allow the solar wind plasma to flow directly into the magnetospheric cavity. At the same time the outer parts of these lines remain embedded in the rapidly moving solar wind and are dragged downstream by the solar wind flow at high speed, as described in chapter 3.

The same thing happens to the lines on the south side of the magnetosphere. Eventually, the north and south lines are brought together far downstream. They end up separated by a very thin layer and reconnection occurs again, producing a new pair of lines, as in figure 14.4. One of these lines then is connected to the earth, while the other is embedded in the solar wind, as a result of this reconnection event. The parts of the north and south lines on the earth side of the reconnection reestablish a single magnetospheric line connected only to the earth on both sides. The solar wind line has been restored to a single line in the solar wind, which then flows away downstream.

The region through which the two lines pass between reconnections forms the magnetotail of the earth. The second reconnection event produces greatly stretched magnetic field lines, which snap back to the earth, producing magnetic substorms and aurora. This picture was proposed theoretically by Dungey (1961) as an explanation for the origin of the magnetic storms and aurora. His model was later confirmed by observations (see Hughes 1995).

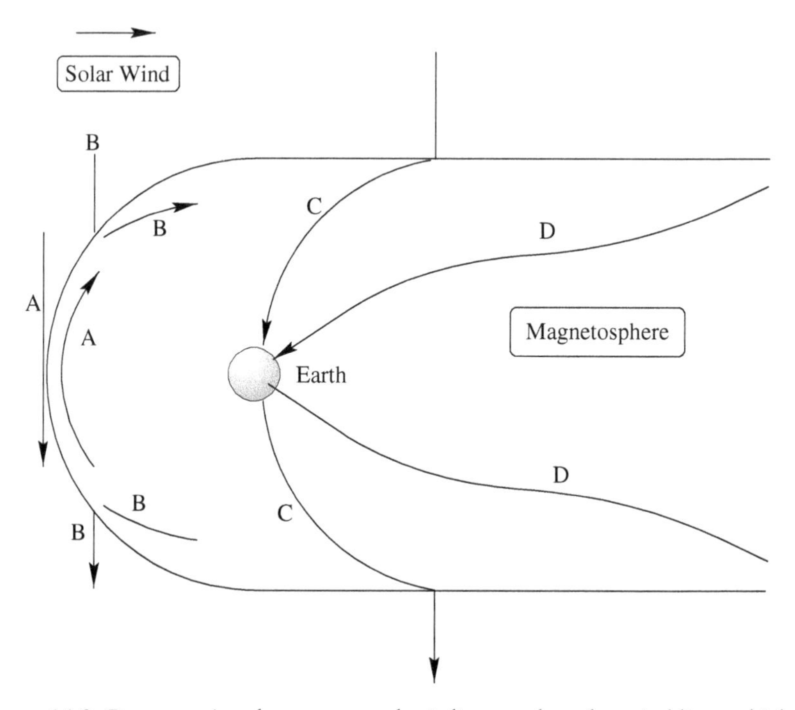

Figure 14.3. Reconnecting the magnetospheric lines to the solar wind lines which drag them down stream

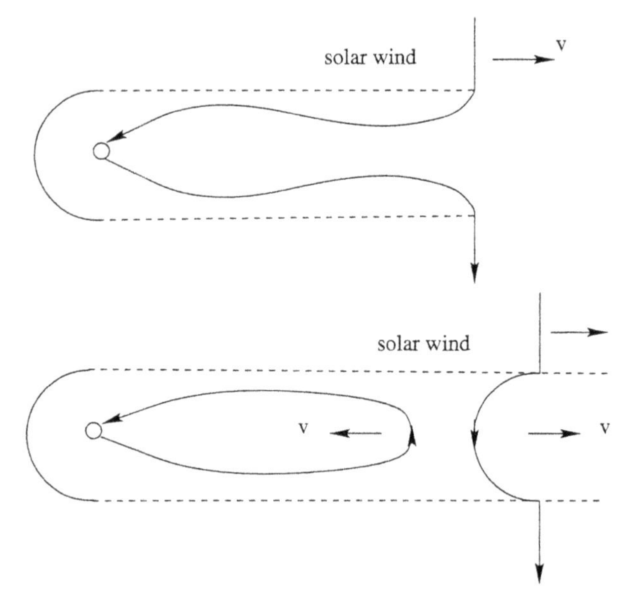

Figure 14.4. Breaking the reconnected field lines and returning the the solar wind lines to the solar wind, and the magnetospheric lines to the magnetosphere

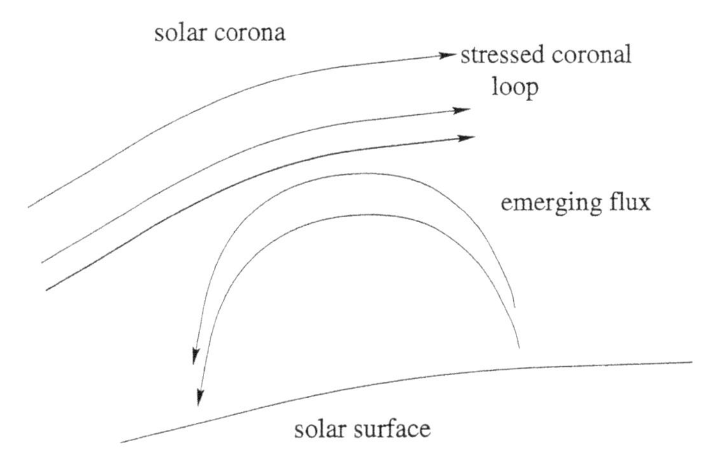

Figure 14.5. Reconnection of emerging flux to a coronal loop

The second example of magnetic reconnection underlies the solar flare phenomena. The solar flare represents a release of energy high up in the solar atmosphere, where there is no visible source of such energy. The origin of this energy, apparently out of nowhere, appeared very mysterious at first, but the mystery was resolved when it was appreciated that there is sufficient free magnetic energy that a rearrangement the sun's magnetic field would account for the emitted solar flare energy.

Using the concept of flux freezing it is imagined that the magnetic field lines are twisted up by the motions of their foot points in the solar photosphere. This motion increases the energy of the magnetic field much above the energy of the sun's vacuum dipole field. Then a bubble of magnetic field could rise up from the photosphere and collide with the twisted field, as in figure 14.5.

The magnetic energy of the twisted field is stored there by flux freezing. The collision with the rising bubble produces a reversed field configuration on a very short scale, δ, through which the field rapidly varies in direction. Reconnection then occurs in this thin region, releasing the twist of the earlier field and lowering its energy. The released energy then provides the observed emitted energy of the flare.

14.2 The Sweet–Parker Model of Magnetic Reconnection

The resistive decay time for the field in the solar atmosphere is very long unless the scale is greatly reduced. The question is how short δ can become. We get slippage of lines through a distance δ in a time t if

$$\delta = \ell = \sqrt{\frac{\eta c t}{4\pi}} \tag{2}$$

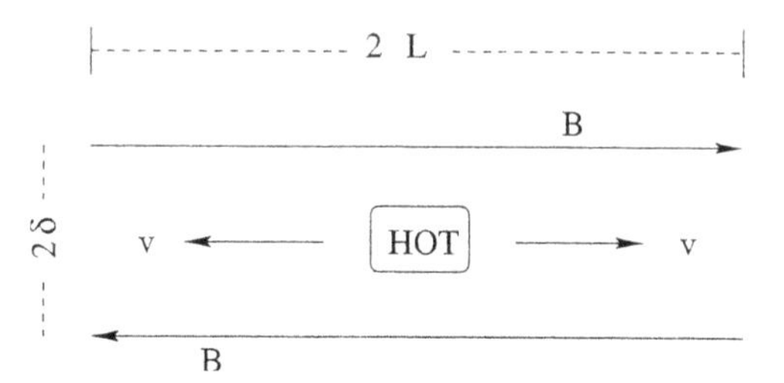

Figure 14.6. The geometry of the current layer

We see that the shorter δ is, the shorter t can be, and the faster the plasma can move through it to reconnect. However, there is a natural limit on how small δ can be. As δ decreases, the current density increases because $j \approx B/4\pi\delta$. As the current density grows, Ohmic heating in the layer also grows. The Ohmic heating

$$\eta c j^2 = \eta c \left(\frac{B}{4\pi\delta} \right)^2 \tag{3}$$

heats the plasma to a high pressure and drives the field lines out of the narrow layer before they can reconnect (see figure 14.6).

To determine the actual amount of heating we have to determine how long any piece of plasma stays in the layer. Now, since $p + B^2/8\pi$ is a constant across the thin layer and $B \to 0$ at the center of the layer, the pressure inside the layer must be greater than that outside the layer by $B_0^2/8\pi$, the magnetic field pressure just outside of the layer. But because the layer is of finite length, there is a pressure gradient along the layer, which causes the plasma to flow along the layer at the Alfven speed $v_A = B_0/\sqrt{4\pi\rho}$. This means that the plasma is Ohmically heated for a time $t_\eta = L/v_A$. Since the Ohmic heating rate is

$$\eta c j^2 = \frac{\eta c B^2}{(4\pi)^2 \delta^2} \tag{4}$$

the plasma is heated to a pressure

$$\Delta p = \frac{\eta c B^2}{(4\pi)^2 \delta^2} t_\eta = \frac{\eta c}{4\pi} \frac{B^2}{4\pi\delta^2} \frac{L}{v_A} = p \tag{5}$$

This pressure increase must not be greater than the outside magnetic pressure $B_0^2/8\pi$, so taking $B^2 \approx B_0^2/2$ we get

$$\frac{\eta c}{4\pi\delta^2} \frac{L}{v_A} < 1$$

or the thickness must be greater than

$$\delta = \sqrt{\frac{\eta c}{4\pi} \frac{L}{v_A}} = \frac{L}{\sqrt{S}} \tag{6}$$

where

$$S = \frac{Lv_A}{\eta c/4\pi} \tag{7}$$

is called the Lundquist number and is generally very large. Equation 6 is the natural lower limit to the thickness of a reconnection layer.

On the other hand, in a steady state the change in pressure $\Delta p \approx B_0^2/8\pi$, so the thickness of the layer should actually be given by equation 6. During steady-state reconnection it must be the case that the flux of outgoing plasma,

$$4\delta v_A$$

must be replaced by the flux of incoming plasma

$$4Lv_R$$

so equating these we have that the velocity of incoming plasma must be

$$v_R = \frac{\delta}{L} v_A \tag{8}$$

or, with equation 6,

$$v_R = \sqrt{\frac{\eta c}{Lv_A}} v_A = \frac{v_A}{\sqrt{S}} \tag{9}$$

This gives the reconnection rate since the incoming plasma brings in lines from either side with this velocity.

Can the lines slip through this plasma to reconnect? They certainly can because, in the time $t = L/v_A$, the lines diffuse through the plasma a distance

$$\ell = \sqrt{\frac{\eta c t}{4\pi}} = \sqrt{\frac{\eta c}{4\pi} \frac{L}{v_A}} = \delta \tag{10}$$

The time L/v_A is the time that the lines have to slip through the plasma before new lines arrive.

This equation is usually expressed as

$$v_R = \frac{\eta c}{4\pi\delta} \tag{11}$$

which comes directly from Ohm's law, with $j = B/4\pi\delta$ (equation 47 in chapter 3). Combining this equation with equation 8 we again get a derivation of the reconnection velocity equation 9.

This argument, or its equivalent, was put forward in 1956 by Sweet (1958), and independently and simultaneously by Parker (1957). The result of their discussions of magnetic reconnection is now called the Sweet–Parker model. The velocity of reconnection, v_R, given in equation 9, is called the Sweet–Parker reconnection velocity, and the thickness δ, given in equation 6, is called the Sweet–Parker layer thickness. We denote this thickness by δ_{SP}.

Let us start a little further back in time before reconnection commences. Imagine that the flux lines, of length L, in a layer of thickness $\delta > \delta_{SP}$ are

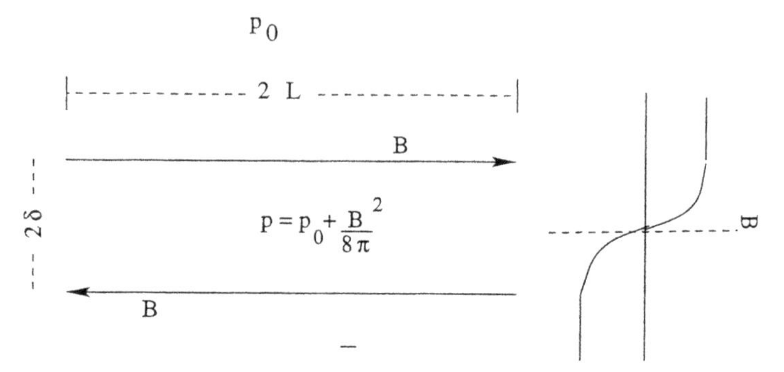

Figure 14.7. The pressure rises by $B_0^2/8\pi$ due to the decreased field

oppositely directed and open to downstream flow, as in figure 14.7. There is an excess pressure $\Delta p \approx B_0^2$ in the layer that drives downstream flow with a velocity v_A/L.

Because $\delta > \delta_{SP}$, the Ohmic heating in the layer is not enough to produce the excess pressure, so δ must decrease to produce some adiabatic heating $-(d\delta/dt)p/\delta$ as well. To keep things simple we will assume that $p = \Delta p$ and that the only pressure in the layer is due to these heating mechanisms. The Ohmic heating is given by equation 4, and the Ohmic heating during the time L/v_A can be written as $p(\delta_{SP}/\delta)^2$. The total heating, adiabatic plus Ohmic, in the layer during the time L/v_A is

$$\frac{L}{v_A}\left(-p\frac{d\delta}{dt}\right) + p\left(\frac{\delta_{SP}}{\delta}\right)^2 \delta$$

and this must equal Δp. Thus, we can write the heat balance equation as

$$-\frac{d\delta}{dt} + \frac{\delta_{SP}^2}{\delta}\frac{v_A}{L} = \frac{v_A}{L}\delta$$

The solution to this equation with an initial value δ_0 for δ is

$$\delta^2 - \delta_{SP}^2 = (\delta_0^2 - \delta_{SP}^2)e^{-2v_A t/L} \tag{12}$$

A plot of δ as a function of time is given in figure 14.8. We see that at first δ decreases exponentially. The heating is purely adiabatic, until δ_{SP} is approached. Then the Ohmic heating becomes important and δ levels off in time and approaches δ_{SP} asymptotically.

At this time the field lines start to slip across the plasma and reconnect as fast as the plasma flows inward. The magnetic field and the plasma velocity become stationary. The fresh plasma entering the reconnection region from $|y| > \delta_{SP}$ is heated to the pressure $B^2/8\pi$ in a time L/v_A. δ_{SP} is given in equation 6.

In summary, prior to magnetic reconnection we expect the layer thickness δ to first decrease exponentially until it approaches δ_{SP}, and then to level off at δ_{SP}.

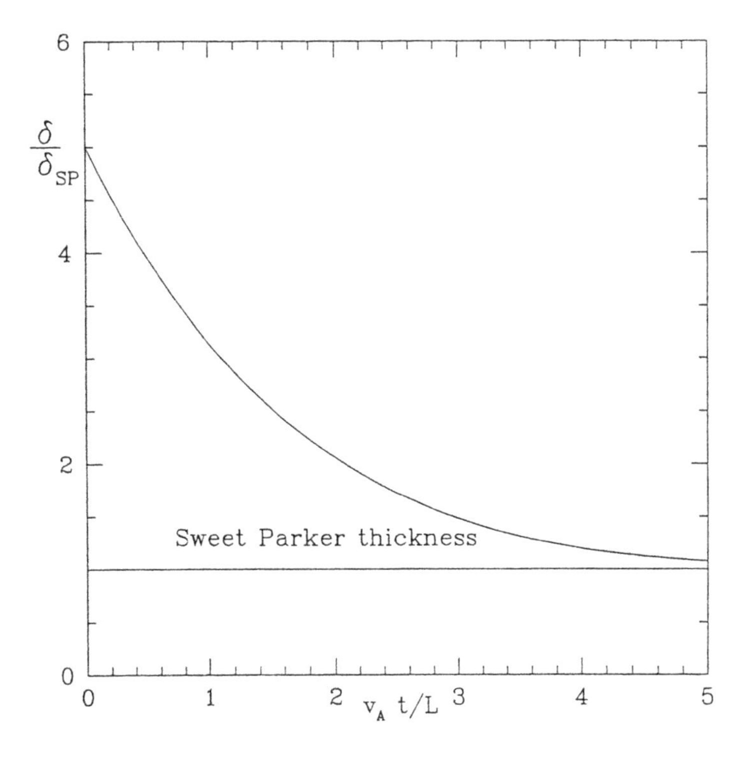

1v

Figure 14.8. The thickness δ versus time

14.3 The Uzdensky Model

How can we make the Sweet–Parker picture more quantitative? Let us examine reconnection in a specific model that was introduced by Uzdensky (Uzdensky and Kulsrud 2000). Let us consider two identical cylinders of plasma, each carrying current into the paper and in near contact, so that their cylindrical fields are in opposite directions, as in figure 14.9.

Let the cylinders be surrounded by a plasma with closed lines of force, which in turn are bounded by a rigid, infinitely conducting wall. For simplicity, we assume that the plasma is incompressible. Let the poloidal flux in each of the cylinders be ψ. We will call the regions of the cylinders the private regions (since they each have their own private flux). The region outside the cylinder but inside the wall will be called the public region since it will contain the combined reconnected flux. (Although the situation naturally starts with no reconnected flux, we discuss the situation at a later time when some reconnection has occurred and there is already some reconnected flux.)

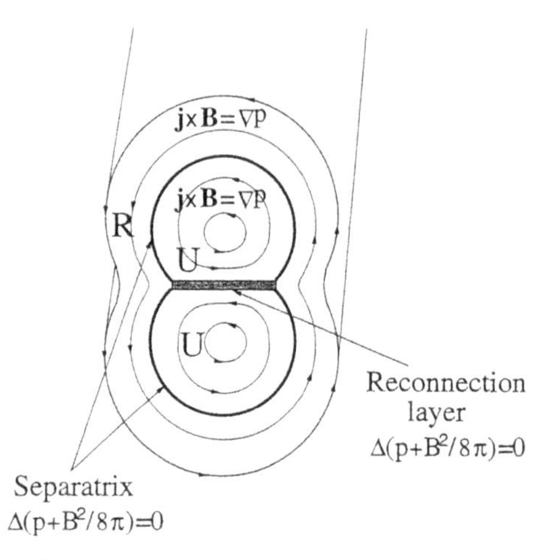

Figure 14.9. The global picture of the reconnection of two cylinders

Let the public region have flux ϕ. ψ and ϕ will change with time as reconnection proceeds, but their sum will remain constant. The volumes of the private regions and their fluxes will decrease, while the public volume and its flux will increase. The solid line between the cylinders is the reconnection region. The solid line between the public and private regions is the separatrix region.

We will also call the private and public regions the global region and the reconnection layer and the separatrix layers the local region. We assume that the resistivity is so small that the local regions can be considered infinitely thin. Thus, to lowest approximation we see only the global regions and the local regions appear as thin lines. We further assume that the reconnection proceeds so slowly that the velocities that it induces in the public and private regions are small. Then at any time the magnetic and pressure forces must be in balance, at every point in the public and private regions, so

$$\mathrm{j} \times \mathbf{B} = \nabla p \tag{13}$$

In the narrow reconnection and separatrix regions there are flows and the forces are not in balance. However, on either side of these surfaces there are magnetic and plasma pressures that must balance. That is, the jump of the total pressure across the surface

$$\left\langle p + \frac{B^2}{8\pi} \right\rangle = 0 \tag{14}$$

Now, it turns out that, as far as the global region is concerned, once we specify ϕ and ψ, the public and private fluxes, and the volume per flux in each region, there is a unique solution to equation 13 satisfying the jump condition equation 14. The volume per flux $V(\psi)$ that must be specified in

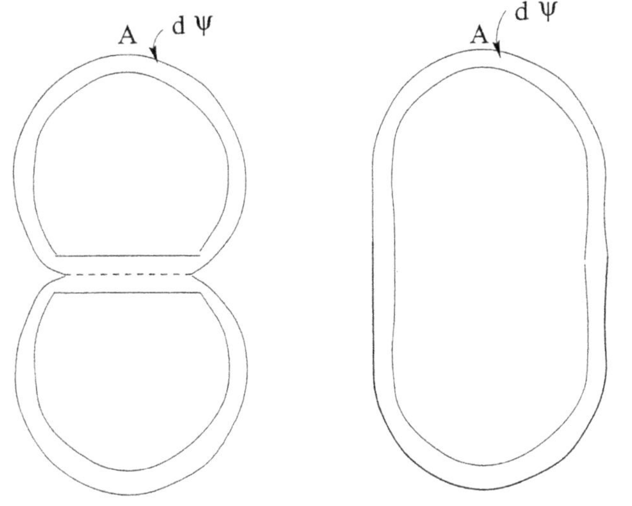

Figure 14.10. Two flux tubes before reconnection and the combined tube after reconnection

each private region is a function given by

$$dV(\psi) = \int dA d\ell = d\psi \int \frac{dA}{d\psi} d\ell = d\psi \int \frac{d\ell}{B} \tag{15}$$

which gives $dV/d\psi$ in the private region. There is a similar specification of volume per flux in the public region.

As flux is reconnected in the local region, the global equilibrium slowly changes, but, by incompressibility and flux freezing, $\int d\ell/B$ must be constant following the plasma as a function of ψ. (The volume of the plasma with flux $d\psi$ must stay constant and $d\psi$ must stay constant following a particular piece of plasma.)

When the plasmas on the two ψ tubes from the private regions merge at the reconnection surface they turn into a single public tube with the same flux $\phi = \psi$ and this tube must have twice the volume of the individual private flux tubes, as we see from figure 14.10. This is true to lowest order in the resistivity. The tube must have the twice the volume to lowest order since the slippage velocity in the local region is of order $v_R \ll v_A$ as the plasma passes through the reconnection region and the separatrix, and the amount that the flux slips relative to the plasma during the Alfven time, L/v_A, is $v_R(L/v_A)B_x \ll \psi$. After the plasma emerges from these narrow regions and enters the public region, it must continue to keep $\int d\ell/B$ equal to its new value, but now as a function now of ϕ. The constant, of course, varies in time so as to match the value of $V(\psi)$ to the value in the private region of the flux just being reconnected (see figure 14.10). Thus, measuring ψ from

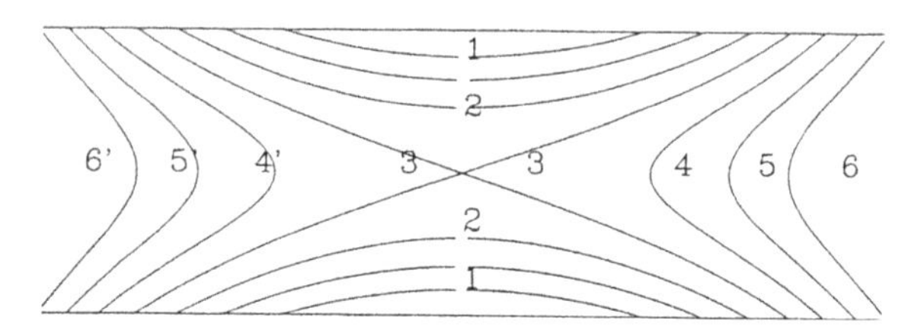

Figure 14.11. The field lines in the reconnection region

the central axis of the cylindrical region and ϕ from the wall we can write the conditions in the form

$$\frac{\partial V(\psi, t)}{\partial t} = 0$$

$$\frac{\partial V(\phi, t)}{\partial t} = 0$$

$$V[\phi_r(t), t] = 2V[\psi_r(t), t] \tag{16}$$

where $\phi_r(t)$ corresponds to the reconnected flux in the private regions at time t, and $\psi_r(t)$ is the unreconnected private flux at time t.

From figure 14.10 we see that the part of a reconnecting line that crosses the separatrix is essentially undisturbed. The part passing through the reconnection region breaks in two at an X point and each half moves in opposite directions horizontally.

So far we have discussed only the behavior in the global region for which the local region is considered infinitely small. In fact, the only way the local region affects the global region is through the rate of reconnection. In every other way, the sequence of states of the global region is specified independently and only the time it takes to progress through these states is controlled by the physics of the local region.

The proper way to treat the local region is by a standard boundary layer treatment in which its thickness is scaled up to a finite value in terms of the infinitely small resistivity. In figure 14.11 we sketch the field lines in the scaled up reconnection region obtained from this boundary layer analysis. The behavior is displayed after blowing up its vertical thickness, which is very thin (of order δ_{SP}). Lines at different positions are numbered. The two lines to be reconnected enter at 1. Later, they are at 2. They break at 3, going through the X point. The right-hand parts of the two lines join to make 4, and the left-hand parts join to make 4′. They then move horizontally to 5 and 5′. Finally, at 6 and 6′ they are in the public region.

Suppose that we know the functions $V(\psi)$ and $V(\phi)$ at some time t and we also know the fluxes ψ_r and ϕ_r at the same time. Then equations 13 and 14 lead to a unique solution of the quasistatic equilibrium. Now, assume that an amount of flux $d\psi_r$ is reconnected so that ψ_r decreases by $d\psi$ and

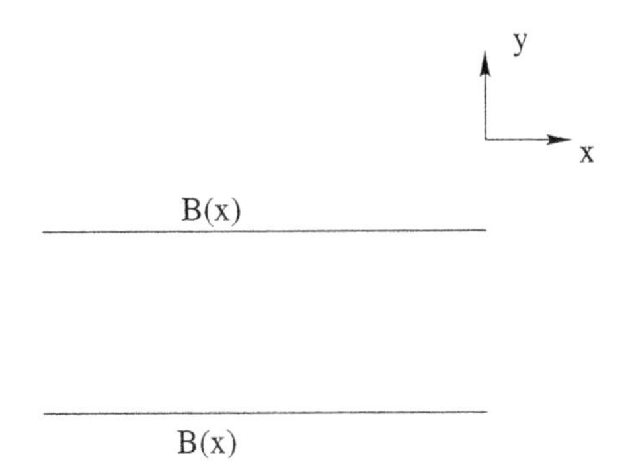

Figure 14.12. The boundary conditions on the incoming surfaces

ϕ_c increases by $d\psi_r$. The functions for the volumes per flux $V(\psi)$, $V(\phi)$ remain the same except for $V(\phi_c)$, which is $2V(\psi_c)$. Again we can invoke equations 13 and 14 to derive a new close by equilibrium solution for p, B outside of the reconnection and separatrix layers.

In fact we can see that there is a unique continuous sequence of equilibria with ψ_r as the parameter characterizing them. ψ_r will vary from the initial amount of private flux down to 0, when all the private flux has been reconnected.

These equilibria all have reconnection layers and separatrix layers and the length of the reconnection layer and the positions of the reconnection and separatrix layers are fixed in each equilibrium of the sequence. Further, the field strength B along each side of these layers is also fixed (see figure 14.12). If $d\psi_r/dt$ is known, then along the reconnection layer we also know

$$v_y = \frac{d\psi_r/dt}{B_x} \tag{17}$$

where we introduce Cartesian coordinates with the origin at the center of the reconnection layer, and x measured along the layer and y measured perpendicular to it. As reconnection proceeds, the unreconnected lines leave the private regions, and the pieces along the reconnection layer merge and are destroyed or, more correctly, are pulled out of the layer, as in figure 14.11, and the rest of the line crosses the separatrix smoothly.

The only unknown quantity is $-d\psi_r/dt$ as a function of ψ_r, $F(\psi_r)$, namely the rate of reconnection of flux. Once $F(\psi_r)$ is known we can summarize the entire reconnection process by the equation

$$\frac{d\psi_r}{dt} = -F(\psi_r) \tag{18}$$

and find the entire time of reconnection from the solution of this equation. This is the proper analytic way to describe the Sweet–Parker reconnection model.

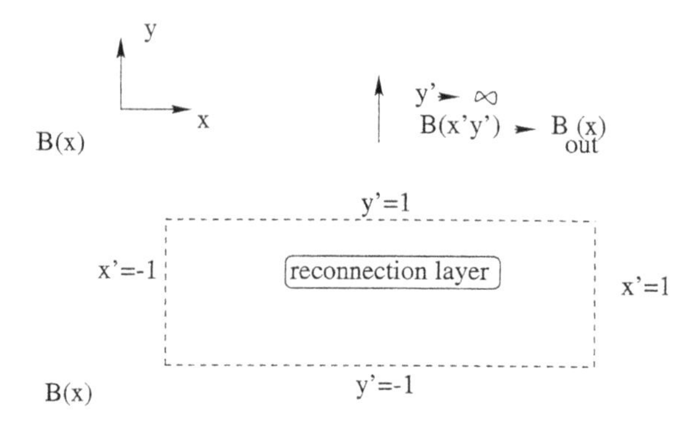

Figure 14.13. The boundary condition on the outgoing surfaces

Before proceeding to the analysis of the reconnection layer, let us consider why the equilibrium is uniquely determined by ψ_r, ϕ_r, $V(\psi)$, and $V(\phi_r)$. Consider the following thought experiment: Start the plasma out of equilibrium with a configuration topologically similar to figure 14.9, with the volume in each layer $d\psi$ given by $V'(\psi)d\psi$ and in $d\phi$ by $V(\phi)d\phi$. Since the situation is out of equilibrium the plasma will start to move. Introduce some friction with some fixed background of artificial material. This will gradually damp the motion. Make sure that the plasma is infinitely conducting and incompressible so that $V(\psi)$ and $V(\phi)$ are preserved. If necessary, separate the three regions by some flexible sheets so that no reconnection occurs during the process. Then the plasma will settle down into a unique equilibrium. In fact, the equilibrium will be the minimum of the magnetic energy over all states with the given V's. It can be shown directly that this is so and that such a minimum is an equilibrium state (see problem 5 in chapter 4). This is actually a special case of a much more general theorem about magnetostatic equilibria.

How do we find $F(\psi_r)$, the reconnection rate? The answer depends on the physics that goes on inside the layer, where flows and resistivity are important. But what we know about the equilibrium, the position of the layers, and the conditions just outside of the reconnection layer $B_x(x)$ on either side and the relative value of v_y are just the conditions for a boundary layer analysis (see figure 14.13). The Sweet–Parker analysis is actually a zero-order approximation to such an analysis.

Let us scale the physical quantities in the layer to conform to the Sweet–Parker estimates. First, set B_0 equal to the value of B_x above the layer at $x = 0$, $y = 0$ and let L be the length of the layer. Then rescale all the quantities by setting

$$x = x'L$$

$$y = y'\delta = y'\sqrt{\frac{L\eta c/4\pi}{v_A}}$$

$$B_x = B'_x B_0$$

$$B_y = B'_y \frac{\delta}{L} B_0$$

$$v_x = v'_x v_A$$

$$v_y = v'_y \frac{\delta}{L} v_A \tag{19}$$

where $v_A = B_0\sqrt{4\pi\rho}$. As $y' \to \infty$ the inner solution should go over smoothly to the outer solution $y > 0$.

We now substitute these new scaled variables into the resistive MHD equations. With Ohm's law instead of the magnetic differential equation and in x, y coordinates, these are

$$E_z + \frac{v_x B_y - v_y B_x}{c} = \eta j = \frac{\eta}{4\pi}\left(\frac{\partial B_y}{\partial x} - \frac{\partial B_x}{\partial y}\right) \tag{20}$$

$$\frac{\partial B_x}{\partial t} = -c\frac{\partial E_z}{\partial y} \tag{21}$$

$$\frac{\partial B_y}{\partial t} = c\frac{\partial E_z}{\partial x} \tag{22}$$

$$\rho\left[\frac{\partial v_x}{\partial t} + (\mathbf{v}\cdot\nabla)v_x\right] = -\frac{\partial}{\partial x}P + \frac{1}{4\pi}(\mathbf{B}\cdot\nabla)B_x \tag{23}$$

$$\rho\frac{\partial v_y}{\partial t} + (\mathbf{v}\cdot\nabla)v_y = -\frac{\partial}{\partial y}P + \frac{1}{4\pi}(\mathbf{B}\cdot\nabla)B_y \tag{24}$$

In the end we want the solution with no time derivatives. However, we still keep the time dependence in order to see how fast the layer will relax to a steady state, and also to facilitate the numerical solution of the equations. $P = p + B^2/8\pi$ is the total pressure, magnetic plus plasma pressure. The electric field E_z at the origin is the rate of evolution of the private flux or the public flux, as can be seen by integrating the induction equation in the global region. We therefore scale it according to the Sweet–Parker theory:

$$E_z = \frac{v_A B_0}{c\sqrt{S}} E' \tag{25}$$

where S is the dimensionless number characterizing the ratio of the dynamo term to the resistive term in Ohm's law

$$S = \frac{v_A L}{\eta c/4\pi} \tag{26}$$

The dynamics in the layer evolves on the Alfven timescale, so we take

$$t = \frac{L}{v_A} t' \tag{27}$$

The rescaled equations in the reconnection layer in terms of the rescaled variables are

$$E' = v'_x B'_y - v'_y B'_x = -\frac{\partial B'_x}{\partial y'} \tag{28}$$

$$\frac{\partial B'_x}{\partial t'} = -\frac{\partial E'}{\partial y'} \tag{29}$$

$$\frac{\partial B'_y}{\partial t'} = \frac{\partial E'}{\partial x'} \tag{30}$$

$$\frac{\partial v'_x}{\partial t'} + \left(\mathbf{v}' \cdot \nabla' v'_x\right) = -\frac{\partial P'}{\partial x'} + \mathbf{B}' \cdot \nabla' B'_x \tag{31}$$

$$0 = -\frac{\partial P'}{\partial y'} \tag{32}$$

$$\frac{\partial v'_x}{\partial x'} + \frac{\partial v'_y}{\partial y'} = 0 \tag{33}$$

where

$$P = \frac{B_0^2}{8\pi} P' \tag{34}$$

$$\mathbf{v}' \cdot \nabla' = v'_x \frac{\partial}{\partial x'} + v'_y \frac{\partial}{\partial y'} \tag{35}$$

$$\mathbf{B}' \cdot \nabla' = B'_x \frac{\partial}{\partial x'} + B'_y \frac{\partial}{\partial y'} \tag{36}$$

Several terms have been dropped because they are of higher order. These are the $\partial B'_y/\partial x'$ term in equation 28 and all the terms in equation 32 except $-\partial P'/\partial y'$.

The total pressure gradient perpendicular to the reconnection layer clearly must dominate over any other force. The result of this is that P' is a function of x' only and is determined by the known total pressure just outside of the layer

$$\frac{B_0^2}{8\pi} P' = \frac{B_{\rm out}^2(x)}{8\pi} + p_{\rm out}(x) \tag{37}$$

But $p'_{\rm out}(x')$ is constant along B, so we have in equation 32

$$\frac{\partial P'}{\partial x'} = \frac{\partial}{\partial x'} \frac{B_{\rm out}^2}{2} \tag{38}$$

and P' can be regarded as known.

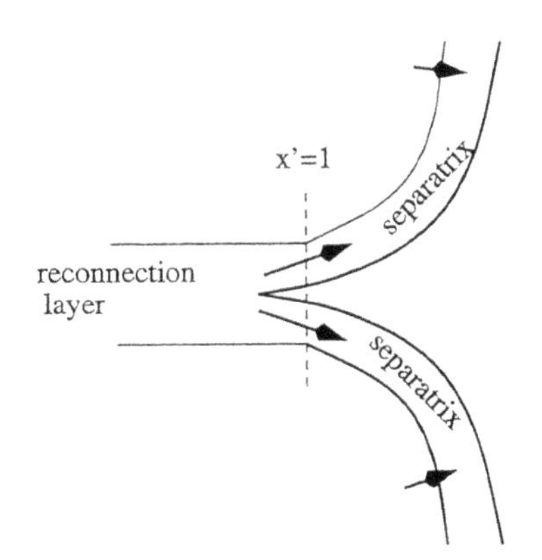

Figure 14.14. Plasma leaving the reconnection region and entering the separatrix region

We have a scaled set of equations that are independent of η and analytically represent the Sweet–Parker solution for any η. We must solve them in the region $-1 < x' < 1$. As y' becomes large they must go over to the outside region, and this gives us the correct boundary conditions from the outside solution, as indicated in figure 14.13. There are 6 equations, equations 28–33, for the six unknowns, E', B'_x, B'_y, v'_x, v'_y, P'. P' can be found as above, so equation 32 can be dropped. As y' goes to ∞ we must have $B'_x \to B^{\text{out}}(x)$ above, and as y' goes to minus infinity $B'_x \to B^{\text{out}}(-x) = -B^{\text{out}}(x)$ below. Also, $v'_x \to 0$ as $y' \to \pm\infty$. since there are no finite flows in the outside region.

The boundary conditions at $x' = \pm 1$ are more subtle. As the flow reaches the end of the reconnection region it must pass into the separatrix region (see figure 14.14). But plasma is already directly crossing the separatrix region from the private to the public region, so that the plasma flowing into it from the reconnection region crashes into this other plasma and accelerates it along the separatrix region. We might expect this to produce a backpressure that would slow down the outflow v'_x from the reconnection region to below the Alfven speed predicted by Sweet and Parker.

However, from a similar scaled set of equations in the separatrix region we can show that the flow is supersonic, so no backpressure is produced until the flow reaches the top of the separatrix (point A in figure 14.10). By this time the tube on which this flow is taking place has pulled out of the reconnection region and so the backpressure acts only on the plasma in the public region, not the plasma in the reconnection region. As a consequence, the downstream boundary conditions at $x' = \pm 1$ are unaffected by backflow and are simply $\partial v'_x/\partial x' = 0$. The boundary condition on the B'_x field is obtained by convecting it inward by the v'_y velocity, at $x' = 1$, neglecting resistivity.

This completes the boundary conditions for the reconnection region problem and we can solve it once we know B_x^{out} along the $y = 0$ line. This comes from the equilibrium solution. (Once the line has entered the reconnected region, the flows on it are rapidly damped by parallel viscosity, and the kinetic energy of the flows is converted to thermal energy. However, we may ignore this latter process because the plasma is incompressible.)

Once we are convinced a solution exists, which has been verified numerically for a several reasonable choices for $B_x^{\text{out}}(x')$, we expect to find E' to be of order unity, since there are no parameters in the scaled problem. This is actually the case. In fact, $E' = 1.07$ for one example. Thus, the Sweet–Parker reconnection result is substantiated and we may conclude that, based on the standard resistive MHD equations, Sweet and Parker have the correct order of magnitude for the rate of magnetic reconnection (and, of course, the correct scaling with η).

We have discussed the reconnection problem for a specific model, merging cylinders, so as to have a well-defined problem to treat with physical precision. With a little imagination, we can see that this model has similar characteristics to most astrophysical situations where reconnection occurs. In particular, it is applicable to reconnection in our two examples.

We now have

$$v_R = \frac{v_A}{\sqrt{S}}; \qquad S = \frac{v_A L}{\eta c / 4\pi} \tag{39}$$

as a valid estimate for reconnection rates, where L and v_A are typical values for any given simulation. The required time to reconnect flux in a region of size L is of order

$$\frac{L}{v_R} = \frac{L\sqrt{S}}{v_A} \tag{40}$$

a time longer than the Alfven time by the factor $\sqrt{S}$. Because S is generally large, this results in a very slow rate of reconnection. Basically, this is due to the thin layer through which the plasma on the reconnected line has to escape (see figure 14.15).

14.4 Comparison of the Sweet–Parker Model with Observations

The Sweet–Parker model leads to a reconnection time much shorter than the resistive decay time for the macroscale L, $T_{\text{decay}} = L^2/(\eta c/4\pi)$, but much longer than the Alfven time $\tau_A = L/v_A$. In fact,

$$\tau_A = \frac{L}{v_A} \ll \tau_{\text{SP}} = \frac{L\sqrt{S}}{v_A} \ll T_{\text{decay}} = \frac{LS}{v_A} \tag{41}$$

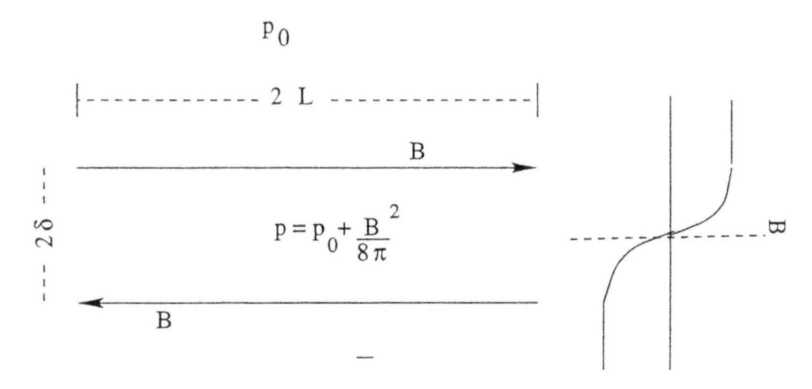

Figure 14.15. Because the reconnection channel is so thin the reconnection is very slow

Let us consider some possible parameters for a solar flare:

$$B = 300 \text{ G}$$

$$\rho = 10^{-15} \text{ g/cm}^3$$

$$v_A = 2.7 \times 10^7 \text{ cm/sec}$$

$$T_e = 100 \text{ eV}$$

$$\frac{\eta c}{4\pi} = 10^4 \text{ cm}^2\text{/sec}$$

$$L = 10^4 \text{ km}$$

$$S = \frac{L v_A}{\eta c/4\pi} = 2.7 \times 10^{12} \tag{42}$$

Then,

$$\tau_A = 37 \text{ sec} : \tau_{SP} = 6 \times 10^7 \text{ sec } (= 2 \text{ years})$$

$$t_{\text{decay}} = 10^{14} \text{ sec } (= 3 \times 10^6) \text{ years} \tag{43}$$

We see that there is an enormous difference between these timescales. The Sweet–Parker theory reduces the reconnection time from a few million years to a few months. However, the observed energy release time for a solar flare ranges from 15 min $\approx 10^3$ sec to several hours $\approx 10^4$ sec. Assuming that this is the same as the reconnection time, we see that there is still a very large discrepancy between the presumed reconnection time from observations and that predicted by Sweet–Parker, which is still a factor 1000 times too long.

Consider our other reconnection example of the solar wind impinging on the earth's magnetosphere. Here, for illustration, we choose the parameters to be $B = 10^{-4}$ G, $n = 10$ cm^{-3}, $T = 10^2$ eV, $L = 10^3$ km, and $\eta c/4\pi =$

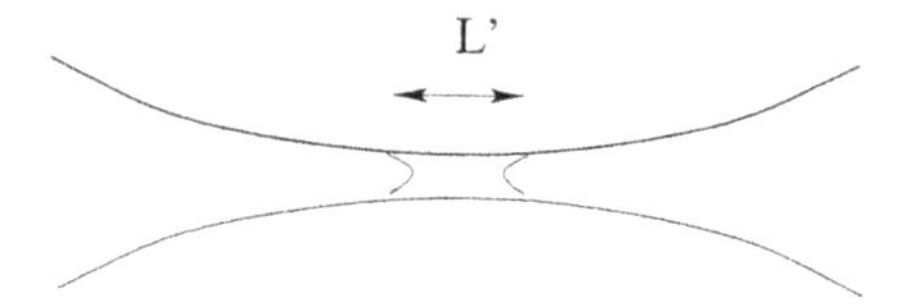

Figure 14.16. Reconnection of two field lines need only be carried out along a short length of the lines

10^7 cm^2/sec. Then $v_A = 7 \times 10^6$ cm/sec, $S = 7 \times 10^7$, and

$$v_R = \frac{v_A}{\sqrt{S}} \approx 10^3 \text{ cm/sec} \tag{44}$$

The incoming velocity behind the shock is $v_0 \approx 100$ km/sec, so only $\approx v_D/v_0 \approx 10^{-4}$ of the lines should reconnect, the other lines flowing without reconnection around the east and west sides of the earth. The fraction of lines that reconnect (obtained from counting the the number of lines crossing the pole) is inferred to be from 5 to 10 percent. In both reconnection examples the predicted and "observed" fractions differ by a factor of a thousand.

14.5 Petschek's Model for Magnetic Reconnection

Clearly, something is wrong. In 1963 Petschek (1964) proposed a possible solution. He carefully analyzed the Sweet–Parker model to see why the reconnection rate is so terribly slow. First, he noted that the Sweet–Parker model requires a very thin layer δ such that the field diffuses as fast as the incoming flow.

Then he observed that for the reconnection to proceed all the incoming mass must flow through this narrow channel of thickness δ, so that more fresh matter on unreconnected can come in. Petschek further made the very important observation that breaking and reconnecting lines was a topological process, and it is only necessary to break and reconnect the lines near one point, as in figure 14.16. The resistivity along the rest of the line is not really necessary for reconnection to proceed. Therefore, he reasoned that the channel has to be so narrow only near the region of reconnection. He suggested that the narrow channel could have a much shorter of length, L'. With this change, then, we get a faster rate of reconnection:

$$v_R = \sqrt{\frac{L}{L'}} v_{\text{SP}} \tag{45}$$

This is not the whole story. The plasma coming in along the length $2L$ but outside the length $2L'$ must still be removed at the velocity v_A to allow the fresh plasma on the next flux line to flow in. But Petschek argued this could be done by the magnetic tension of the reconnected line rather than by a pressure drop, which does not exist beyond L'.

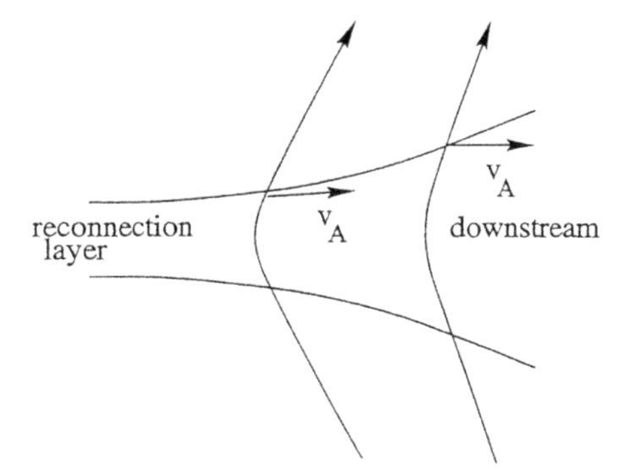

Figure 14.17. The formation of shocks in the down stream region

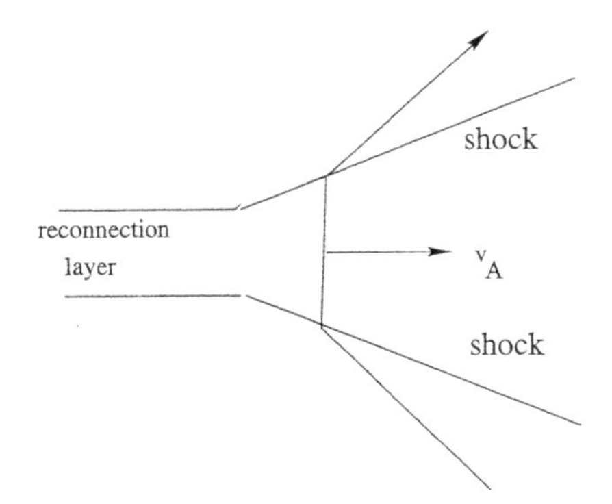

Figure 14.18. Shocks in downstream region

In this case the tension acts as a wave that steepens into a slow shock, as in figures 14.17 and 14.18. From the Rankine–Hugoniot relations for a shock of chapter 6, we have

$$v_R B_x = v_x B_y \qquad \text{(conservation of flux)} \tag{46}$$

and

$$\frac{B_x B_y}{4\pi} = \rho v_x v_R \qquad \text{(conservation of tangential momentum)} \tag{47}$$

The second relation is easy to derive. The horizontal momentum gained by a gram of plasma is $j_z B_y \Delta / v_R$, where Δ is the thickness of the shock. But $4\pi j_z \Delta = B_x$. Dividing equations 46 and 47 we find

$$v_R^2 = \frac{B_y^2}{4\pi\rho} \tag{48}$$

so v_R is the Alfven speed based on the B_y field. Multiplying these equations we get

$$v_x^2 = \frac{B_0^2}{4\pi\rho} \tag{49}$$

Therefore, the downstream speed of the plasma passing through the shock is just v_A as required.

Thus, from the shock conditions we find

$$v_R = \frac{B_y}{\sqrt{4\pi\rho}} \tag{50}$$

But v_R is also given by equation 45. These results must agree. Do they? Indeed, they do. To see this, assume that $L' \ll L$. The pressure gradient is determined by matching to outside conditions and still has the scale L, so the pressure drop is $(L'/L)^2 B_0^2/8\pi$ and cannot accelerate the plasma to v_A. This acceleration must be done by the tension of the field lines in the dissipative region.

Integrate the horizontal component of the equation of motion in y over δ:

$$\frac{\rho\delta}{2}\frac{dv_x^2}{dx} = -\delta j_z B_y = \frac{B_y B_0}{4\pi} \tag{51}$$

But by conservation of mass $v_x = v_R x/\delta$, $x < L'$. Substituting this into equation 51 we find

$$B_y = \frac{\rho\delta 4\pi}{2B_0}\frac{d}{dx}\frac{v_R^2 x^2}{\delta^2} = \frac{4\pi\rho}{B_0}\frac{x v_R^2}{\delta}$$

Thus, B_y increases linearly with x and at $x = L'$:

$$\frac{B_y B_0}{4\pi\rho} = \frac{L' v_R^2}{\delta} = v_R v_x' \tag{52}$$

Now, if $v_x' = v_x = v_A$, which it must, then the two results for v_R, equations 50 and 52, agree. From the second equality in equation 52 we have

$$v_R = \frac{\delta}{L'} v_A \tag{53}$$

and combining this with the condition $v_R = \eta c/4\pi\delta$ we get equation 45.

In Petschek's model, L' is a free parameter and for any choice of it we have a value for the reconnection rate, and this rate increases with decreasing L'. By choosing L' as small as we please, we can get any reconnection rate we please. This is clearly absurd. Petschek pointed out that there is a limit on how small L' can be. He showed that the current densities in the shocks are finite and since they terminate a distance L' from the origin, they produce a large magnetic field upstream proportional to $|\ln L'|$ that tends to cancel the incoming zero-order B_0. Petschek argued that the maximum that this perturbed field could be is $B_0/2$. Beyond this the field would be weakened to zero and the reconnection rate would be reduced to zero.

From these considerations he estimates a maximum value for v_R,

$$v_R = \frac{\pi v_A}{8 \ln(v_R L/v_A L')} = \frac{\pi}{8} \frac{v_A}{\ln\left[(v_R/v_A)^3 S\right]} \tag{54}$$

where in the second expression we use equation 45 for L/L' in the argument of the logarithm. Since $\ln S \gg 3 \ln v_A/v_R \approx 3 \log\log S$, we usually drop the v_R/v_A factor in the log and write simply

$$v_R(\max) \approx \frac{\pi}{8} \frac{v_A}{\ln S} \tag{55}$$

Petschek argued that the reconnection process would adapt itself to the maximum rate. This result has been widely used, since it produces results in much better agreement with observations than the Sweet–Parker model. For example, for our choice of parameters for the solar flare we get

$$\tau_{\text{Petschek}} = \frac{8\tau_A}{\pi} \ln S = 73\tau_a = 30 \text{ sec}$$

which is considerably shorter even than the observed value. Notice, though, that $L' = L(\tau_{\text{Petschek}}/\tau_{SP})^2 = (5 \times 10^{-5})^2 L = 2.5$ cm, which is extremely short. For the example of the solar wind impinging on the magnetosphere, we have for Petschek's reconnection velocity,

$$v_R = 1.5 \times 10^5 \text{ cm/sec} \tag{56}$$

corresponding to 2 percent of the solar wind lines being reconnected. These results were much more comfortable for the observers and made Petschek's result very popular.

However, there are doubts concerning Petschek's model. Petschek's model was actually a boundary layer model that should fit into the Uzdensky formalism. However, shocks do not emerge in Uzdensky's boundary layer analysis, as they should if the Petschek theory were correct.

More importantly, Petschek did not explain where his B_y field came from. He only derived the value it must have to make his model work. It is clear that the Alfven downstream velocity takes the B_y lines away at a rapid rate. What resupplies them? The only possible way to regenerate the B_y field is by nonuniform merging. It is clear that if B_x^{out} were a constant in x, then the fields would merge in such a way that there would be no B_y component to the field in the layer.

Uzdensky attempted to check the Petschek theory by starting with an initial configuration that simulated Petschek's model. He included current layers that were equivalent to the Petschek shocks, and included the transverse B_y field that supported the shock configuration. He chose this field to increase rapidly in x over a distance L'. He chose a value for L' equal to one-fourth of L.

He found that initially the reconnection rate was twice the Sweet–Parker rate, consistent with equation 45. However, the solution was not in steady state. In less than an Alfven time, L/v_A, the rate drops back to the Sweet–Parker rate. Further, he tracked the value of L' with time and found that it increased from $L/4$ to $\approx L$ in less than one Alfven time L/v_A.

Let us estimate the length L' must be, so that the regeneration that arises from nonuniform merging balances the loss of B_y flux by downstream flow at the Alfven speed. First, how fast does B_y develop? As various magnetic lines of force move into the reconnection layer from outside, they will turn to adapt to the pattern of lines seen in figure 14.11. The typical y velocity of the line is $v_R = (\eta c/4\pi)/\delta$. But this is the average velocity. The end nearest $x = 0$ moves faster. We can get an upper limit for how much faster by assuming that the velocity is proportional to the nonuniformity of the field strength $B_x^{\text{out}}(x)$. But this nonuniformity in B_x is set by the macroscopic scale, and near $x = 0$ is

$$B_x = B_0 \left(1 - \frac{x^2}{L^2}\right) \tag{57}$$

If the diffusive region is of length L', then at the two ends of the line B will change by a fraction L'^2/L^2 of itself. This change will induce a change in the y velocity of the two ends of the line by the same fraction, L'^2/L^2, of the mean merging velocity. Thus, an estimate of the upper limit to the rate of regeneration of the B_y flux Φ_y in the region $x < L'$ is

$$\begin{aligned}\left(\frac{d\Phi_y}{dt}\right)_{\text{reg}} &= \frac{v_y(0) - v_y(L')}{L'} B_0 L' \\ &= \frac{L'^2}{L^2} v_R B_0 \end{aligned} \tag{58}$$

In a time L'/v_A all the B_y flux $\Phi_y = L' B_y$ is swept downstream, so we get

$$\left(\frac{d\Phi_y}{dt}\right)_{\text{lost}} = L' B_y \frac{v_A}{L'} = v_A B_y \tag{59}$$

Equating these two rates we get a balance if

$$\left(\frac{d\Phi_y}{dt}\right)_{\text{reg}} - \left(\frac{d\Phi_y}{dt}\right)_{\text{lost}} = \frac{L'^2}{L^2} v_R B_0 - v_A B_y = 0 \tag{60}$$

For this we require

$$B_y = \frac{L'^2}{L^2} \frac{v_R}{v_A} B_0 \tag{61}$$

But

$$\frac{B_y}{B_0} = \frac{v_R}{v_A} \tag{62}$$

so equation 61 reduces to

$$1 = \frac{L'^2}{L^2} \tag{63}$$

and L' is equal to the Sweet–Parker length L. Thus, the balancing of B_y flux in the proper way determines the value of L' in Petschek's theory, and it leads right back to the Sweet–Parker rate. This derivation of L' is more convincing than Petschek's choice of the maximum rate, and is more compelling, since it yields a unique value for L' and explains how the B_y flux is created.

However, if η were larger at the origin than at L' by $\Delta\eta$, then the regeneration rate would be increased. This could make a more important contribution to the differential merging velocity than that due to the variation in B_x, which is small, if $L' \ll L$. Let

$$\eta = \eta_0 \left(1 - \frac{x^2}{L_\eta^2}\right) \tag{64}$$

where L_η is the scale size for η. Then the L'^2/L^2 factor in equations 58, 60, and 61 is replaced by

$$\xi = \frac{L'^2}{L_\eta^2} \tag{65}$$

so from equation 62 the L'^2/L^2 factor in equation 63 is also replaced by ξ and as a consequence $L' = L_\eta$. For inhomogeneous resistivity the Petschek reconnection rate is $v_{\mathrm{Petschek}} = \sqrt{L/L_\eta}\, v_{\mathrm{SP}}$.

However, to get a reconnection rate much faster than Sweet–Parker L_η would have to be very small. We would not expect this to happen in any natural fashion, since there is no reason to expect background resistivity to vary rapidly near any particular point. (On the other hand, when we discuss the role of anomalous resistivity, we will show that it actually can happen, in a very natural way, that L_η can be very small.)

14.6 Non-MHD Reconnection

Up to this point we have based our discussion of magnetic reconnection solely on the basis of resistive MHD physics. Recall that resistive MHD is a macroscopic theory valid only when the scale lengths are sufficiently large. In actual fact, this is not true in either of the two examples of magnetic reconnection that we have considered.

Consider our first example, that of a solar flare. The Sweet–Parker thickness for our choice of parameters is

$$\frac{L}{\sqrt{S}} = \frac{10^9 \text{ cm}}{\sqrt{2.7 \times 10^{12}}} = 600 \text{ cm} \tag{66}$$

The current density in the Sweet–Parker layer is given by

$$j = ne\frac{v_i - v_e}{c} = \frac{B}{4\pi\delta} \approx \frac{300}{4\pi \times 600} = 0.04 \text{ abamps/cm}^2 \tag{67}$$

The relative drift velocity of the electrons and ions is

$$v_i - v_e = \frac{cj}{ne} = 2.5 \times 10^9 \text{ cm/sec} \tag{68}$$

much larger than the ion sound speed of 10^7 cm/sec (if $T = 10^2$ eV). Such a relative drift would certainly drive an ion acoustic instability (see equation 74 in chapter 10).

Further, in MHD physics the assumption is made that the ions and electrons flow at the same velocity in the plane of the reconnection. But if $v_i - v_e = c_s \approx 10^7$ cm/sec, we have a current density of

$$j = \frac{ne(v_i - v_e)}{c} = 1.6 \times 10^{-4} \text{ abamps/cm}^2 \tag{69}$$

If such a current flowed through the Sweet–Parker layer, it would produce a field of only

$$\Delta B = 4\pi j\delta = 1.2 \text{ G} \tag{70}$$

a negligible field compared to B.

Usually, if MHD physics applies, and if such a current, due to a difference of electron and ion flow velocities comparable to c_s, flows through a layer δ, then the perturbed field ΔB would be much larger than the existing field B, and this would force the relative drift to be much smaller (see equation 70 in chapter 8). In our example, we have $\Delta B \sim B$ and the ions and electrons flow with very different velocities the difference being large compared to the ion sound speed, so a two-fluid theory is required.

We see from our solar flare example that MHD theory can fail for the Sweet–Parker application in two ways: First, the relative drift can be large enough to drive microinstabilities, which will make the resistivity much larger. Second, the electrons need not flow at the same velocity as the ions. These failures indicate that the physics in the reconnection process can involve much more than ordinary resistive MHD.

These two conditions for the validity of MHD are identical and reduce to a comparison between the ion skin depth c/ω_{pi} and the thickness of the layer. The condition that resistive MHD applies to the Sweet–Parker model is

$$\delta_{\text{SP}} > 2\frac{c}{\omega_{pi}\sqrt{\beta}} \tag{71}$$

where $\beta = 8\pi p/B_x^2$ is the ratio of plasma pressure in the layer to the magnetic pressure of the reconnecting field measured outside of the layer. In fact, $v_D = \Delta v = v_i - v_e$ is determined by the current in the z direction due to the B_x field. Then,

$$\frac{ne(\Delta v)}{c} = j_z = \frac{B_x}{4\pi\delta} = \frac{\sqrt{8\pi p}}{4\pi\delta\sqrt{\beta}} = \frac{\sqrt{16\pi n M v_i^2}}{\sqrt{\beta}4\pi\delta} \tag{72}$$

where

$$\beta = \frac{8\pi p}{B_x^2} \tag{73}$$

and we have

$$\frac{\Delta v_z}{v_i} = 2\frac{c}{\omega_{pi}\sqrt{\beta}\delta} \tag{74}$$

If the B_z field is zero, then $\beta = 1$. If the B_z field is strong, then its variation between outside and inside the layer takes most of the pressure balance and p is about the same inside and out and β can be determined by outside conditions. (Since our criteria is only an approximate one, it is not very important where p is measured.)

Now, if we want the Sweet–Parker model and resistive MHD to be valid, we must have $\Delta v \ll v_i$ so that the electrons and ions flow with the same velocity, their difference being small compared to v_i. The condition for this is equation 71.

For the solar flare example, with the parameters of equation 42, we find that $c/\omega_{pi}\sqrt{\beta} \approx 7 \times 10^4$ cm, since $\beta \approx 9 \times 10^{-5}$. Since from equation 66 the Sweet–Parker thickness, 600 cm, is considerably thinner than this, we can conclude that the Sweet–Parker model is not applicable and the discrepancy of the Sweet–Parker reconnection time from the observed reconnection time is not really so surprising.

We can also test this condition for the magnetospheric example. Here β is unity, since there is a negligible B_z field. The ion skin depth is 6.3×10^6 cm, while the Sweet–Parker thickness $\delta_{SP} = 10^8/\sqrt{S} = 1.1 \times 10^4$ cm, so again the Sweet–Parker model is not applicable. We see that resistive MHD fails in both of the two examples in which the Sweet–Parker reconnection rate differs so drastically from the observed ratio. New physics must be added.

What is the consequence of having $\Delta v \gg c_s$? Such a large drift velocity gives rise to intense microinstabilities, actually in the form of ion acoustic waves. These waves interact strongly with the electrons and the ions dragging them toward the wave phase velocity. The force due to the wave–particle interaction acts on both the electrons and ions, and is much larger than the Coulomb interaction, so the effective resistivity becomes much larger. This leads to an increased thickness of the current layer. The layer thickness δ will grow to a value δ_c at which the relative drift velocity $v_i - v_e$ drops to just above the critical drift velocity for instabilities. The slight excess above the critical velocity will be sufficient to maintain the waves that give the required diffusion velocity

$$v_R = \frac{\eta' c}{4\pi\delta_c} \tag{75}$$

where η' is the resistivity required to maintain this thicker layer (see equation 11). Since a small increase in current density is sufficient to produce whatever resistivity is required, this equation simply gets replaced by

$$\delta = \delta_c \tag{76}$$

The other Sweet–Parker equation, equation 8, related to mass conservation is still valid but now reads

$$v_R L = v_A \delta_c \tag{77}$$

so the reconnection velocity with anomalous resistivity becomes

$$v_R = \frac{\delta_c}{L} v_A \tag{78}$$

Since δ_c is larger than S_{SP}, the reconnection rate will be correspondingly larger.

For the moment, let us take $v_c \approx v_i$. Then we see from equation 74 that $\delta_c = c/\omega_{pi}\sqrt{\beta}$, so the reconnection rate can be written

$$v_R \approx \frac{c}{\omega_{pi}\sqrt{\beta}} \frac{v_A}{L} \tag{79}$$

and the reconnection time is

$$\tau \approx = \frac{L^2}{\delta_c v_A} = \frac{L^2\sqrt{\beta}}{c/\omega_{pi}} v_A \tag{80}$$

For our solar flare example these lead to

$$\tau = \frac{(10^9)^2}{4.7 \times 10^4 \times 2.7 \times 10^7} = 5 \times 10^5 \text{ sec} \tag{81}$$

which is much closer to the observed value but still a factor of 500 too long. For the magnetospheric case we have

$$v_R = \frac{6.3 \times 10^6}{10^8} v_A = 0.063 v_A = 4.4 \times 10^5 \text{ cm/sec} \tag{82}$$

so with anomalous resistivity, a percent of the incoming lines reconnects.

14.7 Anomalous Resistivity

In our discussion of the onset of the ion acoustic instability we have tentatively taken the critical velocity as $v_D = v_i$, the ion thermal speed. Let us examine this instability in the framework of reconnection. We must distinguish two distinct cases. Take the y–z plane as the plane of reconnection. $\hat{\mathbf{z}}$ is the direction of the reversed fields, and $\hat{\mathbf{y}}$ the direction of the incoming flows. The current $\mathbf{j}$ and the electric field $\mathbf{E}$ are in the $\hat{\mathbf{x}}$ direction (see figure 14.19). We will discuss the simple case in which $B_x = 0$ first. The case where $B_x \neq 0$ is briefly discussed later.

The simpler case, $B_x = 0$, is the case relevant to magnetospheric reconnection both on the front side of the earth and in its tail. In this case $\mathbf{j}$ and $\mathbf{v}_D$ are

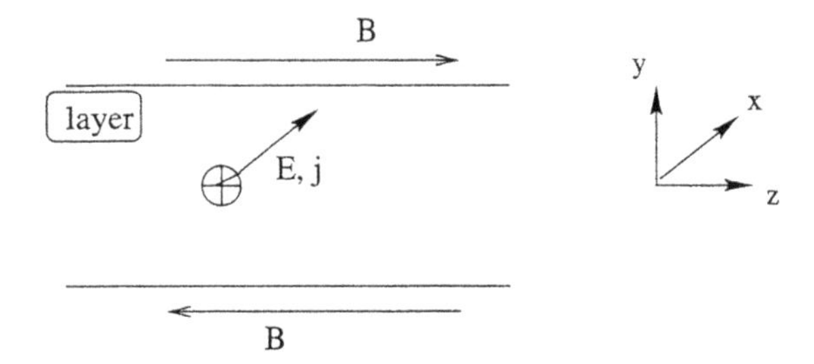

Figure 14.19. The electric field and the current in the x direction

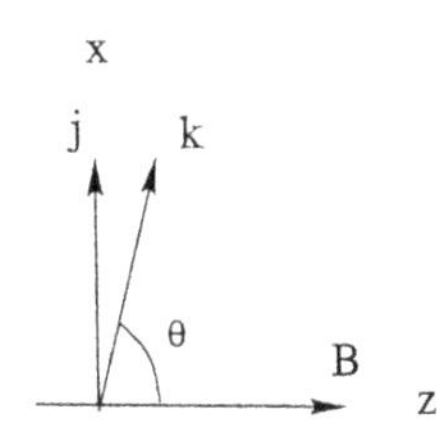

Figure 14.20. The directions of j, k, and B. B is along the z axis.

both in the x direction, and this can make ion acoustic waves propagating nearly perpendicular to the B_x field unstable. It is shown in problem 4 in chapter 10 that above the ion cyclotron frequency, the ion acoustic wave still propagates as an ordinary sound wave. (This is because the electrons are magnetized and respond only to the E_z field.)

First, take $v_D = 0$. Then, for waves propagating near the x axis, the real part of the ion contribution to dielectric tensor ϵ is

$$\epsilon^i_{1xx} = 1 - \frac{\omega^2_{pi}}{\omega^2} \tag{83}$$

as we saw in chapter 9. Thus, if **E** is in the direction of **k** making an angle θ near to 90°, the **B** field, as in figure 14.20, then the real part of the electrostatic equation becomes

$$k\epsilon^i_{1xx} E_x + k_z \epsilon^e_{1zz} E_z = |E|(k\epsilon^i + k\cos^2\theta\epsilon^e_{zz})$$

$$= |E|k\left[\left(1 - \frac{\omega^2_{pi}}{\omega^2}\right) + \frac{\cos^2\theta}{k_z^2\lambda_d^2}\right] = 0 \tag{84}$$

or

$$\omega^2 = \frac{\omega^2_{pi}}{1 + 1/k^2\lambda_D^2} \approx c_s^2 \tag{85}$$

the dispersion relation for a sound wave. The fact that the wave makes a large angle to B makes no difference to the phase velocity. We can see that the electrons adopt a Boltzmann distribution in the potential of the wave.

These results apply only under certain conditions. First, the phase velocity along **B** must not be much greater than $v_e = c_s\sqrt{M/m}$, so $\cos\theta$ must not be much less than $\sqrt{m/M}$. We take the propagation angle close to, but not too close to, 90°. Second, the perpendicular wavelength must be larger than ρ_e. This means that

$$k\rho_e = kc_s\frac{\rho_e}{c_s} = \omega\frac{v_e}{\Omega_e c_s} = \frac{\omega}{\Omega_e}\sqrt{\frac{M}{m}} < 1 \tag{86}$$

and

$$\omega < \Omega_e\sqrt{\frac{m}{M}} = \omega_{\text{LH}} \tag{87}$$

so that the frequency of the wave must be below the lower hybrid frequency.

Now, the most important such waves are the ones with highest frequencies and with $\cos\theta \sim \sqrt{m/M}$. Such waves have a parallel phase velocity of order

$$\frac{c_s}{\cos\theta} = \frac{c_s}{\sqrt{m/M}}\left(\frac{\sqrt{m/M}}{\cos\theta}\right) = v_e\frac{\sqrt{m/M}}{\cos\theta}$$

so when $\cos\theta$ ranges about $\sqrt{m/M}$, the phase velocity ranges over all the electron velocities. On the other hand, their perpendicular phase velocity is near the ion thermal velocity. Such waves can simultaneously be in resonance with the ions and electrons, and are therefore particularly effective in producing an anomalously large friction between them.

Now, let $v_D \neq 0$. Then these waves in the range $\cos\theta \sim \sqrt{m/M}$ are excited when the perpendicular drift velocity of the ions with respect to the electrons exceeds v_i (Davidson and Gladd 1975). This is the case even when $T_i \sim T_e$. The reason is that the resonant electrons are near the peak of the gradient of the electron distribution so that inverse electron Landau damping is comparable with the ion Landau damping.

To see this in more detail let us write the dispersion relation in the frame of the ions:

$$0 = \frac{4\pi e^2}{k^2 m}\int\frac{E_z\partial f/\partial v_z}{\omega - k_z v_z}d^3v + \frac{4\pi e^2}{k^2 M}\int\frac{E\partial F/\partial v}{\omega - kv}d^3v \tag{88}$$

where f is the electron distribution function and F is the ion distribution function. In this frame the electron distribution is

$$f(v) = \delta(v_x - v_D)\delta(v_y)\frac{e^{-mv_z^2/v_e^2}}{\sqrt{2\pi}\,v_e}$$

In the Vlasov equation we neglect the $E_x\partial f/\partial v_x$ term, since the electron are strongly anchored to the magnetic field and have no effective v_x motion.

This assumption is too simple to rigorously justify instability calculation since for a narrow current layer the magnetic field gradient is actually very large and the grad B drifts are important (Carter et al. 2003). However, the results from this assumption do illustrate the possibilities for generating anomalous resistivity by microinstabilities. Therefore, we will neglect these drifts and proceed with the calculation under the simplified assumption.

As in chapter 9 we find, after substitution for f and F, that equation 88 becomes

$$0 = -\epsilon_1 - i\epsilon_2$$

where

$$\epsilon_1 = -\frac{\omega_{pi}^2}{\omega^2} + \frac{1}{k^2\lambda_D^2}$$

$$\epsilon_2 = \left[\pi \left.\frac{\partial F}{\partial v}\right|_{\omega/k} + \pi\frac{M}{m}\left.\frac{\partial f}{\partial v_z}\right|_{(\omega - kv_D)/k_z}\right]\frac{4\pi e^2}{k^2 M} \tag{89}$$

To lowest order $\epsilon_1 = 0$, and we have

$$\omega \approx kc_s$$

where $c_s \approx v_i$ for $T_e = T_i$. To next order, setting $\omega = \omega_0 + i\gamma$, we have

$$i\gamma = \frac{-\epsilon_2}{\partial\epsilon_1/\partial\omega} = -i\frac{\omega^3}{\omega_{pi}^2}\epsilon_2$$

so the growth rate with $T_e = T_i$ is

$$\frac{\gamma}{\omega} = \pi c_s^2\left[\frac{v_D - c_s}{\mu' v_i^3}e^{-(v_D - c_s)^2/2\mu' v_i^2} - \frac{c_s}{v_i^3}e^{-c_s^2/2v_i^2}\right] \tag{90}$$

where $\mu' = \mu\sqrt{M/m} = (k_z/k)\sqrt{M/m}$ characterizes how close the propagation is to perpendicular propagation. To get a rough idea of the growth rate we may take the first exponential of order unity. The second exponential is then smaller, and we drop it. The critical drift velocity is of order c_s.

We see that because the resonant electrons are in the heart of the distribution where $\partial f/\partial v_z$ is large, the electron and ion Landau damping coefficients are comparable. Thus, a drift velocity of order v_i drives waves unstable. There is a critical drift velocity at which the growth rate is zero. For a drift velocity twice this value the growth is of order ω. Thus, we write

$$\gamma \approx \omega\frac{v_D - v_{Dc}}{v_{Dc}} = \omega\frac{j - j_c}{j_c} \tag{91}$$

The waves do not grow indefinitely. They are balanced by a nonlinear damping, similar to the nonlinear damping of Alfven waves (see Kadomtsev (1965) and equation 18 in chapter 14):

$$\gamma_{NL} = -\left(\frac{\mathcal{E}_{\text{waves}}}{nT}\omega\right) \tag{92}$$

Comparing equations 91 and 92 we see that

$$\left(\frac{\mathcal{E}_{\text{waves}}}{nT}\right) \approx \frac{j - j_c}{j_c} \tag{93}$$

(A closer examination of the nonlinear behavior shows that the bulk of the wave energy is at the largest frequency.) Thus, when $j = 2j_c$ all the thermal energy is in the form of waves.

What is the force on the electrons due to these waves? The growth rate of the waves consists of two parts: a part due to the electrons γ^e and a part due to the ions γ^i. The first is positive and the last negative, so

$$\gamma = |\gamma^e| - |\gamma^i| \tag{94}$$

If the electrons acted alone, they would amplify the waves, giving them momentum at the rate

$$-\frac{dP^e_{\text{waves}}}{dt} = 2\gamma^e \frac{\mathcal{E}_{\text{waves}}}{c_s} \tag{95}$$

The ions take momentum from the waves at the rate

$$\frac{dP^i_{\text{waves}}}{dt} = (2\gamma^i + \gamma_{\text{NL}}) \frac{\mathcal{E}_{\text{waves}}}{c_s}$$

or

$$\frac{dP^i_{\text{waves}}}{dt} = -\left|2\gamma^i + \gamma_{\text{NL}}\right| \frac{\mathcal{E}_{\text{waves}}}{c_s} \tag{96}$$

since both γ^i and γ_{NL} are negative.

In a steady state

$$\gamma^e + \gamma^i + \gamma_{\text{NL}} = 0 \tag{97}$$

so all the momentum given to the waves by the electrons is immediately given by the waves to the ions. There is effectively a direct transfer of momentum from the electrons to the ions in this steady state. Making use of equations 91 and 95 and the fact that γ^e is roughly a constant times ω, we have for the force on the electrons

$$F = 2\omega \frac{\mathcal{E}_{\text{wave}}}{c_s} = 2\frac{\omega}{c_s}(nT)\left(\frac{j - j_c}{j_c}\right) \tag{98}$$

The largest effect comes from waves with $\omega \approx \omega_{\text{LH}} = \sqrt{M/m}\Omega$. At $j \approx 2j_c$, $v_D = 2c_s$ we can find the effective collision frequency due to these waves from

$$F = mn\nu(v_D - c_s) = 2\frac{\omega_{\text{LH}}}{c_s} nT \tag{99}$$

or

$$\nu^* \approx 2\sqrt{\frac{M}{m}}\Omega_e$$

where we have set $\omega_{\text{LH}} = \sqrt{\Omega_e \Omega_i}$. This is an enormously large collision frequency. The anomalous resistivity is then of order

$$\frac{\eta^* c}{4\pi} = \frac{\nu^* c^2}{\omega_{\text{pe}}^2} = \sqrt{\frac{M}{m}} \frac{c^2 \Omega_e}{\omega_{\text{pe}}^2} \tag{100}$$

Thus, we expect j to remain very close to j_c and δ very close to δ_c, as discussed above.

14.8 Petschek Reconnection Revisited

A striking result that emerges from this formula for η^* is the very strong sensitivity of the anomalous resistivity to the current j. Let us write

$$\eta = \eta_S + \eta^* \frac{j - j_c}{j_c} \tag{101}$$

where η^* is given by equation 100. Such a large sensitivity is just what is needed to regenerate the B_y flux of Petschek (see equation 64). Equation 58 should now be replaced by

$$\begin{aligned}\left(\frac{d\Phi_y}{dt}\right)_{\text{reg}} &= \frac{c}{4\pi}\frac{\eta(0) - \eta(L')}{\delta_c L'} \\ &= \frac{\eta^* c}{4\pi\delta_c}\frac{j(0) - j(L')}{j_c}B_0 \\ &= \frac{\eta^* c}{4\pi\delta_c}[B(0) - B(L')] \\ &= \left(\frac{\eta^* c}{4\pi\delta_c}\right)\frac{L'^2}{L^2}B_0\end{aligned} \tag{102}$$

Setting this equal to the rate of loss of flux

$$\left(\frac{d\Phi_y}{dt}\right)_{\text{loss}} = v_A B_y \tag{103}$$

we find

$$B_y = \frac{\eta^* c}{4\pi\delta_c}\left(\frac{L'}{L}\right)^2 \frac{B_0}{v_A} \tag{104}$$

Now, this B_y supports shocks and we get from Petschek's theory the relation

$$v_R = \frac{B_y}{\sqrt{4\pi\rho}} = \frac{\eta^* c}{\delta_c}\frac{L'^2}{L^2} \tag{105}$$

But we have from mass conservation,

$$v_R L' = v_A \delta_c \tag{106}$$

so eliminating L' from these two equations yields

$$v_R = \sqrt[3]{\frac{\eta^* c \delta_c}{4\pi v_A L^2}}\, v_A = \sqrt[3]{\frac{\delta_c}{L}\frac{1}{S^*}}\, v_A \tag{107}$$

where $S^* = v_A L/(\eta^* c/4\pi)$ is an effective Lundquist number based on η^* instead of η (see Kulsrud 2001). This is a modified form of Petschek's formula for the reconnection velocity v_R based on anomalous resistivity theory rather than normal resistivity MHD.

Now, η^* is enormously larger than η. This is partially compensated for by the $\sqrt[3]{4\pi\delta_c/L}$ factor, which is fairly small. Thus, anomalous resistivity, because it is so sensitive to the current density j, can reestablish most of the Petschek theory. Notice that the normal η, which enters in a logarithm in Petschek's η_S, does not enter into the present formula for v_R. Further, L' is determined by our different condition, the balance of the loss and gain of the B_y field on which the shocks are supported, rather than Petschek's choice of the length L' that would lead to a maximum reconnection velocity. Finally, the logarithm that came from a different condition to determine L' is gone.

14.9 Which Is the Correct Reconnection Velocity?

The question arises under different conditions as to which of the various reconnection formulas we should use: (a) the Sweet–Parker velocity, equation 9; (b) the Sweet–Parker velocity with anomalous resistivity, equation 79; or (c) the modified Petschek velocity, equation 107. (We have justified the latter two formulas only for the simple case $B_z = 0$.) We can establish a simple qualitative criterion to decide which formula is applicable.

Let us take a given situation and calculate the Sweet–Parker thickness, $\delta_{SP} = L/\sqrt{S}$, and the ion skin depth, c/ω_{pi}. If $\delta_{SP} > c/\omega_{pi}$, then resistive MHD is applicable with the Sweet–Parker model and formula (a) is the correct one. On the other hand, if $\delta_{SP} < c/\omega_{pi}$, then resistive MHD breaks down, and either formula (b) or (c) applies. But for (c) to apply the resistivity at the center of the reconnection layer must be considerably larger than that at $x = L'$. It is easy to see that if $x = \delta_{SP}/(c/\omega_{pi})$ does not differ from one by an order of magnitude, then $\eta_{anom}/\eta_{spitzer} \approx 1/x$, and does not range by an order of magnitude over the reconnection layer and therefore does not have a large enough range to regenerate the cross-field and support the Petschek shocks. Then the applicable reconnection rate is given by the second formula. Finally, if $x \ll 1$, we expect the shocks to develop and then the third formula is the correct one.

Thus, in summary we have three cases:

1. If $1 < x$, we have Sweet–Parker reconnection:

$$v_R = \frac{v_A}{\sqrt{S}} \tag{108}$$

2. If $0.2 < x < 1$, we have normal anomalous resistivity reconnection:

$$v_R = v_A\left(\frac{c}{\omega_{pi}L}\right) \tag{109}$$

3. If $x < 0.2$, we have the new form of Petschek reconnection:

$$v_R = v_A\sqrt[3]{\frac{\delta_c}{LS^*}} \tag{110}$$

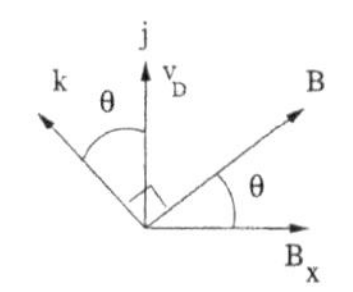

Figure 14.21. The case with $B_z \neq 0$

Normally, x is much larger than one in most regions of the interstellar medium, and Sweet–Parker gives the correct reconnection rate, but it is so slow that we can neglect it. However, occasionally there are situations where x is actually smaller than one and fast Petschek reconnection can occur.

We see that the reconnection rate depends sensitively on x. A moderate change of x can cause a transition from the slow reconnection rate 1 to the fast reconnection 3. This could conceivably trigger a rapid onset of fast reconnection and energy release such as is observed in a solar flare.

These results for reconnection rates are based directly on equation 101 for the dependence of η on j. But the rigorous establishment of this relation depends on finding the correct instability and this is currently being sought.

14.10 The Case When the Guide Field Is Nonzero

We have deferred discussion of a non-zero guide field B_z because it was easier to present the basic physics of non-MHD reconnection in the simpler $B_z = 0$ case. We can see that if $B_z \neq 0$ but is of order B_x, then since $\mathbf{k}$ must be nearly perpendicular to the total $\mathbf{B}$ it can no longer be nearly parallel to $\hat{\mathbf{z}}$ but must make an angle of at least θ to it, where θ is the angle between $\mathbf{B}$ and the direction of the reversing magnetic field, $\hat{\mathbf{x}}$ (see figure 14.21). Since j is still in the $\hat{\mathbf{z}}$ direction, the projection of $\mathbf{v}_D$ on $\mathbf{k}$ is reduced by $\cos\theta$. Thus, now the critical v_D must be larger than the critical velocity v_c for the simple case. That is, the critical drift velocity is $v_c/\cos\theta = v_c B/B_x$. When $B \gg B_x$ the critical value for v_D approaches the electron thermal velocity v_e, and even parallel propagating waves become unstable. However, this condition corresponds to a very large current density and is difficult to achieve. In short, it gets more and more difficult to trigger anomalous resistivity and faster reconnection when there is a substantial magnetic guide field perpendicular to the reversing field.

14.11 Hall Reconnection

The other correction to resistive MHD physics is the separation of the electron and ion flows when $\delta_{SP} < c\omega_{pi}$. This is called the Hall effect. It is clear that this separation should considerably relieve some of the difficulties in magnetic reconnection. One requirement of reconnection physics in the

MHD framework of Sweet–Parker is a very small layer thickness, δ, to allow lines to diffuse. But the diffusion of lines is controlled by electrons and not ions. On the other hand, the narrow layer does not allow enough flux of ions for them to get away from the incoming flux. The ions are forced to flow in this narrow channel because they are constrained by MHD to flow with the same velocity as the electrons.

However, the relief of this condition when δ is as small as c/ω_{pi} allows the ions to flow in the wider channel of width c/ω_{pi}, while the electrons can move through a much narrower channel of width c/ω_{pe} and cause much more rapid diffusion of the field lines. This leads to reconnection at the speed

$$v_R = \frac{c}{\omega_{pi} L'} v_A \tag{111}$$

This is actually the same as the reconnection velocity obtained from anomalous resistivity equation 79, except the length L' is here undetermined. Most of the important results on reconnection involving the Hall effect have emerged from numerical simulations of considerable power and elegance (Drake et al. 1997). The one important question remaining is how to scale up the simulation results from the short lengths L' they employed, of order 30 ion skin depths, to the much longer lengths required to understand space and astrophysical reconnection.

It is not yet clear whether the Hall physics allows the efficient regeneration of the B_y field that Petschek would require to carry out his extension. At the moment this part of the problem is still an open one.

One final remark is worth making. In the two examples we used to consider the breakdown of MHD physics, we found that the Sweet–Parker theory led to a layer thickness much thinner than the ion skin depth, so that the two approximations of MHD were violated. However, there are examples we could think of on larger scales where the opposite is true. For instance, in many of the galactic dynamo theories in which we would like to invoke reconnection, we find that the Sweet–Parker thickness is considerably larger than the ion skin depth (see problem 1). Under these circumstances, the Sweet–Parker theory should be valid. However, on galactic scales it leads to such long reconnection times that reconnection should not produce any effects. On really large scales reconnection does not seem to occur even during a Hubble time.

14.12 Problems

1. (Sweet–Parker reconnection in the interstellar medium) In a plasma of one electron volt temperature, what is the distance of resistive diffusion in a Hubble time?

Consider Sweet–Parker reconnection in the interstellar medium. If we take a magnetic field of 2 μG, a density of one per cubic centimeter, a temperature of 1 eV, and a length L of 1 pc, what is the reconnection velocity v_R? Evaluate τ_R, the time to reconnect a region of magnetic field of thickness L, where $\tau_R = L/v_R$. What is the thickness of the Sweet–Parker reconnection layer δ_{SP}? How does it compare in size to the ion skin depth c/ω_{pi}? Is ordinary MHD applicable for the Sweet–Parker model? What is the value of the Lundquist number $S = Lv_A/(\eta c/4\pi)$?

2. (The critical thickness for anomalous resistivity) Suppose the Sweet–Parker thickness δ_{SP} were one-tenth of the critical thickness for microinstabilities, $\delta_c = c/\omega_{pi}$. Assuming that anomalous resistivity is large enough to keep $\delta = \delta_c$, what is the ratio of the reconnection velocity to that predicted by Sweet and Parker?

3. (The production of a sheet current) Consider a sheared magnetic field

$$\mathbf{B} = B_0 \frac{x}{a} \hat{\mathbf{y}}$$

between infinitely conducting walls at $x = \pm a$. Change the position of the walls to

$$x = \pm a(1 - \delta \cos ky)$$

where δ is small compared to one and keep the magnetic field tangent to the wall. Show that the changed magnetic field is

$$\mathbf{B} + \mathbf{b} = \mathbf{B} + \hat{\mathbf{z}} \times \nabla\{[\psi(x) \cos ky]\}$$

where $\psi(\pm a) = B_0 \delta a$. The perturbed equilibrium equations are

$$\mathbf{j}_1 \times \mathbf{B} + \mathbf{j}_0 \times \mathbf{b} = \nabla p$$

where p is at our disposal, and $\mathbf{j}_0$ and $\mathbf{j}_1$ are the unperturbed and perturbed currents, respectively. Take the curl of this equation to show that ψ satisfies

$$B_0 \frac{x}{a} \left(\frac{\partial^2 \psi}{\partial x^2} - k^2 \psi \right) = 0$$

Show that the solution of this equation for which the field lines have the same topology is

$$\psi(x) = \pm B_0 \delta a \frac{\sinh(kx)}{\sinh(ka)}, \qquad x \neq 0,$$

where we take the plus sign for $x > 0$ and the minus sign when $x < 0$. Evaluate the jump in B_y at $x = 0$ and calculate the surface current.

Show that another solution that has no surface current but has a different topology and reconnected flux is

$$\psi(x) = B_0 \delta a \frac{\cosh(kx)}{\cosh(ka)}$$

Calculate the number of reconnected field lines per period, $\int b_x dy$, and show that it is

$$\frac{2 B_0 \delta a}{\sinh(ka)}$$

Estimate how long it would take to reconnect this number of lines according to Sweet–Parker. Proceed as follows: Show that this number of lines occupies a distance $x_1 = 2a\sqrt{\delta/\sinh(ka)}$. Find the time $t_{SP} = x_1/v_R$, where $v_R = v_A/S$ is the Sweet–Parker reconnection velocity. Take $v_A = b_y(0)/\sqrt{4\pi\rho}$ and $L = \pi/k$.

To express your answer in physical quantities, introduce the resistive time τ_R and the Alfvenic time τ_A by

$$\tau_R = \frac{4\pi a^2}{\eta c}, \qquad \tau_A = \frac{a}{v_{A0}}$$

where $v_{A0} = B_0/\sqrt{4\pi\rho}$ is the zero-order Alfven speed. Show that this Sweet–Parker time is given, up to a constant of order unity, by

$$\tau_{SP} \approx \frac{\sqrt{\tau_R \tau_A}}{ka}$$

The actual reconnection time t_R is given in Hahm and Kulsrud (1985) and is $t_R \approx \tau_R^{3/5}\tau_A^{2/5}$. The difference in the dependence on η is due to some changes from the Sweet–Parker model and also due to the equilibrium changing as the field reconnects.

4. (Ohmic heating in the reconnection layer) Calculate the amount by which the plasma in the reconnection layer is heated in the Sweet–Parker model when B_0, the transverse field, equals zero. Take the heating rate as $\eta c j^2$ with $j \approx B/4\pi\delta$. The time during which any fluid element in the layer is heated is the Alfven time L/v_A. Show that the fluid will thus have its pressure raised by of order $B_0^2/8\pi$, as assumed in the Sweet–Parker model. Repeat the calculation for the case of an anomalous resistivity, assuming that the heating rate is $\eta' c j^2$.

5. (The obliquely propagating ion acoustic wave instability) Consider the obliquely propagating ion acoustic wave driven unstable by the relative drift velocity of the ions and electrons associated with the reconnection current, but analyzed in the electron frame of motion instead of the ion frame. Again take the electrons constrained to move along the magnetic field in the z direction, and take the ions as unmagnetized. Note that the resonant electrons satisfying $\omega = k_z v_z$ have $v_z > 0$, so that $\partial f/\partial v_z < 0$ is negative. From equation 54 in chapter 10, the contribution of the electrons to ϵ_2 is positive and, thus, note from equation 64 in chapter 10 that the electrons reduce the energy of the wave. Explain how this can lead to instability. For your explanation use the fact that in this frame the dielectric constant is

$$\epsilon_1 = 1 - \frac{1}{(\omega - k v_D)^2} + \frac{1}{k^2 \lambda_d^2}$$

See the second problem of chapter 10. Thus, show from equation 62 in chapter 10 that the energy of the ion acoustic wave is negative in the electron frame.

Show that $\gamma > 0$ directly by expanding the dispersion relation

$$\epsilon(\omega) = \epsilon_1(\omega_1 + i\gamma) + i\epsilon_2(\omega_1) = 0$$

using $\epsilon_2 > 0$.

6. (Reconnection in a solar flare) Evaluate the reconnection velocity for a solar flare from the modified Petschek formula equation 107 using the parameters given in equation 42 and determine the reconnection time. (You may take $\delta_c = 10^5$ cm.) Do the same for magnetospheric reconnection determining the fraction of lines reconnected. Use the parameters given just before equation 44. Take δ_c equal to the ion skin depth. What fraction of the incoming solar wind lines are reconnected to the magnetospheric lines?

References

Biskamp, D. 1994. *Nonlinear Magnetohydrodynamics*. Cambridge University Press, Cambridge, UK.

Carter, T. A., H. Ji, F. Trintchouk, M. Yamada, and R. Kulsrud. 2003. *Physics of Plasmas*, **10**, 319.

Davidson, R. C., and N. T. Gladd. 1975. *Physics of Fluids* **18**, 1327.

Drake, J., D. Biskamp, and E. Schwarz. 1997. *Physics of Plasmas* **4**, 1002.

Dungey, J. W. 1961. *Physical Review Letters* **6**, 47.

Hahm, T. S., and R. M. Kulsrud. 1985. *Physics of Fluids* **28**, 2412.

Hughes, W. 1995. In *Introduction to Space Science*, p. 227, ed. M. G. Kivelson and C. T. Russell, Cambridge University Press, Cambridge, UK.

Kadomtsev, B. B. 1965. *Plasma Turbulence*, Chapter IV, p. 68, trans. L. C. Ronson (translation ed. W. G. Rusbridge), Academic Press, London.

Kulsrud, R. M. 1998. *Physics of Plasmas* **5**, 1599.

Kulsrud, R. M. 2001. *Earth, Planets and Space* **53**, 417–422.

Parker, E. N. 1957. *Journal of Geophysical Research*, **62**, 509.

Petschek, H. P. 1964. In *Physics of Solar Flares*, p. 425, ed. W. N. Ness, NASA SP-50, Washington, DC.

Priest, E., and T. Forbes. 2000. *Magnetic Reconnection*, Cambridge University Press, Cambridge, UK.

Sweet, P. A. 1958. In *Electromagnetic Phenomena in Cosmical Physics*, IAU Symposium **6**, p. 123, ed. B. Lehnert, Cambridge University Press, London.

Uzdensky, D., and R. M. Kulsrud. 2000. *Physics of Plasmas* **7**, 4018.

Vasyliunas, V. M. 1975. *Reviews of Geophysics and Space Physics* **13**, 303.

SUGGESTED FURTHER READING

The following is a list of books that I have not quoted in previous references but that I found very useful in learning plasma physics. They are in the order that I think would be the most useful.

P. A. Sturrock, *An Introduction to Theory of Astrophysical and Laboratory Plasmas*, Cambridge University Press, New York, 1994.

D. Melrose, *Plasma Astrophysics: Non-Thermal Processes in Diffuse Magnetized Plasmas*, 2 vols., Gordon & Breach, New York, 1980.

F. Chen, *Introduction to Plasma Physics*, Plenum Press, New York, 1983.

R. J. Goldston and P. H. Rutherford, *Introduction to Plasma Physics*, Institute of Physics Publications, Philadelphia, 1995.

D. Melrose, *Instabilities in Space and Laboratory Plasmas.* Cambridge University Press, New York, 1986.

S. Chandrasekhar, *Plasma Physics*, University of Chicago Press, Chicago, 1960.

G. Schmidt, *Physics of High Temperature Plasmas*, Academic Press, New York, 1966.

J. Huba, *NRL Plasma Formulary*, NRL Publication 6790.98.358.

T. J. M. Boyd and J. I. Sanderson, *The Physics of Plasmas*, Cambridge University Press, Cambridge, UK, 1967.

J. Friedberg, *Ideal MHD*, Plenum Press, New York, 1989.

D. R. Nicholson, *Introduction to Plasma Theory*, Wiley, New York, 1993.

R. O. Dendy, *Plasma Dynamics*, Oxford Clarendon Press, New York, 1990.

D. C. Montgomery and D. A. Tidman, *Plasma Kinetic Theory*, McGraw Hill, New York, 1964.

A. G. Sitenko, *Electromagnetic Fluctuations in a Plasma* (translated by M. Friedman), Academic Press, New York, 1967.

S. Ichimaru, *Statistical Plasma Physics*, Addison Wesley, Redwood City, CA, 1992.

S. Ichimaru, *Plasma Physics: An Introduction to Statistical Physics of Charged Particles*, Benjamin-Cummings, Menlo Park, CA, 1983.

S. Ichimaru, *Basic Principles of Plasma Physics, a Statistical Approach*, Benjamin (Frontiers of Physics), Reading, MA, 1973.

H. Alfven, *Cosmic Physics*, D. Reidel, Dordrecht, 1981.

W. Thompson, *An Introduction to Plasma Physics*, Addison Wesley, Reading, MA, 1962.

D. A. Tidman and W. A. Krall, *Shock Waves in Collisionless Plasmas*, Wiley Interscience, New York, 1971.

G. Bekefi, *Radiation Processes in Plasmas*, Wiley, New York, 1966.

G. Ecker, *Theory of Fully Ionized Plasmas*, Academic Press, New York, 1972.

A. Hasegawa, *Plasma Instabilities*, Springer-Verlag, New York, 1975.

D. Biskamp, *Non Linear Magnetohydrodynamics*, Cambridge University Press, New York, 1993.

T. Tajima and K. Shibata, *Plasma Astrophysics*, Addison-Wesley, Reading, MA, 2002.

INDEX

A *P* after a number indicates the reference is to a problem.

GPSR Authorized Representative: Easy Access System Europe - Mustamäe tee 50, 10621 Tallinn, Estonia, gpsr.requests@easproject.com

www.ingramcontent.com/pod-product-compliance
Ingram Content Group UK Ltd.
Pitfield, Milton Keynes, MK11 3LW, UK
UKHW021902310325
456954UK00008B/236

* 9 7 8 0 6 9 1 1 2 0 7 3 7 *